T0400943

GENE SILENCING: THEORY, TECHNIQUES AND APPLICATIONS

GENETICS – RESEARCH AND ISSUES

Additional books in this series can be found on Nova's website under the Series tab.

Additional E-books in this series can be found on Nova's website under the E-books tab.

GENE SILENCING: THEORY, TECHNIQUES AND APPLICATIONS

ANTHONY J. CATALANO
EDITOR

Nova Science Publishers, Inc.
New York

Library of Congress Cataloging-in-Publication Data

Gene silencing : theory, techniques, and applications / editor, Anthony J. Catalano.
p. ; cm.
Includes bibliographical references and index.
ISBN 978-1-61728-276-8 (hardcover)
1. Gene silencing. I. Catalano, Anthony J.
[DNLM: 1. Gene Silencing. QU 475 G3264 2010]
QH450.G4637 2010
572.8'65--dc22
2010016741

Published by Nova Science Publishers, Inc. ✦ *New York*

Contents

Preface

Gene silencing is a general term describing the epigenetic processes of gene regulation. The term gene silencing is generally used to describe the "switching off" of a gene by a mechanism other than genetic modification. This new book reviews research in the study of gene silencing including RNA silencing in transgenic plants and mycorrhizal research, gene silencing in the CNS and on the most extensively studied systems to mediate siRNA and shRNA delivery into the brain, siRNA delivery strategies as a therapeutic tool in gene therapy, galectin-3 epigenetics and effective methods for selecting siRNA sequences by using the average silencing probability and a hidden Markov model.

Chapter I - This review aims to describe the state of and progress in the knowledge of RNA silencing in transgenic plants, including the experimental applications and new perspectives opened by the most recent studies. Modern plant breeding involves new technical approaches, and genetic transformation is undoubtedly a powerful tool in plant biology and plant pathology. However, genetic engineering does not always result in efficient transgene expression, and often transgene copy number does not correlate with transgene expression level. Research in the past decade has shed light on the importance of RNA silencing as a mechanism of virus resistance in transgenic plants. Several plants that are resistant to viruses have been obtained, and some have commercially been applied for crop protection on fields. Transgene silencing is part of a broad host defence system called RNA silencing, a process that leads to homologous RNA degradation, which has widely been observed in animals, plants, and fungi. A key feature of RNA silencing is the presence of small RNAs, such as microRNAs (miRNAs) and small interfering RNAs (siRNAs), which are processed by a member of the RNAse III-like enzyme family, known as DICER. In plants, several distinct RNA silencing pathways operate to repress gene expression at transcriptional or post-transcriptional level. Transcriptional silencing is associated with DNA methylation, in which DNA homologous to a dsRNA is methylated *de novo*. In addition to defence responses against viruses and transposons, short RNAs have been demonstrated to have a role in a diverse range of functions, including regulation of gene expression, development and chromatin structure. RNA silencing is also a powerful tool for functional genomic studies in several species. Transgene-mediated gene silencing through tissue-specific, partial and/or total gene inactivation is a convenient approach to study target genes functions, particularly in species for which mutant collections are not available. The authors review various strategies for small RNA-based gene silencing: viral expression vectors (virus-induced gene silencing,

VIGS), transgenes containing hairpin RNA structures and a recently introduced approach, based on artificial microRNAs (amiRNAs).

Chapter II - Mycorrhiza is a mutualistic association between fungi and the roots of the vast majority of terrestrial plants. In natural ecosystems the plant nutrient uptake from the soil takes place via the extraradical mycelia of these mycosimbionts. While most herbaceous plants and tropical trees form endomycorrhiza-type interactions, trees of boreal and temperate ecosystems are typically ectomycorrhizal (ECM). These species include the majority of ecologically and economically important trees and the fungal symbionts are predominantly filamentous basidiomycetes.

The symbiotic phase in the life cycle of ECM basidiomycetes is the dikaryon. Hence, studies on symbiotic relevant gene functions would require the inactivation of both gene copies in the dikaryotic mycelium.

RNA silencing is a sequence homology-dependent degradation of target mRNAs based on an ancient cellular mechanism believed to have evolved as protection of eukaryotic cells against alien nucleic acids. In different eukaryotic organisms, including fungi, the RNA silencing pathway can be artificially triggered to target and degrade gene transcripts of interest, resulting in gene knock-down. Most importantly, RNA silencing can act at the cytosolic level affecting mRNAs originating from several gene copies and different nuclei, and it can thus offer an efficient way for altering gene expression in dikaryotic organisms.

Laccaria bicolor, the first symbiotic fungus with its genome sequenced, has rapidly turned into a model fungus in ectomycorrhizal research. *Laccaria* possesses a complete set of genes known to be needed for RNA silencing in eukaryotic cells. The authors have demonstrated that RNA silencing is functional in *L. bicolor* and that it can be triggered via *Agrobacterium*-mediated transformation. Moreover, targeted gene knock-down in dikaryotic mycelium can result in functional phenotypes altered in the symbiotic capacity confirming that RNA silencing is a powerful way to study symbiosis- regulated genes. These findings have now initiated the RNA silencing era in mycorrhizal research, a field that has been hindered by the lack of proper genetic tools.

Chapter III - RNA interference (RNAi) has recently emerged as a powerful tool in functional genomic studies, allowing dissection of entire signalling pathways and elucidation of the molecular mechanisms of neurobiological processes, thereby facilitating rapid identification and validation of possible therapeutic targets. Moreover, RNAi holds great therapeutic potential since application of small interfering RNAs (siRNAs) and short hairpin RNAs (shRNAs) may allow specific knockdown of selected toxic proteins, even when allele-specific silencing is needed, as in the case of dominantly inherited disorders.

Nevertheless, the development of RNAi-based therapeutics for *in vivo* application faces the same challenge common to all classes of drugs: achieving an efficient and sustained distribution into the target tissue at sufficient concentrations to accomplish a therapeutic effect. Although significant progress has been made regarding the safety and stability of siRNAs and shRNAs, a major limitation for the *in vivo* application of RNAi technology concerns the inability of these molecules to cross cellular membranes. Multiple delivery methodologies, including viral and non-viral vectors, have been developed with different degrees of success for the introduction of siRNAs and shRNAs into cells, both *in vitro* and *in vivo*.

This review is focused on the available strategies to achieve gene silencing in the CNS and on the most extensively studied systems to mediate siRNA and shRNA delivery into the brain. Moreover, the authors summarize the most important studies concerning RNAi application in the context of neurodegenerative diseases and other neurological disorders.

Chapter IV - Small interfering RNAs (siRNA) are emerging as promising therapeutic agents for the treatment of inherited and acquired diseases, as well as research tools for the elucidation of gene function. Since the molecules undergo rapid enzymatic degradation and have poor cellular uptake, there is a need to design a delivery system which can protect and efficiently transport siRNA to the target cells. Polymeric nanoparticles have emerged as systems of choice with reduced cytotoxicity and enhanced efficacy. These systems not only protect siRNA from enzymatic degradation by forming condensed complexes but also leads to tissue and cellular targeting, improve cellular penetration, release the siRNA in the right intracellular compartment. Nanoparticles prepared from polycationic polymers like polyethylenimine, chitosan have been widely investigated due to ease of manipulatibility, stability, low immunogenicity, low cost and high flexibility regarding the size of transgene delivered. This chapter presents an overview of siRNA delivery strategies employing polymeric nanoparticles, with emphasis on self-assembled polymeric nanoparticles with promising potential to evolve as therapeutic tool in gene therapy.

Chapter V - With the aim in view to improve physicochemical and biological properties of natural oligonucleotides, several types of DNA analogues and mimics were designed, particularly negatively charged PNA-like mimics. Among them, two types of DNA mimics representing hetero-oligomers constructed from alternating monomers of phosphono peptide nucleic acids and monomers on the base of *trans*-1-acetyl-4-hydroxy-L-proline (HypNA-pPNAs) as well as oligomers constructed from chiral analogues of peptide nucleic acids with a constrained trans-4-hydroxy-N-acetylpyrrolidine-2-phosphonic acid backbone (pHypNAs) were developed. Their physico-chemical and biological properties were evaluated in the comparison with natural oligonucleotides, classical peptide nucleic acids and morpholino phosphorodiamidate oligonucleotide analogues. The results obtained in a set of experiments revealed a high potential of these phosphonate-containing PNA derivatives for a number of biological applications, such as diagnostic, nucleic acids analysis and inhibition of gene expression. HypNA-pPNA and pHypNA mimics combine high hybridization and mismatch discrimination characteristics with good water solubility and biological stability as well as the ability to penetrate cell membranes. Their effectiveness to provide the specific knockdown of a target protein production was demonstrated in research involving *in vitro* systems, living cells and intact organisms. As their effect lasts over a period of several days, due to their high stability in living cells, it represents a very potent technology for administrating antisense- or antigene-based drugs for future therapeutic applications.

Chapter VI - Insects are organisms of considerable interest for comparative biology and medicine, therefore it is not surprising that several publications referred to them as model organisms. Insect and vertebrate evolution diverged more than 500 million years ago, but the molecular bases of several fundamental biological functions, including innate immune response, were already established in their common progenitor and have been conserved. Consequently, starting from information collected in insects, new insights into human biology and pathology were gained. Gene silencing includes several powerful methods, such as the production of loss-of-function mutants and RNA interference. These procedures, in particularly when performed in models for which molecular databases are already available, allow the genetic dissection of several immune-related processes and pathways. In the present review, we will concentrate our attention on the information derived from gene silencing techniques on insect immune signalling with particular attention for *Drosophila melanogaster* and *Anopheles gambiae*.

Chapter VII - Sequencing of plant genome and expressed sequence tag (EST) have provided abundant sequence information in several plant species. Elucidating function(s) of all of these genes is a huge undertaking. Even in well-studied plants like Arabidopsis, function is not known for majority of genes. Hence, a powerful tool that can be widely used to understand gene function is necessary. Several functional genomics tools were developed in the recent past to achieve this goal. RNA interference (RNAi) is one such tool widely used to analyze gene function. RNAi is also proved to be a tool for plant researchers to produce improved crop varieties.

First part of this review is focused on three RNAi based concepts that has potential applications in plant functional genomics and agriculture. These concepts are tissue specific silencing, inducible silencing and host delivered RNAi (hdRNAi) during plant-pest interaction. Tissue specific promoters driving RNAi constructs can induce gene silencing in a particular organelle or tissue. Also, RNAi constructs with stress or chemical inducible promoters can be used to induce gene silencing only when required. These two concepts together can be used to achieve temporal and spatial control of gene silencing in plants. In the hdRNAi, dsRNA generated in an RNAi transgenic plant is delivered to interacting target organism (pest), activating gene silencing in the target organism. A comprehensive review pertaining to these areas is presented.

Second part of the review deals with applications of RNAi in agriculture, animal husbandry and biofuel industry. As suppression of gene expression by RNAi is inheritable, this has been a tool for developing transgenic crop plants for resistance against disease, pests, drought and in other areas of agriculture. This review summarizes developments in these areas with major emphasis on application of RNAi for development of biotic stress tolerant crops. We also note limitations of RNAi technology and ways to overcome the same.

Chapter VIII - Protein-carbohydrate interactions play significant role in modulating cell-cell and cell-extracellular matrix interactions, which, in turn, mediate various biological processes such as growth regulation, immune function, cancer metastasis, and apoptosis. Galectin-3, a member of the β-galactoside-binding protein family, is found multifunctional and is involved in normal growth development as well as cancer progression and metastasis, but the detailed mechanisms of its functions or its transcriptional regulations are not well understood. Besides, several regulatory elements such as GC box, CRE motif, AP-1 site, and NF-κB sites, the promoter of galectin-3 gene (*LGALS3*) contains several CpG islands that can be methylated during tumorigenesis leading to the gene silencing. This review discusses the galectin-3 epigenetics, which represents a novel regulatory mechanism of its transcription.

Chapter IX - Chronic infection with hepatitis B virus (HBV) occurs in approximately 6% of the world's population and is often complicated by cirrhosis and hepatocellular carcinoma (HCC). Existing therapy rarely has durable effects and improving treatment to counter the infection remains an important medical priority. Although harnessing the RNA interference (RNAi) pathway to achieve therapeutic HBV gene silencing holds promise, precise regulation of the expression of silencing sequences is critically important for safe application of this approach. Earlier work from our laboratory demonstrated that pri-miR-31- and pri-miR-122-based anti HBV shuttles were capable of potent, safe antiviral activity and can be used in modular multimeric arrangements. To advance this approach, and limit the potential problems caused by extrahepatic expression of anti HBV RNAi activators, these sequences were placed under control of liver specific transcription control elements, viz. the human Factor VIII (FVIII), alpha-1-antitrypsin (A1AT), HBV preS2 and HBV basic core (BCP) promoters. Using a luciferase reporter gene assay optimal liver-specific transcription control was observed with A1AT and BCP regulating sequences. These elements were then incorporated into pri-miR-expression cassettes and were tested for antiviral efficacy in cell culture and a murine model of HBV replication. Results showed that silencing of HBV replication was achieved. Importantly there was no evidence for disruption of endogenous miR function, which is a significant advantage over use of stronger and constitutively active RNA polymerase (Pol) III promoter RNAi expression cassettes. The use of anti HBV pri-miR shuttles in the context of liver-specific Pol II promoters is likely to have usefulness for therapeutic HBV knockdown, and should also have general applicability to silencing of pathology causing genes in the liver.

Chapter X - Short interfering RNA (siRNA) has been widely used for studying gene functions in mammalian cells but varies markedly in its gene silencing efficacy. Although many design rules/guidelines for effective siRNAs based on various criteria have been reported recently, there are only a few consistencies among them. This makes it difficult to select effective siRNA sequences in mammalian genes. This chapter first reviews the recently reported siRNA design guidelines and then proposes a new method for selecting effective siRNA sequences from many possible candidates by using the average silencing probability on the basis of a large number of known effective siRNAs. It is different from the previous score-based siRNA design techniques and can predict the probability that a candidate siRNA sequence will be effective. The results of evaluating it by applying it to recently reported effective and ineffective siRNA sequences for various genes indicate that it would be useful for many other genes. The evaluation results indicate that the proposed method would be useful for many other genes. It should therefore be useful for selecting siRNA sequences effective for mammalian genes. The chapter also describes another method using a hidden Markov Model (HMM) to select the optimal functional siRNAs and discusses the frequencies of the combinations for two successive nucleotides as important characteristics of effective siRNA sequences.

Chapter XI - Small RNA-mediated gene silencing as a natural defense mechanism against viruses, transposons, and other invading nucleic acids or a means of regulating plant endogenous genes is a powerful tool and is being employed to down-regulate the expression of the targeted genes. Such a small RNA-mediated gene silencing has many different applications in a variety of organisms including humans and animals to control disease as a therapeutic agent, as well as plants to alter plant phenotypes. This silencing platform works through RNA-directed degradation or translational repression of target mRNA and has been devised towards a high-throughput approach for the gene suppression. In particular, sequence-specific control of gene expression by these non-coding RNAs has gained a significant amount of importance in plant biotechnology to influence specific plant phenotypes over the past years. It has been demonstrated that crops that were transformed with RNAi constructs, introduced stable modifications to the biochemical pathways. This can open new avenues in the improvement of crop productivity and quality. Here, the authors review the role of small RNA-directed gene silencing in plant biotechnology. The review will focus on the application of a gene silencing approach mediated by three subclasses of small RNAs for improved oil quality, reduced allergen, virus resistance, and other agronomical traits. The advantages and drawbacks of each gene silencing approach are also discussed with regard to crop improvement.

Chapter XII - Several abiotic stress specific functional and regulatory genes have been cloned, and a number of EST databases representing stress specific genes are available for many plant species. These sequences have to be translated into functional information, necessitating the need for potential functional genomic approaches. Post transcriptional gene silencing (PTGS) is one of approaches to characterize functional relevance of stress responsive genes. Virus-induced gene silencing (VIGS) and developing stable gene knock down plants using hairpin RNA interference (hpRNAi) constructs (referred here as RNAi) are two important PTGS methods. Over a period, these methods are becoming integral part of plant stress functional genomics. Among these two methods, use of VIGS for characterizing abiotic stress responsive genes is still an emerging approach while RNAi has been widely used.

This review is focused on VIGS vector resources, brief methodology of VIGS and application of gene silencing to identify/characterize genes involved in drought-, salinity-oxidative-, high light-, and nutrient-stress management. VIGS can be used as fast forward genetic screening method to identify genes involved in stress tolerance and also an effective reverse genetic tool to validate the relevance of genes identified from high-throughput screening. Further, VIGS can be effectively integrated with abiotic stress imposition and response of gene silenced plants can be quantified using suitable techniques. We describe here an comprehensive approach to silence large number of cDNA clones and characterize the silenced plants under abiotic stresses. We also discussed application of other PTGS based methods like RNAi and artificial micro RNA (amiRNA) in abiotic stress functional genomics.

We propose that PTGS is an useful technology for translational genomics to assign function to large number of abiotic stress responsive genes. Even with their current limitations, gene silencing techniques are set to revolutionize plant abiotic stress functional genomics. Limitations and future directions for these techniques are also briefly discussed.

Chapter XIII - Small regulatory RNAs including short interfering RNAs (siRNAs) and microRNAs (miRNAs) are crucial regulators of gene expression at the posttranscriptional level. Recently, additional roles for small RNAs in gene activation and suppression at the transcriptional level were reported; these RNAs were shown to have sequences that closely or completely match to their respective promoter regions. However, no global analysis for identifying target sequences for miRNAs in the promoter region have been carried out in the human genome.

We performed a genome-wide search for upstream sequences of mRNA transcription start sites where miRNAs are capable of hybridizing with high complementarity. We identified 219 sites in the 10-kb upstream regions of transcription start sites with complete complementarity to 94 human mature miRNAs. Furthermore, the mismatched positions and nucleotides in near-completely matched sites were highly biased, and most of them appear to be possible target sites of miRNAs. The expression of downstream genes of miRNA target sites were examined following transfection of each miRNA into three different human cell lines. The results indicate that miRNAs dynamically modulates gene expression depending on the downstream genes and the cell type.

Chapter XIV - Gene silencing is an exciting field of functional genomics. It has been used as a research tool to discover or validate the functions of genes. It involves short sequence of nucleic acid that can bind to RNA of the gene and interferes the process of its expression. It is diverse in occurrence as well as in applications. This phenomenon occurs from nematodes to fungi and can cause gene silencing in plants, animals and human beings. The core aspects of the mechanisms and functions of gene silencing include co-suppression, RNA-mediated virus resistance and RNA-directed DNA methylation (RdDM). The applications of gene silencing cover a wide spectrum in plants, from designer flower colors to plant-produced medical therapeutics. These functions are achieved by two types of approach such as protection of the plant against attack and fine-tuning of metabolic pathways. RNA-mediated gene control mechanism has already provided new platforms for developing molecular tools for gene function studies and crop improvements. We are now exploring this technology for commercially focused applications in plants. Here, we review the theory of gene silencing discovery and the mechanism of this technique in plants. Further, we discuss the potential use of this technique in plant science particularly in crop improvements.

Chapter XV - RNA interference (RNAi) has been utilized in a variety of applications to target specific gene silencing mediated by small-interfering RNA (siRNA) over the last few years. Cell-penetrating peptides (CPPs) were proven to be able to traverse cell membranes and deliver biological macromolecules into living cells. Here, we provided an efficient and safe method for the delivery of siRNA into mammalian cells mediated by CPPs noncovalently. We first established a GC-EGFP cell line stably expressing enhanced green fluorescent protein (EGFP) from human gastric cells. CPPs were demonstrated to interact with and deliver siRNA into GC-EGFP cells, and the internalized dsRNA tended to localize in the perinuclear region within cells. The sulforhodamine B (SRB) assay further confirmed CPPs were nontoxic to cell viability. Finally, our results showed that siRNA fulfilled its targeted *egfp* gene silencing. In the future, CPPs may provide a useful and nontoxic tool for the delivery of siRNA into mammalian cells.

Chapter XVI - RNA interference (RNAi) has become an indispensible technology for biomedical research and promises to usher in a brand new class of therapeutics that work by silencing disease genes. Until recently, the paradigm for gene silencing in mammalian cells has relied on a small symmetrical RNA structures containing a 19-base-pair duplex with 2 nucleotide overhangs at each 3' end: the standard siRNA structure. The standard siRNA scaffold is based on structures generated by Dicer digestion of a double stranded RNA, and is considered to be the fundamental template for designing RNAi inducers. In fact, early studies suggested there was only very limited flexibility regarding the length and symmetry of the siRNA structure in order to maintain optimal gene silencing. Recent studies, however, have demonstrated that gene silencing siRNAs with duplex lengths shorter than 19 bp or asymmetric structures can trigger specific gene silencing in mammalian cells. Importantly, asymmetric siRNA structures can ameliorate several sequence-independent, nonspecific effects triggered by the canonical siRNA structure. These findings demonstrate the structural flexibility of RNAi inducers in mammalian cells.

In: Gene Silencing: Theory, Techniques and Applications
Editor: Anthony J.Catalano

ISBN: 978-1-61728-276-8
© 2010 Nova Science Publishers, Inc.

Chapter I

Transgene Silencing in Plants: Mechanisms, Applications and New Perspectives

Chiara Pagliarani[1], Irene Perrone[1],
Andrea Carra[2] and Giorgio Gambino[3], *

[1]Department of Arboriculture and Pomology, University of Torino. Via L. da Vinci 44,
I-10095, Grugliasco (TO), Italy
[2]Research Unit for Wood Production Outside Forests, CRA, Strada Frassineto 35,
I-15033, Casale Monferrato (AL), Italy
[3]Plant Virology Institute, National Research Council, Grugliasco Unit,
Via L. da Vinci 44, I-10095 Grugliasco (TO), Italy

Abstract

This review aims to describe the state of and progress in the knowledge of RNA silencing in transgenic plants, including the experimental applications and new perspectives opened by the most recent studies. Modern plant breeding involves new technical approaches, and genetic transformation is undoubtedly a powerful tool in plant biology and plant pathology. However, genetic engineering does not always result in efficient transgene expression, and often transgene copy number does not correlate with transgene expression level. Research in the past decade has shed light on the importance of RNA silencing as a mechanism of virus resistance in transgenic plants. Several plants that are resistant to viruses have been obtained, and some have commercially been applied for crop protection on fields. Transgene silencing is part of a broad host defence system called RNA silencing, a process that leads to homologous RNA degradation, which has widely been observed in animals, plants, and fungi. A key feature of RNA

* Author for correspondence. Phone: +39-011-6708666; Fax: +39-011-6708658; E-mail: g.gambino@ivv.cnr.it

silencing is the presence of small RNAs, such as microRNAs (miRNAs) and small interfering RNAs (siRNAs), which are processed by a member of the RNAse III-like enzyme family, known as DICER. In plants, several distinct RNA silencing pathways operate to repress gene expression at transcriptional or post-transcriptional level. Transcriptional silencing is associated with DNA methylation, in which DNA homologous to a dsRNA is methylated *de novo*. In addition to defence responses against viruses and transposons, short RNAs have been demonstrated to have a role in a diverse range of functions, including regulation of gene expression, development and chromatin structure. RNA silencing is also a powerful tool for functional genomic studies in several species. Transgene-mediated gene silencing through tissue-specific, partial and/or total gene inactivation is a convenient approach to study target genes functions, particularly in species for which mutant collections are not available. We review various strategies for small RNA-based gene silencing: viral expression vectors (virus-induced gene silencing, VIGS), transgenes containing hairpin RNA structures and a recently introduced approach, based on artificial microRNAs (amiRNAs).

Introduction

RNA silencing, a process leading to the degradation of homologous mRNAs, has been widely observed in animals, plants, and fungi [1]. In the early '90s, gene silencing phenomena were first noted through a surprising observation that occurred during the course of plant transformation experiments: the introduction of transgenes inside the genome led to silence both transgenes and their homologous endogenes. Napoli and collaborators [2], with experiments on a chimeric chalcone synthase (*chs*) gene over-expressed in petunia (*Petunia hybrida*) petals, discovered a co-suppression process of both the transgene and the homologous mRNA endogenous sequence. In fungi, Romano and Macino [3] described transient gene expression inactivation events in *Neurospora crassa* transformants and called the discovered silencing process "quelling"; in animals, Lee and colleagues [4] carried out experiments on the nematode *Caenorhabditis elegans*, finding that the gene *lin-4* encodes for a small RNA that strongly regulates larval developmental transitions. Since those results, several improvements have been made to achieve a whole comprehension of RNA silencing pathways; today, most functions involved in this biological process have been described. A key feature of RNA silencing is the presence of small RNAs, which were first observed in plants [5], such as microRNAs (miRNAs) and small interfering RNAs (siRNAs), which are processed from double-stranded RNAs (dsRNAs) by a family of the RNAse III-like enzyme, known as DICER [6]. RNA silencing is an important mechanism used by plants to defend themselves against viral infections and transposons. In addition, short RNAs have been demonstrated to have a role in a diverse range of functions, including regulation of gene expression, development and chromatin structure.

As in co-suppression, genetic engineering does not always result in efficient transgene expression levels. Several cases have been reported, where the transgene copy number does not correlate with the level of transgene expression [7]. Two types of events contribute to the realization of transgene silencing. The first is linked to the specific genomic site in which the T-DNA integrates itself [8], whereas the second closely depends on the configuration of the integrated T-DNAs. Indeed, it is possible that multiple T-DNAs, integrated at one locus, bind to each other to form complex T-DNA structures [7]. Over the last decade, transgene silencing has no longer become just an unforeseen consequence of genetic transformation, but it has been induced with high efficiency through various strategies, using viral expression vectors (virus-induced gene silencing, VIGS), transgenes containing hairpin RNA structures, and a recently introduced approach based on artificial microRNAs (amiRNAs).

There are numerous possible applications for transgenic silencing in plants. Initially, the main research field was focused on enhancing disease resistance. Several plants resistant to viruses have been obtained to date, and some have commercially been used for crop protection on fields. Gene silencing has been applied to change the chemical composition of plant products, for industrial use or to improve fruit quality, and to increase the nutritional value. In parallel, other applications have been tested to reduce the production of plant-derived substances, as in the case of allergens, which can be harmful to human health if present in nutritious foods. In recent years, functional genomic studies have been gaining more and more importance, as they can provide interesting insights about genetic regulation of plant functions and support novel tools for isolating and characterizing genes. Transgene silencing through tissue-specific, partial and/or total gene inactivation is a convenient approach to study target gene functions, in particular in species for which mutant collections are not available.

This review focuses on the main molecular mechanisms involved in plant RNA silencing, with particular attention to transgene-mediated gene silencing. We report the major applications and future perspectives of this technology applied to the most important woody and herbaceous crop plants.

Mechanisms of Gene Silencing

RNA silencing evolved in eukaryotes to regulate gene expression, to control transposable elements and to fight against pathogens. In plants, several distinct RNA silencing pathways operate to repress gene expression at the transcriptional (transcriptional gene silencing, TGS) or post-transcriptional level (post-transcriptional gene silencing, PTGS) [9]. RNA silencing pathways are characterized by the production of double stranded RNAs (dsRNAs) from endogenous or exogenous transcripts, or by the production of self-complementary foldback RNAs. These dsRNAs are digested by a family of RNAse III-like enzymes, termed DICER, into short RNA duplexes of 21-24-nt length, referred to as short interfering RNAs (siRNAs) or microRNAs (miRNAs) [10]. Once generated, siRNAs and miRNAs guide the sequence-specific inactivation of a target mRNA via an RNA-induced silencing complex (RISC), which includes a member of the Argonaute (AGO) proteins, a family of enzymes with endonucleasic activity. RISC mediates cleavage or translational repression of target mRNAs. RNA silencing also involves RNA-directed DNA methylation (RdDM), in which DNA homologous to a dsRNA triggering gene silencing is methylated *de novo* [11]. Small RNAs guide the RNA-induced transcriptional gene silencing (RITS) complex to direct chromatin modifications and DNA methylation of the homologous DNA sequences [12]. In mammals, DNA methylation occurs almost exclusively on cytosines in the symmetric di-nucleotides CpG, whereas in plants, cytosine methylation occurs both at symmetric sites (CpG and CpNpG, where N is A, T or C) and at asymmetric sites (CpNpN).

Transgene Silencing in Plants

Transgene silencing is only a small item in the wide RNA silencing system, but it was actually from the study of transgene suppression phenomena that researchers started to uncover step by step functions and roles of this RNA microcosm. TGS was observed by Matzke and colleagues [13], whereas PTGS was phenotypically identified as the direct cause of flower pigmentation loss in transgenic petunia, where it was studied for the first time [2]. The PTGS pathway is always initiated through a dsRNA, which is often directly produced by the transgene loci as a consequence of imperfect integration events inducing juxtaposed sense-antisense transgenes. In early reports, the introduction of antisense (antisense suppression) and sense construct (co-suppression) into plants reduced the corresponding endogenous mRNA with a low frequency of success [14]. If the sense or antisense transgene is integrated as single copy and associated to a high transcription rate, the transgene transcript will be converted in dsRNA by a RNA dependent RNA polymerase (RDR), preferentially a RDR6 [15] [16]. RDR6 recognizes typical transcript aberrant features, as lack of 5' end cap or/and polyA tail [17], working in tight cooperation with the RNA-binding protein SGS3 and the RNA helicase SDE3 to produce and stabilize the dsRNA (Figure 1). In some cases, the antisense transgene can directly hybridize with the endogenous sense transcript and thus contribute to dsRNA formation [18]. When multiple copies of transgenes arrange by chance as inverted repeats (IR), dsRNA is likely produced by read-through transcription resulting in the formation of an hairpin RNA (hpRNA) showing a self-complementary structure. PTGS is greatly enhanced by direct production of dsRNAs from IR transgenes and in the last years several authors directly used transgenes designed to have self-complementary sequences producing hpRNAs (Figure 1) [19] [20].

Two distinct Dicer-like (DCL) enzymes might process the dsRNA. DCL3 probably produces 24-nt length siRNAs, whereas DCL4 is probably the preferred enzyme for the production of 21-nt length siRNAs population [21]. Both these types of siRNAs are methylated at their 3' termini by the small RNA-specific methyltransferase HUA ENHANCER1 (HEN1) to maintain their stability against possible degradation [22]. 24-nt siRNAs generally enter into the RITS complex to guide DNA and histone methylation at homologous loci, whereas the 21-nt siRNAs enter into the RISC complex to guide cleavage of homologous mRNAs (Figure 1) [23].

PTGS mechanisms can be activated by sense (S) or antisense (AS) transgenes, inverted repeats (IR), hairpin RNAs (hp-RNAs) or viral sequences (VIGS). Sense transgene transcripts are converted in dsRNAs by RDR6; whereas antisense transgenes can directly form dsRNAs or can meet the sense transgene pathway at the ssRNA step. The dsRNA is further sliced by DCL4 into ≈21-nt long siRNAs, which are methylated by HEN1. AGO1 binds these 21-nt siRNAs to form the RISC complex and guide target mRNAs degradation. Sometimes, S and AS transgenes can also follow the TGS pathway. Once converted into hairpin stem-loop forms, both IR and hpRNAs become substrates for DCL4 or DCL3 activity, following the PTGS or TGS pathway respectively, the same happens for VIGS.

In the TGS pathway, a transgene harbouring a sequence complementary to a specific promoter is transcribed and converted into a dsRNA by RDR2. The dsRNA is processed by DCL3 to form a ≈24-nt siRNAs population, always methylated by HEN1. These siRNAs associate with AGO4 to form the RITS complex and activate DNA methylation events.

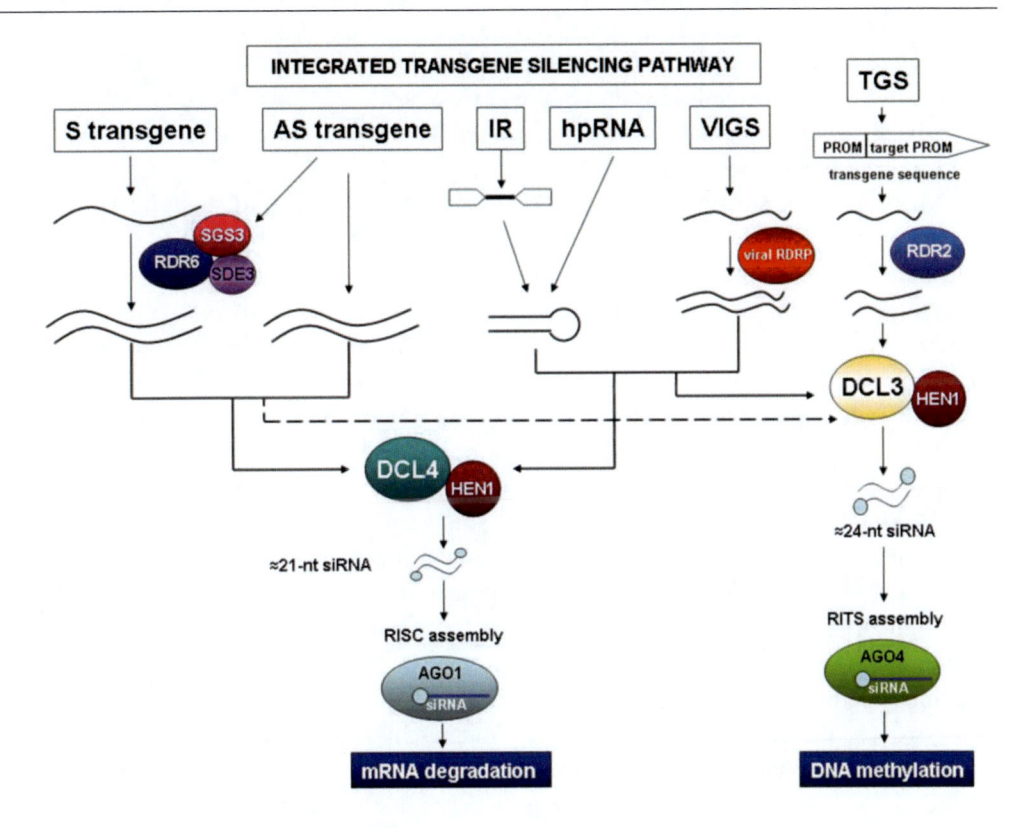

Figure 1. Transgene silencing pathways in plants. See the text for a detailed explanation of the pathways shown in figure.

Endogenous Short Non-coding RNAs in Plants

Different endogenous RNA silencing pathways operate in plants to regulate biological processes. Mostly in the last decade, several of these RNA silencing pathways were deepened, especially thank to loss of function mutants in *Arabidopsis*. Small endogenous RNAs are classified into two categories on the base of their biogenesis: siRNAs are processed from long, perfectly dsRNA precursors, and miRNAs from single-stranded RNA transcripts folding themselves up to produce imperfect double-stranded stem loop precursors. Both these categories are the effectors of RNA silencing phenomena and are involved in several developmental processes, in defence responses to biotic and abiotic stresses and in genome stability maintenance, as reviewed in many works [24] [25] [26] [27] [28] [29] [30].

miRNAs

First evidences about miRNAs and their relative biological roles were obtained in nematodes during the early 90's through forward genetic screenings [4]. Since then, direct cloning and bioinformatic approaches helped to identify miRNAs, their functions and their related silencing pathway both in animals and plants. In plants, miRNAs are mainly involved in negative regulation of genes linked to plant development and cell fate determination [31], responses to abiotic stimuli [30] and biotic stresses [32], and in hormone signal transduction [33]. Functions and regulatory activities of plant miRNAs are widely described and well deepened in several reviews [24] [28] [34].

MiRNA genes are transcribed by Pol II into step-loop containing primary transcripts, termed pri-miRNAs. Pri-miRNAs are processed in the nucleus by DCL1, which, with the help of a double stranded RNA binding protein, HYPONASTIC LEAVES 1 (HYL1), and a zinc finger protein, SERRATE (SE) [23] [35], produces a 21-nt long miRNA-miRNA* duplex. The miRNA-miRNA* duplex is further methylated by HEN1 on the 2'OH at the 3' end, to prevent its urydilation and consequent degradation [36], and exported in the cytoplasm by the exportin-5 orthologous HASTY (HST) [37]. Once in the cytoplasm, only one strand, referred to as the mature miRNA, is loaded onto AGO1, the catalytic centre of the RISC complex, which stabilizes it and guides its pairing to the complementary sequence of a target mRNA (Figure 2). Pairing with the target site results in target cleavage or translational inhibition. In plants, miRNA-directed repression of gene expression is predominantly mediated by mRNA cleavage, whereas in animals the predominant mode of action is the translational repression [38]. The mechanism determining the choice between cleavage or translation inhibition is still unclear. One hypothesis is that when the mRNA and the miRNA are only partially complementary to each other, the RISC mediates the translational repression of the target mRNA [39] [40]. Another hypothesis is that miRNAs involved in crucial developmental processes, such as cell fate determination during embryogenesis, direct cleavage of target mRNAs; whereas when a reversible adaptation to external conditions is required, miRNAs negatively affect target proteins levels [41]. Although translational repression was once thought to be an exception, recent studies have suggested that there might be a coexistence between these two silencing pathways [42].

Trans-acting siRNAs

Trans-acting siRNAs (ta-siRNAs) are a class of plant endogenous small RNAs, which are produced from non-coding transcripts of TAS loci. TAS transcripts are targeted by miRNAs at positions that define the starting sites for double strand synthesis by RDR6 and subsequent processing in 21-nt increments by DCL4 [23] [43] [44] [45] [46]. In particular, two miRNAs, miR173 and miR390, have been shown to target three TAS loci in *Arabidopsis* [44] [47] [48] (Figure 2). Ta-siRNAs are negative regulators of gene expression with a mode of action similar to that of miRNAs. In particular, they target a number of important auxin response factors (ARFs) [49] [50] [51].

Natural Antisense Transcript-derived siRNAs

Natural antisense transcript-derived siRNAs (nat-siRNAs) are produced from partially overlapping transcripts [52] of antisense gene pairs encompassing an inducible and a constitutive gene. When both genes are transcribed, a 24-nt long nat-siRNA is produced by a partially characterized pathway that requires DCL2 and RDR6. This nat-siRNA targets and directs cleavage of the transcript of the constitutive gene, which is then converted in a dsRNA by RDR6 and processed into phased 21-nt nat-siRNAs by DCL1, with a reaction similar to that of ta-siRNAs biogenesis (Figure 2). Thus, as other small RNAs, nat-siRNAs are used as guide to direct sequence-specific silencing of homologous mRNAs. The nat-siRNAs known to date are involved in environmental stress responses, for instance cold and salt stresses tolerance [52] and pathogen attacks [53].

Heterochromatic siRNAs

Heterochromatic siRNAs (hc-siRNAs) are 24-nt long small RNAs that are implicated in transcriptional gene silencing (TGS), which is an epigenetic mechanism resulting in silencing of endogenous genes or transgenes through the inactivation of promoter sequences (Figure 2). They are also termed *cis*-acting siRNAs (ca-siRNAs), because they silence the same genomic loci from which they have been transcribed. Their main function is limiting transposons activities and thus maintaining intact the genome stability. Specifically, hc-siRNAs guide the *de novo* methylation of cytosine residues of the homologous DNA region and also induce chromatin-based silencing through histone modifications, including methylation and acethylation, and chromatin remodelling events. The whole of these regulatory events is known in plants as RNA-directed DNA methylation system (RdDM) and is typical of heterochromatic genome regions and in general of repeated sequences [54]. The silencing pathway associated to RdDM has recently been updated by studies on its effector proteins [55] [56]. In brief, methylated DNA is a template for the transcription of aberrant RNA by RNA polymerase IV (Pol IVa) or less frequently by Pol II. Later, RDR2 converts these transcripts into dsRNAs, which are directly processed by DCL3 in 24-nt primary hc-siRNAs, for whose stability the action of HEN1 is required [54] [34]. These primary hc-siRNAs are used by AGO4 to direct the sequence specific DNA methylation step of RdDM, mediated by Pol IVb, the DNA methyltransferase DOMAINS REARRANGED METHYLASE2 (DRM2) [57], the DEFECTIVE IN RNA-DIRECTED DNA METHYLATION (DRD1), an SNF2-like chromatin remodelling protein ATP-dependent [58], and DMS3, a SMC-hinge domain-containing protein [59]. Once established *de novo* DNA methylation, the METHYLTRANSFERASE1 (MET1) is required for replicative maintenance of methylation at CG sites, and the CHROMETHYLASE3 (CMT3) is required for the maintenance at CNG and asymmetrical CNN sites [60].

As described above, the interactive regulatory network triggered by hc-siRNAs functions to maintain genome stability and avoid both transposons movements and genomic rearrangements phenomena. However, hc-siRNAs can also act to silence protein-coding genes. For instance, the expression of flowering locus C (FLC) in *Arabidopsis thaliana*, a negative regulator of flowering, is regulated both by DNA methylation and chromatin modification events. This occurs on one side thank to the presence of a transposable element in one of the FLC introns, which seems to attract DNA and histone methyltransferases [61], and on the other side thank to the siRNAs biogenesis from antisense transcription of the 3' FLC gene UTR region [62], which directs repressive histone structural remodelling, thus providing the floral transition.

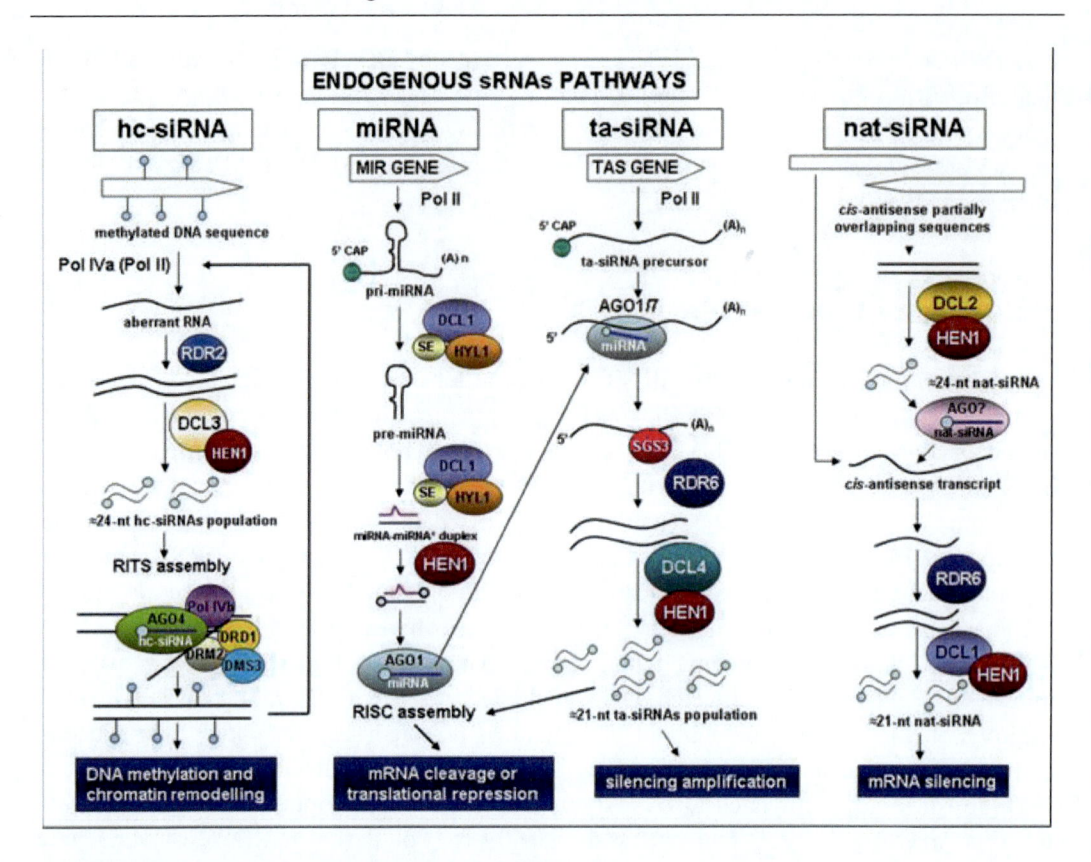

Figure 2. Endogenous short RNAs (sRNAs) pathways.
See the text for a detailed explanation of the pathways shown in figure.

Amplification of Silencing Signals

To achieve a more comprehensive vision of the RNA silencing picture, it is important not to think about it as a cell restricted phenomenon. Several experiments carried out in *Nicotiana* plants witness the existence of a systemic amplification of silencing in plants [63]. Transitive silencing is a mechanism discovered both in nematodes and in plants, which occurs when silencing is spread from the source mRNA sequence along the 5' and 3' adjacent non-target sequences [64] [65]. This phenomenon is typical of transgene induced silencing, even though cases of transitivity have been reported for highly transcribed endogenes [66]. In transitive silencing, primary siRNAs derived from dsRNA recruit cellular RDRs to homologous ssRNA, with the production of new dsRNA and secondary siRNAs that are not related to primary siRNAs. The amplification of silencing signals is related to the spreading of silencing effects, which can either occur through the plasmodesmata, resulting in a local spreading, or through the phloematic system thus determining a systemic amplification of silencing. More in detail, the local spreading can be based either on short-range movements of silencing effectors, as dsRNAs and siRNAs, in a layer of 10-15 cells surrounding the site source of RNA silencing, thus without a true amplification, or on an extensive local spreading outside the 10-15 cells layer, which actually operates to amplify silencing in sink tissues through a RDR6-directed mechanism [67]. The systemic spreading of RNA silencing via vascular tissues is essentially an antiviral strategy, which acts to prevent virus spreading outside the site source of infection to the whole organism [67]. The silencing signal is indeed carried in phloematic sap from the source leaf to all sink leaves, in order to block viral diffusion. Systemic silencing was widely ascertained through grafting experiments [68] and by studying the strategies activated by viral silencing suppressor proteins (see below) to react against host defences [69].

RNA Silencing: A Host Defence System

Eukaryotic organisms have evolved defence strategies against invading nucleic acids, such as those deriving from endocellular pathogens, as viruses, and from genomic mobile elements. Virus diseases are significant threats to modern agriculture and their control remains a challenge to the management of cultivation. Virus infestation of cultivated areas results in a range of effects, from reduced crop quality to complete plant devastation. Plant antiviral defence strategies were hypothesized for the first time observing phenotypes of tobacco plants inoculated with less virulent *Tobacco ringspot virus* (TRSV) strains [70]. Among host reactions, the RNA silencing was recognized as an antiviral mechanism that protected organisms against RNA viruses [71]. There are two major systems in which RNA silencing prevents virus multiplication in plants. The first is the limitation of viral RNA accumulation in the initially infected cells, achieved by processing the viral dsRNA into viral siRNAs, which are further used to guide the destruction of target viral genomes. The second is the spreading of silencing to the whole plant via systemic signalling. This systemic response to viral infections is connected to phenomena known for decades by plant virologists. Scientists had described as 'recovery' the fact that virus-infected plants were able to grow new symptom-free leaves, which could not be infected by the same (or a related) virus [9]. Linked to recovery is the 'cross-protection' phenomenon, which occurs when a pre-existing viral infection protects the plant from a more severe strain of the same virus or from a related virus species. On the basis of the nucleotide sequence similarity, the first virus primes the silencing mechanism to target the second virus and prevent its accumulation [72]. Another classical phenomenon is the plant viral 'synergism'. When two viruses are simultaneously inoculated, synergism results in enhanced disease symptoms and virus accumulation, especially in systematically infected leaves. This mechanism is an evident consequence of the effective suppression of silencing caused by one virus, resulting in an increase in the accumulation of the co-infecting virus [73]. More in detail, a member of the potyvirus group is involved in several cases of synergisms and provides potent suppression of silencing (see below), resulting in the accumulation of a broad range of viruses. A classical example of this is the increase in *Potato virus X* (PVX) accumulation, when PVX is inoculated together with the *Potato virus Y* (PVY) encoding for the P1/HC-Pro, a strong silencing suppressor [74].

Viral siRNAs, which are key molecules in the establishment and spreading of silencing, arise from dsRNA that can be produced during replication of viral RNA genomes in the cytoplasm by viral RNA-dependent RNA polymerases (RDRs) or by bi-directional transcription of DNA viruses in the nucleus [75]. Alternative sources of dsRNAs are the activity of host RDRs and the formation of secondary RNA structures as intramolecular hairpins in viral genomes. High-throughput sequencing of small RNAs from virus infected plants showed approximately equal ratios of (+) and (-) viral siRNAs strands derived from (+) ssRNA viruses [76] [77]. Probably, these siRNAs derived from the replication of long dsRNA precursors. However, more abundant (+) siRNAs were detected in plants infected with tombusviruses and carmoviruses [78] [77] and many of them mapped to hairpin secondary structures within viral genomes. Similarly, siRNAs from the genomic (+) strand of viroid RNA seem to accumulate much higher than those deriving from the replicating (-) strand [79]. However, other authors have reported abundant (-) siRNAs from viroids infecting tomato and grapevine [80] [81] [82]. These (-) siRNAs could derive from DCL targeting dsRNAs generated during viroid replication or resulting from the activity of host RDRs on viroid ssRNA template. RNA silencing can be amplified and spread by cellular RDRs through the mechanisms analyzed in the above sections. The mobile silencing signal moves along the same route of the virus infection: cell-to-cell through the plasmodesmata and over long distances through the phloem. This signal moves out of the initially infected cell range to anticipate and prevent the movement of viral RNAs, thus it helps to immunize sink tissues, establishing antiviral silencing ahead at the viral infection front [75].

Viral Suppressor of RNA Silencing (VSR)

Further support for RNA silencing as an antiviral mechanism comes from the evolution of viral proteins that suppress RNA silencing (VSR). VSRs inhibit the generation and the spreading of silencing signal by a range of diverse mechanisms, including block of siRNAs generation, inhibition of their incorporation into effector complexes and direct interaction with effector complexes [83]. Over 30 VSRs have been identified from different types of viruses, including RNA and DNA viruses [84]. As no sequence homology has been detected between distinct silencing suppressors, they probably evolved independently in different virus groups. VSR activity has been identified in structural (coat protein) as well as non-structural proteins involved in almost every viral function, including movement proteins, viral replicases, replication enhancers and transcriptional activators [85].

One of the first VSR identified, the potyvirus-helper component proteinase P1/HC-Pro, was known to affect aphid transmission, polyprotein processing, genome amplification and the long distance movement of the virus [86]. Nowadays, it is more precisely known that it strongly suppresses RNA silencing, most likely acting on a maintenance step affecting assembling and/or targeting of the RISC complex [87]. The role of P1/HC-Pro in genome amplification and long distance movement is correlated with its silencing suppressor activity. P1/HC-Pro interacts with rgs-CaM, a calmodulin-like protein that negatively regulates silencing [88]. The 2b protein of cucumoviruses was recognized as a silencing suppressor at about the same time [89], and it was found out that it enhances the long distance movement of *Cucumber mosaic virus* (CMV) in a host-dependent manner. The 2b VSR also interferes with the RISC complex: it binds AGO1 directly interfering with its slicing activity [90]. Moreover, 2b prevents any other plant attempt to stop the viral infection blocking the systemic amplification of RNA silencing signals, as demonstrated through grafting experiments in 2b-expressing transgenic plants [68]. In different tombusviruses, including *Cymbidium ringspot virus* (CymRSV), and *Tomato bushy stunt virus* (TBSV), the p19 gene functions to assist systemic spreading and symptom development in host plants [91]. Several groups independently demonstrated p19 silencing suppressor activity studying different tombusviruses [92] [93]. P19 was the first protein demonstrated to directly bind siRNAs through a dimerization step, presumably functioning to prevent the siRNA assembly with the RISC complex [93] [94]. P19 constitutes the first suppressor for which a target in the silencing pathway has been identified.

Voinnet *et al.* [95] demonstrated that p25, a protein previously shown to be important in the cell-to-cell movement of PVX, was indeed a silencing suppressor that avoided silencing amplification out of the primary-infected cells, where silencing initiation took place. However, the PVX p25 is a relatively weak suppressor of RNA silencing when compared to the p25 equivalents of other potexviruses, which were effectively able to re-activate a silenced transgene [96]. *Citrus tristeza virus* (CTV) is an RNA virus encoding at least 11 open reading frames. It has recently been established that three different proteins, p20, p23 and the CP of this complex virus, have silencing suppression activities. Specifically, it was suggested that the RNA silencing suppression, occurring at multiple steps of the silencing pathway, may be essential for viruses with large RNA genomes, such as CTV [97]. The VSR P38 is the coat protein of *Turnip crinkle virus* (TCV), and competes with DCL4 to bind dsRNAs in order to avoid digestion of viral genome into short RNAs [98]. Other viruses are able to prevent RISC assembling: the polerovirus protein P0, after associating with AGO1 [99], directs its degradation through a ubiquitin-based way [100], thus affecting AGO proteins steady-state level and counterattacking the host-defence silencing system [101]. V2 protein, encoded by the viral DNA of *Tomato yellow leaf curl virus* (TYLCV), binds the tomato orthologous of the *Arabidopsis* protein SGS3, a factor typically required for RDR6 proper functionality [6], preventing SGS3 association with dsRNAs [102] [103]. Other two geminivirus VSRs, L2 and AL2, alter the RdDM pathway in the plant nucleus by negative regulating the adenosine kinase ADK, which is tightly associated with the methylation machinery. Therefore, this causes defects in proper genome methylation management and cut out any plant tolerance to the viral infection [104] [105]. Also in the nucleus, the *Cauliflower mosaic virus* (CaMV) P6 and the *Tobacco mosaic virus* (TMV) P122 proteins alterate two factors necessary for siRNAs production and stability, DRB4 and HEN1 respectively [106] [107] [108]. The above cited suppressor proteins and other identified VSRs have been described in detail in several recent reviews [109] [110] [111].

It is well established that antiviral and endogenous silencing pathways share common elements (see above) and VSRs often interact with these common elements. Thus, an attractive hypothesis is that viral infection induces disease symptoms by perturbation of the endogenous small RNA pathways. When expressed from a stably integrated transgene, some of the VSRs (for instance, P1/HC-Pro that inhibits both endogenous mi- and siRNA-mediated gene regulations) cause developmental abnormalities in transgenic plants similar to those exhibited in infected plants [112] [113]. However, recent studies have further shown that VSRs function to suppress antiviral silencing, but at the same time they are dispensable for disease induction, which thus does not support a direct role of VSRs in eliciting disease symptoms [111].

In addition to RNA silencing, another antiviral defence adopted by plants is based on resistance (R) genes, which can trigger highly effective defence programs, known as hypersensitive response (HR) and systemic acquired resistance (SAR) [109]. R genes confer resistance to specific viral strains possessing an avirulence (Avr) determinant. Avr proteins are usually necessary for successful infection and are almost invariably virulence factors in a susceptible host. It seems probable that plants could develop a system allowing RNA silencing and R genes to communicate each other and limit viral infections. In some cases, this cross-talk already exists: the TCV-CP P38 is at the same time the Avr determinant of HR and an RNA silencing suppressor [114].

Transgene Silencing Techniques

Since the 80s, when the inhibitory effect of long antisense RNAs on corresponding protein-coding (sense) mRNAs in animal cells was discovered [115], different transgene silencing techniques have been developed in plants. The achievement of sense or antisense transgenic plants, which belongs to classical PTGS approaches, has widely been used to investigate plant gene functions, but proved to be labour- and time-consuming, and not always successful [116] [117]. Constructs directly harbouring dsRNA structures bypass the requirement for dsRNA synthesis via RDRs, which is probably the rate limiting step in sense and antisense techniques [118]. In the hairpin RNA (hpRNA) method, which was first applied to rice [18], RNA hybridizes with itself to form a hairpin structure comprising a single-strand loop region and a base-paired stem region that triggers the RNA silencing process. The efficiency of silencing increases when the loop region is replaced by an intron that gives stability *in vivo* [119] [120]. As shown by various works, hairpin constructs are one of the most efficient strategies for triggering gene silencing at whole plant level or in specific tissues [121] [122] [123] [124] [125] [126] [127].

Virus-Induced Gene Silencing

The virus-induced gene silencing (VIGS), besides being a host defence system (see the 'cross-protection' described above), was used as strategy to introduce and silence target sequences in plant. The VIGS technique is based on the introduction of exogenous sequences in a virus genome. During plant infection, foreign sequences induce and become the target of the RNA silencing response activated by the host [123]. This system allows a transient transformation of plants, even in fully developed plants, and is useful for those species that are difficult to transform stably (see the following sections). Shortly after its discovery, VIGS was refined to develop a technology for functional genomic studies by directing the silencing response against endogenous plant sequences [128]. VIGS was first demonstrated by silencing a tobacco phytoene desaturase gene (*pds*, which encodes an enzyme of the carotenoid biosynthetic pathway), inserted into an infectious strain of *Tobacco mosaic virus* (TMV): infected plants showed marked photobleaching under strong light conditions [129]. Later, the Baulcombe's research group developed the first VIGS constructs, using PVX [130] or *Tobacco rattle virus* (TRV) [131]. Nowadays, different VIGS system are available. Burch-Smith *et al.* [132] developed an efficient TRV-based VIGS method to silence the *A. thaliana* genes; Pflieger *et al.* [133] showed how a viral vector derived from *Turnip yellow mosaic virus* (TYMV) was able to induce VIGS in *Arabidopsis*. The *Tomato golden mosaic virus*-based VIGS system seems an effective initiator of targeted RNA silencing in *N. benthamiana*

[134]. In the satellite-virus-induced silencing system (SVISS), the target sequence is inserted into the satellite RNA, which is further inoculated with the associated virus [135].

TRV-VIGS is probably the best system to date, because it bypasses different strong drawbacks of VIGS system. First, TRV-VIGS seems to infect almost all tissues, also meristems and floral organs, whereas many other viruses do not colonize the meristem region. In addition, TRV shows mild symptoms and this is very important, because many VIGS have strong symptoms that might mask the phenotype resulting from the gene silencing event [123].

Among disadvantages of VIGS system, there are host range limitations (virus has to be infective for the species of interest) [123] and, furthermore, complete loss-of-function may not be achieved [133]. On the other side, several advantages can be referred. Indeed, VIGS-based vectors are high-throughput (they can be delivered by simply rubbing an infective transcript on a plant leaf or they can be injected as Agro-VIGS); they can be applied to mature plants; they are characterized by a generation of rapid phenotypes (2-3-weeks from infection) and, finally, VIGS is a very good system for those species difficult to transform [123].

More detailed descriptions of VIGS vectors and their applications have recently been described in several reviews [123] [136] [137].

Artificial miRNAs

Artificial microRNAs (amiRNAs) have been developed in a recent strategy to trigger RNA silencing. This technology, which was first employed in human cell lines [138] and later in *Arabidopsis* [139], follows the pathway of endogenous miRNAs biogenesis starting from miRNA precursor transcription. Both sequences of the stem region of miRNA precursors can be altered without changing structural features. This is indispensable to guide DCL1-mediated processing and obtain a high-level accumulation of the desired miRNA sequence. Engineered *Arabidopsis* precursors have successfully been used in tomato and tobacco [140], even though further tests on endogenous miRNA precursors from the species of interest are recommended.

Ossowski and collaborators [141] developed a platform aimed to automate and accelerate the amiRNA design. The WMD (Web MicroRNA Designer) platform was initially implemented for *A. thaliana*, but at present it has been extended to more than 90 plants (http://wmd3.weigelworld.org) [142]. This design tool selects small RNAs on the base of their effectiveness and highest target specificity using optimization criteria. For instance, amiRNAs are required to resemble natural miRNAs: they have to start with a U, to display 5' instability relative to their amiRNA* and to show an A or an U as 10th nucleotide. The 21-mer amiRNA selected is engineered into a miRNA precursor through an overlapping PCR to replace the endogenous miRNA sequence. Finally, the obtained amiRNA precursor can be transferred into a binary vector under the control of either a strong or an inducible promoter.

This RNA silencing approach is used both to silence endogenous genes and to confer virus resistance [143]. AmiRNAs have several advantages in generating viral immunity: unlike VIGS, they do not have host range limitations, a long viral cDNA fragment is not required [144] and the environmental concern of hypothetical complementation/recombination of viral sequences with non-target viruses is overcome [145]. The success of the amiRNA strategy is not only dependent from a good construct design, but it also relies on the accessibility of the target selected region. This was confirmed by studies involving tobacco and *Arabidopsis* transgenic plants engineered to acquire resistance to CMV [144]. The 3'-UTR CMV genome region was selected as target, since it showed a high conserved nucleotide sequence among different CMV strains, suggesting that this region could harbor sequences required for biological functions. In this way, CMV could not evade the amiRNA mechanism through mutations. Moreover, the 3'-UTR CMV region presents a TLS (tRNA-like structure) segment very important for virus replication. However, the presence of this segment confers a low accessibility to miRNA-RISC-mediated cleavage, which was experimentally confirmed: transgenic plants expressing amiRNAs targeting the TLS region showed no protection against CMV infection. On the contrary, *Arabidopsis* and tobacco plants engineered with amiRNAs targeting RISC accessible sites acquired high resistance to CMV attack [144].

In 2006, Niu and colleagues [143] developed a dimeric amiRNA strategy: the expression of more than one type of amiRNA, designed for different target RNAs, aimed to confer resistance to different viruses. Two viral silencing suppressors were selected as amiRNA target sequences: P69, a 69-kDa movement protein of TYMV, and HC-Pro, a proteinase of *Turnip mosaic virus* (TuMV). The expression of a dimeric amiRNA[159] precursor, which can generate both amiR-P69[159] and amiR-HC-Pro[159], conferred resistance to both viruses. The resistance is also maintained at low temperatures, which usually inhibit PTGS [146]. More recently, the attention has been focused on the identification of critical positions leading this resistance type within the target site and on the virus counter defensive response [147]. Researchers employed a heterologous-virus system using a TuMV-GFP viral vector to carry on a dispensable 21-nt sequence of the P69 gene targeted by amiR[159]-P69. Different viral vectors were generated, each containing a different nucleotide mutation in 21-nt sequence. The 21-mutant viruses were inoculated on amiR[159]-P69 plants and the proportion of infected transgenic plants was used to determine the importance of the mutated nucleotide position within the amiRNA target site. The authors concluded that more critical positions are localized on those sequences complementary to the 5' portion of the amiRNA, whereas moderate critical positions are localized in the central region of the target. This can help to design effective amiRNAs. Mutation in certain critical positions reduces amiRNA-RISC-mediated cleavage efficiency; certain mutant viruses can escape from the amiRNA mechanism and replicate themselves. During subsequent virus replications, additional mutations or deletions of the target sequence could be positively selected.

The amiRNA technology was also employed to silence endogenous genes of interest. For instance, Warthmann *et al.* [148] demonstrated that this phenomenon naturally occurs in different rice cultivars expressing 3 different genes: *eui1*, elongated uppermost internode1, *pds* and *spl11*, Spotted leaf 11.. Interestingly, the amiRNA strategy allowed to discriminate among several homologous of *spl11* in the rice genome. Recently, the amiRNA strategy has allowed to deepen current knowledge about the regulation of ethylene response in *Arabidopsis* [149]. EIN2 (ETHYLENE-INSENSITIVE2) is an integral membrane protein that works as central regulator of all ethylene responses. Two F-box proteins (ETP1, EIN2 TARGETING PROTEIN1; ETP2, EIN2 TARGETING PROTEIN2) interact with EIN2 modulating its responses. amiRNA-mediated down-regulation of ETP1 and ETP2 impairs the interaction and leads to constitutive ethylene response phenotypes.

Transitive RNA Silencing

An efficient system to induce RNA silencing is based on the transitive silencing mechanisms described above. Brummell and colleagues [120] demonstrated that, within the construct, the presence of a 3'-inverted repeat untranslated region (3'-UTR), located at the 3'-end of the transgene, leads to high-frequency and high-efficiency of PTGS: up to 90% of the tomato transformants population, modified with polygalacturonase (PG) transgenes driven by constitutive promoter, showed a 98% reduction of endogenous PG mRNAs in ripened fruits. This system does not require inverted repeat sequence of target gene, but a 3'-UTR region of a heterologous gene is used (SHUTR, silencing by heterologous 3'-UTR method). This method represented an interesting improvement for RNA silencing techniques, because it allowed to gain an easier, more rapid and efficient preparation than traditional constructs, indeed the target sequence is inserted through a single cloning step inside an already prepared vector. High-throughput methods for gene silencing are preferred when the purpose is characterizing the function of hundreds of genes in sequenced genomes.

Recently, Petsch and collaborators [150] have used transitive RNA silencing approaches to develop a peculiar forward mutagenesis technique. They combined an RNA silencing-based method with the laser microdissection technology. More in detail, bi-directional (i.e. sense and antisense) cDNA libraries from laser microdissected mesophyll cells of *Arabidopsis* leaves were prepared, directionally cloned into a transitive silencing vector and subsequently used to transform wild-type *Arabidopsis* plants by *Agrobacterium tumefaciens*. In this way, a selective population of transcripts was chosen and analyses on mutant phenotypes allowed to deepen information about genes involved in specific biological processes (as, in this case, photosynthetic process). Similarly, this target silencing strategy could help to identify those genes expressed in response to external stimuli (i.e. pathogen, abiotic stresses) or during a specific developmental phase.

Screenings on mutant phenotypes and tests on the transgenic lines obtained require a large investment in terms of both work and time to set up an RNA silencing study. Computer-based RNA folding predictions or target site predictions, based on statistical or structural motifs, are aimed to minimize these troubles and to select accessible target sites in structured mRNA [151]. *In silico* predictions had limited success, probably due to the foreseeable influence of intracellular factors [152]. *In vivo* validation of the efficiency of candidate RNA silencing-trigger molecules was preferred and a transient co-expression assay involving target::luciferase fusions and RNA silencing-trigger constructs was developed [153] [154]. Effectiveness of different constructs against the target sequence is quantified by the reduction of luciferase protein activity. This approach has recently been named "InVITE assay" (*in vivo* transient expression analysis) by Birch and collaborators [152] and has been tested in plants with various target genes. The authors concluded that this method should actually help to identify those constructs effective for specific target silencing in plants.

Transgene Silencing Applications in Crop Plants

Genetic transformation has emerged as a powerful tool for the genetic improvement of all plants, particularly for woody species, for which breeding programs are hindered by long juvenile periods, high levels of heterozygosity and, in some cases, sexual incompatibility. Furthermore, cultivation of woody crops is based on a few cultivars multiplied by vegetative propagation. Genetic engineering techniques allow to introduce target genes avoiding modifications of other important characteristics of the cultivar itself, while vegetative propagation potentially provides unlimited production of the desired transgenic lines.

In the last years, significant advances have been made in the development of various components of transgenic technology, including transformation techniques and protocols for plants regeneration. Combined with these protocols, gene silencing has largely been used as a tool to knockout and/or down-regulate genes for functional genomic studies. Furthermore, this technology has been used for commercial applications mainly focused on enhancing disease resistance against biotic agents and on modification of metabolic pathways for quality improvement, new substances production and allergens reduction. The perceived negative aspects, derived from modern biotechnology development and its direct application in agriculture, have soon increased scepticism among customers and public authorities. However, the use of PTGS, as a mechanism based on the expression of small amounts of non protein-coding RNAs, would be preferable to strategies based on protein expression.

Transgene Silencing for Plant Disease Resistance

Plants are generally susceptible to virus diseases, particularly perennial crops are more exposed to infections because of their long-term life span. The protection of plants from viral infection was one of the first and most important purposes of transgenic research in plants and it was also one of the first commercial applications of PTGS. However, early attempts to get virus resistance through the introduction of virus-derived sense or antisense sequences were not always successful. In retrospect, some of the unexplained effects observed in coat protein (CP)-mediated resistance techniques seem actually caused by PTGS through the accidental formation of dsRNA.

Among first applications of PTGS, there was the transgenic papaya (*Carica papaya* L.) with resistance to *Papaya ringspot virus* (PRSV). PRSV is a potyvirus non-persistently transmitted by aphids, which destroys the photosynthetic capacity of the canopy, leading to fruit quality and yield reductions, loss of vegetative vigour, and eventual mortality. During the 90's, this threat invaded the major papaya production areas of Hawaii islands, causing extensive damages with the risk of totally destroying this industry. Fitch and colleagues [155] [156] obtained transgenic plants expressing the sense CP gene of PRSV, and showed that lines 55-1 and R1, which derived from a cross between a female 55-1 and a non-transgenic 'Sunset', were highly resistant to PRSV in greenhouse [156] and field experiments [157]. Later, it was shown that this resistance was an homology-dependent resistance via PTGS [157]. Two cultivars developed from line 55-1 ('SunUp', homozygous for the CP gene, and 'UH Rainbow', a F1 hybrid between 'SunUp' and the non-transgenic 'Kapoho' cultivar) were produced, successfully cultivated and commercialized, contributing to save the papaya industry in Hawaii. At present, transgenic papaya is the only transgenic fruit tree authorized for field cultivation and commercialization. Greenhouse experiments showed that line 55-1 deriving from hemizygous plants are susceptible to a number of PRSV isolated from other geographic regions [158]. This prompted several laboratories to develop more transgenic lines effective against others PRSV isolates [159]. The complete genome sequence of the papaya

'SunUp' variety, recently reported by Ming *et al.* [160], represents the first example of a transgenic organism completely sequenced. From a biosafety perspective, the papaya sequencing project provided molecular evidences that the transgene structurally and functionally remains intact in distant descendents of the original integration event [161]. In addition to PRSV, another potyvirus, the *Papaya leaf-distortion mosaic virus* (PLDMV), causes one of the major constraint to papaya production in Japan. Lately, Kung *et al.* [162] have transferred into papaya an untranslatable chimeric construct, containing both the truncated CP coding region of the PLDMV and the truncated CP coding region with the complete 3' untranslated region of PRSV. The results of greenhouse evaluation have indicated that several transgenic lines showed double virus resistance to PLDMV and PRSV occurring via a PTGS mechanism.

After this first report, many other transgenic fruit plants protected against viruses via PTGS mechanism were produced. For instance, *Citrus* was transformed with the CP gene of CTV [163] [164] [165]. CTV, an aphid-transmitted closterovirus, is the causal agent of one of the most economically important diseases of *Citrus*. This virus is restricted to *Citrus* species and causes two main diseases: decline and death of most *Citrus* species grafted on sour orange (*Citrus aurantium* L.), and stem pitting, stunting and reduced fruit yields and quality rates in some *Citrus* varieties regardless of the rootstock. Domínguez *et al.* [166] demonstrated the PDR (pathogen-derived resistance) in transgenic Mexican lime (*C. aurantifolia* (Christ.) Swing.) carrying the CTV-CP gene. Transgenic Mexican limes expressing the p23 gene of CTV displayed characteristics typical of PTGS: multiple copies of the transgene, low levels of the corresponding mRNA, methylation of the silenced transgene, and accumulation of p23-specific siRNAs. When graft- or aphid-inoculated with CTV, some propagations of these silenced lines were immune: they expressed neither symptoms nor accumulated virions [167]. Later, Febres *et al.* [168] obtained some lines of grapefruit transformed with the 3' end of CTV sequence, which showed partial or complete resistance to CTV with mechanisms involving PTGS and TGS. Mexican lime plants were transformed with a region of CTV genome in sense, antisense and hairpin forms, in order to acquire virus resistance [127]. All sense, antisense and empty-vector transgenic lines were susceptible to CTV, whereas 9 on 30 hairpin lines showed CTV resistance.

Plum pox virus (PPV), the causal agent of the Sharka disease, is one of the major constraints to *Prunus* production. PPV is naturally spread through a non-persistent manner by several aphid species; symptoms of virus infection include chlorotic spots and ring patterns on fruits and seeds, and mild to severe fruit deformations. Infected fruits are often acidic with low sugar content and unsuitable for fresh consumption or processing. Scorza *et al.* [169] transformed plum (*Prunus domestica* L.) and apricot (*Prunus armeniaca* L.) with the CP of PPV. The transgenic clone C5 showed high resistance to PPV whether inoculated by aphids and chip budding [170]. This transgenic clone displayed characteristics typical of PTGS, including a high level of transgene transcription in the nucleus, low levels of transgenic mRNAs in the cytoplasm, a complex multicopy transgene insertion with aberrant copies, and methylation of the silenced PPV-CP [171]. Non-transformed plum trees react to viral infections by initiating PTGS-like mechanisms involving the production of 21-nt siRNAs, but high-levels of virus resistance in C5 transgenic plum were characterized by the production of a long-size class of 24-nt siRNAs [172]. Field tests on C5 and other transgenic lines performed in Poland, Romania and Spain, demonstrated that C5 trees exposed for several years to natural infection did not become infected, while susceptible transgenic and untransformed trees developed severe symptoms within the first year [173]. Although highly resistant in field tests, C5 trees could be artificially infected by chip budding or via susceptible rootstock. Even 8 years after the virus inoculation, infected C5 trees showed only a few mild symptoms on single isolated shoots, indicating the long-term nature and high level of resistance to PPV [174]. A different strategy was adopted by Hily *et al.* [175], which transformed plum through a construct with self complementary sequences of PPV-CP, separated by an intron, which produced hpRNAs efficiently eliciting PTGS. In 2009, the USA Food and Drug Administration (FDA) approved the use of transgenic plum resistant to PPV as food, concluding that transgenic plum is not materially different in composition, safety, or any relevant parameter from the other non-transformed plums (AGBIOS Biotech Crop Database, http://www.agbios.com/dbase.php?action=ShowForm).

Despite these positive results, induction of PTGS in transgenic plants does not always result in effective protection against the target virus. In grapevine (*Vitis* spp.), at least sixty different documented viruses, which infect these species and cause global losses estimated for over 1 billion US dollars, have been detected to date [176]. Despite the considerable efforts made in recent years, and with the partial exception of some rootstocks resistant to *Grapevine fanleaf virus* (GFLV) [177], transgenic virus-resistant grapevines have not been obtained yet, confirming the complexity of this plant-pathogen relationship. Transgenic grapevine containing the GFLV-CP gene and showing transgene silencing were unable to contrast the virus spread simply by graft inoculation [178]. In these transgenic grapevine lines the correlation between accumulation of siRNAs, transgene methylation and RNA silencing could not be confirmed. Highly methylated cytosine residues were detected in the GFLV-CP transgene, in the terminator and in the 35S promoter of grapevines without transgene expression, but no detectable level of siRNAs was recorded. However, it is possible that RNA signalling molecules were responsible for the DNA methylation patterns observed in these lines, even though siRNAs concentrations were below detection level [179]. The susceptibility to GFLV could be due both to the high viral inoculum and to the constant viral pressure from the rootstock applied to relatively young and small plants: under these conditions, transgenic grapevines might be unable to suppress GFLV replication [178].

Potato (*Solanum tuberosum* L.) is considered one of the most important crop worldwide and has been the subject of many breeding efforts. Potatoes are generally multiplied by vegetative propagation, so the related viral infections are very destructive: viruses persist in the tubers and the tuber-borne secondary infections are in some cases more severe than primary infections. PVY is one of the most damaging viruses of potato (transmitted both mechanically and through aphids), and in some parts of Europe, it is the most widespread virus infecting potatoes. In the past, diverse attempts to engineer PVY-resistance were made in potato plants [180] [181] [182], but the resistance was fairly weak and the protection seemed to be restricted to a few transgenic lines. Missiou *et al.* [183] obtained transgenic potato lines containing a part of the PVY-CP with a typical hairpin-like RNA structure. The great majority of the analyzed lines showed very strong resistance with broad strain specificity, which could not be overcome by simultaneous or prior infection with the PVX, another widespread potato virus. Jorgensen and Albrechtsen [184] discovered that the silencing pathway triggered by an inverted repeat construct of PVY-CP persisted in presence of the repressor of RNA silencing HC-Pro, when the silencing induced by a single sense CP construct had failed. Marker gene-free transgenic plants resistant to both PVX and PVY have recently been obtained by Bai *et al.* [185], using an RNA silencing-mechanism. The authors introduced in potato a chimeric gene derived from the PVX-CP, using a marker-free expression vector and the nuclear inclusion protein sequence of PVY in the form of an hpRNA. *Potato leafroll virus* (PLRV) is another damaging virus infecting potatoes, thus productivity yields may decrease up to 80-90% in susceptible cultivars and greater losses may be expected when the viral infection occurs in association with PVY and PVX. Resistance to PLRV was engineered in potato and commercialized by Monsanto Company (http://www.monsanto.com/monsanto_today/for_the_record/newleaf_potato.asp), and gained via a PTGS-based method developed by Vazquez-Rovere *et al.* [186] through the introduction of replicase gene sequences. In sweetpotato (*Ipomoea bataas* L.), Kreuze *et al.* [187] obtained RNA silencing-mediated resistance to *Sweetpotato chlorotic stunt virus* (SPCSV). SPCSV is one of the most important pathogens of sweetpotato, which can reduce yields up to 50% and causes diverse synergistic disease complexes when co-infecting with other viruses. No sources of true resistance and only low levels of tolerance to SPCSV are available in sweetpotato germplasm. In addition, sweetpotato is a clonally propagated, highly heterozygous, polyploid and out-crossing species, whose low fertility makes the introgression of gene traits challenging. When a sweetpotato was transformed with an hairpin construct targeting the replicase of SPCSV, none of the transgenic events was immune to SPCSV, but the resulting transgenic plants showed a reduction of virus titres and weaker symptoms [187].

RNA silencing is an effective defence tool against RNA viruses, however the application of this strategy against DNA viruses have only produced inconsistent results to date. *Tomato yellow leaf curl Sardinia virus* (TYLCSV) and TYLCV are two species of geminiviruses (viruses with a ssDNA genome) causing tomato yellow leaf curl, one of the most important worldwide tomato (*Solanum lycopersicum* L.) diseases. Numerous investigations were published to prevent TYLCV infection in tomato plants. Most approaches involved the replication-associated protein (Rep) sequences in dysfunctional forms. Brunetti *et al.* [188] first reported transgenic tomato plants resistant to TYLCSV by a truncated replicase gene. This resistance involves an RNA silencing process [189] and is overcome by the virus [190]. Yang *et al.* [191] demonstrated resistance and immunity against TYLCV in transgenic tomato carrying variations of the TYLCV *Rep* gene. The resistance effect was gained through a mechanism involving PTGS, which is able to interfere with TYLCV transcription. Fuente *et al.* [192] transformed tomato with an intron-hpRNA directed against invading TYLCV C1 gene coding for the Rep protein. These authors reported for the first time resistance to a plant DNA virus obtained by the use of the hpRNAs approach. Bian and colleagues [193] engineered tomato with a hpRNA of the *C2* open reading frame of *Tomato leaf curl virus* (TLCV). The transgenic plants produced a significant level of siRNAs, but they only exhibited delayed symptoms and were susceptible to TLCV infection [193].

CMV, the type member of the genus cucumovirus, family Bromoviridae, has been found in most countries of the world and its host range exceeds 800 plant species, making it one of the most important viruses for its economic impact. CMV was one of the primary targets for development of transgene-mediated resistance, especially because of its importance and the absence of CMV resistance genes in the germplasm of most crops. Several powerful strategies, based on PDR, have been applied to produce transgenic plants resistant to CMV, as resistance mediated by the viral CP, the viral replicase, and PTGS, which has recently been reviewed [194]. In the case of plants expressing a CMV-CP gene, there is an apparent lack of evidence for PTGS-based resistance and some findings suggest that interference with virus movement is primarily involved. The CMV 2b gene is a potent suppressor of PTGS, and the absence of PTGS-mediated resistance in CP constructs may be strictly attributed to the activity of this VSR. In squash (*Cucurbita pepo* spp. *ovifera* var. *ovifera*), the efficacy of the CMV-CP approach was demonstrated by the creation of the triply resistant line CZW-3 (deregulated in 1996 for commercial use in the United States), in which the expression of three CP genes confers resistance to CMV, to *Zucchini yellow mosaic virus* and *Watermelon mosaic virus* [195]. Resistance to CMV mediated by PTGS was shown by transforming tobacco plants with a construct containing an inverted-repeated fragment of CMV-CP [196] [197]. In the transgenic lines siRNA concentrations were positively correlated to resistance levels. Transgenic tomato plants expressing a benign variant of CMV satellite RNA (CMV Tfn-satRNA) do not produce symptoms when challenged with a satRNA-free strain of CMV [198]. The same transgenic plants were initially susceptible to necrosis, when inoculated by a CMV strain supporting a necrogenic variant of satRNA. However, a total recovery from the necrosis was observed in the newly developing leaves through a mechanism involving the RNA silencing [198].

Gene silencing was initially studied in dicotyledonous plants, however during the late 90's, several reports showed that transgene silencing occurs also in monocots. For instance, in rice, resistance to *Rice yellow mottle virus* (RYMV) was gained using a transgene encoding the RNA-dependent RNA polymerase of the virus [199]. RYMV causes a major limiting disease in African rice production, with yellowing, mottling, and stunting of infected plants with sterile or unfilled grains. In the field, the infection with RYMV reduces crop production up to 97%, and up to 54% in a tolerant cultivar. In addition, it was not possible to achieve introgression of this resistance into cultivated rice, because resistance is a recessive, polygenic trait. Pinto and colleagues [199] obtained several lines resistant to virus and in particular in one transgenic line was observed a complete suppression of virus multiplication. The resistance derives from an RNA-based mechanism associated to PTGS and was stable over at least three generations. Wang and colleagues [200], working on barley genetic transformation (*Hordeum vulgare* L.), used an hpRNA construct and obtained resistance to *Barley yellow dwarf virus* (BYDV), which is a virus of global importance, since it infects and reduces yields of several worldwide cereal species. In particular, the authors enhanced PTGS-mediated resistance using an inverted-repeat transgene encoding the 5' end of the viral genome, driven by the maize ubiquitin promoter. In transgenic sugarcane (*Saccharum* spp. L.), an untranslatable form of the CP gene, derived from the *Sorghum mosaic virus* (SrMV), was responsible for virus resistance [201]. The epidemic spread of this virus, together with the *Sugarcane mosaic virus* (SCMV), caused in the 20's of the last century a near collapse of the sugar industry in South America and was the main factor in replacing the susceptible cultivars with interspecific hybrids tolerant to SCMV. However, even today SrMV and SCMV cause the major losses to the sugarcane industry. Ryegrass (*Lolium perenne* L.) is an agronomically important perennial monocot in areas with a temperate climate. *Ryegrass mosaic virus* (RgMV), a virus belonging to the family potyvirideae, frequently reduces yield and persistence of perennial ryegrasses. Xu *et al.* [202] introduced an untranslatable RgMV-CP gene into perennial ryegrass using particle bombardment and found resistance PTGS-mediated in transgenic plants. Resistance against high-dose virion inoculum was demonstrated in primary transformants and in the progeny, after inoculation with different RgMV-strains.

RNA silencing is also a tool to protect plants against infective organisms other than viruses. For instance, *Agrobacterium tumefaciens* is a soilborne bacterium effector of the crown gall disease, which causes economically significant damages in fruit and nut orchards, vineyards, and nurseries all around the world. Tumors on stems and leaves result from excessive production of the phytohormones auxin and cytokinin in plant cells genetically transformed by *Agrobacterium*, during its natural infection cycle. *Arabidopsis* and tomato plants were transformed with constructs containing direct inverted repeats of indolacetamide hydrolase/tryptophan monooxygenase (*iaaM*) and isopentenyl transferase (*ipt*), the oncogenes responsible for auxin and cytokinin production [203]. In transgenic tomato and in *Arabidopsis*, *Agrobacterium*-mediated transformation not simply prevented, but completely abolished gall formation. Similar results, using self-complementary constructs containing the *iaaM* and *ipt* genes, were reported by Escobar *et al.* [204] in walnut (*Juglans regia* L.) and by Viss *et al.* [205] in apple (*Malus.domestica* Borkh.).

Recently, Baum *et al.* [206] have transformed maize (*Zea mais* L.) using a hpRNA against a subunit of the midgut enzyme vacuolar ATPase. The authors grew transgenic plants protected against western corn rootworm infection at a level comparable with that provided by the transgenic maize containing the *Bacillus thuringiensis* (Bt) toxin. This approach could provide an alternative strategy to Bt protection both in maize and cotton (*Gossypium hirsutum* L.), where insects have gradually been developing resistance to Bt.

Transgene Silencing to Improve Fruit Quality and for Functional Genomics Studies

In addition to the examples described above about transgene silencing-based strategies to protect plants from pathogen attacks, another widely use of this technology has been tested to reshape metabolic pathways. In the post genome era of plant biology, a pivotal challenge is to determine functions of all genes within the plant genome. Gene silencing is an effective tool to understand and modify the organization of biochemical pathways in order to better quality and yields. PTGS has also been employed to prevent or substantially reduce the production of plant-derived substances, as in the case of allergens that can become harmful to human health in otherwise nutritious foods.

For fruit quality improvement, Murata *et al.* [207] transformed apples with the polyphenol oxidase (*ppo*) gene in antisense orientation and ascertained a reduction of potential browning in transgenic plants. Transgenic apples with the aldose 6-phosphate reductase (*a6pr*) gene in antisense orientation were built to highlight the regulation of photosynthesis and the carbon partitioning directed by sorbitol synthesis. Indeed, in these plants were observed significant reductions in sorbitol concentrations, increment in sucrose concentrations in the source leaves, and no alterations in CO_2 assimilation and plant vegetative growth [208]. Moreover, in shoot tips of the transgenic apple trees with decreased sorbitol synthesis, the sorbitol dehydrogenase was down-regulated, whereas the sucrose synthase was up-regulated leading to homeostasis of vegetative growth [209]. Transgenic apple fruits obtained both from plants silenced for either ACS (ACC synthase; ACC-1-aminocyclopropane-1-carboxylic acid) or ACO (ACC oxidase), key enzymes responsible for ethylene biosynthesis, showed reduced autocatalytic ethylene production, as expected. No significant difference was revealed in sugar or acid accumulation, but an evident suppression of the synthesis of volatile esters was detected in fruit silenced for ethylene biosynthesis [210]. Johnston *et al.* [211] characterized the ethylene response curves for individual ripening characters in transgenic apples in which the expression of the *mdaco1* gene was suppressed. The same research group blocked anthocyanin biosynthesis by silencing the key enzyme anthocyanidin synthase (*mdans*) in transgenic plants of a red-leaved apple cultivar. The authors observed a shift in the profile of flavonoids and related polyphenols, and a severe reduction of viability by necrotic leaf lesions [212].

In *Citrus*, Wong *et al.* [213] controlled the ethylene biosynthesis by introducing a chilling-inducible *acs* gene in antisense orientation. Ethylene production in *Citrus* plants may be managed to enhance resistance to environmental stresses and improve post-harvest quality of fruits. Reduction in ethylene production in the abscission zone of leaves and fruits at the onset of the natural fruit-degreening stage may prolong the 'on-tree storage' of fruits. Pear (*Pyrus communis* L.) was transformed with the endogenous gene *aco* both in sense and antisense orientations. Abnormal phenotypes, including *in vitro* flowering and abnormal rooting, were shown by some antisense transgenic pear lines obtained, but the same lines followed a normal growth and shared wild type phenotypes under greenhouse conditions. Finally, transgenic plants producing low ethylene levels were proposed to have great value in the production of fruits with an improved shelf life [214]. Botella *et al.* [215] engineered pineapple (*Ananas comosus* L.) with specific constructs to inactivate *acs* gene and the resulting transgenic plants had delayed flowering [216]. Another objective of pineapple transformation was the management of blackheart disorder, a fruit defect caused by exposure of pineapples to lower than 20°C temperatures, which stimulates PPO activity. *ppo* gene, whose sequence was isolated from pineapple fruits under conditions that typically produce blackheart disease, was silenced in transformed plants to restrain blackheart disorder [217].

In the last years, strawberry (*Fragaria* x *ananassa* Duch.) has been selected as a genomic model for the *Rosaceae* family, both because this species is an herbaceous perennial plant, characterized by rapid growth and a small genome, and because it can be transformed and regenerated in a time scale of weeks or months rather than years. To modify strawberry fruit softening, transgenic plants were produced silencing two endo β-1,4-glucanase genes (*cel1* and *cel2*) [218]. In plants transformed with antisense pectate lyase gene, a decrease in postharvest softening was observed [219] together with the improvement of several quality traits of the berry [220]. Strawberry transformed with an antisense cDNA of ADP-glucose pyrophosphorylase (AGPase) small subunit put under the control of the ascorbate peroxidase (APX) promoter, showed increased sugar and decreased starch contents [221]. Transgenic strawberry plants containing an antisense *chs* gene were used to analyze the role of CHS in metabolic channelling between the flavonoid and the phenylpropanoid pathways [222]. Constructs encoding self-complementary hpRNA of *chs* gene was agro-infiltrated with a syringe into the growing fruit receptacles. The authors demonstrated the down-regulation of the target gene and conveyed that this technique, combined with metabolite profiling analyses, may be useful to study unknown gene functions during strawberry fruit development and ripening [223]. Furthermore, walnuts transgenic plants, containing the antisense *chs* gene, allowed to uncover a correlation between flavonoid content and rooting capacity [224].

Transgenic plum containing an hpRNA structure of a *pds* gene deriving from peach resulted in knockout of the endogenous *pds* gene. This system may provide a rapid high throughput strategy for functional genomic studies in *Prunus* [225]. Reduced activity of phytoene desaturase, a key enzyme of the carotenoid biosynthetic pathway, results in inhibition of the carotenoid biosynthesis and causes chlorophyll photo-oxidation. This gene is also often used to evaluate VIGS efficiency from the resulting plant phenotype [226].

An RNA silencing approach was used to down-regulate in apple the allergen Mal d 1, a member of a family of at least 18 genes displaying IgE cross-reactivity to the birch pollen allergen Bet v 1, a PR-protein. Transgenic plantlets showed up to a 10-folds reduction in Mal d 1 leaf expression, with no detectable phenotypic abnormalities when comparing to wild type [227]. Transgenic apple plants were also produced harbouring extra copies of the endogenous S-gene controlling self-incompatibility (SI). SI prevents self-fertilization in many tree fruit crops, and suboptimal pollination through insects activity may lead to low yields. Self-pollination could ensure more consistently high production yields compared to cross-pollination. Controlled self-pollination of the flowers of apple trees over a 3-years-period showed that the silenced lines for SI produced normal levels of fruits and seeds after selfing [228].

In woody species an important theme of genetic improvement is the production of plant material with lower contents of lignin for the paper industry, because the separation of lignin from cellulose and hemicellulose is an expensive and environmentally hazardous process. One of the approaches directed to obtain reduced lignin contents in forest trees was the down-regulation of lignin biosynthetic pathway by silencing the 4-coumarate coenzyme A ligase (*pt4cl1*) [229]. In the transgenic *Populus tremuloides* Michx. trees the authors found a 45% reduction of the lignin content, however the plant growth was substantially enhanced, and structural integrity maintained both at the cellular and at the whole-plant levels. Jouanin *et al.* [230] silenced the caffeate/5 hydroxyferulate O-methyltransferase (COMT - enzyme involved in syringyl lignin synthesis) in *Populus tremula x Populus alba*, obtaining a lignin reduction in transgenic plants. Lignin biosynthesis was also reduced in *Pinus radiata* D. Don by silencing the cynnamyl alcohol dehydrogenase (CAD - the final enzyme in the biosynthesis of lignin monomers) [231].

Tomato is a broadly used model system for studying plant microbe interactions and fleshy fruit biology in *Solanaceae*, owing to its short generation time and the routine protocol for production of transgenic plants. In one of the first reports about gene silencing in tomato, introduction of truncated polygalacturonase (*pg*) transgenes, driven by the CaMV 35S promoter, caused a strong co-suppression of both the transgene and the endogenous ripening-specific *pg* gene [232]. Fruits expressing a truncated pectin methylesterase (*pme*) sense transgene showed developmentally regulated co-silencing of the endogenous *pme* and sense transgene [233]. The absence of *pme* transcripts accumulation in transgenic fruits was correlated to the accumulation of the unspliced *pme* transcript in the nucleus. Thus, ectopic expression of the sense transgene impairs processing of the transcript of the homologous endogenous gene. Pectinesterase (PE) is a ubiquitous cell wall-associated enzyme responsible for galacturonyl residues demethylation in high-molecular weight pectins and is believed to play an important role in cell wall metabolism. Transgenic tomato plants expressing *pmeu*1, a ubiquitously expressed *pe* gene, in antisense orientation showed reduced PE activity levels in both green fruits and leaf tissues. In leaf tissues, silencing of *pmeu*1 did not result in any detectable phenotype, whereas silencing in fruits resulted in an enhancement of softening rates during ripening [234]. Expansins are cell wall-localized proteins highly expressed in expanding and ripening green tomato fruit. In particular, the *LeExp*1 was found to accumulate specifically during fruit ripening. Suppression of *LeExp*1 in transgenic fruit reduced softening during ripening, whereas over-expression of *LeExp*1 increased fruit softening [235]. The fruit for which the *LeExp1* expression was suppressed improved shelf life and processing properties, but they did not result in increased resistance to necrotrophyic pathogens, as *Botrytis cinerea* and *Alternaria alternate* [236]. Tomato allergen Lyc e 1 (profilin) was inhibited through expression of silencing vector. The plants silenced for this gene showed different phenotypes from wild type (for instance, dwarfing under greenhouse conditions, delayed flowering, and yield reductions) due to down-regulation of profilin, which plays a regulative role in the cytoskeleton organization [237]. Recently, Fernandez *et al.* [238] have characterized five promoters expressed at defined time windows during tomato fruit development, which can be used for regulating expression or targeting gene silencing. The promoter sequences were inserted as entry clones compatible with the versatile MultiSite Gateway format, and their activity was confirmed in transgenic tomato lines. Efficient silencing of the endogenous *pds* gene was demonstrated in transgenic tomato lines, producing a matching amiRNA under the phosphoenolpyruvate carboxylase promoter, which is present in developing tomato fruits. The availability of several fruit and flower promoters is an important asset for future functional analyses in tomato and related crops [238].

One of the first examples of gene silencing in rice involved the study of the waxy (*wx*) gene,which encodes granule-bound starch synthase (GBSS), an enzyme that catalyzes a key step in amylose synthesis and that is specifically expressed in pollen and endosperm. Introduction of *wx* gene in antisense orientation into Japonica rice causes a reduction of amylose in the endosperm [239] [240]. Itoh *et al.* [241] showed that rice plants transformed with *wx* transgene exhibited silencing in the pollen grain (but not in endosperm), and this silencing was meiotically transmitted to the progenies by selfing and outcrossing. When the same *wx* transgene was introduced into the waxy mutant, no gene silencing was observed, so the mechanism appeared to require the expression of the endogenous gene. A similar situation was previously observed in tomato plants, in which polygalacturonase is expressed when fruits start to ripen, and silencing of this gene was only initiated in ripening fruits [232]. Transgenic rice plants were obtained to reduce phytic acid in seeds by silencing the 1 D - myo-inositol 3-phosphate synthase gene *rino 1* [242]. Phytic acid in cereal grains and oilseeds forms insoluble mixed salt phytate with various minerals, thus reducing the bioavailability of phosphorus and minerals to monogastric animals. Breeding programs, involving mutants with less phytic acid and more inorganic phosphate, have been frustrated by undesirable agronomic characteristics associated with the phytic acid-reducing mutations. Antisense *rino1* cDNA was expressed under the control of the major storage protein glutelin GluB-1 promoter, however the effects of this transgene were not as robust as those found in low phytic acid (*lpa*) mutants [242]. Recently, Kuwano *et al.* [243] have reported that strong *lpa* transgenic rice was generated when the *rino1* antisense construct was expressed under the control of the oleosin promoter, which specifically directs expression in the seed embryo at early developmental stages. These results clearly indicate that the expression patterns of the promoter and target gene strongly affect gene silencing exploitation.

In wheat (*Triticum aestivum* L.), the starch composition has been modified by altering its amylose-amylopectin ratio, in order to face the incidence of cardiovascular diseases and colon cancers. Regina and colleagues [244], using hpRNA constructs, silenced an isoform of a starch-branching enzyme to produce a high-amylose transgenic wheat line. Gil-Humanes *et al.* [245] demonstrated that the RNA silencing can also be used to down-regulate groups of proteins encoded by multigene families in wheat. These authors performed a silencing experiment, inserting a hpRNA construct with the γ-gliadins, which is responsible for the extensibility and viscosity of gluten and dough, and with the polymeric glutenins, which instead are responsible for elasticity. In the same context, Yue and colleagues [246] silenced one of the genes (*1dx5*) codifying for the high molecular weight glutenin subunits (HMW-GSs), the major components of gluten protein in wheat. Gluten proteins confer characteristic visco-elastic properties, which allow wheat to be processed into food products. Although gluten comprises a large number of proteins, the HMW fraction appears to be particularly important in the formation of high M_r polymers, which are highly elastic. These researchers showed that the silencing of *1dx5* caused a substantial decrease in the flour processing quality based on Farinograph and gluten tests. Basing on these approaches, it will be possible in the future to elucidate roles held by specific groups of proteins, especially for the determination of flours functional properties and the related activities triggering celiac disease. Xia *et al.* [247] obtained transgenic wheat, introducing the puroindoline a (*pina*) gene, which, together with puroindoline b, is the major constituent of friabilin, a starch granule-associated protein. Hard wheat has little or no friabilin present on the surface of water-washed starch granules, whereas in soft wheat friabilin levels are much higher. In the resulting transgenic lines, as a consequence of the *pina* transgene over-expression, the endogenous *pina* genes were co-suppressed. These lines gave interesting clues to further understand puroindolines functions, to underline mechanisms about the kernel texture formation, and finally they provided insights for modifying grain texture through genetic engineering in cereal plants.

In maize, Cigan *et al.* [248] obtained male-sterile plants by constitutively expressing inverted repeats of the promoter of gene *ms45* associated to male fertility. This study supported evidences that the constitutive expression of dsRNAs, consisting of promoter sequences, resulted in the TGS of both endogenous genes and transgenes. The described technique contributed to supply a new powerful approach to study gene functions and identify regulatory components unique to transcriptional gene control. Indeed, transgene silencing was soon used for a large-scale of functional genomics projects in maize. McGinnis and colleague [249] realized transgenic lines containing short gene segments inserted in inverted repeat orientation, properly designed to cut down the expression of several chromatin-related genes involved in most epigenetic phenomena. The authors observed that, in some cases, the IR construct was able to cause a reduction in the mRNAs levels, not only of the target gene, but also of another closely related gene. Since maize does not possess great nutritional qualities for human and animal consumption, because of its low lysine content, two recent works [250] [251] developed a biotechnological approach to face this problem. Indeed, they developed a solution to increase corn lysine content through genetic engineering. An inverted-repeat sequence of gene for the lysine degradation enzyme, lysine-ketoglutarate reductase/saccharophine dehydrogenase (LKR/SDH) suppressed the lysine catabolism, thus transgenic maize kernels accumulated a significant amount of lysine [250]. The same group showed that the LKR/SDH silencing combined with the CordapA (the lysine biosynthetic enzyme) expression led to the accumulation of high levels of free lysine in transgenic corn grains [251].

Cotton is the sixth largest source of vegetable oil in the world, but its oil profile shows relatively high levels of palmitic acid, which is responsible for the low-density of lipoprotein cholesterol-raising properties in humans. Liu *et al.* [252] used hpRNA constructs to silence the D9 and D12 desaturases, which catalyze the biosynthesis of fatty acids. The transformed plants produced seed oil with a lower palmitic acid content and a richer concentration both of oleic and stearic acids, thus making them more suitable for human consumption. Sunilkumar *et al.* [253] produced gossypol toxin free oil by targeting a gene involved in its biosynthesis in cotton.

Peanut (*Arachis hypogaea* L.) is responsible for the majority of food-induced anaphylactic shocks. The principal allergen Ara h 2 is a seed storage protein belonging to the prolammin family, which shows a typical trypsin inhibitor activity. Recently, in two reports, the use of constructs inducing silencing [254] [255] for engineering plants helped to limit allergen production and decrease IgE binding capacity.

Cassava (*Manihot esculenta* L.) is one of the most important food for people in the tropical and subtropical regions of the world. However, cassava contains plant toxins with high human toxicity, which leads to permanent disability when the food is not properly processed before consumption. Cassava roots are the primary plant organ consumed and are the major source of calories (starch), but it is especially the cassava root cortex that contains high concentrations of cyanogenic glycosides (as linamarin and lotaustralin). Transgenic strategies have been used to modify cyanogens levels in cassava plants and processed foods via the suppression of two cytochromes P450s (CYP79D1 and CYP79D2) catalyzing the first step of cyanogen synthesis [256] [257]. The expression of both *cyp79d1* and *cyp79d2* genes was selectively inhibited either in leaves or roots driving antisense constructs. Reduced *cyp79d1* and *cyp79d2* expressions in leaves resulted in decreased leaf linamarin contents and also in roots the authors found a 99% reduction in linamarin [257].

Besides being triggered by transgenes, RNA silencing can naturally occur as a consequence of certain genetic changes. For instance, most commercial varieties of soybean (*Glycine max* L.) produce yellow seeds, in which the loss of pigmentation in seed coats is due to PTGS of the *chs* genes [258]. Wild soybean, an ancestor of the cultivated soybean, exclusively produced seeds with pigmented seed coats. The non pigmented phenotype was probably generated after soybean domestication. Humans chose to maintain plant lines with *chs* RNA silencing, due to a structural change in the *chs* gene cluster, with the production of inverted repeat *chs* [259]. Koseki *et al.* [260] demonstrated that the phenotype of petunia with flowers having a star-type red and white bicolor pattern was caused by a natural RNA silencing phenomenon, targeting *chs* genes in white sectors. Finally, in rice, a dominant mutation, called LGC-1 (Low Glutelin Content-1), produces hpRNA from an inverted repeat for glutelin, which lower glutelin contents through an RNA silencing mechanism [261].

Conclusions and New Perspectives

Transgene silencing techniques have largely been improved. They are contributing to the development of several applications of commercial and agronomical interest, and they have been employed to characterize and identify genes and their functions. Applications focused on enhancing disease resistance, controlling metabolic pathways to change chemical composition of plant products for industrial use or to improve fruit quality or nutritional value, reducing production of plant-derived substances as allergens and functional genomic studies, are described in this chapter. Transgene silencing approaches, and in particular the amiRNA strategy, seem to be a promising method of modulating important traits in cultivars used in modern breeding programs. Through functional studies mediated by transgene silencing approaches, it was possible not only to infer functions of genes, but also to find examples where loss of gene function results in improved plant performances, such as increased yields or biotic and abiotic stress tolerance [148] [262] [263].

The development of high-throughput transgene silencing for genome functional studies, and the acquisition of techniques for rapid construct preparations, have allowed to activate projects for high-throughput plant genomic research in plant model system. The CATMA group (Complete *Arabidopsis* Transcriptome MicroArray; www.catma.org [264]) is now generating a set of PCR products representing each *Arabidopsis* gene. The AGRIKOLA consortium (*Arabidopsis* genomic RNAi knock-out line analysis; www.agrikola. org/index.php?o=/agrikola/main; [265]) is using this set of PCR products to generate a gene-specific RNA silencing construct for each *Arabidopsis* gene. These constructs can be used for large-scale RNA silencing studies and the gene fragments cloned inside these vectors can also be recombined inside VIGS vectors.

Recently, the genomes of two economically important woody plants, poplar [266] (http://genome.jgi-psf.org/Poptr1_1/Poptr1_1.home.html) and grape [267] (http://www.genoscope.cns.fr/externe/GenomeBrowser/Vitis/) [268] (http://genomics.research.iasma.it/cgi-bin/gbrowse/dasrelease3/), have been sequenced. These last two works represent a pivotal starting point for the future activation of a large-scale platform for RNA silencing studies also in woody species, especially for those having a major agronomic and commercial interest for humans.

References

[1]   Hannon, GJ. RNA interference. *Nature,* 2002 418, 244-251.

[2]   Napoli, C; Lemieux, C; Jorgensen, R. Introduction of a chimeric *chalcone synthase* gene into petunia results in reversible co-suppression of homologous genes in trans. *Plant Cell,* 1990 2, 279-289.

[3]   Romano, N; Macino, G. Quelling: transient inactivation of gene expression in *Neurospora crassa* by transformation with homologous sequences. *Mol Microbiol,* 1992 6, 3343-3353.

[4]   Lee, RC; Feinbaum, RL; Ambros, V. The *C. elegans* heterochronic gene *lin-4* encodes small RNAs with antisense complementarity to lin-14. *Cell,* 1993 75, 843-854.

[5]   Hamilton, AJ; Baulcombe, DC. A species of small antisense RNA in post-transcriptional gene silencing in plants. *Science,* 1999 286, 950-952.

[6]   Vaucheret, H. Post-transcriptional small RNA pathways in plants: mechanisms and regulations. *Genes Dev,* 2006 20, 759-771.

[7]   Gelvin, SB. *Agrobacterium*-Mediated Plant Transformation: the Biology behind the "Gene-Jockeying" Tool. *Microbiol Mol Biol R,* 2003 67, 16-37.

[8]   Tzfira, T; Li, J; Lacroix, B; Citovsky, V. *Agrobacterium* T-DNA integration: molecules and models. *Trends Genet,* 2004 20, 375-383.

[9]   Baulcombe, D. RNA silencing in plants. *Nature,* 2004 431, 356-363.

[10]   Bernstein, E; Caudy, AA; Hammond, SM; Hannon, GJ. Role for a bidentate ribonuclease in the initiation step of RNA interference. *Nature,* 2001 409, 363-366.

[11] Mette, MF; Aufsatz, W; Van der Winden, J *et al*. Transcriptional silencing and promoter methylation triggered by double-stranded RNA. *EMBO J*, 2000 19, 5194-5201.

[12] Verdel, A; Jia, S; Gerber, S *et al*. RNAi-mediated targeting of heterochromatin by the RITS complex. *Science*, 2004 303, 672-676.

[13] Matzke, MA; Primig, M; Trnovsky, J; Matzke, AJM. Reversible methylation and inactivation of marker genes in sequentially transformed tobacco plants. *EMBO J,*1989 8, 643-649.

[14] Meyer, P; Saedler, H. Homology-dependent gene silencing in plants. *Ann Rev Plant Physiol Plant Mol Biol*, 1996 47, 23-48.

[15] Dalmay, T; Horsefield, R; Braunstein, TH; Baulcombe, DC. SDE3 encodes an RNA helicase required for post-transcriptional gene silencing in *Arabidopsis*. *EMBO J*, 2000 20, 2069-2078.

[16] Wassenegger, M; Krczal, G. Nomenclature and functions of RNA-directed RNA polymerases. *Trends Plant Sci*, 2006 11, 142-151.

[17] Gazzani, S; Lawrenson, T; Woodward, C *et al*. A link between mRNA turnover and RNA interference in *Arabidopsis*. *Science*, 2004 306, 1046-1048.

[18] Waterhouse, PM; Graham, MW; Wang, MB. Virus resistance and gene silencing in plants can be induced by simultaneous expression of sense and antisense RNA. *Proc Natl Acad Sci USA*, 1998 95, 13959-13964.

[19] Wassenegger, M. Gene silencing-based disease resistance. *Transgenic Research*, 2002 11, 639-653.

[20] Wang, MB; Metzlaff, M. RNA silencing and antiviral defence in plants. *Curr Opin Plant Biol*, 2005 8, 216-222.

[21] Brodersen, P; Voinnet, O. The diversity of RNA silencing pathways in plants. *Trends Genet*, 2006 22, 268-280.

[22] Li, J; Yang, Z; Yu, B *et al*. Methylation protects miRNAs and siRNAs from a 3'-end uridylation activity in Arabidopsis. *Curr Biol*. 2005 15, 1501-1507.

[23] Vazquez, F; Vaucheret, H; Rajagopalan, R; Lepers, K *et al*. Endogenous *trans*-acting siRNAs regulate the accumulation of *Arabidopsis* mRNAs. *Mol Cell*, 2004 16, 69-79.

[24] Bartel, DP. MicroRNAs: Genomics, Biogenesis, Mechanism, and Function. *Cell*, 2004 116, 281-297.

[25] Chen, X. MicroRNA biogenesis and function in plants. *FEBS Letters*, 2005 579, 5923-5931.

[26] Voinnet, O. Induction and suppression of RNA silencing: insights from viral infections. *Nat Rev Genet*, 2005 6, 206-220.

[27] Vaucheret, H. Post-transcriptional small RNA pathways in plants: mechanisms and regulations. *Genes Dev*, 2006 20, 759-771.

[28] Jones-Rhoades, MW; Bartel, DP; Barte, B. MicroRNAs and their regulatory roles in plants. *Annu Rev Plant Biol*, 2006 57, 19-53.

[29] Chapman, EJ; Carrington, JC. Specialization and evolution of endogenous small RNA pathways. *Nat Rev Genet*, 2007 8, 884-896.

[30] Sunkar, R; Chinnusamy, V; Zhu, J; Zhu, JK. Small RNAs as big players in plant abiotic stress responses and nutrient deprivation. *Trends Plant Science*, 2007 12, 301-309.

[31] Chuck, G; Candela, H; Hake, S. Big impacts by small RNAs in plant development. *Curr Opin Plant Biol,* 2009 12, 81-86.

[32] Navarro, L; Dunoyer, P; Jay, F *et al.* A plant miRNA contributes to antibacterial resistance by repressing auxin signaling. *Science,* 2006 312, 436-439.

[33] Wang, JW; Wang, LJ; Mao, YB *et al.* Control of root cap formation by microRNA-targeted auxin response factors in *Arabidopsis. Plant Cell,* 2005 17, 2204-2216.

[34] Chen, X. Small RNAs and their roles in plant development. *Annu Rev Cell Dev Biol,* 2009 35, 21-44.

[35] Yang, L; Liu, Z; Lu, F *et al. SERRATE* is a novel nuclear regulator in primary microRNA processing in *Arabidopsis. Plant J,* 2006 47, 841-850.

[36] Yu, B; Yang, Z; Li, J; Minakhina, S *et al.* Methylation as a crucial step in plant microRNA biogenesis. *Science,* 2005 307, 932-935.

[37] Park, MY; Wu, G; Gonzalez-Sulser, A *et al.* Nuclear processing and export of microRNAs in *Arabidopsis. Proc Natl Acad Sci USA,* 2005 102, 3691-3696.

[38] Mallory, AC; Vaucheret, H. Functions of microRNAs and related small RNAs in plants. *Nat Genet,* 2006 38, Suppl: S31-S36.

[39] Hutvagner, G, Zamore, PD. RNAi: nature abhors a double-strand. *Curr Opin Genet Dev,* 2002 12, 225-232.

[40] Chen, X. A microRNA as a translational repressor of APETALA2 in Arabidopsis flower development. *Science,* 2004 303, 2022-2025.

[41] Voinnet, O. Origin, Biogenesis, and Activity of Plant MicroRNAs. *Cell,* 2009 136, 669-687.

[42] Brodersen, P; Sakvarelidze-Achard, L; Bruun-Rasmussen, M *et al.* Widespread translational inhibition by plant miRNAs and siRNAs. *Science,* 2008 320, 1185-1190.

[43] Peragine, A; Yoshikawa, M; Wu, G *et al. SGS3* and *SGS2/SDE1/RDR6* are required for juvenile development and the production of trans-acting siRNAs in *Arabidopsis. Genes Dev,* 2004 18, 2368-2379.

[44] Allen, E; Xie, Z; Gustafson, AM; Carrington, JC. MicroRNA-directed phasing during trans-acting siRNA biogenesis in plants. *Cell,* 2005 121, 207-221.

[45] Yoshikawa, M; Peragine, A; Park, MY; Poethig, RS. A pathway for the biogenesis of trans-acting siRNAs in Arabidopsis. *Genes Dev,* 2005 19, 2164-2175.

[46] Axtell, MJ; Jan, C; Rajagopalan, R; Bartel, DP. A Two-Hit Trigger for siRNA Biogenesis in Plants. *Cell,* 2006 127, 565-77.

[47] Montgomery, TA; Yoo, SJ; Fahlgren, N *et al.* AGO1-miR173 complex initiates phased siRNA formation in plants. *Proc Natl Acad Sci USA,* 2008 105, 20055-20062.

[48] Montgomery, TA; Howell, MD; Cuperus, JT *et al.* Specificity of ARGONAUTE7-miR390 interaction and dual functionality in TAS3 trans-acting siRNA formation. *Cell,* 2008 133, 128-141.

[49] Fahlgren, N; Montgomery, TA; Howell, MD *et al.* Regulation of *AUXIN RESPONSE FACTOR3* by *TAS3* ta-siRNA affects developmental timing and patterning in *Arabidopsis. Curr Biol,* 2006 16, 939-944.

[50] Adenot, X; Elmayan, T; Lauressergues, D *et al. DRB4*-dependent *TAS3* trans-acting siRNAs control leaf morphology through *AGO7. Curr Biol,* 2006 16, 927-932.

[51] Garcia, D; Collier, SA; Byrne, ME; Martienssen, RA. Specification of leaf polarity in *Arabidopsis* via the trans-acting siRNA pathway. *Curr Biol,* 2006 16, 933-938.

[52] Borsani, O; Zhu, J; Verslues, PE *et al*. Endogenous siRNAs derived from a pair of natural *cis*-antisense transcripts regulate salt tolerance in *Arabidopsis*. *Cell,* 2005 123, 1279-1291.

[53] Katiyar-Agarwal, S; Morgan, R; Dahlbeck, D *et al*. A pathogen-inducible endogenous siRNA in plant immunity. *Proc Natl Acad Sci,* 2006 103, 18002-18007.

[54] Matzke, M; Kanno, T; Daxinger, L *et al*. RNA-mediated chromatin-based silencing in plants. *Curr Opin Cell Biol,* 2009 21, 367-376.

[55] Zhang, X; Henderson, IR; Lu, C; Green, PJ; Jacobsen, SE. Role of RNA polymerase IV in plant small RNA metabolism. *Proc Natl Acad Sci USA*, 2007 104, 4536-4551.

[56] Daxinger, L; Kanno, T; Bucher, E *et al*. A stepwise pathway for biogenesis of 24-nt secondary siRNAs and spreading of DNA methylation. *EMBO J,* 2009 28, 48-57.

[57] Cao, X; Aufsatz, W; Zilberman, D *et al*. Role of the DRM and CMT3 methyltransferases in RNA-directed DNA methylation. *Curr Biol,* 2003 13, 2212-2217.

[58] Pikaard, CS; Haag, JR; Ream, T; Wierzbicki, AT. Roles of RNA polymerase IV in gene silencing. *Trends Plant Science,* 2008 13, 390-397.

[59] Kanno, T; Bucher, E; Daxinger, L *et al*. A structural-maintenance-of chromosomes hinge domain-containing protein is required for RNA-directed DNA methylation. *Nat Genet,* 2008 40, 670-75.

[60] Chan, SW; Henderson, IR; Jacobsen, SE. Gardening the genome: DNA methylation in *Arabidopsis thaliana*. *Nat Rev Genet,* 2005 6, 351-360.

[61] Liu, J; He, Y; Amasino, R; Chen, X. siRNAs targeting an intronic transposon in the regulation of natural flowering behaviour in Arabidopsis. *Genes Dev,* 2004 18, 2873-2878.

[62] Swiezewski, S; Crevillen, P; Liu, F *et al*. Small RNA-mediated chromatin silencing directed to the 3' region of the Arabidopsis gene encoding the developmental regulator, FLC. *Proc Natl Acad Sci USA,* 2007 104, 3633-3638.

[63] Fagard, M; Vaucheret, H. Systemic silencing signal(s). *Plant Mol Biol,* 2000 43, 285-293.

[64] Vaistij, FE; Jones, L; Baulcombe, DC. Spreading of RNA targeting and DNA methylation in RNA silencing requires transcription of the target gene and a putative RNA-dependent RNA polymerase. *Plant Cell,* 2002 14, 857-867.

[65] Himber, C; Dunoyer, P; Moissiard, G *et al*. Transitivity-dependent and -independent cell-to-cell movement of RNA silencing. *EMBO J,* 2003 22, 4523-4533.

[66] Bleys, A; Van Houdt, H; Depicker, A. Down-regulation of endogenes mediated by a transitive silencing signal. *RNA,* 2006 12, 1633-1639.

[67] Kalantidis, K; Schumacher, HT; Alexiadis, T; Helm, JM. RNA silencing movement in plants. *Biol Cell,* 2008 100, 13-26.

[68] Guo, HS; Ding, SW. A viral protein inhibits the long range signalling activity of the gene silencing signal. *EMBO J,* 2002 21, 398-407.

[69] Ruiz-Ferrer, V; Voinnet, O. Viral suppression of RNA silencing: 2b wins the Golden Fleece by defeating Argonaute. *Bioessays,* 2007 29, 319-323.

[70] Wingard, SA. Hosts and symptoms of ring spot, a virus disease of plants. *J Agric Res,* 1928 37, 127-153.

[71] Waterhouse, PM; Wang, MB; Lough, T. Gene silencing as an adaptive defence against viruses. *Nature,* 2001 411, 834-842.

[72]   Ratcliff, FG; MacFarlane, SA; Baulcombe, DC. Gene silencing without DNA: RNA-mediated cross protection between viruses. *Plant Cell,* 1999 11, 1207-1215.

[73]   Marathe, R; Anandalakshmi, R; Smith, TH *et al.* RNA viruses as inducers, suppressors and targets of post-transcriptional gene silencing. *Plant Mol Biol,* 2000 43, 295-306.

[74]   Vance, VB; Berger, PH; Carrington, JC *et al.* 5' proximal potyviral sequence mediates potato virus X/ potyviral synergistic disease in transgenic tobacco. *Virology,* 1995 206, 583-590.

[75]   Mlotshwa, S; Pruss, GJ; Vance, V. Small RNAs in viral infection and host defense. *Trends Plant Sci,* 2008 13, 375-382.

[76]   Yoo, BC; Kragler, F; Varkonyi-Gasic, E *et al.* A systemic small RNA signalling system in plants. *Plant Cell,* 2004 16, 1979-2000.

[77]   Ho, T; Pallett, D; Rusholme, R *et al.* A simplified method for cloning of short interfering RNAs from *Brassica juncea* infected with Turnip mosaic potyvirus and Turnip crinkle carmovirus. *J Virol Methods,* 2006 136, 217-223.

[78]   Molnar, A; Csorba, T; Lakatos, L *et al.* Plant virus-derived small interfering RNAs originate predominantly from highly structured single-stranded viral RNAs. *J Virol,* 2005 79, 7812-7818.

[79]   Itaya, A; Zhong, X; Bundschuh, R *et al.* A structured viroid RNA serves as a substrate for dicer-like cleavage to produce biologically active small RNAs but is resistant to RNA-induced silencing complex mediated degradation. *J Virol,* 2007 81, 2980-2994.

[80]   Machida, S; Yamahata, N; Watanuki, H *et al.* Successive accumulation of two size classes of viroid-specific small RNA in potato spindle tuber viroid-infected tomato plants. *J Gen Virol,* 2007 88, 3452-3457.

[81]   Carra, A; Mica, E; Gambino, G *et al.* Cloning and characterization of small non-coding RNAs from grape. *Plant J,* 2009 59, 750-763.

[82]   Navarro, B; Pantaleo, V; Gisel, A *et al.* Deep Sequencing of Viroid-Derived Small RNAs from Grapevine Provides New Insights on the Role of RNA Silencing in Plant-Viroid Interaction. *PLOS ONE,* 2009 4, e7686.

[83]   Chapman, EJ; Prokhnevsky, AI, Gopinath, K *et al.* Viral RNA silencing suppressors inhibit the microRNA pathway at an intermediate step. *Genes Dev,* 2004 18, 1179-1186.

[84]   Li, F; Ding, SW. Virus counterdefense: diverse strategies for evading the RNA silencing immunity. *Annu Rev Microbiol,* 2006 60, 503-531.

[85]   Voinnet, O. Induction and suppression of RNA silencing: Insights from viral infections. *Nature Rev Genet,* 2005 6, 206-220.

[86]   Kasschau, KD; Cronin, S; Carrington, JC. Genome amplification and long-distance movement functions associated with the central domain of tobacco etch potyvirus helper component-proteinase. *Virology,* 1997 228, 251-262.

[87]   Mallory, AC; Ely, L; Smith, TH *et al.* HC-Pro suppression of transgene silencing eliminates the small RNAs but not transgene methylation or the mobile signal. *Plant Cell,* 2001 13, 571-583.

[88]   Anandalakshmi, R; Marathe, R; Ge, X *et al.* A calmodulin-related protein that suppresses post-transcriptional gene silencing in plants. *Science,* 2000 290, 142-144.

[89]   Brigneti, G; Voinnet, O; Li, WX *et al.* Viral pathogenicity determinants are suppressors of transgene silencing in Nicotiana benthamiana. *EMBO J,* 1998 17, 6739-6746.

[90]   Zhang, X; Yuan, YR; Pei, Y *et al*. Cucumber mosaic virus-encoded 2b suppressor inhibits Arabidopsis Argonaute1 cleavage activity to counter plant defence. *Genes Dev,* 2006 20, 3255-3268.

[91]   Russo, M; Burgyan, J; Martelli, GP. Molecular biology of Tombusviridae. *Adv. Virus Res,* 1994 44, 381-428.

[92]   Qu, F; Morris, TJ. Efficient infection of Nicotiana benthamiana by Tomato bushy stunt virus is facilitated by the coat protein and maintained by p19 through suppression of gene silencing. *Mol Plant-Microbe Interact,* 2002 15, 193-202.

[93]   Silhavy, D; Molnar, A; Lucioli, A *et al*. A viral protein suppresses RNA silencing and binds silencing-generated, 21-to 25-nucleotide double-stranded RNAs. *EMBO J,* 2002 21, 3070-3080.

[94]   Vargason, JM; Szittya, G; Burgyan, J; Hall, TMT. Size selective recognition of siRNA by an RNA silencing suppressor. *Cell,* 2003 115, 799-811.

[95]   Voinnet, O; Lederer, C; Baulcombe, D. A viral movement protein prevents spread of the gene silencing signal in Nicotiana benthamiana. *Cell,* 2000 103, 157-167.

[96]   Voinnet, O; Pinto, YM; Baulcombe, DC. Suppression of gene silencing: a general strategy used by diverse DNA and RNA viruses of plants. *Proc Natl Acad Sci USA,* 1999 96, 14147-14152.

[97]   Lu, R; Folomonov, A; Shintaku, M *et al*. Three distinct suppressors of RNA silencing encoded by a 20-kb viral RNA genome. *Proc Natl Acad Sci USA,* 2004 101, 15742-15747.

[98]   Deleris, A; Gallego-Bartolome, J; Bao, J *et al*. Hierarchical action and inhibition of plant Dicer-like proteins in antiviral defense. *Science,* 2006 313, 68-71.

[99]   Bortolamiol, D; Pazhouhandeh, M; Marrocco, K *et al*. The Polerovirus F box protein P0 targets ARGONAUTE1 to suppress RNA silencing. *Curr Biol,* 2007 17, 1615-1621.

[100] Baumberger, N; Tsai, CH; Lie, M *et al*. The Polerovirus silencing suppressor P0 targets ARGONAUTE proteins for degradation. *Curr Biol,* 2007 17, 1609-1614.

[101] Mangwende, T; Wang, ML; Borth, W *et al*. The P0 gene of Sugarcane yellow leaf virus encodes an RNA silencing suppressor with unique activities. *Virology,* 2009 384, 38-50.

[102] Glick, E; Zrachya, A; Levy, Y *et al*. Interaction with host SGS3 is required for suppression of RNA silencing by tomato yellow leaf curl virus V2 protein. *Proc Natl Acad Sci USA,* 2008 105, 157-61.

[103] Fukunaga, R; Doudna, JA. dsRNA with 50 overhangs contributes to endogenous and antiviral RNA silencing pathways in plants. *EMBO J,* 2009 28, 545-555.

[104] Wang, H; Buckley, KJ; Yang, X *et al*. Adenosine kinase inhibition and suppression of RNA silencing by geminivirus AL2 and L2 proteins. *J Virol,* 2005 79, 7410-7418.

[105] Raja, P; Sanville, BC; Buchmann, RC; Bisaro, DM. Viral genome methylation as an epigenetic defense against geminiviruses. *J Virol,* 2008 82, 8997-9007.

[106] Love, A; Laird, J; Holt, J *et al*. Cauliflower mosaic virus protein P6 is a suppressor of RNA silencing. *J Gen Virol,* 2007 88, 3439-3444.

[107] Haas, G; Azevedo, J; Moissiard, G *et al*. Nuclear import of CaMV P6 is required for infection and suppression of the RNA silencing factor DRB4. *EMBO J,* 2008 27, 2102-2112.

[108] Lozsa, R; Csorba, T; Lakatos, L; Burgyan J. Inhibition of 3' modification of small RNAs in virus infected plants requires spatial and temporal coexpression of small RNAs and viral silencing-suppressor proteins. *Nucleic Acids Res,* 2008 36, 4099-4107.

[109] Soosaar, LM; Burch-Smith, TM; Dinesh-Kumar, SP. Mechanisms of plant resistance to viruses. *Nat Rev,* 2005 3, 789-798.

[110] Burgyan, J. Virus induced RNA silencing and suppression: defence and counter defence. *J Plant Pathol,* 2006 88, 233-244.

[111] Díaz-Pendón, JA; Ding, SW. Direct and indirect roles of viral suppressors of RNA silencing in pathogenesis. *Annu Rev Phytopathol,* 2008 46, 303-326.

[112] Kasschau, KD; Xie, Z; Allen, E; *et al.* P1/HC-Pro, a viral suppressor of RNA silencing, interferes with Arabidopsis development and miRNA action. *Dev Cell,* 2003 4, 205-217.

[113] Dunoyer, P; Lecellier, CH; Parizotto, EA; *et al.* Probing the microRNA and small interfering RNA pathways with virus encoded suppressors of RNA silencing. *Plant Cell,* 2004 16, 1235-1250.

[114] Qu, F; Ren, T; Morris, TJ. The coat protein of turnip crinkle virus suppresses post-transcriptional gene silencing at an early initiation step. *J Virol,* 2003 77, 511-522.

[115] Izant, JG; Weintraub, H. Inhibition of thymidine kinase gene expression by anti-sense RNA: a molecular approach to genetic analysis. *Cell,* 1984 36, 1007-1015.

[116] Jorgensen, RA; Doetsch, N; Müller, A *et al.* A paragenetic perspective on integration of RNA silencing into the epigenome and its role in the biology of higher plants. *Cold Spring Harb Symp Quant Biol,* 2006 71, 481-485.

[117] Mishiba, K; Nishihara, M; Nakatsuka, T *et al.* Consistent transcriptional silencing of 35S-driven transgenes in gentian. *Plant J,* 2005 44, 541-556.

[118] Travella, S; Keller, B. Down-regulation of gene expression by RNA-induced gene silencing. In: Jones HD, Shewry, PR. Methods in Molecular Biology, Transgenic Wheat, Barley and Oats. *Humana Press;* 2009 478; 185-199.

[119] Smith, NA; Singh, SP; Wang, MB *et al.* Total silencing by intron-spliced hairpin RNAs. *Nature,* 2000 407, 319-320.

[120] Brummell, DA; Balint-Kurti, PJ; Harpster, MK *et al.* Inverted repeat of a heterologous 3'-untranslated region for high-efficiency, high-throughput gene silencing. *Plant J,* 2003 33, 793-800.

[121] Chuang, CF; Meyerowitz, EM. Specific and heritable genetic interference by double-stranded RNA in *Arabidopsis thaliana. Proc Natl Acad Sci USA,* 2000 97, 4985-4990.

[122] Wesley, SV; Helliwell, CA; Smith, NA *et al.* Construct design for efficient, effective and high-throughput gene silencing in plants. *Plant J,* 2001 27, 581-590.

[123] Waterhouse, PM; Helliwell, CA. Exploring plant genomes by RNA-induced gene silencing. *Nat Rev Genet,* 2003 4, 29-38.

[124] Watson, JM; Fusaro, AF; Wang, M; Waterhouse, PM. RNA silencing platforms in plants. *FEBS Letters,* 2005 579, 5982-5987.

[125] Dalakouras, A; Moser, M; Zwiebel, M *et al.* A hairpin RNA construct residing in an intron efficiently triggered RNA-directed DNA methylation in tobacco. *Plant J,* 2009 60, 840-851.

[126] Reyes, CA; Peña ,EJ; Zanek, MC *et al.* Differential resistance to *Citrus psorosis virus* in transgenic *Nicotiana benthamiana* plants expressing hairpin RNA derived from the coat protein and 54K protein genes. *Plant Cell Rep,* 2009 28, 1817-1825.

[127] López, C; Cervera, M; Fagoaga, C *et al*. Accumulation of transgene-derived siRNAs is not sufficient for RNAi-mediated protection against *Citrus tristeza virus* in transgenic Mexican lime. *Mol Plant Path,* 2010 11, 33-41.

[128] Baulcombe, DC. Fast forward genetics based on virus-induced gene silencing. *Curr Opin Plant Biol,* 1999 2, 109-113.

[129] Kumagai, MH; Donson, J; della-Cioppa, G *et al*. Cytoplasmic inhibition of carotenoid biosynthesis with virus-derived RNA. *Proc Natl Acad Sci USA,* 1995 92, 1679-1683.

[130] Ruiz, MT; Voinnet, O; Baulcombe, DC. Initiation and maintenance of virus-induced gene silencing. *Plant Cell,* 1998 10, 937-946.

[131]. Ratcliff, F; Martin-Hernandez, AM; Baulcombe, DC. Tobacco rattle virus as a vector for analysis of gene function by silencing. *Plant J,* 2001 25, 237-245.

[132] Burch-Smith, TM; Schiff, M; Liu, Y; Dinesh-Kumar SP. Efficient virus-induced gene silencing in *Arabidopsis. Plant Physiol,* 2006 142, 21-27.

[133] Pflieger, S; Blanchet, S; Camborde, L *et al*. Efficient virus induced gene silencing in *Arabidopsis* using a 'one-step' TYMV-derived vector. *Plant J,* 2008 56, 678-690.

[134] Kjemtrup, S; Sampson, KS; Peele, CG *et al*. Gene silencing from plant DNA carried by a geminivirus. *Plant J,* 1998 14, 91-100.

[135] Gosselé, V; Faché, I; Meulewaeter, F *et al*. SVISS-a novel transient gene silencing system for gene function discovery and validation in tobacco plants. *Plant J,* 2002 32, 859-866.

[136] Burch-Smith, TM; Anderson, JC; Martin, GB; Dinesh-Kumar, SP. Applications and advantages of virus induced gene silencing for gene function studies in plants. *Plant J,* 2004 39, 734-746.

[137] Unver, T; Budak, H. Virus-Induced Gene Silencing, a Post Transcriptional Gene Silencing Method. *Int J Plant Genomics,* 2009 Article ID 198680.

[138] Zeng, Y; Wagner, EJ; Cullen, BR. Both natural and designed micro RNAs can inhibit the expression of cognate mRNAs when expressed in human cells. *Mol Cell,* 2002 9, 1327-1333.

[139] Parizotto, EA; Dunoyer, P; Rahm, N *et al*. In vivo investigation of the transcription, processing, endonucleolytic activity, and functional relevance of the spatial distribution of a plant miRNA. *Genes Dev,* 2004 18, 2237-2242.

[140] Alvarez, JP; Pekker, I; Goldshmidt, A *et al*. Endogenous and synthetic microRNAs stimulate simultaneous, efficient, and localized regulation of multiple targets in diverse species. *Plant Cell,* 2006 18, 1134-1151.

[141] Ossowski, S; Schwab, R; Weigel, D. Gene silencing in plants using artificial microRNAs and other small RNAs. *Plant J,* 2008 53, 674-690.

[142] Schwab, R; Ossowski, S; Warthmann, N; Weigel, D. Directed Gene Silencing with Artificial MicroRNAs. In: Meyers BC, Green PJ. *Plant MicroRNAs: Methods in Molecular Biology. Humana Press;* 2009 592, 71-88.

[143] Niu, QW; Lin, SS; Reyes, JL *et al*. Expression of artificial microRNAs in transgenic *Arabidopsis thaliana* confers virus resistance. *Nat Biotechnol,* 2006 24, 1420-1428.

[144] Duan, CG; Wang, CH; Fang, RX; Guo, HS. Artificial MicroRNAs Highly Accessible to Targets Confer Efficient Virus Resistance in Plants. *J Virology,* 2008 82, 11084-11095.

[145] Garcia, JA; Simon-Mateo, C. A micropunch against plant viruses. *Nat Biotechnol,* 2006 24, 1358-1359.

[146] Szittya, G; Silhavy, D; Molnár, A *et al.* Low temperature inhibits RNA silencing-mediated defence by the control of siRNA generation. *EMBO J,* 2003 22, 633-640.

[147] Lin, SS; Wu, HW; Elena, SF *et al.* Molecular Evolution of a Viral Non-Coding Sequence under the Selective Pressure of amiRNA Mediated Silencing. *PLoS Pathog,* 2009 5, e1000312.

[148] Warthmann, N; Chen, H; Ossowski, S *et al.* Highly Specific Gene Silencing by Artificial miRNAs in Rice. *PLoS ONE,* 2008 3, e1829.

[149] Qiao, H; Chang, KN; Yazaki, J; Ecker JR. Interplay between ethylene, ETP1/ETP2 F-box proteins, and degradation of EIN2 triggers ethylene responses in *Arabidopsis. Genes & Dev,* 2009 23, 512-521.

[150] Petsch, KA; Ma, C; Scanlon, MJ; Jorgensen, RA. Targeted forward mutagenesis by transitive RNAi. *Plant J,* 2010 61, 873-882.

[151] Klingelhoefer, JW; Moutsianas, L; Holmes, C. Approximate Bayesan feature selection on a large meta-dataset offers novel insights on factors that effect siRNA potency. *Bioinformatics,* 2009 25, 1594-1601.

[152] Birch, RG; Shen, B; Sawyer, BJB *et al.* Evaluation and application of a luciferase fusion system for rapid *in vivo* analysis of RNAi targets and constructs in plants. *Plant Biotech J,* 2010 8, 1-11.

[153] Kawasaki, H; Ohkawa, J; Tanishige, N *et al.* Selection of the best target site for ribozyme-mediated cleavage within a fusion gene for adenovirus E1A-associated 300 kDa protein (p300) and luciferase. *Nucleic Acid Res,* 1996 24, 3010-3016.

[154] Sawyer, BJB; Birch, RG. A generic assay for antisense and rybozime inhibition *in vitro. Proc Aust Soc Biochem Mol Biol,* 1996 28, POS-320-302.

[155] Fitch, MMM; Manshardt, RM; Gonsalves, D *et al.* Stable transformation of papaya via microprojectile bombardment. *Plant Cell Rep,* 1990 189, 189-194.

[156] Fitch, MMM; Manshardt, RM; Gonsalves, D *et al.* Virus resistant papaya plants derived from tissues bombarded with the coat protein gene of *Papaya ringspot virus. Nature Biotechnol,* 1992 10, 1466-1472.

[157] Lius, S; Manshardt, RM; Fitch, MMM *et al.* Pathogen-derived resistance provides papaya with effective protection against *Papaya ringspot virus. Mol Breeding,* 1997 3, 161-168.

[158] Tennant, EE; Gonsalves, C; Ling K-S, *et al.* Differential protection against *Papaya ringspot virus* isolates in coat protein gene transgenie papaya and classically cross-protected papaya. *Phytopathology,* 1994 84, 1359-1366.

[159] Bau, H-J; Cheng, Y-H; Yu, T-A *et al.* Broad-spectrum resistance to different geographic strains of *Papaya ringspot virus* in coat protein gene transgenic papaya. *Phytopathology,* 2003 93, 112-120.

[160] Ming, R; Hou, S; Feng, Y *et al.* The draft genome of the transgenic tropical fruit tree papaya (*Carica papaya* Linnaeus). *Nature,* 2008 452, 991-997.

[161] Kohli, A; Christou, P. Stable transgenes bear fruit. *Nature Biotechnol,* 2008 26, 653-654.

[162] Kung, YJ; Bau, HJ; Wu, YL; *et al.* Generation of transgenic papaya with double resistance to Papaya ringspot virus and Papaya leaf-distortion mosaic virus. *Phytopathology,* 2009 99,1312-1320.

[163] Gutiérrez-E, MA; Luth, D; Moore, GA. Factors affecting *Agrobacterium*-mediated transformation in Citrus and production of sour orange (*Citrus aurantium* L.) plants

expressing the coat protein gene of *Citrus tristeza virus*. *Plant Cell Rep,* 1997 16, 745-753.

[164] Domínguez, A; Guerri, J; Cambra, M *et al*. Efficient production of transgenic citrus plants expressing the coat protein gene of *Citrus tristeza virus*. *Plant Cell Rep,* 2000 19, 427-433.

[165] Ghorbel, R; Domínguez, A; Navarro, L; Peña, L. High efficiency genetic transformation of sour orange (*Citrus aurantium*) and production of transgenic trees containing the coat protein gene of *Citrus tristeza virus*. *Tree Physiol,* 2000 20, 1183-1189.

[166] Domínguez, A; Mendoza, AH; Guerri, J *et al*. Pathogen-derived resistance to *Citrus tristeza virus* (CTV) in transgenic mexican lime (*Citrus aurantifolia* (Christ.) Swing.) plants expressing its *p25* coat protein gene. *Mol Breeding,* 2002 10, 1-10.

[167] Fagoaga, C; López, C; Mendoza, AH *et al*. Post-transcriptional gene silencing of the p23 silencing suppressor of *Citrus tristeza virus* confers resistance to the virus in transgenic Mexican lime. *Plant Mol Biol,* 2006 60, 153-165.

[168] Febres, VJ; Lee, RF; Moore, GA. Transgenic resistance to *Citrus tristeza virus* in grapefruit. *Plant Cell Rep,* 2008 27, 93-104.

[169] Scorza, R; Ravelonandro, M; Callahan, AM *et al*. Transgenic plums *(Prunus domestica)* express the plum pox virus coat protein gene. *Plant Cell Rep,* 1994 14, 18-22.

[170] Ravelonandro, M; Scorza, R; Bachelier, JC *et al*. Resistance to transgenic *Prunus domestica* to plum pox virus infection. *Plant Dis,* 1997 81, 231-235.

[171] Scorza, R; Callahan, A; Levy, L *et al*. Post-transcriptional gene silencing in plum pox virus resistant transgenic European plum containing the plum potyvirus coat protein gene. *Transgenic Res,* 2001 10, 201-209.

[172] Hily, JM; Scorza, R; Webb, K; Ravelonandro, M. Accumulation of the Long Class of siRNA Is Associated with Resistance to *Plum pox virus* in a Transgenic Woody Perennial Plum Tree. *Mol Plant Microbe In,* 2005 18, 794-799.

[173] Hily, JM; Scorza, R; Malinowski, T *et al*. Stability of gene silencing-based resistance to *Plum pox virus* in transgenic plum (*Prunus domestica* L.) under field conditions. *Transgenic Res,* 2004 13, 427-36.

[174] Malinowski, T; Cambra, M; Capote, N *et al*. Field trials of plum clones transformed with the *Plum pox virus* coat protein (PPV-CP) gene. *Plant Dis,* 2006 90, 1012-1018.

[175] Hily, JM; Ravelonandro, M; Damsteegt, V *et al*. Plum pox virus coat protein gene Intron-hairpin-RNA (ihpRNA) constructs provided resistance to plum pox virus in *Nicotiana benthamiana* and *Prunus domestica*. *J Am Soc Hortic Sci,* 2007 132, 850-858.

[176] Martelli, GP. Grapevine virology highlights 2006-09. *Extended abstract of 16th ICVG Meeting,* Dijon, France, 2009, 15-23.

[177] Vigne, E; Komar, V; Fuchs M. Field safety assessment of recombination in transgenic grapevines expressing the coat protein gene of Grapevine fanleaf virus. *Transgenic Res,* 2004 13, 165-179.

[178] Gambino, G; Perrone, I; Carra, A *et al*. Transgene silencing in grapevines transformed with GFLV resistance genes: analysis of variable expression of transgene, siRNAs production and cytosine methylation. *Transgenic Res,* 2010 19, 17-27.

[179] Haque, AKMN; Yamaoka, N; Nishiguchi, M. Cytosine methylation is associated with RNA silencing in silenced plants but not with systemic and transitive RNA silencing through grafting. *Gene*, 2007 396, 321-331.

[180] Farinelli, L; Malnoe, P; Collet, G. Heterologous encapsidation of potato virus Y strain O PVY-O with the transgene coat protein of PVY strain N PVY-N in *Solanum tuberosum* cv. Bintje. *Nature Biotechnol*, 1992 10, 1020-1025.

[181] Malnoe, P; Farinelli, L; Collet, G; Reust, W. Small-scale field tests with transgenic potato, cv. Bintje, to test the resistance to primary and secondary infections with potato virus Y. *Plant Mol Biol*, 1994 25, 963-975.

[182] Okamoto, D; Nielsen, SVS; Albrechtsen, M; Borkhardt, B. General resistance against potato virus Y introduced into a commercial potato cultivar by genetic transformation with PVYN coat protein. *Potato Res*, 1996 39, 271-282.

[183] Missiou, A; Kalantidis, K; Boutla, A *et al.* Generation of transgenic potato plants highly resistant to potato virus Y (PVY) through RNA silencing. *Mol Breed*, 2004 14, 185-197.

[184] Jorgensen, B; Albrechtsen, M. Stability of RNA silencing-based traits in potato after virus infection. *Mol Breed*, 2007 19, 371-376.

[185] Bai, Y; Guo, Z; Wang, X *et al.* Generation of double-virus-resistant marker-free transgenic potato plants. *Prog Nat Sci*, 2009 19, 543-548.

[186] Vazquez-Rovere, C; Asurmendi, S; Hopp, HE. Transgenic resistance in potato plants expressing potato leafroll virus (PLRV) replicase gene sequences is RNA-mediated and suggests the involvement of post-transcriptional gene silencing. *Arch Virol*, 2001 146, 1337-1353.

[187] Kreuze, JF; Klein, IS; Lazaro, MU *et al.* RNA silencing-mediated resistance to a crinivirus (Closteroviridae) in cultivated sweetpotato (*Ipomoea batatas* L.) and development of sweetpotato virus disease following co-infection with a potyvirus. *Mol Plant Pathol*, 2008 9, 589-598.

[188] Brunetti, A; Tavazza, M; Noris, E *et al.* High expression of truncated viral rep protein confers resistance to tomato yellow leaf curl virus in transgenic tomato plants. *Mol Plant-Microbe Interact*, 1997 10, 571-579.

[189] Lucioli, A; Noris, E; Brunetti, A *et al.* Tomato yellow leaf curl Sardinia virus rep-derived resistance to homologous and heterologous geminiviruses occurs by different mechanisms and is overcome if virus-mediated transgene silencing is activated. *J Virol*, 2003 77, 6785-6798.

[190] Noris, E; Lucioli, A; Tavazza, R *et al.* Tomato yellow leaf curl Sardinia virus can overcome transgene-mediated RNA silencing of two essential viral genes. *J Gen Virol*, 2004 85, 1745-1749.

[191] Yang, Y; Sherwood, TA; Patte, CP *et al.* Use of tomato yellow leaf curl virus Rep gene sequences to engineer TYLCV resistance in tomato. *Phytopathology*, 2004 94, 490-496.

[192] Fuentes, A; Ramos, PL; Fiallo, E *et al.* Intron–hairpin RNA derived from replication associated protein C1 gene confers immunity to Tomato Yellow Leaf Curl Virus infection in transgenic tomato plants. *Transgenic Res*, 2006 15, 291-304.

[193] Bian, XY; Rasheed, MS; Seemanpillai, MJ; Rezaian MA. Analysis of Silencing Escape of *Tomato leaf curl virus*: An Evaluation of the Role of DNA Methylation. *Mol Plant Microbe In*, 2006 19, 614-624.

[194] Morroni, M; Thompson, JR; Tepfer, M. Twenty Years of Transgenic Plants Resistant to *Cucumber mosaic virus*. *Mol Plant Microbe In*, 2008 21, 675-684.

[195] Tricoli, DM; Carney, KJ; Russel, PF *et al*. Field evaluation of transgenic squash containing single or multiple virus coat protein gene constructs for resistance to Cucumber mosaic virus, watermelon mosaic virus 2, and zucchini yellow mosaic virus. *Nature Biotechnol*, 1995 13, 1458-1465.

[196] Kalantidis, K; Psaradakis, S; Tabler, M; Tsagris, M. The occurrence of CMV-specific short RNAs in transgenic tobacco expressing virus-derived double-stranded RNA is indicative of resistance to the virus. *Mol Plant-Microbe In*, 2002 15, 826-833.

[197] Chen, Y-K; Lohuis, D; Goldbach, R; Prins, M. High frequency induction of RNA-mediated resistance against Cucumber mosaic virus using inverted repeat constructs. *Mol Breed*, 2004 14, 215-226.

[198] Cillo, F; Finetti-Sialer, MM; Papanice, MA; Gallitelli D. Analysis of Mechanisms Involved in the *Cucumber mosaic virus* Satellite RNA-mediated Transgenic Resistance in Tomato Plants. *Mol Plant Microbe In*, 2004 17, 98-108.

[199] Pinto, YM; Kok, RA; Baulcombe, DC. Resistance to rice yellow mottle virus(RYMV) in cultivated African rice varieties containing RYMV transgenes. *Nature Biotechnol*, 1999 17, 702-707.

[200] Wang, MB; Abbot, DC; Waterhouse, PM. A single copy of a virus-derived transgene encoding hairpin RNA gives immunity to *Barley yellow dwarf virus*. *Mol Plant Pathol*, 2000 1, 347-356.

[201] Ingelbrecht, IL; Irvine, JE; Mirkov, TE. Post-transcriptional gene silencing in transgenic sugarcane. Dissection of homology dependent virus resistance in a monocot that has a complex polyploidy genome. *Plant Physiol*, 1999 119, 1187-1197.

[202] Xu, J; Schubert, J; Altpeter, F. Dissection of RNA-mediated ryegrass mosaic virus resistance in fertile transgenic perennial ryegrass (*Lolium perenne* L.). *Plant J*, 2001 26, 265-274.

[203] Escobar, MA; Civerolo, EL; Summerfelt, KR; Dandekar, AM. RNAi-mediated oncogene silencing confers resistance to crown gall tumorigenesis. *Proc Natl Acad Sci USA*, 2001 98, 13437-13442.

[204] Escobar, MA; Leslie, CA; McGranahan, GH; Dandekar, AM. Silencing crown gall disease in walnut (*Juglans regia* L.). *Plant Sci*, 2002 163, 591-597.

[205] Viss, WJ; Pitrak, J; Humann, J *et al*. Crown-gall-resistant transgenic apple trees that silence *Agrobacterium tumefaciens* oncogenes. *Mol Breed*, 2003 12, 283-295.

[206] Baum, JA; Bogaert, T; Clinton, W *et al*. Control of coleopteran insect pests through RNA interference. *Nat Biotechnol*, 2007 25, 1322-1326.

[207] Murata, M; Haruta, M; Murai, N *et al*. Transgenic apple (*Malus* x *domestica*) shoot showing low browning potential. *J Agric Food Chem*, 2000 48, 5243-8.

[208] Cheng, L; Zhou, R; Reidel, EJ *et al*. Antisense inhibition of sorbitol synthesis leads to up-regulation of starch synthesis without altering CO_2 assimilation in apple leaves. *Planta*, 2005 220, 767-76.

[209] Zhou, R; Cheng, L; Dandekar, AM. Down-regulation of sorbitol dehydrogenase and up-regulation of sucrose synthase in shoot tips of the transgenic apple trees with decreased sorbitol synthesis. *J Exp Bot*, 2006 57, 3647-57.

[210] Dandekar, AM; Teo, G; Defilippi, BG *et al*. Effect of down-regulation of ethylene biosynthesis on fruit flavor complex in apple fruit. *Transgenic Res*, 2004 13, 373-384.

[211] Johnston, JW; Gunaseelan, K; Pidakala P *et al*. Co-ordination of early and late ripening events in apples is regulated through differential sensitivities to ethylene. *J Exp Bot,* 2009 60, 2689-2699.

[212] Szankowski, I; Flachowsky, H; Li, H *et al*. Shift in polyphenol profile and sublethal phenotype caused by silencing of anthocyanidin synthase in apple (*Malus* sp.). *Planta,* 2009 229, 681-692.

[213] Wong, WS; Li, GG; Ning, W *et al*. Repression of chilling-induced ACC accumulation in transgenic citrus by over-production of antisense 1-aminocyclopropane-1-carboxylate synthase RNA. *Plant Sci,* 2001 161, 969-977.

[214] Gao, M; Matsuta, N; Murayama, H *et al*. Gene expression and ethylene production in transgenic pear (*Pyrus communis* cv. 'La France') with sense or antisense cDNA encoding ACC oxidase. *Plant Sci,* 2007 173, 32-42.

[215] Botella, JR; Cavallaro, AS; Cazzonelli, CI. Towards the production of transgenic pineapple to control flowering and ripening. *Acta Hort,* 2000 529, 115-122.

[216] Trusov, Y; Botella, JR. Silencing of the ACC synthase gene ACACS2 causes delayed flowering in pineapple [*Ananas comosus* (L.) Merr.]. *J Exp Bot,* 2006 57, 3953-3960.

[217] Ko, HL; Campbell, PR; Jobin-Décor, MP *et al*. The introduction of transgenes to control blackheart in pineapple (*Ananas comosus* L.) cv. *Smooth Cayenne by microprojectile bombardment Euphytica,* 2006 150, 387-395.

[218] Palomer, X; Llop-Tous, I; Vendrell, M *et al*. Antisense down-regulation of strawberry endo-β-(1,4)-glucanase genes does not prevent fruit softening during ripening. *Plant Sci,* 2006 171, 640-646.

[219] Jiménez-Bermúdez, S; Redondo-Nevado, J; Muñoz-Blanco, J *et al*. Manipulation of strawberry fruit softening by antisense expression of a pectate lyase gene. *Plant Physiol,* 2002 128, 751-759.

[220] Sesmero, R; Quesada, MA; Mercado, JA. Antisense inhibition of pectate lyase gene expression in strawberry fruit: characteristics of fruits processed into jam. *J Food Eng,* 2007 79, 194-199.

[221] Park, JI; Lee, YK; Chung, WI *et al*. Modification of sugar composition in strawberry fruit by antisense suppression of an ADPglucose pyrophosphorylase. *Mol Breed,* 2006 17, 269-279.

[222] Lunkenbein, S; Coiner, H; Ric de Vos, CH *et al*. Molecular characterization of a stable antisense chalcone synthase phenotype in strawberry (*Fragaria×ananassa*). *J Agric Food Chem,* 2006 54, 2145-2153.

[223] Hoffmann, T; Kalinowski, G; Schwab, W. RNAi-induced silencing of gene expression in strawberry fruit (*Fragaria×ananassa*) by agroinfiltration: a rapid assay for gene function analysis. *Plant J,* 2006 48, 818-826.

[224] El Euch, C; Jay-Allemand, C; Pastuglia M *et al*. Expression of antisense chalcone synthase RNA in transgenic hybrid walnut microcuttings. Effect on flavonoid content and rooting ability. *Plant Mol Biol,* 1998 38, 467-79.

[225] Petri, C; Webb, K; Hily, JM *et al*. High transformation efficiency in plum (*Prunus domestica* L.): a new tool for functional genomics studies in *Prunus* spp. *Mol Breed,* 2008 22, 581-591.

[226] Holzberg, S; Brosio, P; Gross, C; Pogue, GP. Barley stripe mosaic virus induced gene silencing in a monocot plant. *Plant J,* 2002 30, 315-327.

[227] Gilissen, LJWJ; Bolhaar, STHP; Matos, CI *et al*. Silencing the major apple allergen Mal d 1 by using the RNA interference approach. *J Allergy Clin Immunol*, 2005 115, 364-369.

[228] Broothaerts, W; Keulemans, J; Van Nerum, I. Self-fertile apple resulting from S-RNase gene silencing. Plant Cell Rep, 2004 22, 497-501.

[229] Hu, W; Harding, SA; Lung, J *et al*. Repression of lignin biosynthesis promotes cellulose accumulation and growth in transgenic trees. *Nature Biotechnol*, 1999 17, 808-812.

[230] Jouanin, L; Goujon, T; De Nadai, V *et al*. Lignification in transgenic poplars with extremely reduced caffeic acid O-methyltransferase activity. *Plant Physiol*, 2000 123, 1363-1373.

[231] Moller, R; Steward, D; Phillips, L *et al*. Gene silencing of cinnamyl alcohol dehydrogenase in *Pinus radiata* callus cultures. *Plant Physiol Biochem*, 2005 43, 1061-1066.

[232] Smith, CJS; Watson, CF; Bird, CR *et al*. Expression of a truncated tomato polygalacturonase gene inhibits expression of the endogenous gene in transgenic plants. *Mol Gen Genet*, 1990 224, 477-481.

[233] Mishra, KK; Handa, AK. Post-transcriptional silencing of pectin methylesterase gene in transgenic tomato fruits results from impaired pre-mRNA processing. *Plant J*, 1998 14, 583-592.

[234] Phan, TD; Bo, W; West, G *et al*. Silencing of the major salt-dependent isoform of pectinesterase in tomato alters fruit softening. *Plant Physiol*, 2007 144, 1960–1967.

[235] Brummell, DA; Harpster, MH; Civello, PM *et al*. Modification of expansin protein abundance in tomato fruit alters softening and cell wall polymer metabolism during ripening. *Plant Cell*, 1999 11, 2203-2216.

[236] Brummell, DA; Howie, WJ; Ma, C; Dunsmuir, P. Postharvest fruit quality of transgenic tomatoes suppressed in expression of a ripening-related expansin. *Postharvest Biol Tech*, 2002 25, 209-220.

[237] Le, LQ; Mahler, V; Lorenz, Y *et al*. Reduced allergenicity of tomato fruits harvested from Lyc e 1-silenced transgenic tomato plants. *J Allergy Clin Immunol*, 2006 118, 1176-1183.

[238] Fernandez, AI; Viron, N; Alhagdow, M *et al*. Flexible tools for gene expression and silencing in tomato. *Plant Physiol*, 2009 151, 1729-1740.

[239] Shimada, H; Tada, Y; Kawasaki, T; Fujimura, T. Antisense regulation of the rice WAXY gene expression using a PCR-amplified fragment of the rice genome reduces the amylose content in grain starch. *Theor Apple Genet*, 1993 86, 665-672.

[240] Terada, R; Nakajima, M; Isshiki, M *et al*. Antisense waxy genes with highly active promoters effectively suppress waxy gene expression in transgenic rice. *Plant Cell Physiol*, 2000 41, 881-888.

[241] Itoh, K; Nakajima, M; Shimamoto, K. Silencing of waxy genes in rice containing Wx transgenes. *Mol Gen Genet*, 1997 255, 351-358.

[242] Kuwano, M; Ohyama, A; Tanaka, Y *et al*. Molecular breeding for transgenic rice with low-phytic-acid phenotype through manipulating *myo* –inositol 3-phosphate synthase gene. *Mol Breed*, 2006 18, 263-272.

[243] Kuwano, M; Mimura, T; Takaiwa, F; Yoshida , KT. Generation of stable 'low phytic acid' transgenic rice through antisense repression of the *1 D -myo-inositol 3 -phosphate synthase* gene using the 18-kDa oleosin promoter. *Plant Biotechnol J,* 2009 7, 96-105.

[244] Regina, A; Bird, A; Topping, D *et al.* High amylose wheat generated by RNA interference improves indices of large-bowel health in rats. *Proc Natl Acad Sci USA,* 2006 103, 3546-3551.

[245] Gil-Humanes, J; Pistón, F; Hernando, A *et al.* Silencing of g-gliadins by RNA interference (RNAi) in bread wheat. *J Cereal Sci,* 2008 48, 565-568.

[246] Yue, SJ; Li, H; Li, YW *et al.* Generation of transgenic wheat lines with altered expression levels of 1Dx5 high-molecular weight glutenin subunit by RNA interference. *J Cereal Sci,* 2008 47, 153-161.

[247] Xia, L; Geng, H; Chen, X *et al.* Silencing of puroindoline a alters the kernel texture in transgenic bread wheat. *J Cereal Sci,* 2008 47, 331-338.

[248] Cigan, AM; Unger-Wallace, E; Haug-Collet, K. Transcriptional gene silencing as a tool for uncovering gene function in maize. *Plant J,* 2005 43, 929-940.

[249] McGinnis, K; Murphy, N; Carlson, AR *et al.* Assessing the efficiency of RNA interference for maize functional genomics. *Plant Physiol,* 2007 143, 1441–1451.

[250] Hournard, NM; Mainville, JL; Bonin, CP. High-lysine corn generated by endosperm-specific suppression of lysine catabolism using RNAi. *Plant Biotechnol J,* 2007 5, 605-614.

[251] Frizzi, A; Huang, S; Gilbertson, LA *et al.* Modifying lysine biosynthesis and catabolism in corn with a single bifunctional expression/silencing transgene cassette. *Plant Biotechnol J,* 2008 6, 13-21.

[252] Liu, Q; Singh, SP; Green, AG. High-stearic and high-oleic cottonseed oils produced by hairpin RNA-mediated post-transcriptional gene silencing. *Plant Physiol,* 2002 129, 1732–1743.

[253] Sunilkumar, G; Campbell, LM; Puckhaber, L *et al.* Engineering cottonseed for use in human nutrition by tissue-specific reduction of toxic gossypol. *Proc Natl Acad Sci USA,* 2006 103, 18054-18059.

[254] Dodo, HW; Konan, KN; Chen, FC *et al.* Alleviating peanut allergy using genetic engineering: the silencing of the immunodominant allergen Ara h 2 leads to its significant reduction and a decrease in peanut allergenicity. *Plant Biotechnol J,* 2008 6, 135-145.

[255] Chu, Y; Faustinelli, P; Ramos, ML *et al.* Reduction of IgE binding and non promotion of Apergillus flavus fungal growth by simultaneously silencing Ara h 2 and Ara h 6 in peanut. *J Agric Food Chem,* 2008 56, 11225-11233.

[256] Siritunga, D; Sayre, RT. Generation of cyanogen-free transgenic cassava. *Planta,* 2003 217, 367-373.

[257] Siritunga, D; Sayre RT. Engineering cyanogen synthesis and turnover in cassava (*Manihot esculenta*). *Plant Mol Biol,* 2004 56, 661-669.

[258] Senda, M; Masuta, C; Ohnishi, S *et al.* Patterning of virus-infected *Glycine max* seed coat is associated with suppression of endogenous silencing of chalcone synthase genes. *Plant Cell,* 2004 16, 807-818.

[259] Kasai, A; Kasai, K; Yumoto, S; Senda, M. Structural features of GmIRCHS, candidate of the I gene inhibiting seed coat pigmentation in soybean: implications for inducing

endogenous RNA silencing of chalcone synthase genes. *Plant Mol Biol,* 2007 64, 467-479.

[260] Koseki, M; Goto, K; Masuta, C; Kanazawa, A. The star-type color pattern in *Petunia hybrida* 'Red Star' flowers is induced by sequence-specific degradation of chalcone synthase RNA. *Plant Cell Physiol,* 2005 46, 1879-1883.

[261] Kusaba, M; Miyahara, K; Iida, S *et al.* Low glutelin content1: a dominant mutation that suppresses the *glutelin* multigene family via RNA silencing in rice. *Plant Cell,* 2003 15, 1455-1467.

[262] Song, XJ; Huang, W; Shi, M *et al.* A QTL for rice grain width and weight encodes a previously unknown RING-type E3 ubiquitin ligase. *Nat Genet,* 2007 39, 623–630.

[263] Leshem, Y; Melamed-Book, N; Cagnac, O *et al.* Suppression of Arabidopsis vesicle-SNARE expression inhibited fusion of H_2O_2- containing vesicles with tonoplast and increased salt tolerance. *Proc Natl Acad Sci USA,* 2006 103, 18008–18013.

[264] Sclep, G; Allemeersch, J; Liechti, B *et al.* CATMA, a comprehensive genome-scale resource for silencing and transcript profiling of *Arabidopsis* genes. *BMC Bioinformatics,* 2007 8, 400.

[265] Hilson, P; Allemeersch, J; Altmann, T *et al.* Versatile gene-specific sequence tags for Arabidopsis functional genomics: transcript profiling and reverse genetics applications. *Genome Res,* 2004 14, 2176-2189.

[266] Tuskan, GA; Difazio, S; Jansson, S *et al.* The genome of black cottonwood, *Populus trichocarpa* (Torr. & Gray). *Science,* 2006 313, 1596-1604.

[267] Jaillon, O; Aury, JM; Noel, B *et al.* The grapevine genome sequence suggests ancestral hexaploidization in major angiosperm phyla. *Nature,* 2007 449, 463-467.

[268] Velasco, R; Zharkikh, A; Troggio, M *et al.* A high quality draft consensus sequence of the genome of a heterozygous grapevine variety. *PLoS One,* 2007 2: e1326.

In: Gene Silencing: Theory, Techniques and Applications ISBN: 978-1-61728-276-8
Editor: Anthony J.Catalano © 2010 Nova Science Publishers, Inc.

Chapter II

RNA Silencing in the Ectomycorrhizal Fungus *Laccaria bicolor*

Minna J. Kemppainen and Alejandro G. Pardo[*]

Laboratorio de Micología Molecular, Departamento de Ciencia y Tecnología,
Universidad Nacional de Quilmes, and Consejo Nacional de Investigaciones Científicas y
Tecnicas (CONICET). Roque Sáenz Peña 352, (B1876BXD) Bernal,
Provincia de Buenos Aires, Argentina

Abstract

Mycorrhiza is a mutualistic association between fungi and the roots of the vast majority of terrestrial plants. In natural ecosystems the plant nutrient uptake from the soil takes place via the extraradical mycelia of these mycosimbionts. While most herbaceous plants and tropical trees form endomycorrhiza-type interactions, trees of boreal and temperate ecosystems are typically ectomycorrhizal (ECM). These species include the majority of ecologically and economically important trees and the fungal symbionts are predominantly filamentous basidiomycetes.

The symbiotic phase in the life cycle of ECM basidiomycetes is the dikaryon. Hence, studies on symbiotic relevant gene functions would require the inactivation of both gene copies in the dikaryotic mycelium.

RNA silencing is a sequence homology-dependent degradation of target mRNAs based on an ancient cellular mechanism believed to have evolved as protection of eukaryotic cells against alien nucleic acids. In different eukaryotic organisms, including fungi, the RNA silencing pathway can be artificially triggered to target and degrade gene transcripts of interest, resulting in gene knock-down. Most importantly, RNA silencing can act at the cytosolic level affecting mRNAs originating from several gene copies and different nuclei, and it can thus offer an efficient way for altering gene expression in dikaryotic organisms.

---

[*] Corresponding author. Phone: 54-11-4365-7100 (ext 4205); Fax: 54-11-4365-7132; E-mail: apardo@unq.edu.ar

Laccaria bicolor, the first symbiotic fungus with its genome sequenced, has rapidly turned into a model fungus in ectomycorrhizal research. *Laccaria* possesses a complete set of genes known to be needed for RNA silencing in eukaryotic cells. We have demonstrated that RNA silencing is functional in *L. bicolor* and that it can be triggered via *Agrobacterium*-mediated transformation. Moreover, targeted gene knock-down in dikaryotic mycelium can result in functional phenotypes altered in the symbiotic capacity confirming that RNA silencing is a powerful way to study symbiosis- regulated genes. These findings have now initiated the RNA silencing era in mycorrhizal research, a field that has been hindered by the lack of proper genetic tools.

Ectomycorrhiza

The mycorrhizal symbiosis is an ancient mutualistic association between fungi and the roots of the vast majority of terrestrial plants. In natural ecosystems the plant nutrient uptake (N, P and several micronutrients) from soil happens via the extra-radical mycelia of these symbiotic fungi. This association also improves plant fitness by increasing water acquisition, heavy metal tolerance and resistance to pathogens. While most herbaceous plants and tropical trees form arbuscular mycorrhiza (AM), forest trees of boreal and temperate zones are typically ectomycorrhizal (ECM) plants. These species include the majority of economically important trees such as pines, spruces, birches, poplars, oaks, etc. On the other hand, the fungal symbionts are predominantly filamentous basidiomycetes (Figure 1).

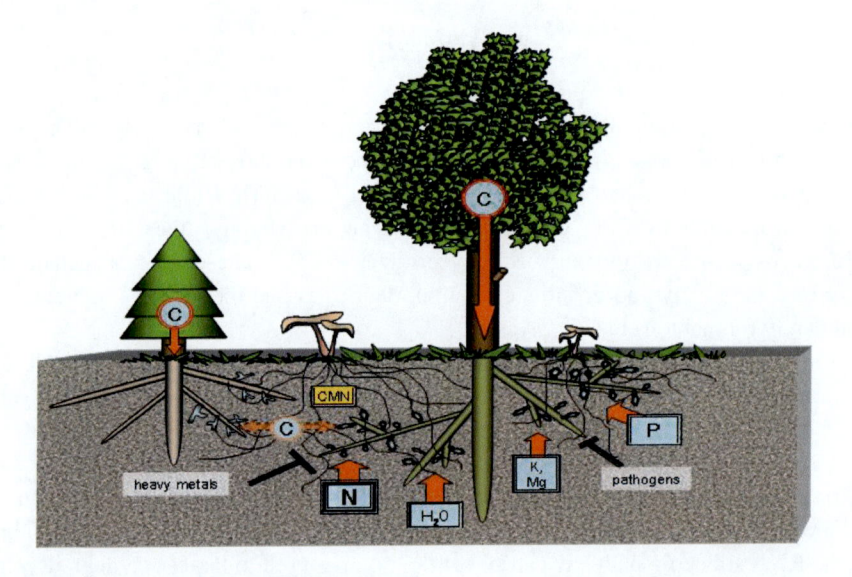

Figure 1. ECM improves plant fitness by increasing nutrient and water uptake as well as by protecting against pathogens and toxic concentrations of heavy metals. The extraradical mycelium multiplies the root nutrient absorption area in soil. ECM fungi have also access to nutrient sources not directly available to host plants (organic N, P as well as rock-minerals). In the symbiosis the fungus is fed with plant photosynthates. Due to the low host specificity of ECM fungi different plant individuals and species are also shown to be connected by common mycelial network (CMN). Horizontal carbon transfer between plants via CMNs has been demonstrated. C, carbon; N, nitrogen; P, phosphorus; K, potassium and Mg, magnesium.

The ECM is formed predominantly by tree lateral roots. Fungal infection results in micro- and macroscopical modification of the roots making the ECM organs detectable with the naked eye. These visual features are dependent on the plant species and the fungal partner involved. For instance the branching pattern, texture and color have led to the identification of different morphotypes which have been used as keys for identification of fungal symbionts (Agerer 1987–2002; Ingleby et al., 1990). Even though considerable variation exists between different morphological classes, ECM is defined by three fundamental structures:

1. A fungal mantle (or sheath) that can cover the root to different extent.
2. Hyphal penetration between, but not within, root cells forming a branched structure called Hartig net.
3. Extra-radical mycelium that extends into the soil from the symbiotic organ.

The Hartig net creates the wide contact surface between the intra-radical mycelium and plant root cells and is the physical interface where the interchange of nutrients takes place between the two symbionts. Both the plant and the fungal cell wall become modified in the symbiotic organ. The structure of the Hartig net shows significant variation between angiosperm and gymnosperm hosts. In conifers it involves both epidermal and cortical cell layers while in angiosperms usually only epidermal cells are surrounded by fungal mycelium (Peterson et al., 2004). No AM-like penetrations of root cells are observed in a functional ECM. It has been reported however that in senescent symbiotic organs ECM fungi can behave in a pathogenic manner and penetrate plant cells (Smith and Read, 2008). The anatomical structures characteristic for an ECM root are presented in Figure 2.

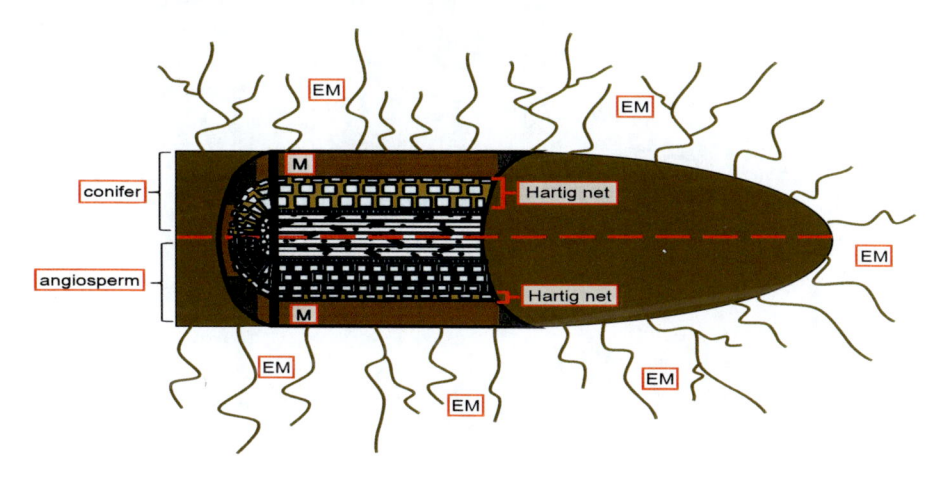

Figure 2. ECM structures. The upper half of the root represents anatomical structures characteristic of conifers where the Hartig net involves both root epidermal and cortical cells. The lower half represents angiosperm ECM where the Hartig net surrounds the epidermal root layer only. M: mantle, EM: extraradical mycelium.

The EMC symbiosis is formed by a vast number (more than 5500) of filamentous basidiomycetes (Smith & Read, 2008). These species typically form macroscopic fruiting bodies and several are also edible, of economical value and therefore collected and commercialized by man (boletales, *Cantarellus spp.*, *Tricholoma matsutake*, etc.).Therefore, this group of mycorrhizal fungi has gained general public and scientific interest for centuries.

Ectomycorrhizal basidiomycetes are clearly a polyphyletic group and many of them belong to large basidiomycete families such as Amanitaceae, Boletaceae and Russulaceae (Brundrett, 2002). They dominate in acidic, nitrogen-limited forest soils with accumulation of complex nitrogen in organic humus layer. ECM species are predominantly homobasidiomycetes with 5-6 μm wide hyphae and they inhabit upper organic soil horizons. They, unlike AM fungi, are not biotrophic and dependent on host plant for their growth capacity and can be thus cultivated as axenic cultures in laboratory. Despite the growth as saprotrophs *in vitro* the possible role of ECM fungi as facultative soil saprotrophs in nature is however under scientific debate. The ligno- and cellulytic capacity of these fungi is low when compared to saprotrophic basidiomycetes. They also occupy soil horizons where the availability of carbon compounds of high energy value is low. These facts highlight the limited capacity of ECM fungi to acquire carbon directly from the soil matrix and support the idea of their dependency on carbon supply from the host trees in nature (Baldrian, 2009).

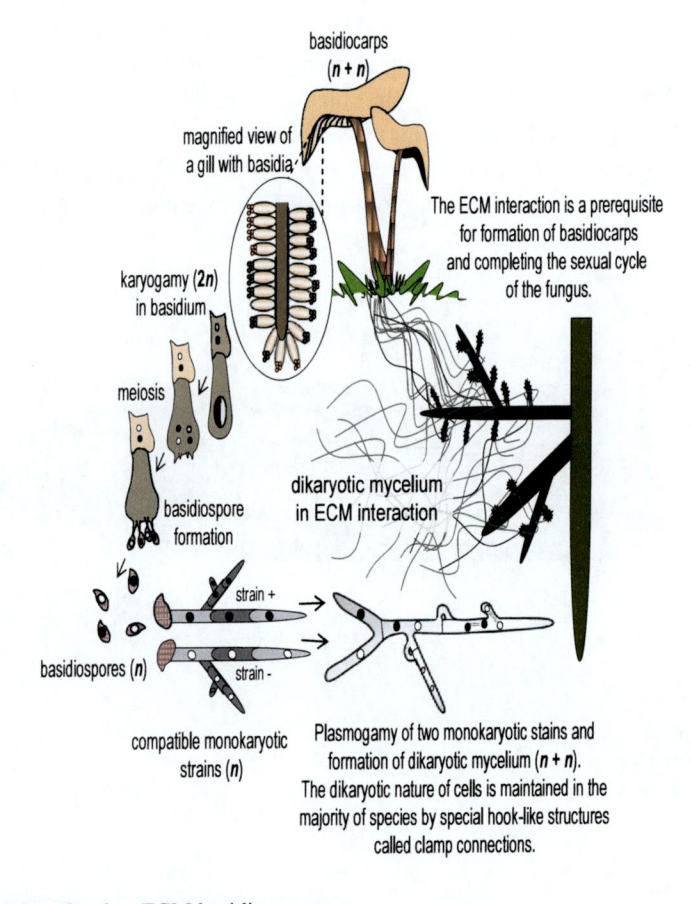

Figure 3. Typical lifecycle of an ECM basidiomycete.

Most basidiomycetes are heterothallic with a mating system controlled by one till four multiallelic mating loci (Aα, Aβ, Bα, Bβ) (Kües & Casselton, 1992). The sexual structures, often macroscopic fruiting bodies, are formed for production of basidiospores via meiosis. These basidiospores germinate into monokaryotic mycelium which needs to anastomose with another compatible monokaryotic strain in order to produce a dikaryotic mycelium. In the dikaryotic mycelium karyogamy is postponed and the two compatible nuclei co-exist separately in a common cell. This dikaryotic condition is maintained in the mycelium usually by special structures called clamp connections which assure the sorting of two separate nuclei in each hyphal compartment separated by septa. It is the dikaryotic mycelium that can further complete the sexual cycle of the fungus. This requires the formation of ECM with the host plant and therefore basidiocarp and basidiospore formation are generally not possible to obtain in axenic fungal cultures. On the other hand, monokaryotic strains are usually unable to form ECM though some exceptions to this rule are known to exist (Kropp & Fortin, 1988; Lamhamedi et al., 1990). Also intraspecific variation in mycorrhization capacity, host specificity, as well as in metabolic profiles is well reported for different ECM basidiomycetes (Smith & Read, 2008). A typical lifecycle of an ECM basidiomycete is presented in Figure 3.

Despite the huge ecological and economical importance of ECM the current comprehension of host-mycosymbiont recognition, establishment of dual-organs and functions of ECM is still rather limited.

The Molecular Revolution of ECM Research

The majority of molecular information on ECM function comes from a relative limited set of fungal-plant interactions considering mostly basidiomycete species easily cultivable in axenic cultures: *Pisolithus spp. - Eucalyptus spp.*, *Suillus bovinus - Pinus sylvestris*, *Hebeloma cylindrosporum - Pinus pinaster*, *Amanita muscaria - Picea abies*, *Paxillus involutus - Betula pendula*, *Laccaria bicolor - Pseudotzuga menziesii / Populus spp.* and *Piloderma croceum - Quercus robur*. Another widely studied EMC pair is the ascomycete *Tuber borchii - Tilia platyphyllos*. Molecular data on ECM function and ontogenic processes involved in the establishment of the symbiotic interaction traditionally originated from targeted studies on precise gene functions. However, ten years ago a molecular biology revolution arrived also in research of symbiotic plant-fungus interactions. This new era in ECM research was initiated by the use of molecular tools such as the creation of mycorrhiza specific EST libraries and boutique cDNA micro-arrays. These were used for studying expression profiles of limited sets of fungal and plant genes both in their free living forms and at different stages of ECM development. Later on, and with the use of more high throughput technologies such as cDNA macro-arrays, the activity status of thousands of genes has been resolved in single experiments (Voiblet et al., 2001; Polidori et al, 2002; Peter et al., 2003; Krüger et al., 2004; Menotta et al., 2004; Johansson et al., 2004; Duplessis et al., 2005; Le Quéré et al., 2005; Morel et la., 2005; Wright, 2005). During the last five years these studies have multiplied the knowledge on genetic processes active in ECM symbiosis. The expression of both fungal and plant host genes has been studied and numerous genes have been detected as symbiosis-regulated. This means that their transcript levels are either increased or decreased during the ECM interaction. However, no ectomycorrhiza-specific fungal or plant genes have been detected yet and their true existence has been therefore questioned.

Laccaria Bicolor and the First Genome of an ECM Fungus

The sequencing of the first tree genome, *Populus trichocarpa*, gave access to the full gene repertoire of a mycorrhizal host plant (Tuskan et al., 2006). *Populus* is an example of a tree species which is able to form both AM and ECM. With the full genome sequence available this tree has naturally turned into a reference host in molecular ECM research. Most importantly, the decision of the Department of Energy Joint Genome Institute (JGI) to further sequence the genomes of *Populus* mycobionts, the AM fungus *Glomus intraradices* and the ECM fungus *Laccaria bicolor* (Martin et al., 2004), took mycorrhizal research into the genomic era.

Laccaria bicolor (common name; bicoloured deceiver) is a homobasidiomycete of the orden Agaricales (gill mushrooms) and a member of the family Tricholomataceae (Figure 4A). It is a widespread fungus both in temperate and boreal forests where it forms ECM symbiosis with several tree species such as birches, pines and poplars (Figure 4C). The strain S238N of *Laccaria bicolor* (Maire) Orton (Figure 4B) was isolated by J. Trappe (Oregon State University, Corvallis, Or., USA) and R. Molina (USDA Forest Service, Corvallis, Or., USA) from a fruiting body collected under *Tsuga mertensiana* (Bong.) Carr. at Crater Lake, Or., USA in 1976 (Di Battista et al., 1996) The strain S238N was later on transferred to the Forest Microbiology Laboratory at INRA-Nancy Center, France. This dikaryotic strain has since been introduced as an inoculant into tree nurseries and plantations in France. In 1988 a set of monokaryons were obtained from *in vitro* germination of spores isolated from one basidiocarp of S238N engaged in ECM with *Pseudotzuga menziesii* (Selosse et al., 1996). One of these monokaryotic strains was S238N-H82 which was chosen for resolving *L. bicolor* genome.

Laccaria genome was sequenced by using the whole-genome shotgun approach by the JGI and the *Laccaria* Genome Consortium which joined research efforts of several ECM laboratories around the world. The genome was publicly released in 2006 (http://genome.jgi-psf.org/Lacbi1/Lacbi1.home.html) and the genome scale analysis was published by the *Laccaria* Genome Consortium in 2008 (Martin et al., 2008). This was the first symbiotic fungal genome sequenced. The whole-genome sequence of *Laccaria* has made possible for the first time to analyze the metabolic traits present in a fungus specialized to exploit both soil nutrient resources and the symbiotic niche in host plant roots. Comparisons with genomes of saprotrophic and pathogenic fungi have offered valuable insights in those evolutive steps leading to different fungal life-styles. Also genomic scale transcriptomic analyses of mycorrhizal development in combination with the full genomic sequence of poplar host are now technically possible. The access to *Laccaria* genome has dramatically changed the possibilities and the way molecular ECM research is and will be carried out in future. The New Phytologist volume 180 (2008) was dedicated to *Laccaria*.

Figure 4. (A) *Laccaria bicolor* (Maire) Orton sporocarps. The name bicolor refers to the violet coloration of the gills and the base of the stipe in contrast to the rest of the salmon-colored fruiting body. (B) Vegetative mycelium of *L. bicolor* dikaryotic strain S238N growing on a Petri-dish with medium P5. (C) *In vitro* mycorrhization between *L. bicolor* strain S238N and micropropagated *Populus tremula* x *Populus alba*. Photo of Laccaria fruiting bodies courtesy of Mr. José Mª. Ausín.

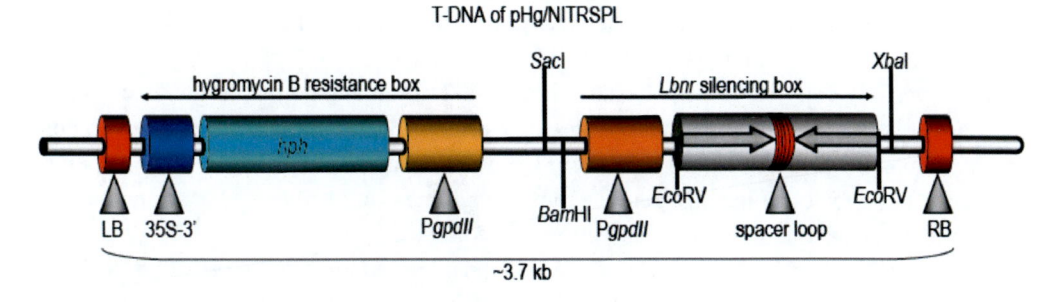

Figure 5. A schematic representation of pHg/NITRSPL binary vector's T-DNA used for launching RNAi on *Laccaria* nitrate reductase gene (*Lbnr*). P*gpdII*: *Agaricus bisporus* glyceraldehyde-3-phosphate dehygrogenase promoter, *hph*: *E. coli* aminocyclitol phophotransferase gene that confers resistance to hygromycin B, 35S-3´: cauliflower mosaic virus 35S 3´UTR polyA signal. Relevant restriction sites within the T-DNA are indicated. *Sac*I and *Bam*HI sites are unique.

Laccaria genome, organized in 12 chromosomes, is 64.9 Mb in size and it is the largest fungal genome sequenced this far. Its size is not due to large-scale duplications but an expansion of several multi-gene families. *Laccaria* genome encodes for 20614 predicted proteins and it is surprisingly rich in transposons (21 % of the genome) and repeated sequences. It has an important number of small effector-type secreted proteins (SSPs) of unknown function. Interestingly, many of these are also highly expressed in symbiotic organs. Expression of over 80 % of predicted *Laccaria* genes has been detected in vegetative mycelium, fruiting bodies or mycorrhizal organs. With the whole genome information of *Laccaria* NimbleGen whole-genome expression oligoarrays are now for the first time available also to the mycorrhizal research community.

Laccaria genome has also revealed the truth of the cellulytic capacity of ECM fungi. *Laccaria* lacks or shows strong reduction in genes needed for plant cell wall degradation which are typical of saprotrophic filamentous fungi (exocellobiohydrolases, xylanases, peroxidases, hemicellulases, pectinases, *etc.*). This agrees well with *Laccaria* co-existence within host roots. This lack of cellulytic capacity also implicates that plant root entry and Hartig net formation in ECM is not accompanied by plant cell wall degradation by fungal enzymes as originally proposed, at least not in the case of *Laccaria* ECM. However, *Laccaria* genome shows a strong capacity to degrade non-cell wall polysaccharides. Also its proteolytic enzyme battery is potent indicating an adaptation to growth in forest soil rich in organic complex forms of nutrients and a strong capacity to feed especially on animal-origin nitrogen. Furthermore, *Laccaria* shows increased nutrient transporter capacity compared to saprotrophic basidiomycetes such as *Coprinus cinereus* and *Phanerochaete chrysosporium*. This is concordant with the role of *Laccaria* as an ECM fungus and with its need for active sequestration of nutrients from soil substrates for maintaining the symbiotic interaction. No invertase genes are present in *Laccaria* genome thus further confirming that sucrose can not be directly used by the fungus and that sugar transport at the ECM interface must depend on the plant invertase activity. Also genomes of other ectomycorrhizal fungi are currently being sequenced and the one of the model ascomycete *Tuber melanosporum* will be sortly released (Bohannon, 2009). Future comparisons of *Tuber* and *Laccaria* genomes will offer valuable information on how adaptation to the symbiotic lifestyle has been orchestrated in two phylogenetically distant fungal species.

However, the capacity of ECM research to take full advantage of the genome sequence and genome scale gene expression profiles and turn *Laccaria* into a true model organism faces a serious obstacle. In future more detailed gene-to-function studies in ECM depend on the access to reverse molecular genetic tools functional in *Laccaria*. These include both efficient transformation and gene inactivation technologies.

We established a high throughput *Agrobacterium*-mediated transformation technique (AMT) based on hygromycin B resistance for *Laccaria* dikaryotic and monokaryotic strains. This gene transfer allows manipulation of intact vegetative mycelium of the fungus circumventing protoplast preparation (Kemppainen et al., 2005). The transgene integration via AMT shows a simple pattern, generating predominantly single integrations which propose high compatibility of this genetic tool for random gene tagging of the fungus. A plasmid rescue method for recovering the transgene integration sites in the fungal genome was also optimized for *Laccaria*. The AMT introduced transgenes seem to integrate with no evident nucleotide sequence preference, but the dikaryotic transformants recovered under selection pressure show a transgene integration bias towards coding sequences of putative genes (Kemppainen et al., 2008). Interruption of one of the gene copies in a dikaryotic fungus did not result in remarkable reduction on the cellular transcript level but suggested that haploinsufficiency phenotypes can be detected under growth liming conditions.

RNA Silencing in *Laccaria Bicolor*

Based on previous knock-out trials the homologous recombination (HR) rate in the monokaryotic strain S238N-H82 of *L. bicolor* was concluded very low. Therefore, the possibility to use RNA silencing for altering gene expression (gene knock-down) directly in dikaryotic mycelium of the fungus was investigated.

Post-transcriptional gene silencing (PTGS) represents an alternative mechanism for reducing the mRNA level in a sequence specific manner in eukaryotes. Even though gene knock-down does not completely abolish the target gene activity a remarkable reduction of mRNA can be achieved resulting in functional phenotypes. Furthermore, strongly silenced phenotypes of different eukaryotes have been demonstrated to be comparable with gene null-mutants. As PTGS acts at cytosolic level especially this category of RNA silencing represents an attractive and straightforward approach for modifying gene expression both in diploid and dikaryotic organisms. Besides its use in monokaryotic filamentous ascomycetes, RNA silencing launched from one transformed nucleus has been shown to efficiently reduce target endogene mRNA levels in dikaryotic saprotrophic basidiomycetes (Namekawa et al., 2005; Wälti et al., 2005; de Jong et al., 2006; Eastwood et al., 2008; Matityahu et al., 2008). These studies have led to the analysis of functional silenced phenotypes demonstrating the power of this genetic tool in resolving fungal gene functions directly in dikaryotic strains.

Although based on an evolutionary conserved cellular mechanism the successful use and initiation of RNA silencing depends on the presence of active RNA silencing protein machinery in the eukaryotic organism under study. In the fungal kingdom the lost of the RNA silencing pathway has been reported in *Saccharomyces cerevisiae, Candida spp.* and *Ustilago maydis* demonstrating that this lost of RNA silencing is not a taxonomically conserved phenomenon. Hence, the activity of RNA silencing mechanisms must be assessed on a species basis. A preliminary study of *Laccaria* full genome sequence revealed that it encodes for a minimum set of genes with predicted protein products shown to be fundamental for RNA silencing. Two putative Dicer-like, and several Argonaute (AGO) and RNA-dependent RNA polymerase (RdRP) proteins were detected (table 1). *Laccaria* Dicer-like proteins were searched based on their sequence homology with *Neurospora crassa* DCL1 (Q7S8J7.1) and DCL2 (Q7SCC1.3) RNAse III domains a and b, AGO proteins with their homology to *N. crassa* PIWI domains from two AGO proteins (NSU04730 and NCU09434) and RdRPs based on their homology with *N. crassa* RdRP domains of two RdRP proteins (NCU07534 and EAA34169.1).

Interestingly only one of the putative *Laccaria* Dicers (ID 294420) contains all the protein domains typical of Dicer proteins (N-terminal DEAD/DEATH-box ATP binding domain, RNA helicase and C-terminal ribonuclease III domains IIIa and IIIb). The second one (ID 311711) is demoted apparently lacking the DEAD/DEATH-box. *Laccaria* Dicer-like protein (ID 294420) seems to be homologous to the fungal group of Dicer-like proteins which also harbors *N. crassa dcl2* and *M. grisea mdl2*. The activity of the first one is involved in quelling, transgene-induced silencing, in *Neurospora* (Catalanotto et al., 2004) and the latter is demonstrated to be needed for hpRNA-induced transgene silencing and siRNA accumulation in *Magnaporthe* (Kadotani et al, 2004). *Laccaria* sequence is also related to DCL2 of *Arabidopsis*. This protein has recently been demonstrated to be fundamental for transitive silencing response in production of secondary siRNAs from various RNAi triggers. It is also proposed to be important in viral defense of plants (Mlotshwa et al., 2008). The other demoted *Laccaria* Dicer-like protein seems to be related to Dicer-1 class proteins. All six *Laccaria* AGOs show conserved Paz and Piwi domains of functional Argonaute proteins and these genes have been manually cured. Of six putative RdRPs three have been manually cured at the moment (ID 141500, 315790 and 316322). One of putative RdRPs (ID 323128) shows a partial RdRP domain and thus might not encode a functional protein.

Table 1. *Laccaria* **RNAi proteins: two Dicer-like, six Argonautes, and six RdRPs. The Dicer-like protein (ID 311711) is demoted lacking a DEAD/DEAH-box. The RdRP protein (ID 323128) has only a partial RdRP-domain. RdRP: RNA-dependent RNA polymerase**

Protein ID	Predicted function	Best Blast Hit	
294420	Dicer	ref|XP_001840952.1| Hypothetical protein CC1G_03181 [Coprinopsis cinerea okayama7#130]. Score (Bits): 857	E Value: 0.0
311711	Dicer	ref|XP_001837094.1| Predicted protein [Coprinopsis cinerea okayama7#130]. Score (Bits): 1056	E Value: 0.0
295925	Argonaute	ref|XP_001838344.1| Hypothetical protein CC1G_04788 [Coprinopsis cinerea okayama7#130]. Score (Bits): 1093	E Value: 0.0
304374	Argonaute	ref|XP_001833415.1| Hypothetical protein CC1G_05115 [Coprinopsis cinerea okayama7#130]. Score (Bits): 494	E Value: 0.0
311445	Argonaute	ref|XP_001840986.1| hypothetical protein CC1G_04830 [Coprinopsis cinerea okayama7#130]. Score (Bits): 832	E Value: 0.0
311854	Argonaute	ref|XP_001840986.1| Hypothetical protein CC1G_04830 [Coprinopsis cinerea okayama7#130]. Score (Bits): 832	E Value: 0.0
315429	Argonaute	ref|XP_001840986.1|Hypothetical protein CC1G_04830 [Coprinopsis cinerea okayama7#130]. Score (Bits): 914	E Value: 0.0
317035	Argonaute	ref|XP_001837864.1|Hypothetical protein CC1G_09846 [Coprinopsis cinerea okayama7#130]. Score (Bits): 1197	E Value: 0.0
141500	RdRP	ref|XP_001828874.1| Hypothetical protein CC1G_03668 [Coprinopsis cinerea okayama7#130]. Score (Bits): 924	E Value: 0.0
301565	RdRP	ref|XP_001829472.1| Predicted protein [Coprinopsis cinerea okayama7#130]. Score (Bits): 921	E Value: 0.0
311867	RdRP	ref|XP_001837206.1| predicted protein [Coprinopsis cinerea okayama7#130]. Score (Bits): 746	E Value: 0.0
315790	RdRP	ref|XP_001833131.1| Hypothetical protein CC1G_01193 [Coprinopsis cinerea okayama7#130]. Score (Bits): 889	E Value: 0.0
316322	RdRP	ref|XP_001838853.1| Hypothetical protein CC1G_09230 [Coprinopsis cinerea okayama7#130]. Score (Bits): 1912	E Value: 0.0
323128	RdRP	ref|XP_001837224.1| predicted protein [Coprinopsis cinerea okayama7#130]. Score (Bits): 534	E Value: 2e-149

This high number of putative RdRPs (6) and AGOs (6) in *Laccaria* reflects the homobasidiomycete specific gene expansion of RNA silencing machinery already reported by Nakayashiki and Kadotani (2006). However, *Laccaria* obviously does not have three Dicer proteins as reported for other homobasidiomycetes by the authors. As a conclusion, at least one putative gene copy encoding each RNAi protein (Dicer, Argonaute and RdRP) with all the domains characteristic of a functional protein is present in *Laccaria* genome. Therefore, artificially triggerd RNAi could result succesfull in modifying gene expression in this fungus.

Laccaria nitrate reductase gene (*Lbnr*, protein ID 254066) was selected as a test gene for assessing functionality of RNAi in the fungus. This gene was considered an optimal target for RNA silencing for several reasons. Its enzyme activity is encoded by a single nuclear gene and it is not member of a multigene-family reducing thus possible silencing off-target effects linked to homologous mRNAs. Moreover, RNAi of *Lbnr* should allow an easy growth phenotype-screening of silenced strains on nitrate medium. As nitrate metabolism is generally repressed in fungi and other eukaryotes by primary N sources, growth of the silenced strains could be expected to be unaffected on ammonium.

The *Lbnr* is also a part of N metabolism of *Laccaria* which makes this gene an interesting target for genetic modification due to its possible role in fundamental functions of ECM symbiosis. Likewise to *Laccaria* the vast majority of ectomycorrhizal fungi studied this far can support growth on nitrate indicating that this metabolic trait is evolutionary conserved (Nygren et al., 2008). However, the true role of nitrate as N source for ectomycorrhizal fungi in forest soils is under current debate. Even though traditionally considered of minor importance for total N pool in forest soils other studies propose that this inorganic N form may have marked temporal and spatial concentration gradients being more available for mycorrhizal fungi than previously believed (Stark & Hart, 1997; Laverman et al., 2000). Moreover, the activity and differential expression of fungal and plant nitrate utilization genes has been connected to functional ectomycorrhizal symbiosis and enhanced host plant N nutrition. The nitrate reductases of plant hosts become down-regulated both in endo- and ectomycorrhizal roots and this repression has been directly linked to the activity of fungal N metabolism in ectomycorrhiza (Kaldorf et al., 1998; Guescini et al., 2003; Bailly et al., 2007). The fungal response to symbiosis can however vary between different ectomycorrhizal species. While in the basidiomycete *Hebeloma cylindrosporum* nitrate reductase expression is lower in mycorrhizal structures than in extraradical mycelium, in ectomycorrhiza formed by the ascomycete *Tuber borchii* fungal nitrate utilization genes are specifically induced in symbiotic structures (Bailly et al., 2007; Guescini et al., 2003; 2007; 2009). Despite of the observed species specific variation increasing data suggest that fungal nitrate metabolism genes can play an important role in establishment and/or function of ectomycorrhizal interactions in nature. Therefore, knocking-down of *Lbnr* could also offer information on how this mutualistic association is regulated.

The functionality of the *Laccaria* RNA silencing pathway was tested by transforming the dikaryotic fungus with a promoter-directed inverted repeated sequence of a partial coding sequence of *Lbnr* gene. Expression of this construct was expected to produce hpRNAs of 417 bp self-complementary dsRNA arms separated by a 98 bp non-intronic spacer sequence. This silencing trigger represents app. 16 % of the total length of the predicted intron-spliced *Lbnr* transcript. The *Lbnr* silencing triggering cassette was constructed in the T-DNA of a pCAMBIA-1300 based transformation vector to form pHg/NITRSPL (Figure 5; Kemppainen et al., 2009a).

Forty seven randomly selected transformants were analyzed for unaffected growth on ammonium and further for their growth capacity on nitrate. Thirty two percent of the tested strains showed a marked reduction of growth when compared to the wild type fungus on nitrate medium. While the colony diameter was not altered these strains produced remarkably less aerial and medium penetrating mycelia (Figure 6A; Kemppainen et al., 2009a). Biomass comparisons showed that these growth phenotypes produced only 5-9 % of wild type mycelial dry weight on nitrate while growth on ammonium was not affected (Figure 6B; Kemppainen et al., 2009a).

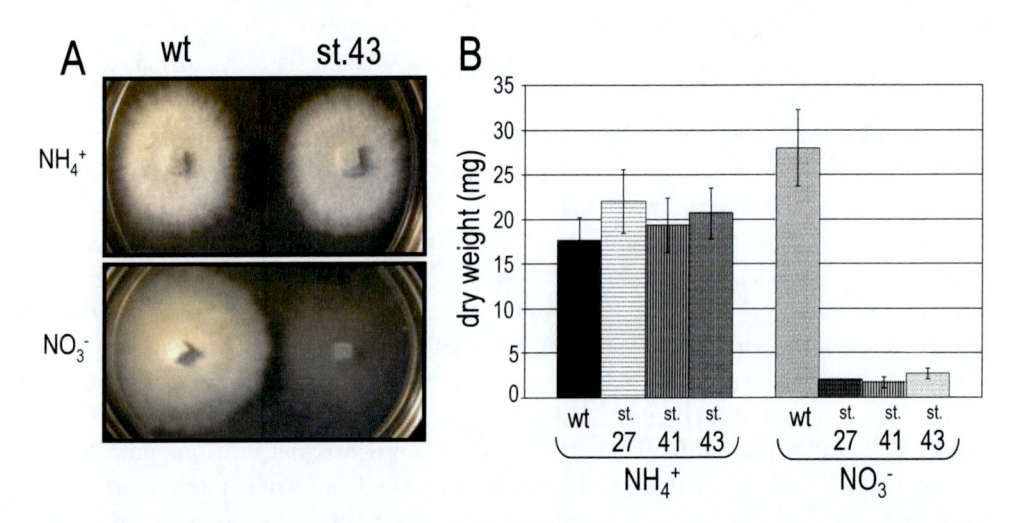

Figure 6. (A) Growth capacity of pHg/NITRSPL transformant (strain 43) on solid ammonium and nitrate media in comparison to *Laccaria* wild type strain. Plate photos show fungal growth after 30 days of culture on modified P5 medium (ammonium medium) and modified P5 medium with 4mM KNO3 as N source. (B) Biomass production (dry weight) of three pHg/NITRSPL transformants (strains 27, 41 and 43) and the wild type fungus in liquid cultures with ammonium or nitrate as N source. The columns represent the mean of triplicates and the bars denote the standard deviation. Ammonium-grown mycelia were harvested after 30 days and nitrate-grown mycelia after 45 days of culture.

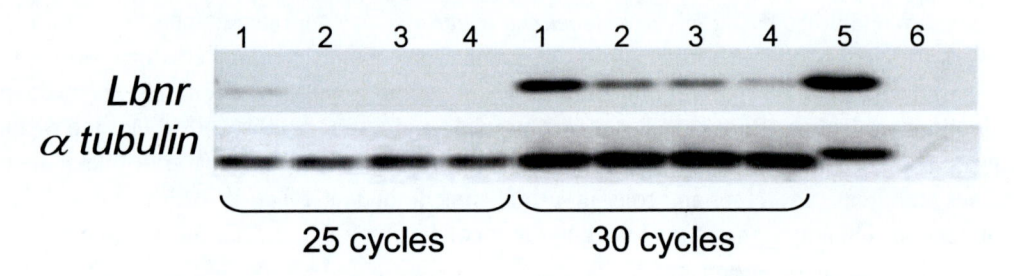

Figure 7. RT-PCR expression patterns of *Lbnr* in dikaryotic *Laccaria* wild type strain S238N and three strongly phenotypic dikaryotic pHg/NITRSPL transformants when growing on nitrate as N source. Detection of *Laccaria* tubulin-1B alpha chain transcript (α tubulin) was used as control amplicon to show even cDNA template loading and lack of gDNA contamination in the samples. The photo shows amplification products after 25 and 30 PCR cycles. 1: wild type; 2: st. 27; 3: st. 41; 4: st. 43; 5: gDNA control; 6: negative PCR control (H₂O template).

Three strains with a remarkable phenotype were used for further molecular analysis (st 27, 41 and 43). Southern blot analysis indicated a random site simple T-DNA integration in two of the strains and a putative double integration in one of them (Kemppainen et al., 2009a). TAIL-PCR and iPCR were used for obtaining transgene integration site information. One genomic integration site could be resolved for each of the strains. Integration site sequence analysis further confirmed that all three strains had incorporated the T-DNA in putative ORFs of different identity (protein ID 311599, ID 332022 or 296689, and ID 311879). None of these ORFs was *Lbnr* or other loci with a predicted role in N metabolism. This demonstrated that the observed affected growth on nitrate was not a haploinsufficiency phenotype caused by T-DNA integration. A semi-quantitative RT-PCR analysis confirmed that the studied strains had remarkable reduced *Lbnr* mRNA level and that target gene specific RNA silencing had been successfully triggered in *Laccaria* (Figure 7; Kemppainen et al., 2009a). This result was later also re-confirmed by northern blot (Figure 8; Kemppainen et al., 2009b).

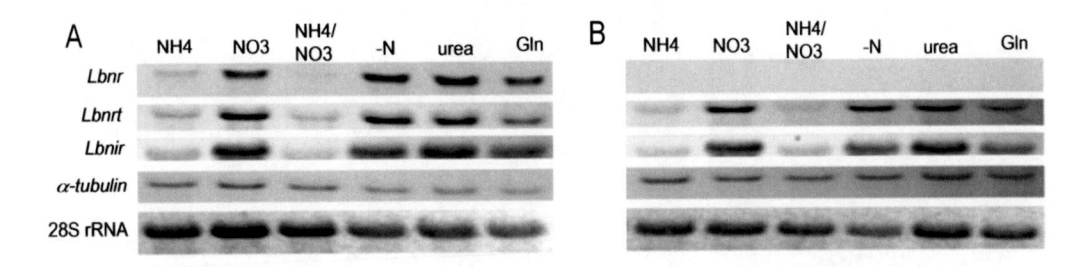

Figure 8. (A) Northern blot detection of *L. bicolor* wild type and (B) *Lbnr*-silenced strain (pHg/NITRSPL st.43) fHANT-AC transcript levels on variable N regimens after 48 h of induction. The N conditions were adjusted to 4 mM of total N (except for NH4/NO3 for which total N was 8 mM). NH4: ammonium (as ammonium sulphate); NO3: nitrate (as potassium nitrate); NH4/NO3: ammonium + nitrate; N-: no nitrogen; Gln: L-glutamine. Loading controls of 28S ribosomal RNA before membrane blotting and control hybridization with α-tubulin probe are presented.

The strain with strongest phenotype on solid nitrate growth medium (st 43) had less than 10 % of wild type mRNA level and it carried a single T-DNA integration in an ORF encoding for a putative DNA polymerase epsilon subunit B (protein ID 311879). The silencing phenotype of this strain was stronger than of strain 27 which carried a double T-DNA integration. This finding indicates that the integration copy number of the silencing trigger-construct is not the main regulator of silencing strength in *Laccaria*. Also another strain with a single T-DNA integration (st 41) presented a silencing level comparable to st 27 further suggesting that some other characteristics of the genomic RNA triggering locus, such as the transcriptional activity would be affecting the silencing strength in *Laccaria*.

Heterochromatin has been shown to be a source of small non-coding transcribed RNAs that participate in function and assembly of heterochromatin itself. These RNAs form a link between the epigenetic control and the RNA silencing machinery (for a review see Bühler & Moazed, 2007; Bühler, 2009). Epigenetic modifications can be long lasting, heritable and affect gene expression making epigenetics one of the most blooming bioscientific research fields. Artificially-induced RNA silencing is also frequently linked to epigenetic modifications. These can affect DNA and/or histone proteins causing organism-specific heterochromatization responses. These nuclear epigenetic modifications can affect both the silencing triggering locus (due to its repeated sequence structure) but also the homologous target gene.

Figure 9. *In vitro* mycorrhization between micopropagated *Populus* and *Laccaria* wild type strain S238N (A) and *nr*-silenced strain (B, C) produced on Petri dish-system. The bigger photos above demonstrate the general interaction on nitrate mycorrhization medium after 35 days of co-cultivation. The small photos below show how the wild type fungus efficiently forms mycorrhized lateral root with a visible mantle structures while the silenced strain does not enter to symbiotic interaction. In the case of the silenced strain bare lateral roots can be observed on the top of a mycelia mat which covers the Petri-dish. The extent of this mycelia mat produced on nitrate mycorrhization medium is highly comparable between the wild type and the *nr-* silenced strain but only the wild type fungus forms mycorrhiza. (D) Shift from nitrate mycorrhization medium to nitrate/L-asparagine medium immediately activates interaction between *Populus* and the silenced strain. Photo D present same root section as presented in photo C but after 8 days on nitrate/L-asparagine medium. Mantle formation is evident.

The increased use of RNA silencing as a genetic modification tool has elevated the interest for epigenetic changes involved in transgene induced silencing also in fungi. Data on this subject are still sparse, mostly coming from model fungi, the fission yeast *Schizosaccharomyces pombe* and the filamentous ascomycete *N. crassa*. It is however evident that the epigenetic response to RNA silencing triggering vary between different fungal species. The direct DNA methylation response in *S. pombe* is weak and the formation and maintenance of heterochromatin involves histone methylation in this fungus.The maintenance of natural heterochromatic sites is shown to be a RNA silencing machinery-dependent process mediated via small centromeric-RNAs (cenRNAs) (Volpe et al., 2002). Another proof for the participation of RNA silencing machinery and siRNAs as mediators of epigenetic changes comes from RNA silencing studies. The siRNAs originated from a dsRNA-expressing transgenic inverted repeats may induce transcriptional gene silencing (TGS) as well as chromatin modifications on unlinked homologous sequences in *S. pombe* (Schramke and Allshire, 2003). In *N. crassa* histone modifications and DNA methylation of repetitive genomic sequences are shown to be interconnected (Tamuru and Selker, 2001; Selker et al., 2002). Single transgene coding sequences, when introduced into vegetative mycelium of the fungus, are not modified but if the transgenes are present as inverted tandem repeats they become hypermethylated (Pandit and Russo, 1992). This methylation of repeated trangenes was shown to be complex but not random and reach gene promoter regions as well (Codón et al., 1997). On the other hand, while the histones in these quelling triggering transgene loci are highly methylated the epigenetic status of the target endogene has been demonstrated unchanged (Chicas et al., 2005). The recognition and marking of repetitive sequences in *N. crassa* is shown to be, surprisingly, independent of the RNA silencing pathway (Freitag et al., 2004). How the machinery responsible for DNA and histone modification is targeted to specific repeated DNA regions independently of sequence specific guidance of siRNA is not yet understood.

The role of the RNA silencing pathway in recognition and in epigenetic modification of genomic and transgene repetitions in other filamentous fungi is poorly known. The silencing locus and target gene sequence methylation as a response to RNA silencing in filamentous fungi seem to vary between species. In the ascomycete *Magnaporthe oryzae* RNA silencing of eGFP, triggered by dsRNA expression, does not induce *de novo* cytosine methylation in CCGG sequences of the target transgene or in the silencing triggering locus (Kadotani et al., 2003). Another example of the direct DNA methylation in RNA silenced filamentous fungi comes from the basidiomycete *Coprinopsis cinerea* (syn. *Coprinus cinereus*). In contrast to *M. oryzae,* dsRNA triggered silencing of eGFP transgene is strongly correlated with cytosine methylation in the CCGG motif, but not in GATC sequences (Wälti et al., 2006). The CpG methylation was shown to be strong in the inverted repeated silencing triggering locus and some weak CpNpG type cytosine methylation was detected also in the target gene sequence. The detected DNA methylation was postulated to depend on strong and constitutive dsRNA expression and to be mediated via RNA silencing pathway in *Coprinopsis*. Transgene-triggered cytosine methylation of both the silencing locus and the target gene has been reported also in another filamentous homobasidiomycete, *Schizophyllum commune* (Schuurs et al., 1997).

We studied the CpG DNA methylation status of the target gene and the silencing triggering locus in *nr*-silenced strains of dikaryotic *L. bicolor*. The use of partial gene sequence in silencing triggering allowed us to investigate the sequence specific epigenetic effects of constitutive dsRNA expression in *Laccaria*. These DNA methylation studies were done by using the isoschizomeric enzymes *Hpa*II and *Msp*I which recognize the same restriction site but show different cytosine- methylation sensitivities (*Hpa*II is CpG methylation sensitive unlike *Msp*I but both enzymes are CpNpG methylation sensitive). Wild type genomic DNA and DNA of three silenced strains were digested with both enzymes in parallel and processed into Southern blots. These were successively hybridized with different DNA probes targeted to different sequences of the silencing triggering T-DNA or the target endogene (Kemppainen et al., 2009a). Based on these hybridizations we could conclude that dsRNA-triggered RNA silencing initiates DNA methylation in *L. bicolor* as both the silencing triggering T-DNA and the target endogene showed some moderate cytosine methylation signatures (Kemppainen et al., 2009a).

While the wild type *Lbnr* locus was free of detectable methylation all three tested silenced stains had increased CpG methylation in the inverted repeated structures of the silencing loci. These included *Lbnr* sequences and *A. bisporus gpdII* promoters which were both present as inverted repeats in the silencing and hygromycin B resistance cassettes in the T-DNA of the transformation vector pHg/NITRSPL. Evaluation of the target gene *Lbnr* methylation indicated that this genomic sequence had become also methylated but these modifications affected only the sequences present in the dsRNA trigger leaving the rest of the gene coding sequence unmodified. Moreover, this endogene methylation did not affect, or did not affect to the same extent, the whole genomic DNA. This proposes that the detected RNAi linked CpG methylation of the target endogene locus would be taking place predominantly in the transformed nucleus. The type of RNAi trigger used for launching silencing on *Lbnr* (hpRNA) is expected to act via siRNA-dependent PTGS at cytosolic level. No siRNA detection was carried out but the fact that the silencing was initiated by dsRNA expression against coding and not the promoter sequences, that the detected endogene methylation was partial and that a strong mRNA reduction was achieved in a dikaryotic strain all support the cytosolic nature of the given silencing response in *Laccaria*. In mammalian cells introduced siRNAs have been shown to access the nucleus only by artificially activated import of by permeabilization of the nuclear membrane (Morris et al., 2004). This might also be the case for cytosol processed siRNAs in *Laccaria* explaining why the detected target endogene methylation would be affecting predominantly the transformed nucleus in the dikaryotic mycelium. However, the use of a non-intronic spacer in the dsRNA trigger could also influence this matter. A non-intronic dsRNA is not expected go through intron-splicing and might thus not be directed with high efficiency to nuclear export. Such dsRNAs could therefore have an increased nuclear retention leading to more pronounced epigenetic effects in the transformed nucleus. Interestingly, the recent discovery of cytosolic loading and active movement of siRNA-dependent components of nuclear RNA-induced silencing complex (nRISC) to nucleus in human cells and in *Caenorhabditis elegans* has demonstrated that siRNAs processed in cytosol can return and act in the nuclear environment (Ohrt et al., 2008; Guang et al., 2008; Weinmann et al., 2009). Also the epigenetic effects linked to virus-induced gene silencing (VIGS) in plants can be now explained by the nuclear entry of cytosolic siRNAs further proposing that this nRISC might be a widely conserved eukaryotic RNAi pathway. Even though nRISC is not yet demonstrated in fungi we can not exclude the possibility that the observed moderate *Lbnr* target gene methylation could be initiated via nuclear entry of cytosolic siRNAs in both transformed and and un-transformed nuclei.

The effect of *Lbnr* silencing on interaction with *Populus* was tested on an *in vitro* Petri dish mycorrhization system with two strongly silenced *Laccaria* strains (41 and 43). As low C availability has been demonstrated to stimulate ectomycorrhization (Smith & Read, 2008), these mycorrhization assays were performed under severe C limitation thus making the symbiotic interaction attractive for the fungus. Sucrose was used as C source in the mycorrhization media and as *Laccaria* lacks invertase activity the weak fungal growth supported on this C source is most likely due to low concentration of monosaccharides liberated from sucrose during autoclaving and/or exposure to acidic pH.

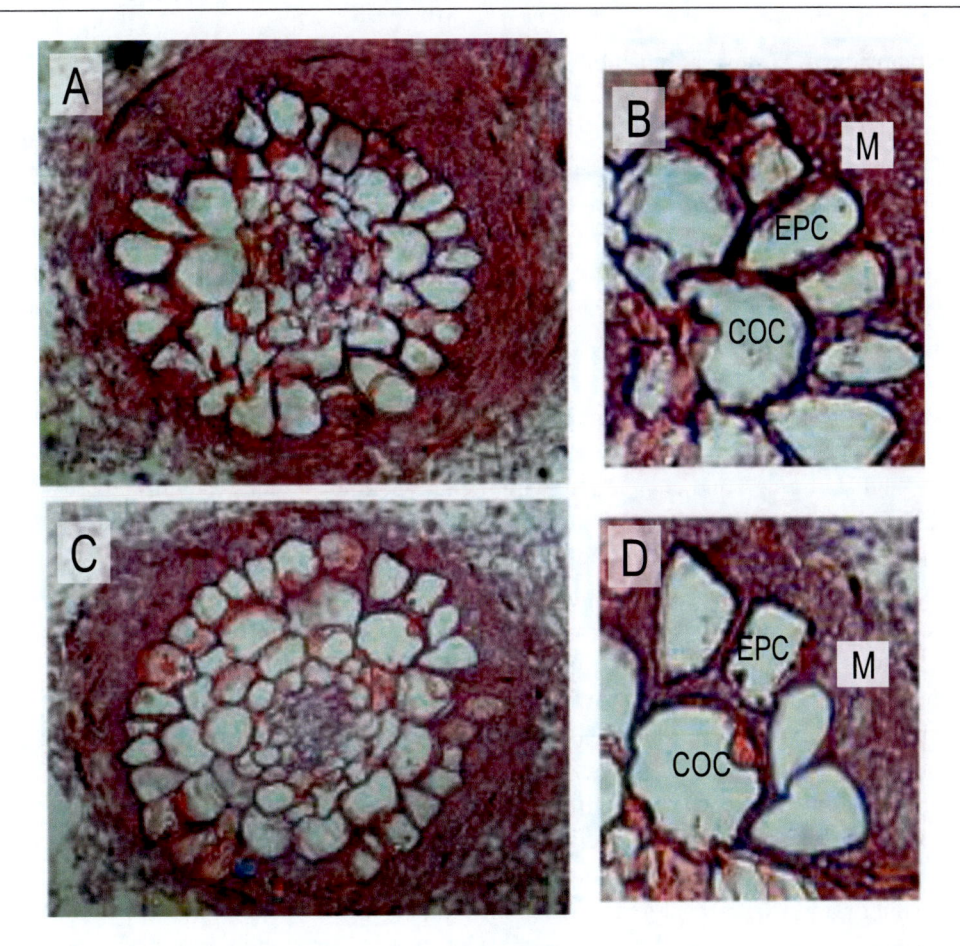

Figure 10. Cross-sections of *Populus-Laccaria* symbiotic organs formed by the wild type fungus (A, B) and a *nr*-silenced strain (C, D). The mycorrhized lateral roots were harvested 8 days after the shift from nitrate medium to nitrate/L-asparagine mycorrhization medium. The samples present a fully developed mantle (M) and Hartig net which can be observed between epidermal (EPC) and cortical cells (COC). Sectioned mycorrhizas were double stained with safranin and Fast Green. Magnification 400x.

While interaction of the silenced strains with the plant did not show alteration on ammonium-mycorrhization medium when compared to the wild type fungus mycorrhization on nitrate gave a dramatically different outcome. Whereas the wild type showed vigorous mycorrhization, (Figure 7A; Kemppainen et al., 2009a) the silenced strains only interacted with the first plant lateral roots forming mantle-like structures (Kemppainen et al., 2009a). This interaction took place only with mycelial inoculum pre-grown on ammonium. The rest of the vast plant root-system did not show signs of mycorrhization as no mantle-like structures were formed. The bare root-haired lateral roots could be observed on the top of mycelia mat of silenced strains (Figure 9B, C; Kemppainen et al., 2009a).

At the observation point (35 days of co-culture) both the wild type and the silenced strains had produced a comparable, thin fungal growth which covered the whole surface area available for plant-fungal interaction (half of the Petri dish) and made contact with the whole plant root-system. The failure of the silenced strains to interact with the plant could thus not be explained by a lower infection capacity linked to a weak growth on the nitrate mycorrhization medium.

In order to associate the observed non-mycorrhizal phenotype on nitrate to the altered N metabolism of the silenced strains, the mycorrhization systems were shifted to a new medium with nitrate and an organic nitrogen source, L-asparagine. This amino acid is an efficiently utilized N source (and also a C source) for the fungus but not for the plant. This medium shift immediately allowed the initiation of a mycorrhization process between plant and *Lbnr* silenced strains. The mycorrhization, observed as mantle formation, occurred both with already existing and newly formed lateral roots of poplar (Figure 7D; Kemppainen et al., 2009a).

Microscopic sections of 8 days old ectomycorrhizas from nitrate/L-asparagine medium showed anatomical structures characteristic of functional and fully developed ectomycorrhiza (mantle and Hartig net) and comparable with the ones formed by the wild type fungus (Figure 10; Kemppainen et al., 2009a). This activation of the symbiotic interaction when the fungus was offered an organic N source strongly suggests that the detected lack of mycorrhization of *Lbnr* silenced strains on nitrate was linked to the altered nitrate utilization and impaired N balance of the fungus. Other *Laccaria* features as a compatible fungal symbiont of *Populus* were obviously not enough for compensating its N metabolism defect and the strongly reduced capacity to offer N compounds to the plant.

This is the first direct genetic proof of the importance of uncompromised fungal N metabolism in formation of the ectomycorrhizal symbiosis. Our results strongly propose that the fungus is not capable of cheating the plant and obtaining C compounds without delivering N to the host. These findings also refute the idea that the symbiontic fungus could act as an occasional parasite as proposed by some researchers. This is also the first time that the altered expression of a precise fungal gene is shown to impair mycorrhization and the first time that RNA silencing has been used in mycorrhizal research. Our results demonstrate that RNA silencing is a suitable genetic tool for studying the ectomycorrhizal symbiotic interaction. They also highlight the two main advantages of gene knock- down when compared to the gene knock-out approach: a modification of gene activity can produce functional phenotypes directly in dikaryotic mycelium and the remnant gene activity allows studying fungal behavior under conditions lethal to the gene knock-out strains. Dikaryotic double *Lbnr* knock-out strains of *Laccaria* would had not been able to grow on the nitrate mycorrhization medium and the effect of altered fungal N balance on the symbiotic interaction could have not been possible to access.

A systematic inhibition of mycorrhization was observed with the *Lbnr* silenced strains under nitrate feeding. This observation raises the questions whether the plant or the fungal partner is the main controller and how this control is regulated. The mycorrhization medium is not especially growth limiting for the plant but the same conditions put the fungus under a severe C starvation. Mycorrhization could therefore offer the fungus an attractive release from its C stress. This mycorrhization response was observed with the wild type strain which is capable of an efficient utilization of nitrate and delivery of N compounds to the plant within the symbotic organs.

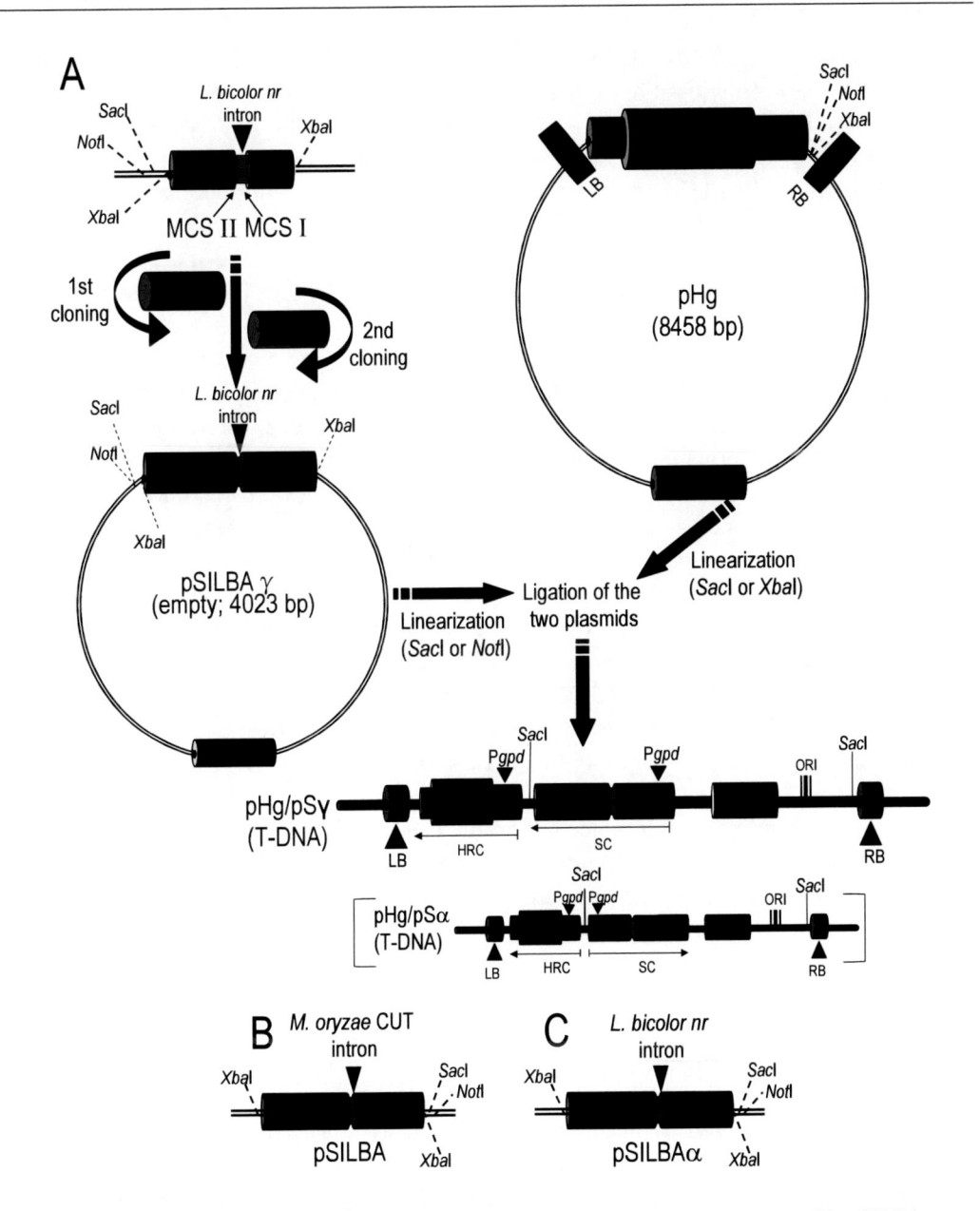

Figure 11. (A) pHg/pSILBAγ vector system. The ihpRNA expression cassette is constructed in pSILBAγ under *A. bisporus* gpdII promoter by two directed cloning steps. *Laccaria Lbnr* intron forms the spacer for the inverted repeats. The hygromycin B resistance cassette is located in the T-DNA of the binary vector pHg. Joining of pSILBAγ and pHg creates the AMT/silencing/rescue-vector with a predicted T-DNA structure (also pHg/pSILBAα T-DNA is presented). (B) pSILBA vector variant has the *M. oryzae* CUT intron and (C) pSILBAα *Lbnr* intron. The orientation of the silencing cassette in these variants differs from the one in pSILBAγ resulting in pHg/pSILBA T-DNAs with inverted repeated P*gpdII* structures.

Even though mycorrhization for obtaining a C source should be highly attractive also to the *Lbnr* silenced fungal strains the benefit gained by the plant in such an interaction can be questioned. The N benefit gained from it would be unfair compared to the C cost as the fungal partner's capacity to obtain N from the growth medium via extraradical mycelium is minimal.

Therefore, the main control over the mycorrhization process under the experimental conditions assayed should be on the plant side and it is the plant which is aborting the development of symbiotic organs. These are formed between the first lateral roots and the ammonium pre-grown inocula of the *Lbnr* silenced strains as the hyphal N resources can be expected shortly camouflage the true nature of affected fungal N metabolism on the nitrate mycorrhization medium. However, when the incapacity of the fungus to deliver N becomes evident no further symbiotic structures are formed. Which would be the plant molecular mechanisms behind such a control and how the N status of the fungal partner could be sensed?

The recognition of compromised N metabolism of a putative fungal symbiont and abortion of symbiosis could theoretically occur at two time-points: (1) the pre-colonization phase or (2) during formation of the nutrient exchange interface. Intercellular penetration leading to Hartig net formation is generally considered posterior to mantle formation. We did not detect mantle on lateral roots of poplar which is pointing towards abortion of symbiosis already during the pre-colonization phase. A strong connection exists between the nutritional status of ectomycorrhizal and other fungi and their expression of surface proteins (Madhani & Fink 1998; Lengeler et al., 2000; Soragni et al., 2001). Fungal cell wall proteins and cell surface polysaccharides have been identified as important molecules in the establishment of symbiosis but the precise roles and regulation of them is not yet understood. Also fungal origin elicitors of still rather uncharacterized nature are proposed to play an important role in the pre-colonization phase of ectomycorrhization (Smith and Read, 2008). Even though environmental N limitation is proposed to stimulate the ectomycorrhization response it is possible that as the silenced fungus is internally compromised for its N metabolism it can not fulfill the extracellular/cell wall molecular profile needed for recognition and/or colonization of the plant partner. This molecular incompatibility of a fungal symbiont with low N delivery potential could be detected by the plant and lead to abortion of mycorrhization at early stages. As a result the development of symbiotic organs potentially disfunctional for the plant partner is avoided.

However, there is no reason to believe that mantle formation and some root-colonization can not happen simultaneously. Also the vigorous mycelial growth which is generally taking place during mantle development under nutrient limiting conditions is better explained if a C source delivery at the symbiotic interface is simultaneously established. No mantle formation was observed but the *Lbnr* silenced hyphae undoubtedly did make contact with surfaces of *Populus* lateral roots and root hairs. Therefore, we can not exclude that some root penetration could have repeatedly been taken place or some other nutrient exchange interface could have been established without root penetration as also inner mantle hyphae, which are not part of the Hartig net but are positioned immediately adjacent to the outer tangential wall of root epidermal cells, are proposed to provide a second possible interface for nutrient exchange in ECM (Dexheimer & Gérard, 1994). Therefore, the detected lack of mantle formation could equally be related to colonization abortion directly due to failing in the establishment of a bi-directional nutrient flow at preliminary nutrient-exchange interfaces. As the silenced fungus is not capable of delivering N to the host the plant neither offers the fungus C and no further mycorrhization takes place due to this metabolic incompatibility.

How the fungal incapacity in offering a source of N compounds could lead to inhibition of C flow from the plant? The answer for this could lay in the regulation of plant monosaccharide transporters in symbiotic organs. In ectomycorrhized roots of *Amanita muscaria / Populus tremula x Populus tremuloides* the expression of both plant and fungal monosaccharide transporters are demonstrated to be strongly enhanced (Nehls, 2008). Also in *Laccaria/ Populus tremula x Populus tremuloides* mycorrhiza expression of putative fungal hexose transporter genes is induced (Fajardo Lopez et al., 2008). This suggests that in functional mycorrhiza both partners are competing for C compounds present at the symbiotic interface. Towards which partner the C flow is proposed to be controlled by the activity of the plant transporter. The PttMST3.1, one of the three mycorrhiza up-regulated monosaccharide transporter genes of *Populus,* is especially highly expressed in mycorrhized roots. Interestingly, it seems that the activity of this protein is post-translationally controlled via phosphorylation (Grunze et al., 2004). The activity of this transporter (an importer) is proposed to be modulated by the amount of nutrients offered by the fungal partner at the symbiotic interface. Insufficient nutrient flow would be detected by the plant leading, via phosphorylation, to activation of this monosaccharide importer and retrieval of C form the apoplast to the host plant. This hypothesis of the plant transporter control over C flow and functionality of ectomycorrhizal organs has been further tested by constitutive over expression of another *Populus* monosaccharide transporter (importer). The transgenic plants show reduced mycorrhizal infection and abnormal termination of symbiosis suggesting that the mycorrhiza specific plant monosaccharide importer can be a key controller in ectomycorrhizal symbiosis (Nehs, 2008). Our detected lack of mycorrhization could thus equally be linked to such mechanisms by which the plant can detect the incapacity of the fungus to establish N flux at the preliminary nutrient exchange interfaces. These interfaces could be formed prior to full developed mantle and lead to the shut-down of the C flux form the plant and abortion of mycorrhization observed as a lack of mantle formation.

Development of the pHg/pSILBAγ-vector System for Efficient RNA Silencing via AMT in *Laccaria Bicolor*

An efficient use of RNA silencing requires an easy-to-use silencing/transformation vector compatible with the organism under study and the transformation method employed. Double-stranded hairpin RNA (hpRNA) expression from stably integrated transgenes or from autoreplicative elements has been shown to be a widely efficient trigger in inducing RNA silencing in fungi. Vectors for plant silencing such as pHANNIBAL and pHELLSGATE (Wesley et al., 2001) and pSTARLING and pOpOFF (CSIRO) have been available already for almost a decade and several silencing vectors for mammalian cells have been reported in last years (Wadhwa et al., 2004; Gou et al., 2007). RNA silencing vectors for fungi however did not exist before the launch of pSilent-1 (Nakayashiki et al., 2005) which has led to the successful use of RNA silencing as a genetic tool in filamentous ascomycetes. Now the second generation of fungal silencing plasmids with high- throughput adaptations such as the incorporation of the GATEWAY technology (Invitrogen), dual promoter sequence expression and use of inducible promoters has been released for ascomycetes (Nguyen et al., 2008; Shafran et al., 2008; Oliveira et al., 2008; Barton & Prade 2008).

The success of RNA silencing in *Laccaria* generated an urgent need for a universal RNA silencing/transformation vector compatible with AMT. While an increasing number of silencing vectors for ascomycetes has been released during the last few years no silencing vectors adapted to basidiomycetes are available. Unfortunately, many commonly used ascomycete promoters are weakly recognized in filamentous basidiomycetes, especially when introduced predominantly as a single copy via AMT, making the use of these vectors difficult in this group of fungi. Neither are these transformation vectors directly compatible with AMT. An efficient use of RNA silencing as a reverse genetic tool in *Laccaria* requires the availability of a universal silencing vector for this fungus. Our aim was to develop a multiuse cloning vector adapted to both gene expression and RNA silencing studies in *Laccaria*. Besides the optimized use in this particular fungus general consideration was taken to offer to the fungal research community a vector system with maximum compatibility with different homobasidiomycetes as well as flexible use via different transformation methodologies.

Our requirements for such a multiuse transformation vector were:

An easy to clone hpRNA expression cassette under a fungal promoter widely recognized by basidiomycetes and with an intronic spacer for RNA silencing studies.

The possibility to use the same RNA silencing cassette also for transgene cloning and gene expression studies.

A hygromycin B resistance cassette under a widely recognized fungal promoter for transformant selection.

Compatibility with *Agrobacterium*-mediated transformation

A plasmid rescue-motif for easy transgene integration site recovery and analysis.

The possibility to use the silencing triggering/gene expression cassette without this plasmid rescue component.

The possibility to launch simultaneous silencing of two target genes from two separate silencing cassettes.

The compatibility of the system with other plasmid-based transformation methods.

The silencing/transformation binary vector system pHg/pSILBAγ fulfilling these requirements was designed and its functionality was evaluated in *L. bicolor* with successful silencing of *Lbnr* in the dikaryotic strain S238N and inositol-1,4,5, triphosphate 5-phosphatase gene (protein ID 306121) in the monokaryotic strain S238N-H82. Two other variants of pSILBAγ (pSILBA and pSILBAα) differing in their intronic sequences and the silencing cassette orientations were also tested for their silencing triggering capacity on *Lbnr*-target gene. The performance of pSILBAγ in producing strongly silenced strains was found superior and this effect was traced back to methylation free promoter sequences in the given construct most probably allowing the maximum silencing trigger production from it.

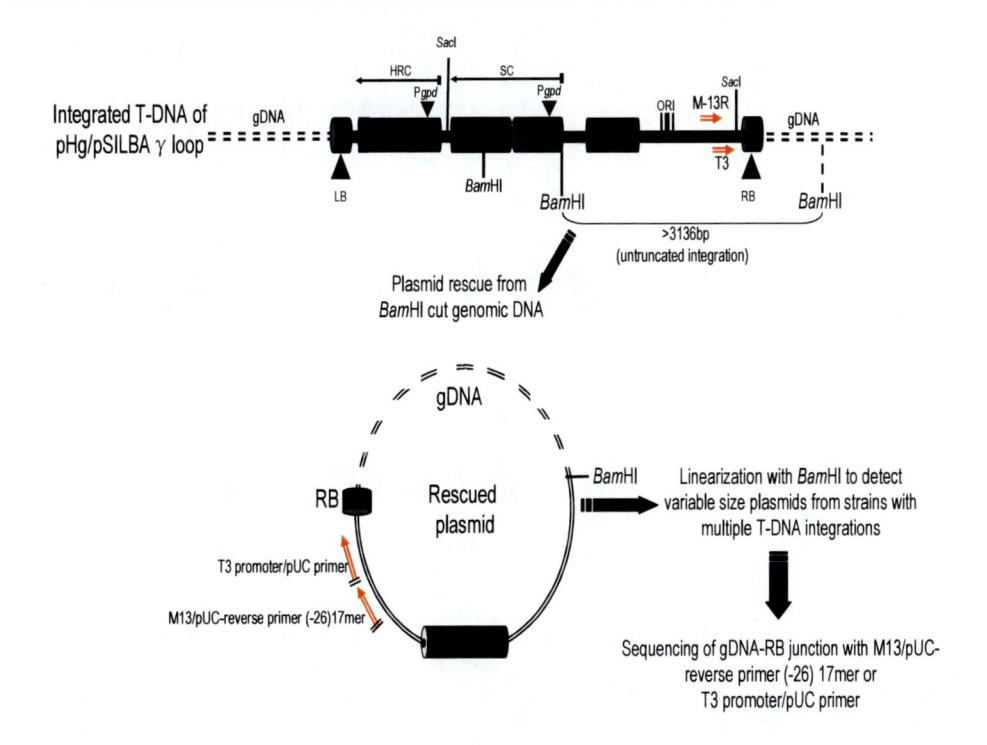

Figure 12. Plasmid rescue of genomic T-DNA integration sites from pHg/pSILBAγ transformed strains can be carried out by RB rescue with *Bam*HI l under ampicilline selection. This *Bam*HI rescue is independent of sequence used in the RNAi triggering as it cuts before P*gpdII* in the silencing cassette. Untruncated integrations generate rescue plasmids of a minimal size of ~ 3.1 kb and linearizable with *Bam*HI. Fungal genomic sequences can be resolved by sequencing with the universal primers T3 promoter/pUC or M13/pUC-reverse (-26)17 mer. HRC: hygromycin B resistance cassette; SC: silencing cassette.

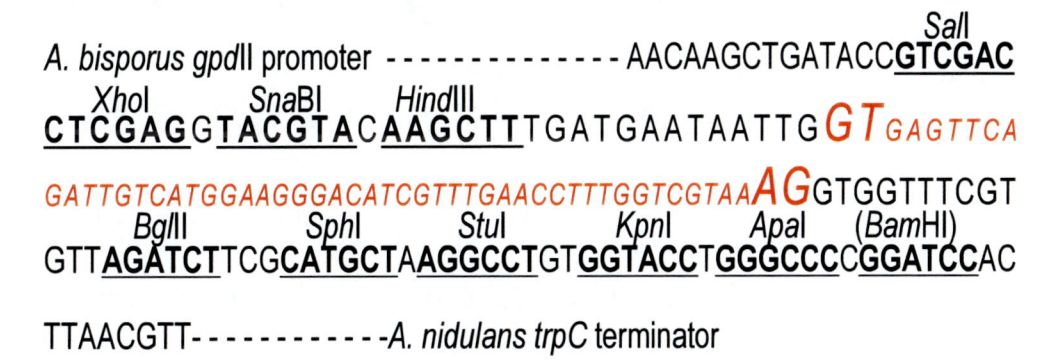

Figure 13. The cloning sites for inverted repeated sequence construction around the *Lbnr* intron in silencing cassettes of pSILBAα and pSILBAγ. All the restriction sites but *Bam*HI are unique in these plasmids. The sequence of *Lbnr* intron is presented in italics, with the donor and acceptor bases in capital letters. The *Sal*I, *Xho*I *Sna*BI and *Hind*II sites form the MCS I, and *Bgl*II, *Sph*I, *Stu*I, *Kpn*I and *Apa*I sites the MCS II.

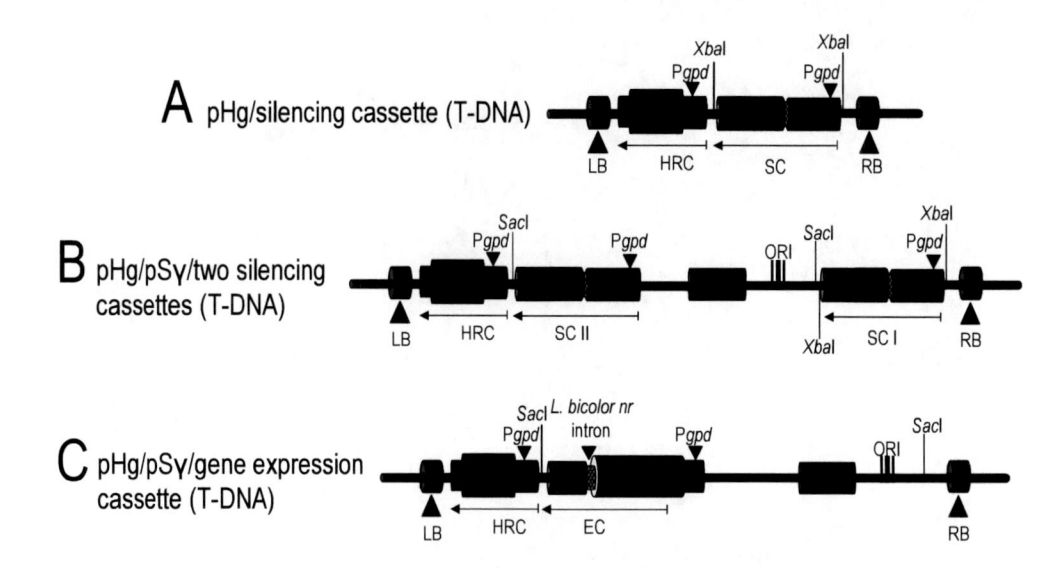

Figure 14. Three alternative uses of the pHg/pSILBAγ vector system. The figure represents the T-DNAs of binary vectors for (A) ihpRNA triggering without the plasmid rescue motif; (B) simultaneous silencing of two different genes or boosting ihpRNA triggering of a single target; (C) the use in transgene expression studies with an intronic sequence included into the expression cassette. HRC: hygromycin B resistance cassette; SC: silencing cassette; EC: expression cassette.

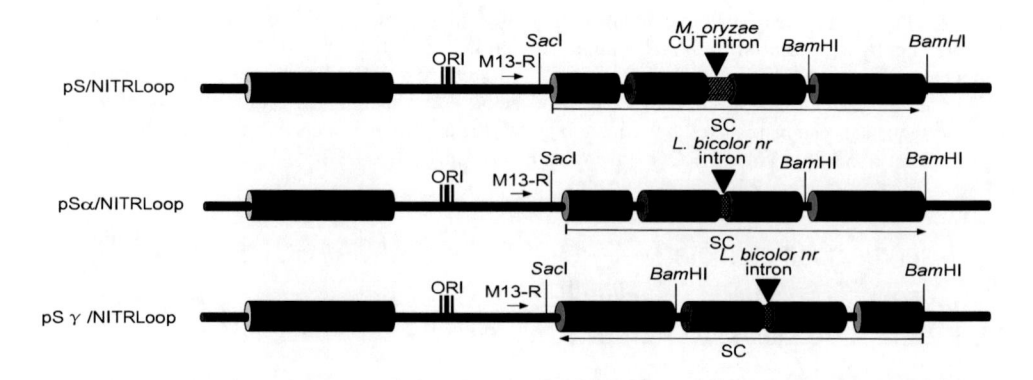

Figure 15. Schematic representations of three pSILBA variants (SILBA: pS; SILBAα: pSα; and SILBAγ: pSγ) tested for *Lbnr* silencing. SC: silencing cassette. The arrows indicate the orientation of the SC. The priming site for M13/pUC-reverse primer(-26)17 mer (M13-R) is indicated. P*gpd*: *A. bisporus* gpdII promoter; T*trpC*: *A. nidulans* tryptophan synthetase terminator; ORI: plasmid origin of replication; Amp[r]: ampicillin resistance cassette; CUT: cutinase intron.

The pHg/pSILBAγ- vector system is based on separating the construction of the RNA silencing trigger cassette in pSILBAγ plasmid (SIL: silencing, BA: basidiomycetes Figure 11A; Kemppainen & Pardo, 2009) and the fungal selection cassette for hygromycin B resistance in the *Agrobacterium* binary plasmid pHg (Figure 9A; Kemppainen et al., 2009a). This separation was done in order to allow easy cloning steps and manipulation of the small size plasmid pSILBAγ and to maintain the maximum number of unique restriction sites for inverted repeated sequence cloning around the intronic sequence of *Lbnr*. Joining these two plasmids together creates a silencing/AMT transformation vector of predicted T-DNA organization with a plasmid rescue motif for T-DNA RB- rescue (with *Bam*HI) under ampicilline resistance in *E. coli*. The rescued T-DNA-gDNA junctions can be sequenced with universal primers M13/pUC-reverse (-26) 17mer or with T3 promoter/pUC (Figure 12; Kemppainen & Pardo, 2009)).

The pSILBAγ is a modification of pSilent-1 (Nakayashiki et al., 2005) where the original hygromycin resistance cassette has been removed, the ascomycete promoter replaced by *Agaricus bisporus* glyceraldehyde-3-phosphate dehygrogenase promoter (P*gpdII*) and the original cutinase intron changed with the sixth intronic sequence (52 bp) of *Lbnr*. The P*gpdII* is widely recognized in homobasidiomycetes. This promoter, besides *A. bisporus and L. bicolor*, has been shown to be functional at least in *Suillus bovinus, Hebeloma cylindsrosporum, Coprinus cinereus, Hypholoma sublateritium* and *Moniliophthora perniciosa* (Hanif et al, 2002; Combier et al., 2003; Burns et al., 2005; Godio et al., 2004; Fagundes Lopes et al., 2008). As the tryptophan synthetase terminator of *Aspergillus nidulans* (T*trpC*) of pSilent-1 was maintained in the cloning cassette of pSILBAγ (Figure 13), it is suited not only for ihpRNA triggering but also for transgene expression studies in basidiomycetes with the possibility to incorporate an intronic sequence before or after the gene coding region of interest (Figure 14C). An intronic sequence is often needed for an efficient transgene expression in homobasidiomycetes (Lugones et al., 1999; Ma et al., 2001; Scholtmeijer et al., 2001; Burns et al., 2005; Yamazaki et al., 2006).

For RNA silencing the PCR amplified target gene sequence repeats are directionally cloned into unique restriction sequences in the multiple cloning sites (MCS) I and II around the *Laccaria* intronic spacer in pSILBAγ (or pSILBAα) (Figure 11). Afterwards the plasmid is linearized with *Sac*I (or if a *Sac*I site is present in the inverted repeated sequences, *Not*I can be used instead). The binary vector pHg has a hygromycin B resistance cassette under the same *A. bisporus gpdII* promoter. This plasmid carries also a unique *Sac*I site in its T-DNA for linearization. The pSILBAγ with the inverted repeated sequence can be cloned as a complete plasmid into the T-DNA of pHg and the joined plasmids are selected under ampicillin and kanamycin in *E. coli*. Due to certain incompatibility between pHg and pSILBA plasmids only one ligation orientation is viable producing T-DNAs with a predicted structure (Figure 11A; Kemppainen & Pardo, 2009). Constructing a RNA silencing/AMT vector can be accomplished by using a minimum of three PCR primers (one blunt end and two sticky end sites) and three ligation reactions. The pSILBA-plasmids tolerate well inverted repeated arms of 700 bp while repeats of over 900 bp were noticed to generate some structural instability and are thus not recommended. Alternatively, if the rescue motif of the pSILBA backbone is not desired, the ihpRNA expression cassette (or a transgene expression cassette) can be excised from pSILBAγ with *Xba*I and cloned into the unique *Xba*I site available in pHg T-DNA (Figure 14A). Also the introduction of two separate RNAi/gene expression cassettes of interests into pHg is possible by first cloning the *Xba*I-liberated cassette and later on the other cassette as a full pSILBAγ-plasmid using the *Sac*I site of pHg (Figure 14B).

As the efficiency of PTGS is also linked to the relative abundance of both the target mRNA and the silencing trigger in cells, making silencing of rare transcripts more difficult than the abundant ones, the double silencing cassette approach could also be used for boosting the silencing triggering of a single target gene. The simultaneous double cassette RNAi triggering of two separate target genes or boosting silencing of a single gene have not been tested yet in *Laccaria*. However, the T-DNA of pHg seems to tolerate well the presence of two inverted repeated sequence cassettes suggesting that such constructs would not cause technical problems in AMT of the fungus.

The presence of an intron as a spacer in hpRNA (ihpRNA) is shown to improve the silencing efficiency in plants and animals (Smith *et al.*, 2000; Kalidas and Smith, 2002). Also in *Neurospora crassa* and recently in the zygomycete *Mucor circinelloides* an intronic spacer has been reported to double the silencing efficiency of an hpRNA trigger (Goldoni et al., 2004; de Haro et al., 2009). This phenomenon has been proposed to be related to a possible higher nuclear accumulation of an intron-free trigger while the intron-containing trigger would be more efficiently exported to the cytoplasm via the nuclear splicing pathway. However, in filamentous fungi and yeasts efficient RNA silencing has been triggered with dsRNAs carrying both non-intronic and intronic spacers and the role of intron in improving silencing efficiency is not completely clear especially in basidiomycete fungi. We had efficiently launched RNA silencing in *Laccaria* with hpRNA carrying a non-intronic spacer sequence (Kemppainen et al., 2009a). In a RNA silencing study conducted in other basidiomycete, *Coprinus cinereus,* no difference in silencing efficiency was observed between a non-intronic and intronic spacer. Unfortunately, the constructs used in this study carried a transcribed intron outside the inverted repeated structure and therefore these results did not rule out the improved nuclear export of the silencing trigger by intron splicing in basidiomycetes (Wälti et al., 2006). Furthermore, even though strong silencing phenotypes have been obtained in fungi with both ihpRNA and hpRNA expression it is not conclusive whether the cellular silencing pathways responsible for the observed phenotypes with these different triggers have been the same. It is possible that both PTGS and TGS mechanisms could contribute to the silencing outcome. Consistently with this idea of simultaneous activity of the two RNA silencing pathways (one active on genomic DNA and affecting gene transcription and the other acting on mRNAs) we had detected moderated epigenetic modification of the target gene sequence during silencing with hpRNA trigger in *Laccaria.* Therefore, despite of best interpreted as an outcome of PTGS some participation of TGS in the silencing phenotypes could not be excluded (Kemppainen et al., 2009a). Possible prolonged dsRNA retention in nucleus due to lack of intron processing could be involved in this activation of a DNA methylation cascade.

We wanted to study the role of intronic sequences in *Laccaria* RNAi triggering with the tree different silencing constructs. Three pSILBA variants: pSILBA, pSILBAα and pSILBAγ were tested for their efficiency in launching RNA silencing using *Lbnr* as a target gene. All plasmids carried inverted sequence repeats of 417 bp corresponding to the protein encoding exonic sequence of the target but they varied in the intronic spacer sequences and in the RNA silencing triggering cassette orientation. In *Laccaria* genome introns are abundant and usually relative short, of about 50-60 bp. This is clearly shorter than the 147 bp *M. oryzae* cutinase intronic spacer (CUT) used in pSilent-1 (Nakayashiki et al., 2005). This size difference could theoretically lead to less efficient recognition and splicing of the introduced heterologous intron in *Laccaria*. This lack of processing could affect especially silencing efficiency of short hpRNAs as a long spacer sequence can hinder their dsRNA formation. Therefore, we tested two intronic sequences: pSILBA had the original pSilent-1 *M. oryzae* CUT intron and pSILBAα an endogenous *Lbnr* intron. The third construct, pSILBAγ, was identical to pSILBAα but carried the silencing cassette inverted with respect to the plasmid backbone (Figure 15; Kemppainen & Pardo, 2009). This cassette switch was done in order to assess whether the inverted repeated promoter structure generated in pHg/pSILBA and pHg/pSILBAα become epigenetically modified in *Laccaria*. Such a promoter region CpG methylation was previously detected when using the silencing trigger construct pHg/NITRSPL which also carried a promoter repetition (Kemppainen et al., 2009a). What we did not known was whether this methylation was due to direct modification of the promoter-repeat *per se* or represented some spread methylation from the silencing triggering repeated sequence. Our concern was that such promoter methylations could be affecting the efficienty of the silencing trigger expression in our plasmid system.

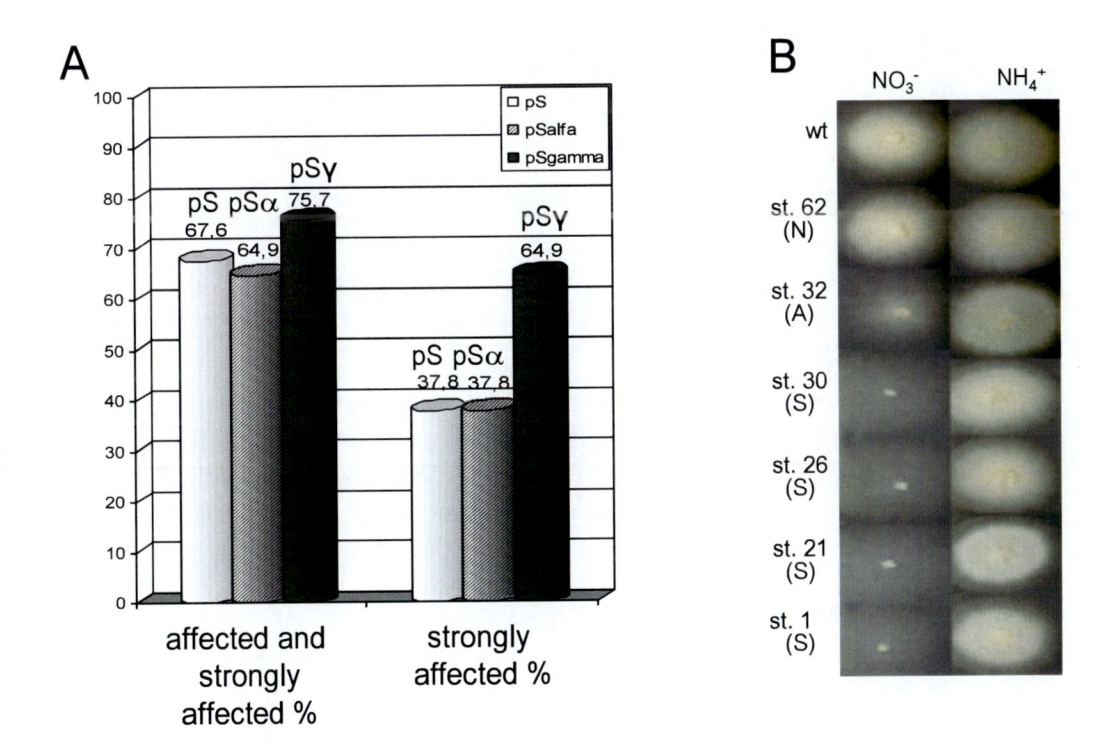

Figure 16. (A) Differences in silencing efficiency of *Lbnr* between pSILBA constructs. The percentages were calculated based on growth of 37 randomly selected pSILBA, pSILBAα and pSILBAγ transformed strains after one month in liquid nitrate medium in the microtitre plate assay. (B) *Laccaria* wild type dikaryon and six pHg/pSILBAγ/NITRLoop transformed strains grown for 23 days on solid medium with ammonium or nitrate as N sources. Growth categories: N, non-affected; A, affected; S, strongly affected.

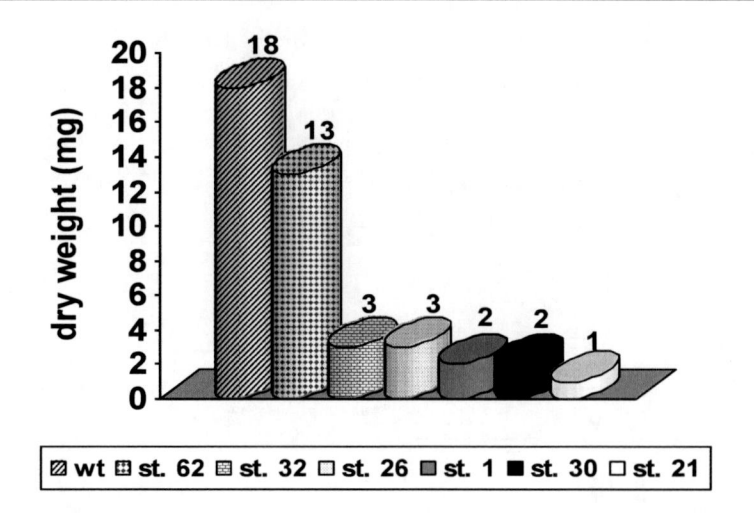

Figure 17. Dry weight of mycelia produced by wild type and six pHg/pSILBAγ/NITRLoop transformed strains in liquid nitrate medium after 22 days. The numbers above the bars refer to milligrams of mycelia produced by the given strain.

The three constructs were equally efficient in *Laccaria* transformation and produced comparable number of fungal strains affected in their growth on nitrate (pSILBA: 67.6 %, pSILBAα: 64.9 % and pSILBAγ: 75.7 %) (Figure 16A; Kemppainen & Pardo, 2009). This demonstrated that the pHg/pSILBA system was functional in silencing triggering in the fungus and there was no difference when the homologous or heterologous intron-spacer was used. It is possible that the *M. oryzae* intron was equally spliced in *Laccaria* or the lack of splicing did not affect the stability of dsRNAs. However, the number of strongly silenced phenotypes was almost double when the fungus was transformed with pHg/pSILBAγ (64.9 %) compared with other pSILBA-variants indicating that the silencing cassette orientation played an important role in the silencing strength. Also the strongly affected growth phenotype category of pSILBAα transformants produced slightly more fungal growth on nitrate than the same category of pSILBAγ transformants (Kemppainen et al., 2009a) indicating that this construct was superior in launching RNA silencing in *Laccaria*.

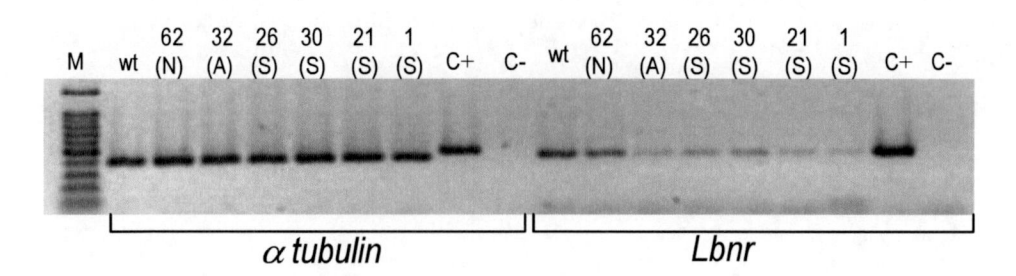

Figure 18. RT-PCR expression analysis of *Lbnr* in dikaryotic *Laccaria* S238N wild type and six pHg/pSILBAγ/NITRLoop transformed strains. The picture shows the amplification products after 30 cycles of PCR. The alpha tubulin (protein ID 192524) was used as a control amplicon for even cDNA template usage. The numbers refer to the transformed fungal strains and the letters to the silencing phenotype categories of growth on nitrate medium. N: not affected; A: affected; S: strongly affected; C+: gDNA positive PCR control; C-: H_2O negative PCR control; M: DNA size marker.

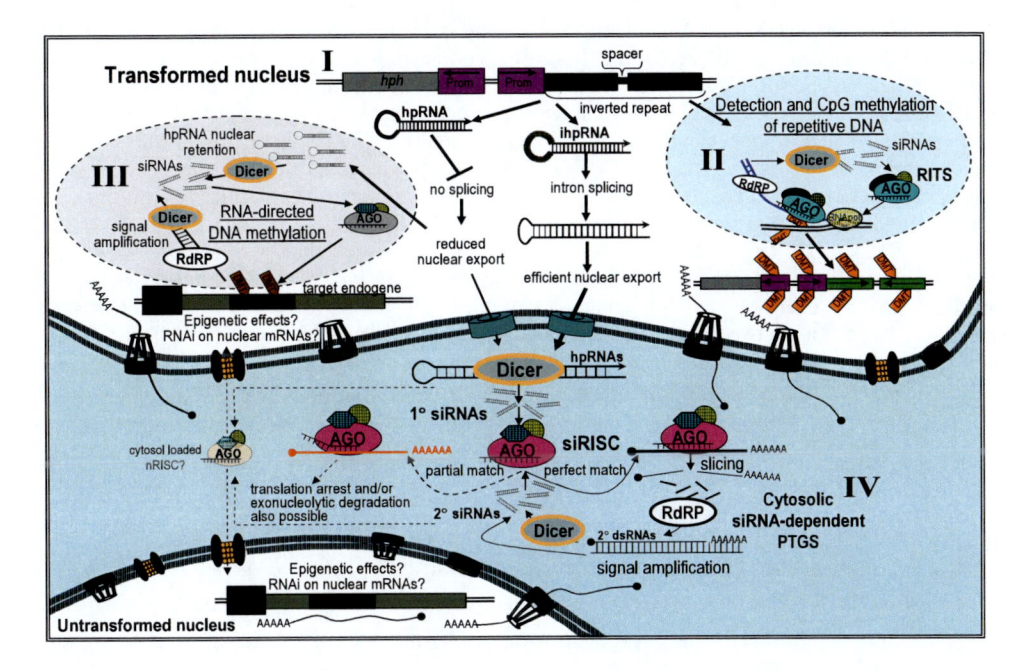

Figure 19.

I An efficient silencing in *Laccaria* requires integration of the silencing triggering DNA into an active genomic site, a coding region. This integration site dependency of silencing strength is believed to be related to more accessible chromatin structure in these active gene regions which also confers a higher transcriptional activity of the silencing triggering cassette and an efficient dsRNA production. Also these active sites might be less susceptible to mechanisms that detect and methylate genomic DNA repeats, as these are present in silencing triggering cassettes, allowing more efficient silencing trigger production.

II Inverted repeated sequences (promoter repetitions or silencing trigger cassette) present in the trans-DNA locus are detected in *Laccaria* by nuclear mechanisms which target repetitive genomic sequences. These sites suffer heterochromatization which involves at least direct DNA CpG methylation but other modifications such as histone methylations can not be excluded either. These two epigenetic modifications of chromatin are shown to work in interconnected manner in many eukaryotes. The detected CpG methylation of repetitive sequences in the silencing triggering locus in *Laccaria* could be carried out by nuclear RNAi-dependent pathway described in other eukaryotes and best characterized in *S. pombe*. According to the nascent transcript model the epigenetic marking of DNA repetitions depends on constant production of RNA transcripts from the repetitive loci. These ssRNAs serve as template for RdRP which produces dsRNAs. DsRNAs are cut by nuclear Dicer into siRNAs and these siRNAs, once incorporated into RITS, target homologous nascent transcripts at their sites of synthesis forming thus a physical link between RITS, DNA-repetitions and DNA and/or histone modifying enzymes.

III Another nuclear epigenetic pathway activated by RNA silencing in *Laccaria* affects the target endogene. A moderate CpG methylation mark was detected in endogene affecting strictly only the sequence used in silencing triggering and proposing a strong link between nuclear siRNAs and DNA-modifications in *Laccaria* according to the RNA-directed DNA methylation model proposed for plants. This pathway is RNAi-dependent and most propably involves siRNA-RISC complex binding to transcripts produced from the targeted locus similar to nascent transcript theory for targeting of genomic repetitions. We detected more pronounced target gene DNA methylation in the case of non-intronic hpRNA versus intronic hpRNA trigger. As ihpRNA is believed to pass through intron splicing and to be more efficiently exported to cytoplasm we propose that the more pronounced nuclear epigenetic effect of hpRNA is related to possible increased nuclear retention of this dsRNA trigger. As the target gene CpG methylation did not seem to affect the whole cellular DNA it proposes that these epigenetic modifications would be taking place predominantly in the transformed nucleus. Other target endogene modifications besides CpG methylation, such as histone modifications, can not be excluded.

IV The dsRNA silencing triggers used in our studies were against gene coding regions and not promoter sequences. Also the detected target gene CpG methylation was moderate and did not affect the whole genomic DNA in the dikaryon and neither reached the gene promoters. These results strongly support that the gene silencing triggered with dsRNA in dikaryotic *Laccaria* is due to a siRNA-dependent cytosolic PTGS mechanisms. This pathway can be expected to follow the conserved RNAi-steps described in different eukaryotes. No cytosolic siRNA detection was performed but the perfect homology nature of the silencing trigger against the target transcript proposes that Dicer cutting, primary siRNA incorporation into siRISC and perfect match-mediated Argonaute mRNA-cut pathway is responsible for the detected gene knockdowns. To which extent this pathway is amplified by RdRPs and what are the characteristic of this transitive silencing in *Laccaria* is not yet known. Even though miRNAs are not detected in fungi it is possible that primary or secondary siRNA could act via imperfect matches in a miRISC manner and cause transcriptional arrest or mRNA degradation of off-target transcripts.

The possible role of cytosol loaded nRISC can not be excluded in *Laccaria*. This could link siRNAs, produced by the PTGS pathway, to RNAi on target transcripts in nucleus as well as mediate epigenetic target gene DNA modifications between the transformed and untransformed nucleus in dikaryotic mycelium.

Six pSILBAγ-transformant strains belonging to different silencing phenotype categories (not affected: N; affected: A; and strongly affected: S) (Figure 16B; Kemppainen & Pardo, 2009) were further analyzed for biomass production on nitrate (Figure 17; Kemppainen & Pardo, 2009), for transgene copy number (Kemppainen & Pardo, 2009) and *Lbnr* mRNA level by semi-quantitative RT-PCR (Figure 18; Kemppainen & Pardo, 2009). These experiments demonstrated that the strongly silenced strains produced only 5.6-16.7 % of wild type biomass, had predominantly single T-DNA integrations and less than 10 % of *Lbnr* mRNA when compared with the wild type fungus. As the pSILBAα and pSILBAγ constructs had showed clear variation in their *Lbnr* silencing capacity and these differed only in the structural organization of their T-DNAs, the CpG methylation status of the *gpdII* promoter regions was studied. This was carried out in two strongly silenced strains triggered with pHg/pSILBAα or pHg/pSILBAγ. The inverted repeated promoter structure in pHg/pSILBAα T-DNA showed a clear methylation signature as suspected while in pHg/pSILBAγ transformants no similar methylation took place (Kemppainen & Pardo, 2009). Most likely a more efficient ihpRNA trigger expression is produced form these methylation free promoters explaining the higher number of strongly silenced phenotypes generated by the pHg/pSILBAγ construct.

The role of a recognizable intron in the hpRNA triggering was further studied by assessing the target gene CpG methylation status. Interestingly, while silencing triggering of *Lbnr* with the non-intronic hpRNA had resulted in epigenetic modification of the target gene sequence (Kemppainen et al., 2009a) the ihpRNA silencing launched from pSILBAα or pSILBAγ constructs did not produce such effect (Kemppainen & Pardo, 2009). This difference could be linked to the processing of the ihpRNA silencing trigger in the transformed nucleus and probably to the more efficient export of dsRNA to fungal cytosol. Similar response has been reported in the fission yeast *Schizosaccharomyces pombe* in which the presence or absence of an intronic sequence in dsRNA homologous to a gene encoding sequence has been connected to different epigenetic effects. While an intron containing dsRNA silencing trigger clearly initiated PTGS (Sigova et al., 2004) a construct with coding sequence repeats separated by a non-intronic spacer was reported to result in TGS (Schramke and Allshire, 2003). Similar kind of mechanisms which interconnect the processing of introns in hpRNA triggers and nuclear epigenetic modification seem to be active during *L. bicolor* RNA silencing as well. Besides, the characteristics of the silencing triggering construct itself and its effect on silencing efficiency, we focused on another well known but less studied aspect of RNA silencing: the silencing strength variation (SSV). This is observed among different strains carrying the same silencing trigger. All RNA silencing studies, in fungi or in other eukaryotes, where gene knock-down is launched from stable silencing trigger expression, show variation on the level of silencing between independent transformants. The reason for this phenomenon is not well understood and different factors have been proposed to be responsible for SSV. These include the nuclear transgene copy number, different epigenetic effects related to transgene integration site (euchromatin vs. heterochromatin) and post-transformational modifications and genomic rearrangements that might affect the transgene integrity. The SSV has been reported in yeast and filamentous fungi and both with integrative and auto-replicative silencing vectors (Liu et al., 2002; Kadotani et al., 2003; Rappleye et al., 2004; Tanguay et al., 2006; Bohse and Woods, 2007). Also non-silenced and silenced fungal strains are often reported to carry full size silencing vectors and the copy number of silencing triggering constructs do not necessarily correlate with higher degree of silencing (Nicolás et al. 2003; Fitzgerald et al. 2004; Nakayashiki et al., 2005; Tanguay et al. 2006; Wälti et al. 2006; Moriwaki et al. 2007; Yamada et al. 2007; Oliveira et al., 2008). It is therefore unlikely that the silencing trigger copy number is a fundamental factor causing SSV in fungi. Also our previous results on RNAi (Kemppainen et al., 2009a) indicated that a single T-DNA transgene integration could produce a silencing phenotype as strong as double transgene integration in *Laccaria*. An analysis of genomic integration sites of three strongly silenced *Laccaria* strains had already revealed that all of them carried the silencing triggering cassette in putative genes. These findings were of special interest because they suggested that the silencing cassette integration in active genomic sites could be one of the main factors causing SSV in *Laccaria*. Further confirmation of the transgene site effect could however only come from resolving more genomic integration sites in silenced *Laccaria* strains. This type of study has not been conducted in fungi yet, and to our knowledge, neither on other silenced eukaryotes.

We resolved genomic integration sites by plasmid rescue form one pHg/pSILBAγ-transformed phenotypically non-silenced (st 62), one intermediately silenced (st 32) and three strongly *Lbnr*-silenced fungal strains (st 30, 26 and 21). Of these strains st 26 harbored a double T-DNA integration both of which could be resolved by plasmid rescue. All the studied strongly silenced strains had their T-DNA integration within genes with EST support, while intermediately silenced and non-silenced strains were the only ones with intergenic T-DNA integrations (Kemppainen & Pardo, 2009). Hence, together with our previous results (Kemppainen et al., 2009a) it becomes clear that for the maximum silencing effect the silencing triggering casset ought to integrate in an actively transcribed genomic region, preferably within a transcriptional unit in *Laccaria* genome. However, the integration orientations of the silencing triggering cassettes in these ORFs did not show conservation with respect to the transcription orientation of the interrupted gene. This suggests that the stronger silencing triggering from these sites is not due to increased ihpRNA transcription from the close-by *Laccaria* promoter. The epigenetic status of the T-DNA integration site itself directly affects the transgene expression from *gpdII* promoter which can be observed as SVV.

The pHg/pSILBAγ vector system was further tested with another target endogene. A putative synaptojanin-like 5-Pase-encoding gene (protein ID 306121) was chosen for this assay due to the remarkable growth phenotype of the monokaryotic S238N-H82 strain with a random T-DNA tag affecting this gene (Kemppainen & Pardo, 2009). The reduced branching and folding of the colony caused by the gene interruption was expected to serve as an easy preliminary screening for monokaryotic strains presenting knock-down of this target gene.

Besides the dikaryotic strain, RNA silencing was functional in a monokaryotic genetic background (Kemppainen & Pardo, 2009).

The following model summarizes the cytosolic and nuclear siRNA-mediated RNAi dependent pathways detected in *Laccaria* (Fig. 19). Our experimental data from hpRNA- and ihpRNA-triggered silencing together with genomic information of RNAi related protein reservoir of the fungus indicates that at least three separate RNAi-dependent pathways are activated by dsRNA expression in *Laccaria*. Two of these are nuclear and produce epigenetic modification at the silencing triggering and the target endogene loci and the third pathway is cytosolic PTGS as it leads to efficient reduction of the target gene mRNA level in a dikaryotic strain.

References

Agerer R, 1987-2002. *Colour Atlas of Ectomycorrhizae.* Einhorn-Verlag, Schwäbisch Gmünd, Germany.

Bailly J, Debaud J-C, Vegner M-C, Plassard C, Chalot M, Marmeisse R, Fraissinet-Tachet L, 2007. How does a symbiotic fungus modulate expression of its host-plant nitrite reductase? *New Phytol* 175:155-165.

Baldrian P, 2009. Ectomycorrhizal fungi and their enzymes in soil: is there enough evidence for their role as facultative soil saprotrophs? *Oecologia* 161:657-660.

Barton LM, Prade RA, 2008. Inducible RNA interference of *brl*Aβ in *Aspergillus nidulans*. *Euk Cell* 7:2004-2007.

Bohannon J, 2009. Rooting around the truffle genome. *Science* 323:1006-1007.

Bohse ML, Woods JP, 2007. RNA interference-mediated silencing of the *YPS3* gene of *Histoplasma capsulatum* reveals virulence defects. *Infect Immun* 75:2811-2817.

Brundrett MC, 2002. Coevolution of roots and mycorrhizas of land plants. *New Phytol* 154:275-304.

Bühler M, 2009. RNA turnover and chromatin-dependent gene silencing. *Chromosoma* 118:141-151.

Bühler M, Moazed D, 2007. Transcription and RNAi in heterochomatic gene silencing. *Nat Struct Mol Biol* 14:1041-1048.

Burns C, Gregory KE, Kirby M, Cheung MK, Riquelme M, Elliott TJ, Challen MP, Bailey A, Foster GD, 2005. Efficient GFP expression in the mushroom *Agaricus bisporus* and *Coprinus cinereus* requires introns. *Fungal Genet Biol* 42:191-199.

Catalanotto C, Pallotta M, ReFalo P, Sachs MS, Vayssie L, Macino G, Cogoni C, 2004. Redundancy of the two dicer genes in transgene-induced posttrancriptional gene silencing in *Neurospora crassa. Mol Cell Biol* 24:2536-2545.

Chicas A, Forrest EC, Sepich S, Cogoni C, Macino G, 2005. Small interfering RNAs that trigger posttranscriptional gene silencing are not required for the histone H3 Lys9 methylation necessary for transgenic tandem repeat stabilization in *Neurospora crassa*. *Mol Cell Biol* 25:3793-3801.

Codón AC, Lee Y-S, Russo VE, 1997. Novel pattern of DNA methylation in *Neurospora crassa* transgenic for the foreign gene *hph*. *Nucl Acids Res* 25:2409-2416.

Combier JP, Melayah D, Raffier C, Gay G, Marmeisse R, 2003. *Agrobacterium tumefaciens*-mediated transformation as a tool for insertional mutagenesis in the symbiotic ectomycorrhizal fungus *Hebeloma cylindrosporum*. *FEMS Micorbiol Lett* 220:141-148.

de Haro JP, Calo S, Cervantes M, Nicolás FE, Torres-Martínez S, Ruiz-Vázquez RM, 2009. A single dicer gene is required for efficient gene silencing associated with two classes of small antisense RNAs in *Mucor circinelloides*. *Euk Cell* 8:1486-1497.

de Jong JF, Deelstra HJ, Wösten HAB, Lugones LG, 2006. RNA-mediated gene silencing in monokaryons and dikaryons of *Schizophyllum commune*. *Appl Environ Mircrobiol* 72:1267-1269.

Dexheimer J, Gérard J, 1994. The cell wall as an essential component of the ectomycorrhizal symbiosis. *J Trace Microprobe Tech* 12:185-199.

Duplessis S, Courty PE, Tagu D, Martin F, 2005. Transcript patterns associated with ectomycorrhiza development in *Eucalyptus globulus* and *Pisolithus microcarpus*. *New Phytol* 165:599-611.

Eastwood DC, Challen MP, Zhang C, Jenkins H, Henderson J, Burton KS, 2008. Hairpin-mediated down-regulation of the urea cycle enzyme argininosuccinate lyase in *Agaricus bisporus*. *Mycol Res* 112:708-716.

Fagundes Lopes FJ, Vieira de Queiroz M, Oliveira Lima J, Oliveira Silva VA, Fernandes de Araújo E, 2008. Restriction enzyme improves the efficiency of genetic transformations in *Moniliophthora perniciosa*, the causal agent of witches broom disease in *Theobroma cacao*. *Braz Arch Biol Technol* 51:27-34.

Fajardo-López M, Dietz S, Grunze N, Bloschies J, Weiß M, Nehls U, 2008. The sugar porter gene family of *Laccaria bicolor* function in ectomycorrhizal symbiosis and soil-growing hyphae. *New Phytol* 180:365-378.

Fitzgerald A, van Kan JAL, Plummer KM, 2004. Simultaneous silencing of multiple genes in the apple scab fungus *Venturia inaequalis* by expression of RNA with chimeric inverted repeats. *Fungal Genet Biol* 41:963-971.

Freitag MD, Lee DW, Kothe GO, Pratt RJ, Aramayo R, Selker EU, 2004. DNA methylation is independent of RNA interference in *Neurospora*. *Science* 304:1939.

Godio R, Fouces R, Guniña E, Martin J, 2004. *Agrobacterium tumefaciens*-mediated transformation of an antitumor clavaric acid-producing basidiomycete *Hypholoma sublateritium*. *Curr Genet* 45:287-294.

Goldoni M, Azzalin G, Macino G, Cogoni C, 2004. Efficient gene silencing by expression of double stranded RNA in *Neurospora crassa*. *Fungal Genet Biol* 41:1016-1024.

Gou D, Weng T, Wang Y, Wang Z, Zhang H, Gao L, Chen Z, Wang P, Liu L, 2007. A novel approach for the construction of multiple shRNA expression vectors. *J Gene Med* 9:751-763.

Guang S, Bochner AF, Pavelec DM, Burkhart KB, Harding S, Lachowiec J, Kennedy S, 2008. An Argonaute transports siRNAs from the cytoplasm to the nucleus. *Science* 321:537–541.

Guescini M, Pierloeni R, Palma F, Zeppa S, Vallorani L, Potenza L, Sacconi G, Giomaro G, Stocchi V, 2003. Characterization of the *Tuber borchii* nitrate reductase gene and its role in ectomycorrhizae. *Mol Gen Genomics* 269:807-816.

Guescini M, Stocchi L, Sisti D, Zeppa S, Polidori E, Ceccaroli P, Satarelli R, Stocchi V, 2009. Characterization and mRNA expression profile of the *TbNre1* gene of the ectomycorrhizal fungus *Tuber borchii*. *Curr Genet* 55:59-68.

Guescini M, Zeppa S, Pierleoni R, Sisti D, Stocchi L, Stocchi V, 2007. The expression profile of the *Tuber borchii* nitrite reductase suggests its positive contribution to the host plant nitrogen nutrition. *Curr Genet* 51:31-41.

Grunze N, Willmann M, Nehls U, 2004. The impact of ectomycorrhiza formation on monosaccharide transporter gene expression in poplar roots. *New Phytol* 164:147-155.

Hanif M, Pardo AG, Gorfer M, Raudaskoski M, 2002. T-DNA transfer and integration in the ectomycorrhizal fungus *Suillus bovinus* using hygromycin B as a selectable marker. *Curr Genet* 41:183-188.

Ingleby K, Mason PA, Last FT, Fleming LV, 1990. *Identification of ectomycorrhizas. Institute of Terrestrial Ecology,* Research Publication No. 5. HMSO, London.

Johansson T, Le Quéré A, Ahren D, Söderström B, Erlandsson R, Lundeberg J, Uhlén M, Tunlid A, 2004. Transcriptional responses of *Paxillus involutus* and *Betula pendula* during formation of ectomycorrhizal root tissue. *Mol Plant Microbe Interact* 17:202-215.

Kadotani N, Nakayashiki H, Tosa Y, Mayama S, 2003. RNA silencing in the phytopathogenic fungus *Magnaporthe oryzae. Mol Plant Microbe Interact* 16:769-776.

Kadotani N, Nakayashiki H, Tosa Y, Mayama S, 2004. One of the two Dicer-like proteins in the filamentous fungi *Magnaporthe oryzae* genome is responsible for hairpin RNA-triggered RNA silencing and related small interfering RNA accumulation. *J Biol Chem* 279:44467-44474.

Kaldorf M, Schmelzer E, Bothe H, 1998. Expression of maize and fungal nitrate reductase genes in arbuscular mycorrhiza. *Mol Plant Microbe Interact* 11:439-448.

Kalidas S, Smith DP, 2002. Novel genomic cDNA hybrids produce effective RNA interference in adult *Drosophila. Neuron* 33:177-184.

Kemppainen, MJ, Alvarez-Crespo MC, Pardo, AG, 2009b. fHANT-AC genes of the ectomycorrhizal fungus*Laccaria bicolor* are not repressed by L-glutamine allowing simultaneous utilization of nitrate and organic nitrogen sources. *Environ Microbiol Rep* doi:10.1111/j.1758-2229.2009.00111.x

Kemppainen, M, Circosta, A, Tagu, D, Martin, F, Pardo, AG, 2005. *Agrobacterium*-mediated transformation of the ectomycorrhizal symbiont *Laccaria bicolor* S238N. *Mycorrhiza* 16:19-22.

Kemppainen, M, Duplessis, S, Martin, F, Pardo, AG, 2008. T-DNA insertion, plasmid rescue and integration analysis in the model mycorrhizal fungus *Laccaria bicolor. Microb Biotechnol.* 1: 258-269.

Kemppainen, M, Duplessis, S, Martin, F, Pardo, AG, 2009a. RNA silencing in the model mycorrhizal fungus *Laccaria bicolor.* Gene knock-down of nitrate reductase results in inhibition of symbiosis with *Populus. Environ Microbiol* 11: 1878–1896.

Kemppainen, M, Pardo, AG, 2009. pHg/pSILBAγ silencing vector system for homobasidiomycetes: an efficient RNA silencing trigger in the model mycorrhizal fungus *Laccaria bicolor* via *Agrobacterium*-mediated transformation *Microb Biotechnol* doi:10.1111/j.1751-7915.2009.00122.x

Kropp BR, Fortin JA, 1988. The incompatibility system and relative ectomycorrhizal performance of monokaryons and reconstituted dikaryons of *Laccaria bicolor. Can J Bot* 66:289-294.

Krüger A, Peškan-Berghöfer T, Frettinger P, Herrmann S, Buscot F, Oelmüller R, 2004. Identification of premycorrhiza-related plant genes in the association between *Quercus robur* and *Piloderma croceum. New Phytol* 163:149-157.

Kües U, Casselton L, 1992. Fungal mating type genes - regulators of sexual development. *Mycol Res* 96:993-1006.

Lamhamedi MS, Fortin JA, Kope HH, Kropp BR, 1990. Genetic variation in ectomycorrhiza formation by *Pisolithus arhizus* on *Pinus pinaster* and *Pinus banksiana*. *New Phytol* 115:689-697.

Laverman AM, Zoomer HR, van Verseveld HW, Verhoef HA, 2000. Temporal and spatial variation of nitrogen transformations in a coniferous forest soil. *Soil Biol Biochem* 32:1661-1670.

Lengeler KB, Davidson RC, D'souza C, Harashima T, Shen WC, Wang P, Pan X, Waugh M, Heitman J, 2000. Signal transduction cascade regulating fungal development and virulence. *Microbiol Mol Biol Rev* 64:746-785.

Le Quéré A, Wright DP, Söderström B, Tunlid A, Johansson T, 2005. Global patterns of gene regulation associated with the development of ectomycorrhiza between birch (*Betuna pendula* Roth.) and *Paxillus involutus* (Batsch) Fr. *Mol Plant Microbe Interact* 18:659-673.

Liu H, Cottrell TR, Pierini LM, Goldman WE, Doering TL, 2002. RNA interference in the pathogenic fungus *Cryptococcus neoformans*. *Genetics* 160:463-470.

Lugones LG, Scholtmeijer K, Klootwijk R, Wessels JGH, 1999. Introns are necessary for mRNA accumulation in *Schizophyllum commune*. *Mol Microbiol* 32:681–689.

Ma B, Mayweld MB, Gold MH, 2001. The green fluorescent protein gene functions as a reporter of gene expression in *Phanerochaete chrysosporium*. *Appl Environ Microbiol* 67:948–955.

Madhani HD, Fink GR, 1998. The control of filamentous differentiation and virulence in fungi. *Trends Cell Biol* 8:348-353.

Martin F, Aerts A, Ahren D, Brun A, Danchin EGJ, Duchaussoy F, Gibon J, Kohler A, Lindquist E, Pereda V, et al., 2008. The genome of *Laccaria bicolor* provides insights into mycorrhizal symbiosis. *Nature* 452:88-92.

Martin F, Tuskan GA, DiFazio SP, Lammers P, Newcombe G, Podila GK, 2004. Symbiotic sequencing for the *Populus* mesocosms. *New Phytol* 161:330-335.

Matityahu A, Hadar Y, Dosoretz CG, Belinky PA, 2008. Gene silencing by RNA interference in the white-rot fungus *Phanerochaete chrysosporium*. *Appl Environ Microbiol* 74:5359-5365.

Menotta M, Amicucci A, Sisti D, Gioacchini AM, Stocchi V, 2004. Differential gene expression during pre-symbiotic interaction between *Tuber borchii* Vitad. and *Tilia americana* L. *Curr Genet* 46:158-165.

Mlotshwa S, Pruss GJ, Peragine A, Endres MW, Li J, Chen X, Poethig RS, Bowman LH, Vance V, 2008. *DICER-LIKE2* plays a primary role in transitive silencing of transgenes in *Arabidopsis*. *PloS One* 3:e1755.

Morel M, Jacob C, Kohler A, Johansson A, Martin F, Chalot M, Brun A, 2005. Identification of genes differentially expressed in extraradical mycelium and ectomycorrhizal roots during *Paxillus involutus–Betula pendula* ectomycorrhizal symbiosis. *Appl Environ Microbiol* 71:382-391.

Moriwaki A, Ueno M, Arase S, Kihara J, 2007. RNA-mediated gene silencing in the pythopathogenic fungus *Bipolaris oryzae*. *FEMS Microbiol Lett* 269:85-89.

Morris KV, Chan SW, Jacobsen SE, Looney DJ, 2004. Small interfering RNA-induced transcriptional gene silencing in human cells. *Science* 305:1289-1292.

Nakayashiki H, Hanada S, Nguyen BQ, Kadotani N, Tosa Y, Mayama S, 2005. RNA silencing as a tool for exploring gene function in ascomycete fungi. *Fungal Genet Biol* 42:275-283.

Nakayashiki H, Kadotani N, Mayama S, 2006. Evolution and diversification of RNA silencing proteins in fungi. *J Mol Evol* 63:127-135.

Namekawa SH, Iwabata K, Sugawara H, Hamada FK, Koshiyama A, Chiku H, Kamada T, Sakaguchi K, 2005. Knockdown of *LIM15/DMC1* in the mushroom *Coprinus cinereus* by double-stranded RNA-mediated gene silencing. *Microbiology* 151:3669-3678.

Nehls U, 2008. Mastering ectomycorrhizal symbiosis: the impact of carbohydrates. J Exp Bot 59:1097-1108.

Nicolás FE, Torres-Martínez S, Ruiz-Vázquez RM, 2003. Two classes of small antisense RNAs in fungal RNA silencing triggered by non-integrative transgenes. *EMBO J* 22:3983-3991.

Nygren CMR, Eberhardt U, Karlsson M, Parret JL, Lindahl BD, Taylor AFS, 2008. Growth on nitrate and occurrence of nitrate reductase-encoding genes in a phylogenetically diverse range of ectomycorrhizal fungi. *New Phytol* 180:875-889.

Nygren CMR, Edqvist J, Elfstrand M, Heller G, Taylor AFS, 2007. Detection of extracellular protease activity in different species and genera of ectomycorrhizal fungi. *Mycorrhiza* 17:241-248.

Ohrt T, Mütze J, Staroske W, Weinmann L, Höck J, Crell K, Meister G, Schwille P, 2008. Fluorescence correlation spectroscopy and fluorescence cross-correlation spectroscopy reveal the cytoplasmic origination of loaded nuclear RISC in vivo in human cells. *Nucleic Acid Res* 36:6439-6449.

Oliveira JM, van der Veen D, de Graaff LH, Qui L, 2008. Efficient cloning system for construction of gene silencing vectors in *Aspergillus niger. Appl Microbiol Biotechnol* 80:917-924.

Pandit NN, Russo VEA, 1992. Reversible inactivation of a foreign gene, *hph*, during the asexual cycle in *Neurospora crassa* transformants. *Mol Gen Genet* 234:412-422.

Peterson RL, Massicotte HB, Melville LH, (eds) 2004. *Mycorrhizas: Anatomy and cell biology*. NRC Research Press, Ottawa.

Peter M, Courty P-E, Kohler A, Delaruelle C, Martin D, Tagu D, Frey-Klett P, Duplessis S, Chalot M, Podila G, Martin F, 2003. Analysis of expressed sequence tags from the ectomycorrhizal basidiomycetes *Laccaria bicolor* and *Pisolithus microcarpus. New Phytol* 159:117-129.

Polidori E, Agostini D, Zeppa S, Potenza L, Palma F, Sisti D, Stocchi V, 2002. Identification of differentially expressed cDNA clones in *Tilia platyphyllos-Tuber borchii* ectomycorrhizae using a differential screening approach. *Mol Genet Genomics* 266:858–864.

Rappleye CA, Engle JT, Coldman WE, 2004. RNA interference in *Histoplasma capsulatum* demonstrates a role for alpha-(1,3)-glucan in virulence. *Mol Microbiol* 53:153-165.

Scholtmeijer K, Wösten HAB, Springer J, Wessels JGH, 2001. Effect of introns and AT-rich sequences on expression of the bacterial hygromycin B resistance gene in the basidiomycete *Schizophyllum commune. Appl Environ Microbiol* 67:481–483.

Schramke V, Allshire R, 2003. Hairpin RNAs and retrotransposons LTRs effect RNAi and chromatin-based gene silencing. *Science* 301:1069-1074.

Schuurs TA, Schaeffer EAM, Wessels JGH, 1997. Homology-dependent silencing of the *SC3* gene in *Schizophyllum commune. Genetics* 147:589-596.

Shafran H, Miyara I, Eshed R, Prusky D, Sherman A, 2008. Development of new tools for stydying gene function in fungi based on the Gateway system. *Fungal Genet Biol* 45:1147-1154.

Selker EU, Freitag M, Kothe GO, Margolin BS, Rountree MR, Allis CD, Tamuru H, 2002. Induction and maintenance of nonsymmetrical DNA methylation in *Neurospora. Proc Natl Acad Sci USA* 99:16485-16490.

Sigova A, Rhind N, Zamore P, 2004. A single Argonaute protein mediates both transcriptional and posttranscriptional silencing in *Schizosaccharomyces pombe. Genes Dev 18*:2359-2367.

Smith SE, Read DJ, 2008. Mycorrhizal Symbiosis. 3rd edition. Academic Press, London.

Smith NA, Singh SP, Wang MB, Stoutjesdijk PA, Green AG, Waterhouse PM, 2000. Total silencing by intron-spliced hairpin RNAs. *Nature* 407:319-320.

Soragni E, Bolchi A, Balestrini R, Gambaretto C, Percudani R, Bonfante P, Ottonello S, 2001. A nutrient-regulated, dual localization phospholipase A(2) in the symbiotic fungus *Tuber borchii. EMBO J* 20:5079-5090.

Stark JM, Hart SC, 1997. High rates of nitrification and nitrate turnover in undisturbed coniferous forests. N*ature* 385:61-64.

Tamuru H, Selker EU, 2001. A histone H3 methyltransferase controls DNA methylation in *Neurospora crassa. Nature* 414:277-283.

Tanguay P, Bozza S, Breuil C, 2006. Assessing RNAi frequency and efficiency in *Ophiostoma floccosum* and *O. piceae. Fungal Genet Biol* 43:804-812.

Tuskan GA, DiFazio S, Jansson S, Bohlmann J, Grigoriev I, Hellsten U, et al., 2006. The genome of black cottonwood, *Populus trichocarpa* (Torr. & Gray). *Science* 313:1596-1604.

Voiblet C, Duplessis S, Encelot N, Martin F, 2001. Identification of symbiosis-regulated genes in *Eucalyptus globulus-Pisolithus tinctorius* ectomycorrhiza by differential hybridization of arrayed cDNAs. *Plant J* 25:181-191.

Volpe TA, Kinder C, Hall IM, Teng G, Grewal SI, Martienssen RA, 2002. Regulation of heterochromatic silencing and histone H3 lysine-9 methylation by RNAi. *Science* 297:1833-1837.

Wadhwa R, Kaul SC, Miyagishi M, Taira K, 2004. Vectors for RNA interferente. *Curr Opin Mol Ther* 6:367-372.

Wälti MA, Villalba C, Buser RM, Grünler A, Aebi M, Künzler M, 2006. Targeted gene silencing in the model mushroom *Coprinopsis cinerea* (*Coprinus cinereus*) by expression of homologous hairpin RNAs. *Euk Cell* 5:732-744.

Weinmann L, Höck J, Ivacevic T, Ohrt T, Mütze J, Schwille P, Kremmer E, Benes V, Urlaub H, Meister G, 2009. Importin 8 is a gene silencing factor that targets Argonaute proteins to distinct mRNAs. *Cell* 136:496–507.

Wesley SV, Helliwell CA, Smith NA, Wang MB, Rouse DT, Liu Q, Gooding PS, et al., 2001. Construct design for efficient, effective and high-throughput gene silencing in plants. *Plant J* 27:581-590.

Wright DP, Johansson T, Le Quéré A, Söderström B, Tunlid A, 2005. Spatial patterns of gene expression in the extramatrical mycelium and mycorrhizal root tips formed by the ectomycorrhizal fungus *Paxillus involutus* in association with birch (*Betula pendula*) seedlings in soil microcosms. *New Phytol* 167:579-596.

Yamazaki T, Okajima Y, Kawashima H, Tsukamoto A, Sugiura J, Shishido K, 2006. Intron-dependent accumulation of mRNA in *Coriolus hirsutus* of lignin peroxidase gene the product of which is involved in conversion/degradation of polychlorinated aromatic hydrocarbons. *Biosci Biotechnol Biochem* 70:1293-1299.

In: Gene Silencing: Theory, Techniques and Applications
Editor: Anthony J.Catalano

ISBN: 978-1-61728-276-8
© 2010 Nova Science Publishers, Inc.

Chapter III

Gene Silencing as a Promising Strategy to Target CNS and Neurological Disorders

Ana L. Cardoso[1], Sara M. Trabulo[1,2], Sónia P. Duarte[1,2], Pedro M. Costa[1,2], Luís P. Almeida[1,3] and Maria C. Pedroso de Lima[1,2]

[1]Center for Neuroscience and Cell Biology, University of Coimbra, Coimbra, Portugal
[2]Department of Life Sciences, Faculty of Science and Technology,
University of Coimbra, Coimbra, Portugal
[3]Laboratory of Pharmaceutical Technology, Faculty of Pharmacy, University of Coimbra,
Coimbra, Portugal

Abstract

RNA interference (RNAi) has recently emerged as a powerful tool in functional genomic studies, allowing dissection of entire signalling pathways and elucidation of the molecular mechanisms of neurobiological processes, thereby facilitating rapid identification and validation of possible therapeutic targets. Moreover, RNAi holds great therapeutic potential since application of small interfering RNAs (siRNAs) and short hairpin RNAs (shRNAs) may allow specific knockdown of selected toxic proteins, even when allele-specific silencing is needed, as in the case of dominantly inherited disorders.

Nevertheless, the development of RNAi-based therapeutics for *in vivo* application faces the same challenge common to all classes of drugs: achieving an efficient and sustained distribution into the target tissue at sufficient concentrations to accomplish a therapeutic effect. Although significant progress has been made regarding the safety and stability of siRNAs and shRNAs, a major limitation for the *in vivo* application of RNAi technology concerns the inability of these molecules to cross cellular membranes. Multiple delivery methodologies, including viral and non-viral vectors, have been developed with different degrees of success for the introduction of siRNAs and shRNAs into cells, both *in vitro* and *in vivo*.

This review is focused on the available strategies to achieve gene silencing in the CNS and on the most extensively studied systems to mediate siRNA and shRNA delivery into the brain. Moreover, we summarize the most important studies concerning RNAi application in the context of neurodegenerative diseases and other neurological disorders.

Introduction

1. The Challenge of Drug Delivery to the Brain

Developing effective treatment for a range of progressive and untreatable CNS disorders remains one of the major goals of clinical medicine. Due to its complex structure, compartmentalized functions and high sensitivity to insult, the brain is indisputably the most challenging organ for drug delivery. By effectively blocking the passage of most blood components into the intracerebral space, the blood-brain barrier (BBB) allows the establishment of a separate and extremely stable environment, which is crucial for accurate synaptic transmission. Moreover, the BBB has a neuroprotective function, limiting the access of potential neurotoxic substances, by either preventing their passage or actively removing them via specific ATP-dependent transporter proteins. The structure of BBB (Figure 1) is generated by the brain endothelial cells, which form the capillaries of the brain and spinal cord. These cells form tight junctions between themselves, sealing the transfer of polar solutes to the interstitial fluid, unless this occurs via specific membrane-bound receptors[1].

The ideal drug for CNS application should present optimal lipid solubility for BBB penetration, while retaining significant pharmacological activity in the extracellular or intracellular aqueous fluid. Nevertheless, the conjugation of both these properties is often impossible to achieve, and increasing the lipid solubility of a drug molecule may have undesirable effects, such as decrease of its bioavailability and increase the binding by plasma proteins and liver retention. During the last decade, several strategies have been developed to circumvent the BBB when delivering therapeutic agents, such as direct injection into the brain parenchyma or cerebrospinal fluid, delivery of prodrugs, delivery through the olfactory route, transient opening of the BBB and targeting of BBB transporters. Recently, several of these approaches have been explored in non-viral gene delivery to promote efficient delivery of nucleic acids, such as siRNAs and plasmid DNA, across the BBB and into the intracellular compartment of the target cells [1].

2. RNA Interference: A New Therapeutic Approach to Treat Neurological Injury

The discovery of the RNA interference (RNAi) pathway can be considered as one of the most remarkable biological breakthroughs of the last decade. Since its initial description by Fire and colleagues [2] in 1998, RNAi has transformed the genomic field and introduced scientists to a series of new and complex gene regulation pathways, mediated by small RNA molecules previously discarded as irrelevant.

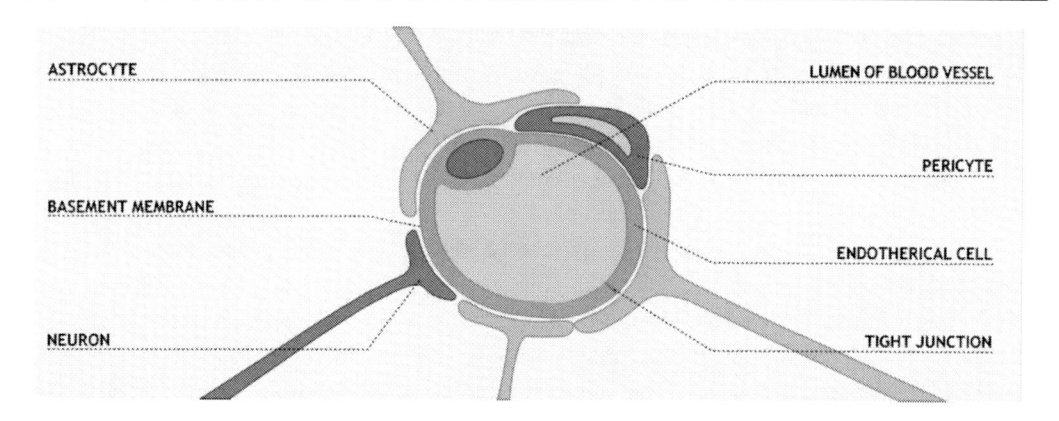

Figure 1. Schematic diagram of the blood brain barrier. The neurovascular unit that constitutes the BBB is composed of cerebral endothelial cells, which form tight junctions and separate the paracellular medium from the blood, and pericytes, which partially surround the endothelium. Astrocytes and microglia, together with axons from neuronal cells form a network which fully surrounds the capillaries and further isolates this structure.

During the last few years, scientists have developed different methods to manipulate RNAi in the laboratory, exploiting its potential as both a research tool and a novel therapeutic approach to treat human diseases. RNAi is currently routinely used to uncover the complexity of intracellular signalling pathways associated with development, cancer, infection and other similar phenomena. Moreover, it holds the promise to be the most selective of all therapeutic drugs, which can seek out and destroy a specific mRNA target without affecting other genes.

[1] The major players of the RNAi process are small RNA molecules, such as small interfering RNAs (siRNAs) and microRNAs (miRNAs), which differ in their cellular biogenesis but converge into the same molecular pathways of gene regulation, albeit with different consequences to the cell. According to their nature, these small RNA species can elicit gene regulation by at least four different types of response: destruction of homologous mRNA [3-5], inhibition of translation [6], *the novo* methylation of genomic regions with consequent blocking of transcription [7, 8] and modifications of heterochromatin with consequent chromosomal rearrangements [9, 10].

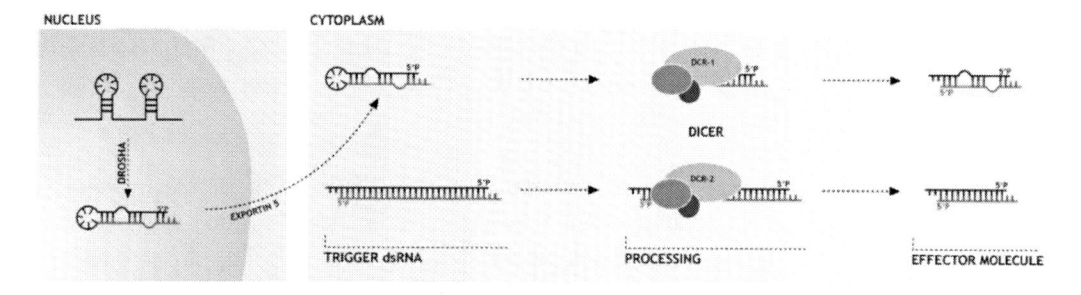

Figure 2. siRNA and miRNA biogenesis. Both siRNAs and miRNAs are derived from large RNA precursors. In the case of miRNAs, the precursor is termed pre-miRNA and adopts a hairpin structure. In the case of siRNAs, the precursor is a long dsRNA molecule. Both precursors are processed by a DICER nuclease into duplexes termed miRNAs or siRNAs, which possess 5'-phosphate and 2-nucleotide 3' overhangs.

In the case of siRNAs, the cellular mechanism of RNAi starts with the production of dsRNA in the cytoplasm by RNA-template RNA polymerization (for example from viral RNA) or by hybridization of overlapping transcripts (for example from repetitive sequences such as transposons). These complementary dsRNA precursors are then cleaved by the dsRNA-specific RNAse type III Dicer, to generate small double-stranded RNAs of 21 to 23 nucleotides (siRNAs), as depicted in Figure 2.

In the case of miRNAs, the process is more complex and starts in the nucleus with the transcription of long primary miRNA precursors (pri-miRNAs) [11, 12]. Most miRNAs genes are expressed individually under the control of specific promoters and regulatory sequences, but others are located in clusters that may be transcribed as polycistrons. The pri-miRNAs, which can be several kilobases long, undergo folding into elaborate stem-loop structures containing complementary or nearly-complementary 20- to 50-base-pair inverted repeats which form dsRNA hairpins. Another RNAse III enzyme called Drosha, with the help of a double stranded RNA-binding protein called DGCR8, cleaves the steem-loop structures away from the pri-miRNAs at specific sites, generating pre-miRNA intermediates [12]. These intermediates are then transported from the nucleus to the cytoplasm by the nuclear export-factor Exportin-5 [13, 14]. Once in the cytoplasm, the pre-miRNAs are further processed by Dicer to generate double stranded miRNAs [15] which usually have a very short half-life. The two chains usually dissociate and only one strands acts as a mature miRNA and the other, the miR*, is eventually degraded.

Although miRNAs and siRNAs share several similarities, such as size, the 5'-phosphate and 3'hydroxil terminals which result from RNAse III cleavage and the mechanism of action on mRNAs through complementary sequences, they also present very important differences. While siRNAs originate from exogenous or endogenous long dsRNAs, miRNAs precursors are always endogenous and typically found in intergenic regions or introns capable of forming hairpin structures. Usually miRNAs are generated from one arm of stem-loop and in vast excess over any complement. Moreover, while the RNA target of a siRNA presents perfect complementarity to one of the strands of the siRNA itself, a single miRNA can regulate hundreds of different genes, which present a miRNA binding site in the 3'-UTR region in their mRNAs, even when the miRNA sequence is not a perfect match to the target mRNA. Several different miRNAs can target the same 3'-region [16, 17].

Both siRNAs and miRNAs are key intervenients in several regulatory pathways through association with various protein partners. Many components of these complexes are yet to be identified, but biochemical and genetic studies have shown that every siRNA or miRNA-binding ribonucleoprotein (RPN) contains a member of the Argonaute (Ago) protein family. After binding of a small RNA molecule, the RPNs are rearranged into the RNA induced silencing complex (RISC) in the case of siRNAs, or into the miRNA-containing effector-complex (miRPN) in the case of miRNAs [18].

Concerning mRNA silencing, the effectiveness and specificity of the RNAi process are dependent upon favouring the loading of the antisense (guide) strand of the siRNA into the RISC complex. Although the sense (passenger) strand confers stability to the dsRNA molecule, it should not be selected by RISC, as it contains the same sequence of the target gene and can induce potential off-target effects. The passenger strand is the initial target of the Ago 2 endonuclease activity and, after this initial cleavage, the strand remnants are discarded and the guide strand is secured, forming the mature RISC complex.

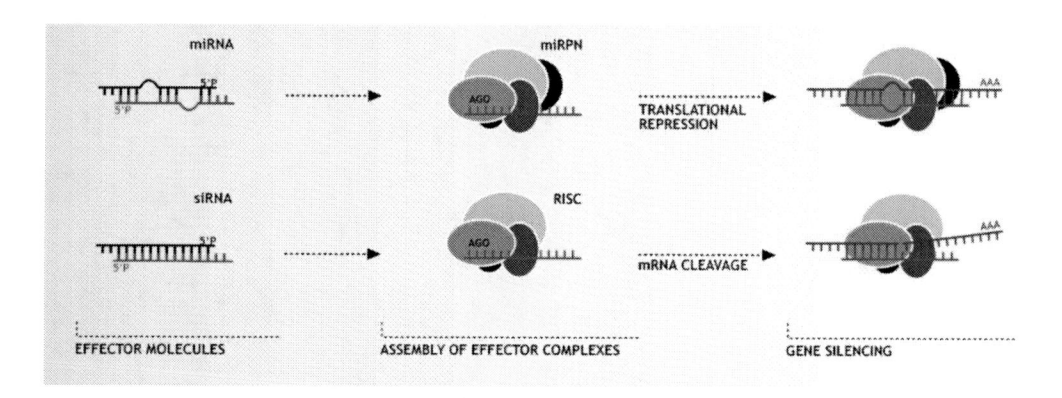

Figure 3. Different mechanisms of siRNA and miRNA action. The mature miRNAs and siRNAs bind to an argonaute protein (Ago) to form the effector complexes known as miRPN and RISC, respectively. The miRNA or siRNA duplex molecules are unwound while being assembled into the miRPN/RISC and one of their strands (guide strand) allows the Ago protein to bind to a specific target RNA. The effector complexes silence gene expression by suppressing the translation of the target mRNA or by promoting its cleavage.

The guide strand of siRNA allows the recognition of the complementary or nearly complementary target mRNA and its cleavage ten nucleotides upstream of the nucleotide paired with the 5' end of the guide siRNA [19, 20]. The cleavage occurs between nucleotides 10 and 11 of the siRNA, at the scissile phosphate of the mRNA.

The complementarity between the small RNA and its target RNA is crucial to determine whether the miRNA or siRNA mediates mRNA cleavage or represses mRNA translation. When a miRNA or siRNA enters the miRPN/RISC complex, an A-form helix must form at the centre of the guide strand, in the so-called "seed region", containing nucleotides 2 to 8, in order for mRNA cleavage to occur. In the absence of this helix, which occurs when the base pairing in the 5' end region of the miRNA is not perfect, translational repression will ensue (Figure 3).

2. siRNAs and shRNAs: Tools for Gene Silencing

Gene silencing mediated by the RNAi technology has been proposed as a new therapeutic strategy for the treatment of several neurodegenerative diseases, of both genetic and sporadic origin. Although the *in vivo* challenges for siRNA and shRNA delivery to the neuronal tissue are very similar to those of conventional drugs, at least two major advantages can be attributed to siRNAs over small drug molecules. The synthesis of siRNA molecules is easy to perform, being similar to all target sequences not requiring complex production or purification methodologies, thereby facilitating large scale production and drastically reducing production costs. The second advantage of RNAi is that, in contrast to conventional drugs, siRNAs can inhibit any protein, including the so-called "non-druggable targets", which have conformations not amenable to small molecule binding. Moreover, allele-specific silencing can be achieved with RNAi by exploring single nucleotide polymorphisms, which opens the perspective of new therapeutic strategies to a group of genetic diseases so far without treatment.

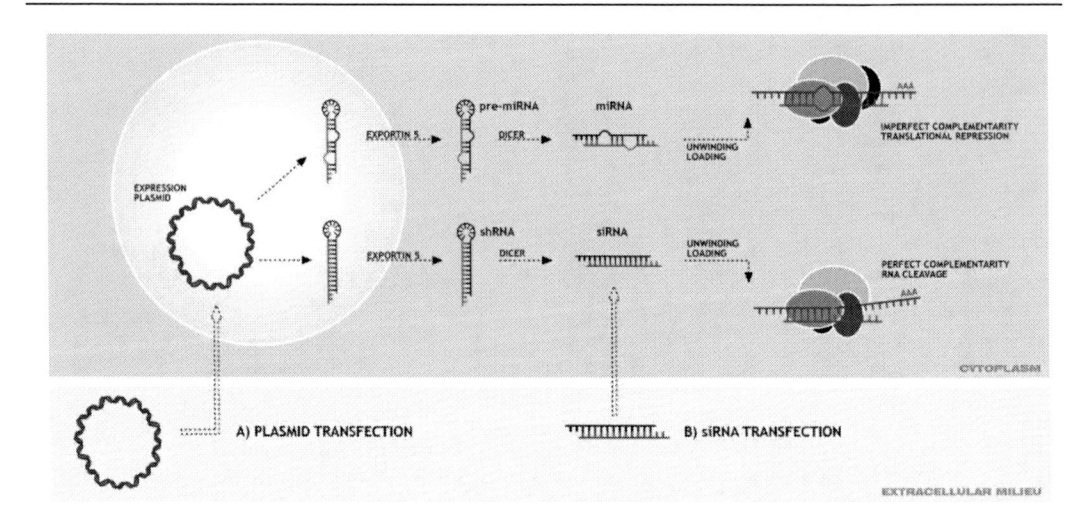

Figure 4. Different pathways for nucleic acid delivery and gene silencing. The exogenous activation of gene silencing can be achieved by transfecting cells **(A)** with plasmid DNA, which is transcribed directly into shRNAs or miRNAs, without requiring Drosha processing, or **(B)** by direct delivery of siRNAs into the cytoplasm. Depending on the degree of complementarity between the miRNA/siRNA and its target, the final result will be translational repression/mRNA decay or RNA cleavage.

Due to its specificity and potency, several strategies have been developed to harness the RNAi machinery and translate it into new tools for research and therapy. Since 2001, several groups have efficiently transfected differentiated mammalian cells with synthetic siRNA [19, 21, 22] and the function of several endogenous genes has also been investigated *in vivo* using this approach. The two most common ways to silence a gene of interest in mammalian cells based on RNAi technology are the delivery of synthetic siRNAs using a variety of non-viral vectors or the delivery of plasmids encoding short hairpin RNAs (shRNAs) by viral or non-viral methods [23] (Figure 4). shRNAs are processed in the nucleus and cytoplasm by the cellular machinery of the miRNA endogenous pathway, originating siRNA oligonucleotides similar to those delivered exogenously. Both these methods allow a highly controlled approach to gene knockdown, in which inducible small RNA vector systems can be used to transiently or permanently decrease the expression of a target protein in selected cells or tissues. These systems have the additional advantage of allowing the titration of protein levels by varying the intracellular concentration of the siRNA or shRNA, thus showing the effects of intermediate levels of the target proteins under study.

The optimization of the design of both siRNAs and shRNAs aiming at increasing their effectiveness involves three major steps: the first is ensuring that the correct siRNA strand (antisense strand) is preferentially loaded by RISC (RNA induced silencing complex), and is fully complementary to the target mRNA, the second is minimizing the complementarity of the designed sequence with other known mRNA targets in order to decrease the potential for "off-target" effects and the last step is to minimize the activation of host cell defence pathways, such as the interferon response. Several rules have been assigned to the design of siRNA sequences in order to increase the efficiency of these molecules. In general, synthesised siRNAs should not target introns, the 5' and 3' ends of untranslated regions (UTR) and sequences within 75 bases of the start codon (ATG). Furthermore, the guanine/cytosine content of the sequence should be between 30 and 50% and the 5' terminal of the sense strand should have higher thermodynamic stability than the 5' terminal of the antisense strand. Finally, a BLAST-search should be performed and sequences with low-stringency should be avoided, to ensure that the potential for targeting unrelated genes is minimal.

While synthetic siRNAs enter the cell directly to the cytoplasm where both RISC and their target mRNA are located, shRNAs can be considered siRNA prodrugs. They need to enter the nucleus in the form of a plasmid, usually under the control of RNA polymerase III promoters, and be expressed using the cellular machinery, and subsequently to enter the miRNA pathway at the level of Exportin-5.

The shRNAs are cleaved in the cytoplasm by Dicer, generating common siRNA duplexes. Transcription of shRNAs is normally performed using the RNA Pol III promoters U6 or H1, which are compact promoters, located 5' to the transcribed sequence. Usually the promoter directs the synthesis of two small inverted repeats separated by a spacer region of varying length. Pol III-mediated transcription terminates after the second or third residue of a "TTTT" stretch and the "UU" 3' terminal of shRNAs is used to form the 3' 2 nucleotide overhangs characteristic of pre-miRNAs. RNAse II promoters can also be used with this purpose, although it has been found that the transcriptional starting site is often less precise in this situation and the termination sequence required is much longer, which limits their applicability. Although there are other possible designs to generate vector-driven expression of siRNAs, the one described above and illustrated in Figure 5 is the most widely employed and has been shown to have higher suppressive activity [24, 25]. This is probably due to a higher yield of annealing and to the presence of the loop in the hairpin system, which plays an important role in the transport of the shRNAs to the cytoplasm. The shRNA expression cassette can be moved into a normal plasmid for non-viral delivery or, most of the times, incorporated in a viral vector which allows long-term, stable expression of the encoded shRNA. More recently, vectors encoding miRNA or shRNA under the control of miRNA RNA polymerase II promoters have been developed to study the function of endogenous miRNAs or induce gene silencing of a specific target, with the additional advantage of being able to produce multiple shRNAs or miRNAs with a single promoter [26].

2.1. siRNAs *versus* shRNAs – Benefits and Disadvantages

Initial RNAi-based studies in neurons focused on the use of synthetic siRNAs with 19-21-nucleotides long, which proved to mediate significant and reproducible gene silencing in both neuronal primary cultures [27] and *in vivo* [28]. Nevertheless, several drawbacks were initially associated with the use of siRNAs in neurons, such as high susceptibility to nuclease degradation, inefficient delivery in the absence of a delivery vector and transient effect. Different strategies have been developed to address these problems, such as the incorporation of chemical modifications in the terminal regions of the siRNA duplexes, conjugation of peptides and cholesterol moieties or complexation with cationic lipid formulations. Application of these new technologies greatly increased siRNA efficiency *in vivo* and resulted in the development of siRNA-based drugs, currently in clinical trials for eye and lung disorders. Although synthetic siRNAs can be very efficient, they do not achieve stable long-term silencing, an important feature for treatment of some CNS disorders, such as neurodegenerative diseases.

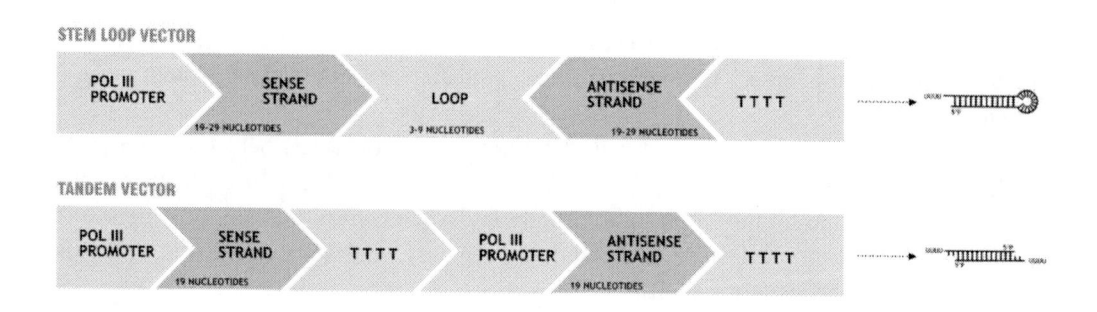

Figure 5 . Plasmid driven expression of shRNAs. Diagrammatic presentation of the tandem and loop type vector-driven shRNA constructs.

Brummelkamp and co-workers [29] first demonstrated the feasibility of using shRNAs as precursors for siRNA-mediated silencing. This strategy was tested shortly after in the brain by direct delivery of an AAV vector [30], and found to allow a long-term efficient decrease of the target protein levels, thus circumventing one of the major limitations for RNAi application. Furthermore, although requiring more effort in the production stage, plasmids encoding shRNAs allow regulated expression by inserting tetracycline response elements [31] or other regulatory systems, which can work as safeguards in chronic disease treatment. At present, shRNA constructs are the models of choice for trials in animal models of neurodegenerative diseases, although transient silencing of target genes mediated by siRNAs cannot be immediately considered as a drawback. In fact, siRNA delivery can be advantageous in some cases when a rapid target knockdown is required, for instance in ischemia or trauma models, or when long-term silencing of a given target may be detrimental to cell function.

3. Strategies for Gene Delivery in the Brain

Although significant progress has been made regarding the safety and stability of siRNA and shRNA molecules, a major limitation for the *in vivo* application of RNAi technology concerns the inability of these molecules to cross biological barriers, such as the BBB and reach the neuronal tissue. The successful clinical application of RNAi to the CNS is therefore dependent on the ability to overcome this particular obstacle.

It is also important to note that siRNA and shRNA target sites are mostly located in the cytoplasm and nucleus, respectively. Therefore, for the success of RNAi-based therapeutics it is essential that these molecules traverse the cellular membrane and/or the nuclear envelope in order to reach their destination and promote gene silencing.

Due to the highly lipophilic nature of the plasma membrane, which excludes the entry of large hydrophilic molecules, siRNA transfer is restricted by its size and anionic charge of the phosphate backbone. In the absence of an appropriate delivery system, which can help to circumvent this problem, the internalization of siRNA and other nucleic acid molecules is always relatively inefficient. Moreover, siRNAs have a half-life of less than one hour in human plasma, and the circulating molecules are quickly excreted by the kidneys [32]. All these factors make the systemic application of siRNAs unlikely to succeed, except when very high doses are employed.

Multiple delivery methodologies, involving both viral and non-viral vectors, have been developed, with different degrees of success, for the introduction of siRNAs and shRNAs into cells, *in vitro* and *in vivo*. The most usual vector classification applied in gene therapy separates virus-based vectors from all other gene delivery vehicles. Due to their intrinsic mechanism for delivering genetic material upon cell infection, viruses have been employed in this field for over a decade and have proven to yield higher transduction efficiencies in most cell types, when compared to standard non-viral delivery systems. The five types of viral vectors currently being used for application of RNAi include those based on retrovirus (RV), adenovirus (AV) and, most importantly, adeno-associated virus (AAV) and lentivirus (LV). Although very efficient, these vectors are known to suffer from unexpected inflammatory reactions and unspecific integration into the host genome, which can lead to tumorogenesis. Nevertheless, significant progress has been made in this field, with the development of tissue-specific serotypes and regulated viral promoters, which allow a safer application.

Regarding non-viral delivery systems, two subclasses can be described: physical delivery methods, such as microinjection, electroporation and hydrodynamic injection, and chemical delivery vehicles, such as liposomes [33], polymers [34], peptide/protein complexes [35], and siRNA conjugates. The most common components used in the design of chemical vectors are: natural or synthetic lipids (typically a mixture of cationic and neutral lipids), cationic polymers, cell penetrating peptides, antibodies and combinations of these elements. Besides increasing siRNA delivery, most non-viral vectors have the additional advantage of protecting the nucleic acids from nuclease-mediated degradation, both outside and inside the cellular milieu. Functional devices can be further introduced in each vector to help overcome different barriers of the transfection process, such as traversing the plasma membrane, escaping lysosomal degradation and translocation to the nucleus. These devices include, for example, the use of targeting ligands to increase the specificity of cellular uptake and fusogenic lipids or peptides to enhance endosomal release. Although usually less efficient than viral vectors, non-viral vectors present several advantages, such as simplicity and ease of large scale production, low price, reduced immunogenicity (in most cases) and unrestricted genetic material size. Table 1 shows the major groups of vectors employed in *in vivo* RNAi studies, as well as their major advantages and disadvantages.

3.1. Non-viral Delivery Systems

Non-viral delivery systems are the vehicle of choice for siRNA delivery, offering several advantages as compared to viral vectors, such as higher safety, dosage control and evasion of immunological surveillance. Also, chemical modifications can be easily incorporated in these vectors to increase siRNA stability *in vivo*. However, most non-viral vectors suffer from relatively low transfection efficiency in primary cells and, until recently, very few non-viral delivery systems have achieved efficient silencing of endogenous genes in the brain.

Table 1. Vectors used for siRNA and shRNA delivery

Type	Vector	Advantages	Disadvantages	References
Viral	RV	- High efficiency - Stable transfer of genes	- Possibility of random Integration in the genome - Transduction limited to dividing cells	[36]
	LV	- Transduction of non-dividing cells - Ability of carrying large genomes - Possibility of pseudotyping	- Possible immunogenicity	[37-39]
	AV	- No risk of insertion in the host's genome - Useful for liver applications	- Dose limiting hepatotoxicity - Transient expression - Inflammatory response - Dependence on specific receptors	[40-44]
	AAV	- Absence of pathogenicity - Possibility of pseudotyping - Transduction of non-dividing cells - Formation of episomal DNA molecules	- Small genome - Presence of antibodies in the majority of the transduced cells - Inflammatory response	[45-48]
Physical	Naked siRNA	- Ease of preparation and application	- Large amounts required - Immunogenicity - Low efficiency	[49-51]
	Hydrodynamic injection	- Low price - Efficiency for liver delivery	- Only useful in mice - No clinical use	[52-54]
	Electroporation	- Reduced immune response - Efficient in vitro	- Local application - Invasive	[55-60]
Chemical	Modified siRNAs	- Increased serum stability - Reduced immune response - Ease of formulation	- Expensive - Carrier requirement - Low efficiency	[61]
	Liposomes	- Low cost - Ease of preparation and application - Delivery of high number of siRNA or shRNA copies - Ease of modification	- Reduced efficiency after systemic delivery - Potential immunogenicity	[33, 62-64]
	Polymer nanoparticles	- Low cost - Ease of preparation - Delivery of high number of siRNA or shRNA copies	- Reduced efficiency after systemic delivery - Potential immunogenicity - Lack of availability	[33, 34, 65]

Table 1. (Continued)

Type	Vector	Advantages	Disadvantages	References
	Cell penetrating peptides	- Reduced immune response - Ease of preparation	- High cost - Limited experience	[35]
	siRNA conjugates	- Efficiency of transfection - Tissue-specificity	- High cost	[49]

Liposomes are an example of a type of non-viral vector which has been successfully used for the delivery of siRNAs to neuronal cells, both *in vitro* and *in vivo* [66-68]. These lipid vesicles have been used as tools for drug delivery since 1964 but it was not until the late 80s, when the cationic lipid called DOTMA was employed to transfer both DNA and RNA [69], that the research field for liposome-mediated delivery of genetic material actually started. Today many non-viral nucleic acid delivery agents are cationic lipid-based and a large number of commercially available reagents, such as Lipofectamine 2000, DOTAP or Oligofectamine, can provide high transfection efficiency in a wide range of mammalian cell types *in vitro*.

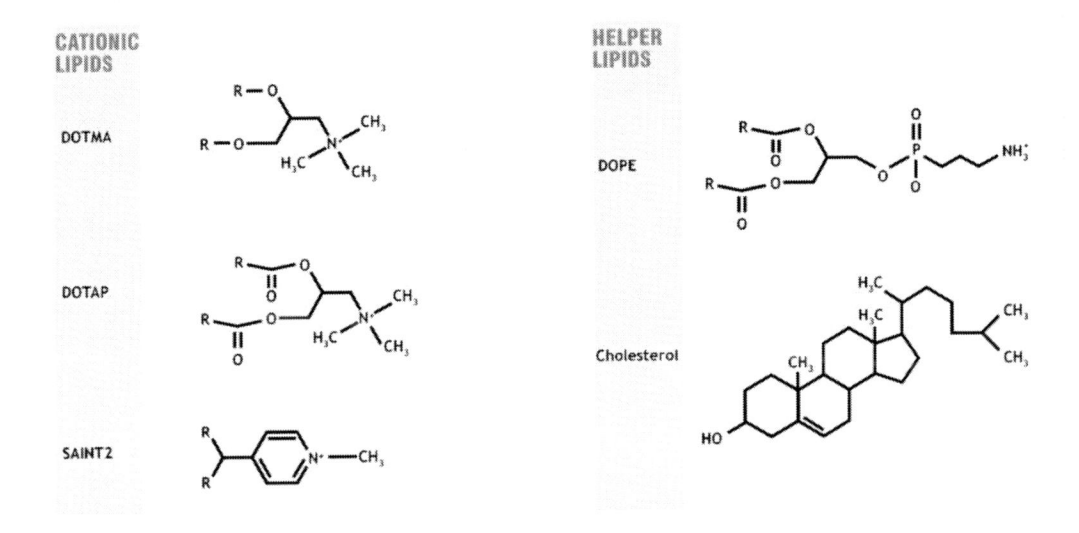

Figure 6. Chemical structures of common cationic and helper lipids. Cationic lipids, such as DOTMA, DOTAP and SAINT 2, are frequently used for nucleic acid complexation and intracellular delivery, usually in the presence of a helper lipid, such as DOPE and cholesterol. Cationic lipids usually present an amphipathic nature, with a hydrophobic core and a hydrophilic head group, which can vary significantly in structure and in the number of positive charges. The head group is linked to the backbone structure through ether (DOTMA), esther (DOTAP) or C-C (SAINT 2) bonds.

Liposomes are small vesicles consisting of an aqueous compartment enclosed by one or more lipid bilayers, which can carry both hydrophobic and hydrophilic molecules. These lipid particles are often very stable and may exhibit favourable pharmacokinetic properties. The lipid bilayer usually contains more than one lipid component and polyethylene glycol (PEG), but functionalizing agents, such as targeting peptides and proteins, can also be present. Cationic liposomes have been the most widely used lipid-based vectors for siRNA delivery and are usually composed of a cationic lipid (such as DOTAP, DOTMA, SAINT 2 and DOGS) and a helper lipid (cholesterol, DOPE or DOPC). Several examples of commonly employed lipids are presented in Figure 6.

Cationic liposomes can form the so-called lipoplexes upon electrostatic interaction of the lipid positive charges with the negatively charged nucleic acids. The excess of positive charges of lipoplexes allows an extensive interaction with the negatively charged proteoglycans present at the cell surface, which makes this lipid-based delivery system usually more efficient than neutral or negatively charged formulations, in the absence of targeting moieties.

Following binding of lipoplexes to the cell surface, endocytosis has been recognized as the major pathway for intracellular nucleic acid delivery. Once inside the endosome, the release of the lipoplexes to the cytoplasm is of crucial importance to avoid siRNA degradation upon fusion with the lysosome. It is believed that, the presence of positively charged lipids inside the endosome will induce the "flip-flop" of the anionic lipids present in the endosomal membrane from the cytoplasmic to the luminal leaflet, which will result in the formation of neutrally charged ion-pairs of cationic and anionic lipids [70]. This will have two major consequences for gene delivery. First, destabilization of the endosomal membrane will occur, with subsequent release of the endosomal content into the cytoplasm, and second, there will be a displacement of the cationic lipid from the lipoplexes, thus freeing the carried nucleic acids and allowing their interaction with the cellular machinery.

Recently, Pedroso de Lima and co-workers provided evidence that a lipid-based formulation, containing DOTAP:Cholesterol liposomes associated with transferrin (Tf), could promote siRNA delivery and efficient silencing of a reporter gene in neurons [68].

As observed in Figure 7a, results from confocal microscopy studies clearly show the efficient uptake of Cy3-labelled siRNAs in neuronal cultures following delivery of transferrin-associated lipoplexes (Tf-lipoplexes) (Figure 7a) [68]. Moreover, transfection of neuronal primary cultures prepared from luciferase-expressing mice with Tf-lipoplexes containing a siRNA sequence against luciferase led to a 50% reduction of luciferase activity in these cells (Figure 7b)[68]. Efficient gene silencing was also observed *in vivo,* following stereotactic injection of the same formulation, containing anti-luciferase siRNAs, in the striatum of luciferase mice [68].

More recently, the same authors have shown that Tf-lipoplexes can be employed to mediate silencing of an endogenous gene in neuronal cells, both *in vitro* and *in vivo*, with a therapeutic purpose [66]. In this study the selected target was the transcription factor c-Jun, a protein with a pro-apoptotic and pro-inflammatory role following brain excitotoxic injury. By selectively silencing c-Jun using Tf-lipoplexes, the authors were able to demonstrate that cell death could be prevented in neuronal primary cultures, following exposure to glutamate or oxygen glucose deprivation, two well described models of excitotoxic injury (Figure 8).

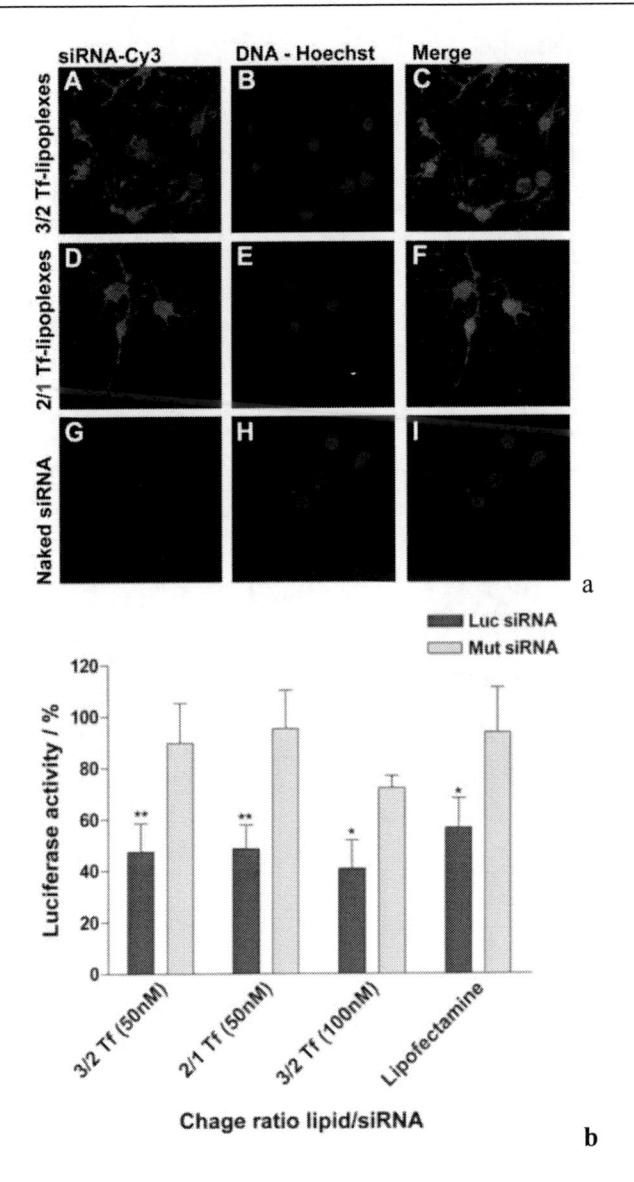

Figure 7. Uptake and gene silencing mediated by Tf-lipoplexes in neuronal primary cultures. Primary cultures of cortical neurons were transfected with Tf-lipolexes prepared with DOTAP:Chol liposomes, transferrin (32 µg/ µg siRNA) and siRNAs, at the indicated lipid/siRNA charge ratios. **a)** Cells were incubated for 4h with Tf-lipoplexes containing Cy3-labelled siRNAs (red), at a final siRNA concentration of 50 nM. In parallel experiments, the same amount of labelled siRNAs was added to the cells. Cells were labelled with the nuclear dye Hoechst 33342 (blue) and observed by confocal microscopy. Efficient siRNA uptake was observed in neurons transfected with Tf-lipoplexes but not in neurons incubated with naked siRNAs. **b)** Cells were incubated for 4h with Tf-lipoplexes containing anti-luciferase siRNAs (Luc siRNA) or non-silencing siRNAs (Mut siRNA), at a final siRNA concentration of 50 or 100 nM. In parallel studies cells were incubated with the Luc or Mut siRNAs complexed with Lipofectamine, at a final siRNA concentration of 50 nM. Luciferase activity was determined for each sample in relative light units (RLU), 48 h after transfection and calculated as a percentage of activity in control (non-transfected) cells. A significant and specific reduction of luciferase activity was observed when neurons were transfected with Luc siRNAs at a final siRNA concentration of 50 nM. * $p<0.5$ and **$p< 0.01$ compared to cells transfected with Mut siRNAs.

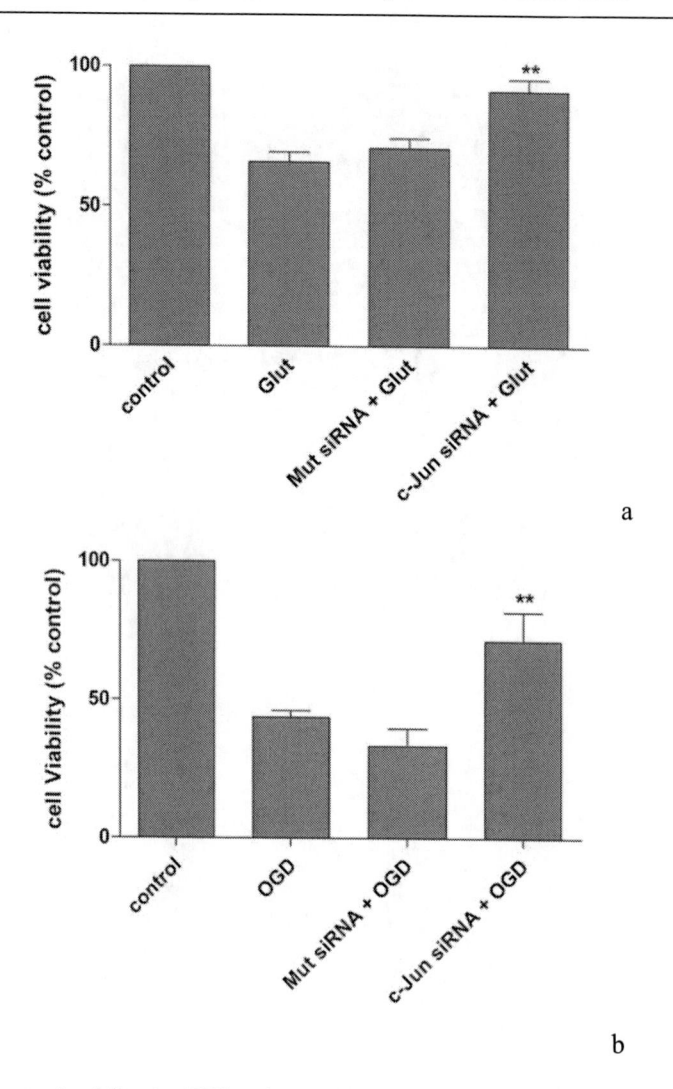

a

b

Figure 8. Neuronal protection following Tf-lipoplex-mediated c-Jun silencing *in vitro*. Tf-lipoplexes containing anti-c-Jun (c-Jun siRNA) or non-silencing (Mut siRNA) siRNAs, prepared at a 2/1 lipid/siRNA charge ratio, were added to primary neuronal cultures at a final siRNA concentration of 50 nM. Twenty four hours after transfection the cultures were incubated with **a)** 125 μM glutamate for 20 min or **b)** exposed for 4 h to glucose-free EBSS medium in the absence of oxygen (OGD). In parallel experiments, non-transfected cells were incubated in the absence (control) or presence of glutamate (Glut) or oxygen/glucose deprivation conditions (OGD). Cells were further incubated for 18 h in fresh medium before analysis of cell viability using the MTT assay. A notorious recovery of neuronal viability was observed in neurons receiving anti-c-jun siRNAs following excitotoxic damage. **p< 0.01 compared to cells exposed to glutamate or OGD.

Moreover, Tf-lipoplex-mediated c-Jun silencing in the CA3 region of the mouse hippocampus led to a decrease of neuronal loss in the treated hemispheres (Figure 9a) and prevent microglia and astrocyte activation following intraventricular injection of kainic acid (KA). No improvement was observed in animals receiving the same formulation carrying

Mut siRNAs, which indicated that the observed effect was related specifically to c-Jun knockdown. This treatment also resulted in a reduction of the number and severity of epileptic seizures induced by KA, as well as in a significant decrease of the production of pro-inflammatory cytokines, such as IL-6, IL-1β and TNF-α (Figure 9b).

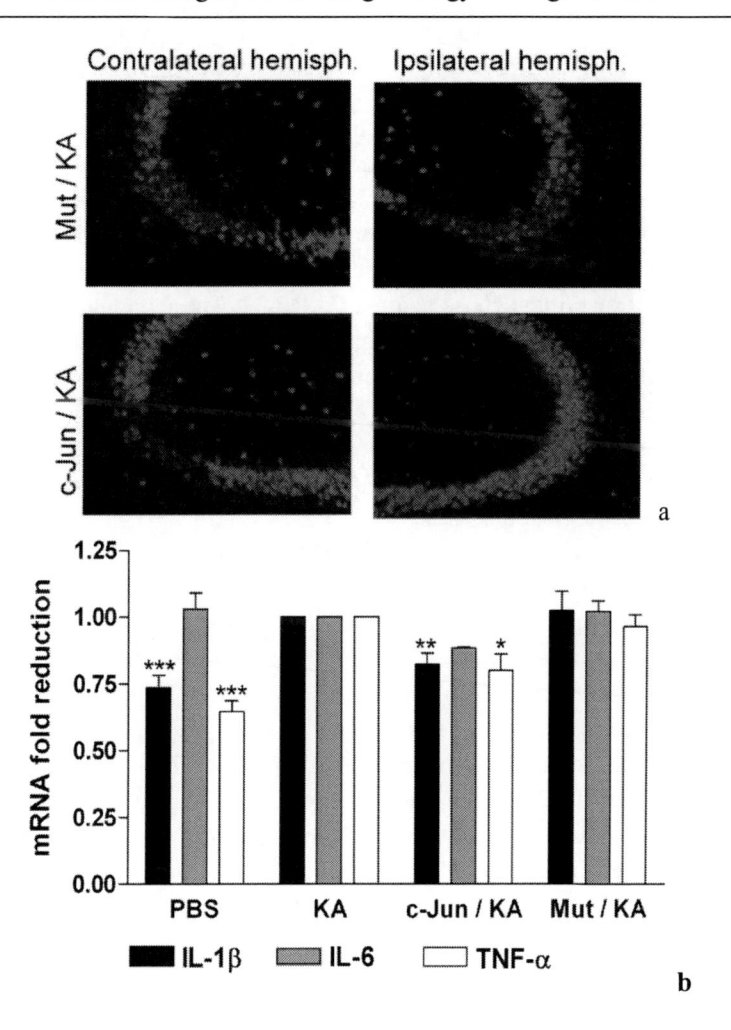

Figure 9. Neuroprotective and anti-inflammatory effect mediated by c-Jun silencing in the hippocampus. Mice were injected stereotactically, in the CA3 region of the hippocampus of the right hemisphere (ipsilateral), with Tf-lipoplexes containing anti-c-Jun siRNAs or Mut siRNAs. Immediately before Tf-lipoplex injection, 0.1 nmol of kainic acid was injected in the lateral ventricle. Animals were sacrificed 3 days postinjection. **a)** Neuronal death in the CA3 region of both hemispheres was evaluated by immunohistochemistry, using an antibody against NeuN, a specific neuronal marker. Representative fluorescence microscopy images of contralateral and ipsilateral hemispheres are shown at 200x magnification. While significant NeuN loss can be observed in the contralateral hemispheres and in the ipsilateral hemisphere injected with Mut siRNAs, no significant change of neuronal morphology or loss of NeuN labelling was observed in the ipsilateral hemisphere injected with anti-c-Jun siRNAs. **b)** mRNA levels of three pro-inflammatory cytokines (IL-1β, IL-6 and TNF-α) were assessed by QRT-PCR and are expressed as fold reduction with respect to cytokine levels in non-treated animals injected with KA. A significant decrease in mRNA levels of the studied cytokines was observed in the animals treated with anti-c-Jun siRNAs delivered by Tf-lipoplexes. Results are presented as mean values ± SD. ***$p < 0.001$, **$p < 0.01$ and *$p < 0.05$ compared to non-treated animals injected with KA.

The use of semiconductor nanocrystals, also called quantum dots (QDs) for biomedical applications, including gene delivery, has also become an area of intense research over the last decade. The surface of these particles can be easily functionalized to incorporate biological molecules and their small size makes them ideal candidates for *in vivo* application. Recently, Bonoiu et al. showed that QDs modified with a cationic polymer coating can be used to prepare QD-siRNA nanoplexes [71]. These nanoplexes were able to silence the metalloproteinase 9 (MMP-9), a protein involved in the permeabilization of the BBB and in the transmigration of inflammatory cells under neuroinflammatory conditions, such as multiple sclerosis, encephalitis, cerebral ischemia and others. Silencing of MMP-9 in brain microvascular endothelial cells resulted in an up-regulation of extracellular matrix proteins and a decrease in endothelial permeability in an *in vitro* model of the BBB.

In another study by the same authors, gold nanorods were employed to generate similar nanoplexes by association with siRNAs against DARP-32. These particles were used to target the dopaminergic signalling pathway, which has been shown to be involved in neurobiological alterations related to drug abuse [72]. Delivery of gold nanorod-DARP-32 siRNA nanoplexes *in vitro* resulted in significant reduction of DARP-32, ERK and PP-1 levels, without any signs of toxicity. Both studies seem to suggest that this kind of nanoparticles may be suited to mediate *in vivo* siRNA delivery, for treatment of drug addiction and other neurological diseases.

An important issue concerning siRNA and shRNA delivery to the brain is related with the intravenous administration of non-viral formulations and their lack of neuronal tropism. To address this problem, several groups have developed strategies to confer cell specificity to their formulations, such as antibodies against insulin and transferrin receptors, highly expressed in BBB endothelial cells and neurons [73], and peptides targeting specific neuronal receptors, such as the acetylcholine receptor [74].

Pardridge and co-workers have developed the concept of Trojan horse liposomes. These particles consist in pegylated immunoliposomes encapsulating a single molecule of DNA (encoding a gene or a shRNA) and conjugated, through the tips of the PEG strands, with a peptide or peptidomimetic antibody (MAb) capable of mediating receptor-mediated transport (RMT) across the BBB. A panel of species-specific MAbs has been developed for brain delivery targeting the mouse transferrin receptor (8D3 MAb TfR), the rat transferrin receptor (OX26 MAb TfR) or the human and primate insulin receptor (83-14 MAb HIR). These targeting MAbs bind to the BBB receptors, triggering RMT from blood to the interstitial fluid and, subsequently, bind the same receptor on brain cells to promote receptor-mediated endocytosis and expression of the gene or shRNA sequence encoded in the cargo DNA.

The delivery of shRNAs using Trojan horse liposomes was first tested in a brain tumor model using a luciferase targeted-shRNA [75]. C6 rat glioma cells stably transfected with the luciferase gene, were stereotactically implanted in the caudate-putamen nucleus of adult rats, resulting in the growth of an intracranial tumor which expressed high levels of luciferase. The Trojan horse liposomes were injected intravenously into the tumor-bearing animals resulting in an impressive 90% knockdown of luciferase expression in the tumor [75]. The same authors presented results using a similar approach to silence the epidermal growth factor receptor (EGFR), which is overexpressed in many brain cancers [73]. In this study a 90% reduction of EGFR expression was observed in a brain tumor model established in severe combined immunodeficient (SCID) mice. Moreover, a significant increase in the survival time of the animals treated weekly with the anti-EGFR shRNA encapsulated in the Trojan horse immunoliposomes was reported [73].

More recently, Kumar and colleagues showed that a short peptide derived from the rabies virus glycoprotein (RVG) was able to interact specifically with the nicotinic acethylcholine receptor (AchR) present at the surface of cholinergic neurons. To enable siRNA binding, a chimaeric peptide was synthesized by adding nine arginine residues at the carboxyl terminus of RVG (RVG-9R). The RVG-9R was able to mediate the transvascular delivery of siRNAs to several brain regions, resulting in specific and efficient gene silencing within the brain and affording robust protection against fatal viral encephalitis in mice [74].

Both the above-described strategies have shown that intravenous RNAi-based therapy in the brain can be achieved with non-viral delivery systems, providing that the RNAi effector molecules are efficiently coupled to brain specific and BBB-permeable vectors.

3.2. Viral Delivery Systems

Concerning viral vectors, which constitute the most common system for shRNA delivery, the major advantages associated with their application are the efficient target silencing and long-term, regulated expression. Probably the most promising viral vectors for shRNA delivery are the last generation of lentivirus and adeno-associated virus. Both types of viruses can be pseudotyped to increase tissue tropism, in the case of AAVs by using one of the naturally occurring viral serotypes and in the case of LVs by taking advantage of the envelope material of other known viruses. Moreover, the risk of insertional mutagenesis is reduced in both cases and both LVs and AAVs are able to transduce primary and non-dividing cells. In general, AAVs are considered to be nonpathogenic in humans and achieve persistence by forming episomal DNA molecules.

Lentiviruses are a subclass of retroviruses which can accommodate a large amount of genetic information in their genome and generate viral vectors less immunogenic than their adenoviral and retroviral counterparts. Due to their ability to persistently transduce non-dividing cells, LVs are the vectors of choice for application to the CNS. Although native viruses show a particular tropism to infect human embryonic and hematopoietic stem cells, pseudotyped viruses may present a particular tropism to the heart and skeletal muscle, liver and even brain. In 2003, Baekelandt and colleagues have shown efficient knockdown of EGFP in the adult mouse brain, following stereotactic injection of lentiviral vectors carrying a shRNA against this protein [76]. LVs have also been employed to downregulate endogenous target genes in the brain, like BACE1 [39], which is involved in the cleavage of the amyloid precursor protein and ataxin-3 [37].

The potential of AAVs for shRNA delivery was assessed in several studies. AAVs modified to present serotype 8 capsid were used to express shRNAs against hepatic targets, involving reporter genes and the hepatitis B virus [46]. Another study reported the use of AAVs with serotype 5 capsid in combination with an eye-specific promoter to deliver shRNAs to the rat retina *in vivo [47]*. Finally, AAVs with serotype 1 have been used for brain delivery of shRNAs in a model of spinocerebellar ataxia [48]. The major disadvantage of this type of vector is related to the presence of anti-AAV antibodies in the vast majority of the cell population, resulting from previous asymptomatic infections, which may hinder a second application of these vectors.

Very recently, a huge breakthrough has been achieved concerning intravenous delivery of viral vectors for gene delivery into the brain. Adeno-associated virus with serotype 9 (AAV-9) were shown to surpass the BBB following intravenous injection both in the facial and tail vein of the mouse, achieving widespread GFP expression in astrocytes and neurons and illustrating the enormous potential of this type of vectors for clinical applications involving both gene delivery and gene silencing [77]. However, although neurons are mostly non-dividing cells, which highly decreases the risk of insertional mutagenesis associated with some integrating viruses, this risk exists and has to be seriously taken into consideration before human application of viral-based therapies. Moreover, several questions have been raised recently concerning the saturation of the cellular RNAi machinery following viral shRNA delivery to the brain, with consequent deregulation of the miRNA pathway, which is particularly important in this organ.

4. RNAi as a Therapeutic Tool in Neuroscience

During the past decade, numerous mediators of neurodegeneration have been identified and validated. These molecules belong to very distinct classes, such as ion channels, neurotransmitters, neurotransmitter receptors, cytokines, growth factors, enzymes and other protein families. Although in some cases substantial pre-clinical validation exists, few new therapeutics have emerged, usually because the molecular targets are not readily amenable to small inhibitor molecules, proteins or antibodies. RNAi technology has just begun to be applied with promising results in drug development, but has already been used efficiently as a research tool to validate new targets in different disease models.

4.1. Genetic Diseases

The most natural candidates for RNAi-based therapies are dominant inherited neurodegenerative disorders, such as the familial amyotrophic lateral sclerosis (ALS) or polyglutamine expansion diseases. In these cases, deletion of the mutated gene should prevent its toxic properties and delay disease onset. Several studies have applied viral vectors encoding shRNAs to effectively silence the mutated protein causing the disease, leading to behavioural and pathological improvements. In the case of Huntington's disease (HD), Machado Joseph's disease (MJD) and spinocerebellar ataxia, a significant reduction in the number of intracellular aggregates and inclusions was observed after protein silencing [48, 78, 79].

Regarding SCA1, it was demonstrated that intracerebellar injection of recombinant adeno-associated virus (AAV) vectors expressing shRNAs resulted in a remarkable improvement of motor coordination, restored cerebellar morphology and resolved characteristic ataxin-1 inclusions in Purkinje cells of SCA1 mice [48]. In the case of HD, which results from a polyglutamine repeat expansion that causes a toxic gain of function in the protein huntingtin (htt), Harper and colleagues showed that RNAi directed against mutant human htt reduced htt mRNA and protein expression in cell culture and in the HD mouse brain[78]. Most importantly, htt gene silencing improved both behavioral and neuropathological abnormalities associated with HD [78]. In a similar study but using a non-viral delivery system, siRNAs directed against the huntingtin gene were used to repress the transgenic mutant huntingtin expression in an HD mouse model [80]. Intraventricular injection of these siRNAs at an early postnatal period inhibited transgenic huntingtin expression in brain neurons and induced a decrease in the number and size of intranuclear inclusions in striatal neurons, prolonging model mice longevity, improving motor function and slowing down the loss of body weight [80].

In a mouse model of ALS, muscle injection of a lentiviral vector mediating the expression of siRNA molecules specifically targeting the human SOD1 gene (SOD1) into various groups of mice engineered to overexpress a mutated form of human SOD1 (SOD1(G93A)), resulted in the silencing of mutated SOD1, preventing neuronal loss and improving motor performance [81]. In addition, a considerable delay in the onset of ALS symptoms by more than 100% and an increase in animal survival by nearly 80% of their normal life span were also observed [81, 82].

An important aspect to take into account is that in several genetic diseases, such as Huntington's disease and Machado Joseph's disease, the function of the disease-causing gene is not yet fully understood. In order to prevent any deleterious effect caused by the silencing of the wild-type protein, several successful strategies have been developed to suppress only the disease-causing allele, preserving the normal protein function [79]. For Machado-Joseph disease/spinocerebellar ataxia type 3, a siRNA molecule that exclusively silences the mutant allele while sparing expression of the WT allele was generated by Miller and coworkers [79]. Allele-specific suppression was achieved with all three approaches currently employed to deliver siRNA: in vitro-synthesized siRNA duplexes as well as plasmid and viral expression of short hairpin RNA [79]. More recently, Alves et al., demonstrated that a single polymorphism present in more than 70% of patients with Machado-Joseph disease could be targeted by a lentiviral vector encoding a shRNA, leading to efficient silencing of mutant ataxin-3, both *in vitro* and *in vivo,* and to a significant decrease in the severity of the neurophatological abnormalities associated with this disease[37]. Importantly, these studies established that siRNA molecules can be engineered to silence disease genes differing by a single nucleotide, highlighting the usefulness of siRNAs in dominantly inherited disorders [79].

In the case of dystonia, a non-degenerative dominant neurological illness, short-time silencing of torsin-A (TOR1A) using an allele-specific approach was proved to be sufficient to revert the disease's phenotype, without undesirable effects caused by long-term silencing of this protein, whose function is yet unknown [83]. Since it is known that torsin-A, carrying the disease-linked mutation torsinA(DeltaE), acts through a dominant-negative effect by recruiting wild-type torsin-A [torsinA(wt)] into oligomeric structures in the nuclear envelope, Gonzalez-Alegre and coworkers designed shRNAs capable of mediating allele-specific suppression of torsinA(DeltaE)[83]. Silencing of torsinA(DeltaE) using this shRNA restored the normal distribution of torsinA(wt), rescuing cells from its dominant-negative effect [83]. Moreover, using a recombinant feline immunodeficiency virus as a delivery vector, this shRNA effectively silenced torsinA(DeltaE) in a neural model of the disease, without triggering an interferon response [83].

4.2. Sporadic Diseases

The most common neurodegenerative diseases, such as Alzheimer's disease (AD) and Parkinson's disease (PD), are usually sporadic in nature, with only few inherited cases. These disorders represent a growing health problem and affect millions of individuals around the world, which makes the development of new therapies a priority. In the last decade, several mutations were identified in the few individuals carrying the genetic form of these diseases, which helped to increase our understanding of their molecular pathogenesis and revealed new potential targets for RNAi-based therapeutics. Since then, several groups have targeted the β-secretase (BACE 1) [39] responsible for the cleavage of the β-amyloid precursor protein, the mutated amyloid precursor protein carrying the Swedish double mutation (APP_{sw}) and also the mutated tau protein [84] in AD models, leading to significant reduction in the number of β-amyloid plaques and improvements in memory associated tasks.

As far as BACE 1 is concerned, it was demonstrated that lowering BACE1 levels using a lentiviral vector expressing a shRNA targeting the enzyme mRNA reduced amyloid production and behavioral deficits in APP transgenic mice [39]. In addition, an efficient method for producing siRNAs against a well-characterized tau mutation (V337M) and the most widely studied APP mutation (APPsw) was also described. The allele-specific RNA duplexes identified by this method then served as templates for constructing shRNA plasmids that successfully silenced mutant tau or APP alleles [84].

The protein α-synuclein has been the major target in studies concerning Parkinson's disease [85]. Using a dual cassette lentivirus that co-expresses an alpha-synuclein-targeting shRNA and GFP as a marker gene, Sapru and colleagues reported an effective silencing of the endogenous human alpha-synuclein gene in the human dopaminergic cell line SH-SY5Y, and also of the overexpressed human alpha-synuclein in the rat brain [85]. Silencing of this protein *in vivo* conferred resistance to a potent and selective dopaminergic neurotoxin [86]. Furthermore, the authors demonstrated for the first time that alpha-synuclein suppression decreased dopamine transport in human cells, reducing the maximal uptake velocity (V(max)) of dopamine and the surface density of its transporter by up to 50% [86].

The knockdown of the tyrosine hydroxylase enzyme, responsible for the rate-limiting step in dopamine synthesis, has also been achieved using shRNAs, leading to behavioural changes, including a motor performance deficit and reduced response to a psychostimulant, similar to those obtained with neurotoxins [30].

Overall, the above-described results clearly show the potential of RNAi technology for both therapeutic application and in the establishment of new disease models. In addition, several other neurological disorders, such as prion diseases and trauma, have been investigated in a clinical perspective using RNAi. Downregulation of the prion protein (PrP^c) was achieved with lentiviral vectors, expressing PrP^c-specific shRNAs, which resulted in a decrease of neuronal accumulation of the infectious form of this protein (PrP^{Sc}) and an increase in the survival time of animals inoculated with PrP^{Sc} [87]. Furthermore, a chimeric mouse model derived from lentiviral vector-transduced embryonic stem cells was generated, and depending on the degree of chimerism, the animals carried the lentiviral shRNAs in a certain percentage of brain cells, expressing reduced levels of PrP^c. In highly chimeric mice, there was an increase in the survival time of animals inoculated with PrP^{Sc} [87].

In a different approach to control neuronal death, silencing of several apoptotic mediators using siRNA in axotomized neurons resulted in significant increase in neuronal survival [88]. For instance, retinas injected with anti-Apaf-1- and anti-c-Jun-siRNA showed significantly more surviving retinal ganglion cells than the untreated controls.

Table 2. RNA interference in neurological disorders

	Disease	Carrier / administration route	Target	Reference
Genetic Diseases	Polyglutamine expansion diseases	AAV-shRNA	CAG repeat ataxin-1	[48]
		AAV-shRNA	Huntingtin	[78]
		siRNA-Lipofectamine	Huntingtin	[80]
		AD-shRNA (allele-specific)	MJD1	[79]
	Amyotrophic lateral sclerosis	LV-shRNA intraspinal	SOD1	[82]
		LV-shRNA intramuscular	SOD1	[81]
	Dystonia	shRNA-LV (allele specific)	TOR1A	[83]
Sporadic Diseases	Parkinson's disease	shRNA-LV	Alpha-synuclein	[85, 86]
		shRNA-LV	Tyrosine hydroxylase	[30]
	Alzheimer's disease	shRNA-LV	BACE1	[39]
		shRNA-LV	Tau, APP$_{sw}$	[84]
	Prion disease	shRNA-LV	PrP^c	[87]
	Pain	Naked siRNA intrathecal	P2X$_3$	[50]
		siRNA – PEI intrathecal	NR2B	[89]
		siRNA-cationic lipid intrathecal	Delta opioid receptor	[90]
	Glioblastoma	Ex-vivo transfected tumour cells	PTPzeta/RPTPbeta	[92]
		siRNA-PEI intratumoral	Pleiotrophin	[34]
		shRNA-Pegylated immunoliposomes	EGF receptor	[73]
	Trauma	Naked siRNA local injection	c-Jun, BAX, Apaf-1	[88]
	Encephalitis	Naked siRNA Hydrodynamic injection	Envelope protein - West Nile Virus	[93]
		shRNA intraperitoneal	M-protein - Japanese encephalitis virus	[94]
		siRNA-peptide intravenous	Japanese encephalitis virus	[74]
	Depression	Naked siRNA intraventricular	Serotonin transporter	[91]

The control of neurophatic pain has been another major area of research in the last decade. The silencing of ion channels and other receptors involved in the modulation of pain sensation, such as the $P2X_3$ ATP-gated cation channel [50], the NR2B subunit of NMDA receptors [89] and the delta opioid receptor (DOR)[90], has been tested through the intrathecal administration of naked siRNAs, with consequent reduction of pain-related behaviour in rats. Regarding P2X3, molecular analysis of tissues from animals receiving P2X3 siRNA revealed that P2X3 mRNA expressed in dorsal root ganglia and P2X3 protein translocated into the dorsal horn of the spinal cord were significantly diminished [50]. The use of siRNA targeting the NR2B subunit not only decreased the expression of NR2B mRNA and its associated protein, but also abolished formalin-induced pain behaviours in rat model [89]. In the case of DOR, a highly effective method for *in vivo* DOR silencing in the spinal cord and dorsal root ganglia by a cationic lipid was developed by Luo et al. The low effective dose of siRNA/i-Fect complex reflected an efficient delivery of the siRNA to peripheral and spinal neurons and produced no behavioral signs of toxicity [50].

Finally, neuropsychiatric disorders, such as depression, which constitute another important problem of modern societies, are also currently being studied using RNAi as a tool. Downregulation of the serotonin transporter (SERT) using an intraventricular infusion of naked siRNAs resulted in an antidepressant-related behaviour in mice, comparable to that obtained with the serotonin reuptake inhibitor citalopran [91]. Infusing the SERT-targeting siRNA significantly reduced the mRNA levels of SERT in raphe nuclei and resulted in a significant, specific and widespread downregulation of SERT-binding sites in the brain. In contrast, infusion of citalopran produced a widespread downregulation of SERT-binding sites, which was found to be independent of any alterations at the mRNA level.

At present, all available treatments to PD, AD and other neurological pathologies are limited to the improvement of the symptoms of the disease, but do not provide a true cure for these pathologies. The above-mentioned studies illustrate the potential of RNAi therapeutics to halt or delay the process of neuronal death at its origin, thus providing a novel approach for the treatment of disorders of the CNS.

Table 2 summarizes the most important studies concerning siRNA and shRNA application in the context of neurodegenerative diseases and other neurological disorders.

Conclusion

Despite being in its infancy, RNA interference is already considered an essential tool for the neuroscientists. RNAi can help to elucidate which are the most relevant proteins to the progression in the cell death program following excitotoxic injury, which may represent potential targets for the development of new neuroprotective therapies. Furthermore, the efficient silencing of these proteins using siRNA or shRNAs may constitute by itself a promising approach for the treatment of several neurological disorders, if the current limitations associated with *in vivo* delivery of these molecules can be overcome. The different properties of non-viral and viral vectors suggest that the design of an universal vector may be unrealistic. Instead, the tailoring of a vector capable of addressing the specific needs of a particular pathology may be the solution for a successful transition of RNAi from bench to the clinic.

References

[1]   D.J. Begley, (2004) Delivery of therapeutic agents to the central nervous system: the problems and the possibilities, *Pharmacology & therapeutics* 104 29-45.

[2]   A. Fire, S. Xu, M.K. Montgomery, S.A. Kostas, S.E. Driver, C.C. Mello, (1998) Potent and specific genetic interference by double-stranded RNA in Caenorhabditis elegans, *Nature* 391 806-811.

[3]   A.J. Hamilton, D.C. Baulcombe, (1999) A species of small antisense RNA in posttranscriptional gene silencing in plants, *Science* (New York, N.Y 286 950-952.

[4]   D.D. Yang, C.Y. Kuan, A.J. Whitmarsh, M. Rincon, T.S. Zheng, R.J. Davis, P. Rakic, R.A. Flavell, (1997) Absence of excitotoxicity-induced apoptosis in the hippocampus of mice lacking the Jnk3 gene, *Nature* 389 865-870.

[5]   P.D. Zamore, T. Tuschl, P.A. Sharp, D.P. Bartel, (2000) RNAi: double-stranded RNA directs the ATP-dependent cleavage of mRNA at 21 to 23 nucleotide intervals, *Cell* 101 25-33.

[6]   P.H. Olsen, V. Ambros, (1999) The lin-4 regulatory RNA controls developmental timing in Caenorhabditis elegans by blocking LIN-14 protein synthesis after the initiation of translation, *Developmental biology* 216 671-680.

[7]   M.F. Mette, W. Aufsatz, J. van der Winden, M.A. Matzke, A.J. Matzke, (2000) Transcriptional silencing and promoter methylation triggered by double-stranded RNA, *The EMBO journal* 19 5194-5201.

[8]   K.V. Morris, S.W. Chan, S.E. Jacobsen, D.J. Looney, (2004) Small interfering RNA-induced transcriptional gene silencing in human cells, *Science* (New York, N.Y 305 1289-1292.

[9]   K. Mochizuki, N.A. Fine, T. Fujisawa, M.A. Gorovsky, (2002) Analysis of a piwi-related gene implicates small RNAs in genome rearrangement in tetrahymena, *Cell* 110 689-699.

[10]   A. Verdel, S. Jia, S. Gerber, T. Sugiyama, S. Gygi, S.I. Grewal, D. Moazed, (2004) RNAi-mediated targeting of heterochromatin by the RITS complex, *Science* (New York, N.Y 303 672-676.

[11]   B.R. Cullen, (2004) Transcription and processing of human microRNA precursors, *Molecular cell* 16 861-865.

[12]   Y. Lee, C. Ahn, J. Han, H. Choi, J. Kim, J. Yim, J. Lee, P. Provost, O. Radmark, S. Kim, V.N. Kim, (2003) The nuclear RNase III Drosha initiates microRNA processing, *Nature* 425 415-419.

[13]   E. Lund, S. Guttinger, A. Calado, J.E. Dahlberg, U. Kutay, (2004) Nuclear export of microRNA precursors, *Science* (New York, N.Y 303 95-98.

[14]   R. Yi, Y. Qin, I.G. Macara, B.R. Cullen, (2003) Exportin-5 mediates the nuclear export of pre-microRNAs and short hairpin RNAs, *Genes & development* 17 3011-3016.

[15]   E. Bernstein, A.A. Caudy, S.M. Hammond, G.J. Hannon, (2001) Role for a bidentate ribonuclease in the initiation step of RNA interference, *Nature* 409 363-366.

[16]   H. Gong, C.M. Liu, D.P. Liu, C.C. Liang, (2005) The role of small RNAs in human diseases: potential troublemaker and therapeutic tools, *Medicinal research reviews* 25 361-381.

[17]   C.D. Novina, P.A. Sharp, (2004) The RNAi revolution, *Nature* 430 161-164.

[18] S.M. Hammond, E. Bernstein, D. Beach, G.J. Hannon, (2000) An RNA-directed nuclease mediates post-transcriptional gene silencing in Drosophila cells, *Nature* 404 293-296.

[19] S.M. Elbashir, W. Lendeckel, T. Tuschl, (2001) RNA interference is mediated by 21- and 22-nucleotide RNAs, *Genes & development* 15 188-200.

[20] J. Martinez, A. Patkaniowska, H. Urlaub, R. Luhrmann, T. Tuschl, (2002) Single-stranded antisense siRNAs guide target RNA cleavage in RNAi, *Cell* 110 563-574.

[21] N.J. Caplen, S. Parrish, F. Imani, A. Fire, R.A. Morgan, (2001) Specific inhibition of gene expression by small double-stranded RNAs in invertebrate and vertebrate systems, *Proceedings of the National Academy of Sciences of the United States of America* 98 9742-9747.

[22] S.M. Elbashir, J. Harborth, W. Lendeckel, A. Yalcin, K. Weber, T. Tuschl, (2001) Duplexes of 21-nucleotide RNAs mediate RNA interference in cultured mammalian cells, *Nature* 411 494-498.

[23] J.Y. Yu, S.L. DeRuiter, D.L. Turner, (2002) RNA interference by expression of short-interfering RNAs and hairpin RNAs in mammalian cells, *Proceedings of the National Academy of Sciences of the United States of America* 99 6047-6052.

[24] M. Miyagishi, H. Sumimoto, H. Miyoshi, Y. Kawakami, K. Taira, (2004) Optimization of an siRNA-expression system with an improved hairpin and its significant suppressive effects in mammalian cells, *The journal of gene medicine* 6 715-723.

[25] M. Miyagishi, K. Taira, (2002) Development and application of siRNA expression vector, *Nucleic acids research* 113-114.

[26] X. Zhu, L.A. Santat, M.S. Chang, J. Liu, J.R. Zavzavadjian, E.A. Wall, C. Kivork, M.I. Simon, I.D. Fraser, (2007) A versatile approach to multiple gene RNA interference using microRNA-based short hairpin RNAs, *BMC molecular biology* 8 98.

[27] A.M. Krichevsky, K.S. Kosik, (2002) RNAi functions in cultured mammalian neurons, *Proceedings of the National Academy of Sciences of the United States of America* 99 11926-11929.

[28] H. Makimura, T.M. Mizuno, J.W. Mastaitis, R. Agami, C.V. Mobbs, (2002) Reducing hypothalamic AGRP by RNA interference increases metabolic rate and decreases body weight without influencing food intake, *BMC neuroscience* 3 18.

[29] T.R. Brummelkamp, R. Bernards, R. Agami, (2002) A system for stable expression of short interfering RNAs in mammalian cells, *Science* (New York, N.Y 296 550-553.

[30] J.D. Hommel, R.M. Sears, D. Georgescu, D.L. Simmons, R.J. DiLeone, (2003) Local gene knockdown in the brain using viral-mediated RNA interference, *Nature medicine* 9 1539-1544.

[31] J. Ohkawa, K. Taira, (2000) Control of the functional activity of an antisense RNA by a tetracycline-responsive derivative of the human U6 snRNA promoter, *Human gene therapy* 11 577-585.

[32] A. Santel, M. Aleku, O. Keil, J. Endruschat, V. Esche, G. Fisch, S. Dames, K. Loffler, M. Fechtner, W. Arnold, K. Giese, A. Klippel, J. Kaufmann, (2006) A novel siRNA-lipoplex technology for RNA interference in the mouse vascular endothelium, *Gene therapy* 13 1222-1234.

[33] S. Zhang, B. Zhao, H. Jiang, B. Wang, B. Ma, (2007) Cationic lipids and polymers mediated vectors for delivery of siRNA, *J Control Release* 123 1-10.

[34] M. Grzelinski, B. Urban-Klein, T. Martens, K. Lamszus, U. Bakowsky, S. Hobel, F. Czubayko, A. Aigner, (2006) RNA interference-mediated gene silencing of pleiotrophin through polyethylenimine-complexed small interfering RNAs in vivo exerts antitumoral effects in glioblastoma xenografts, *Human gene therapy* 17 751-766.

[35] B.R. Meade, S.F. Dowdy, (2007) Exogenous siRNA delivery using peptide transduction domains/cell penetrating peptides, *Advanced drug delivery reviews* 59 134-140.

[36] T.R. Brummelkamp, R. Bernards, R. Agami, (2002) Stable suppression of tumorigenicity by virus-mediated RNA interference, *Cancer cell* 2 243-247.

[37] S. Alves, I. Nascimento-Ferreira, G. Auregan, R. Hassig, N. Dufour, E. Brouillet, M.C. Pedroso de Lima, P. Hantraye, L. Pereira de Almeida, N. Deglon, (2008) Allele-specific RNA silencing of mutant ataxin-3 mediates neuroprotection in a rat model of Machado-Joseph disease, *PLoS One* 3 e3341.

[38] M. Li, H. Li, J.J. Rossi, (2006) RNAi in combination with a ribozyme and TAR decoy for treatment of HIV infection in hematopoietic cell gene therapy, *Annals of the New York Academy of Sciences* 1082 172-179.

[39] O. Singer, R.A. Marr, E. Rockenstein, L. Crews, N.G. Coufal, F.H. Gage, I.M. Verma, E. Masliah, (2005) Targeting BACE1 with siRNAs ameliorates Alzheimer disease neuropathology in a transgenic model, *Nature neuroscience* 8 1343-1349.

[40] Y. Chen, H. Chen, A. Hoffmann, D.R. Cool, D.I. Diz, M.C. Chappell, A.F. Chen, M. Morris, (2006) Adenovirus-mediated small-interference RNA for in vivo silencing of angiotensin AT1a receptors in mouse brain, *Hypertension* 47 230-237.

[41] H. Li, X. Fu, Y. Chen, Y. Hong, Y. Tan, H. Cao, M. Wu, H. Wang, (2005) Use of adenovirus-delivered siRNA to target oncoprotein p28GANK in hepatocellular carcinoma, *Gastroenterology* 128 2029-2041.

[42] H. Osada, Y. Tatematsu, Y. Yatabe, Y. Horio, T. Takahashi, (2005) ASH1 gene is a specific therapeutic target for lung cancers with neuroendocrine features, *Cancer research* 65 10680-10685.

[43] S. Ragozin, A. Niemeier, A. Laatsch, B. Loeffler, M. Merkel, U. Beisiegel, J. Heeren, (2005) Knockdown of hepatic ABCA1 by RNA interference decreases plasma HDL cholesterol levels and influences postprandial lipemia in mice, *Arteriosclerosis, thrombosis, and vascular biology* 25 1433-1438.

[44] H. Xia, Q. Mao, H.L. Paulson, B.L. Davidson, (2002) siRNA-mediated gene silencing in vitro and in vivo, *Nature biotechnology* 20 1006-1010.

[45] D. Boden, O. Pusch, F. Lee, L. Tucker, B. Ramratnam, (2004) Efficient gene transfer of HIV-1-specific short hairpin RNA into human lymphocytic cells using recombinant adeno-associated virus vectors, *Mol Ther* 9 396-402.

[46] C.C. Chen, T.M. Ko, H.I. Ma, H.L. Wu, X. Xiao, J. Li, C.M. Chang, P.Y. Wu, C.H. Chen, J.M. Han, C.P. Yu, K.S. Jeng, C.P. Hu, M.H. Tao, (2007) Long-term inhibition of hepatitis B virus in transgenic mice by double-stranded adeno-associated virus 8-delivered short hairpin RNA, *Gene therapy* 14 11-19.

[47] D.M. Paskowitz, K.P. Greenberg, D. Yasumura, D. Grimm, H. Yang, J.L. Duncan, M.A. Kay, M.M. Lavail, J.G. Flannery, D. Vollrath, (2007) Rapid and stable knockdown of an endogenous gene in retinal pigment epithelium, *Human gene therapy* 18 871-880.

[48] H. Xia, Q. Mao, S.L. Eliason, S.Q. Harper, I.H. Martins, H.T. Orr, H.L. Paulson, L. Yang, R.M. Kotin, B.L. Davidson, (2004) RNAi suppresses polyglutamine-induced neurodegeneration in a model of spinocerebellar ataxia, Nature medicine 10 816-820.

[49] A.R. de Fougerolles, (2008) Delivery vehicles for small interfering RNA in vivo, *Human gene therapy* 19 125-132.

[50] G. Dorn, S. Patel, G. Wotherspoon, M. Hemmings-Mieszczak, J. Barclay, F.J. Natt, P. Martin, S. Bevan, A. Fox, P. Ganju, W. Wishart, J. Hall, (2004) siRNA relieves chronic neuropathic pain, *Nucleic Acids Res* 32 e49.

[51] D.R. Thakker, F. Natt, D. Husken, R. Maier, M. Muller, H. van der Putten, D. Hoyer, J.F. Cryan, (2004) Neurochemical and behavioral consequences of widespread gene knockdown in the adult mouse brain by using nonviral RNA interference, *Proceedings of the National Academy of Sciences of the United States of America* 101 17270-17275.

[52] P. Hamar, E. Song, G. Kokeny, A. Chen, N. Ouyang, J. Lieberman, (2004) Small interfering RNA targeting Fas protects mice against renal ischemia-reperfusion injury, *Proceedings of the National Academy of Sciences of the United States of America* 101 14883-14888.

[53] D.L. Lewis, J.E. Hagstrom, A.G. Loomis, J.A. Wolff, H. Herweijer, (2002) Efficient delivery of siRNA for inhibition of gene expression in postnatal mice, *Nature genetics* 32 107-108.

[54] L. Zender, S. Hutker, C. Liedtke, H.L. Tillmann, S. Zender, B. Mundt, M. Waltemathe, T. Gosling, P. Flemming, N.P. Malek, C. Trautwein, M.P. Manns, F. Kuhnel, S. Kubicka, (2003) Caspase 8 small interfering RNA prevents acute liver failure in mice, *Proceedings of the National Academy of Sciences of the United States of America* 100 7797-7802.

[55] Y. Akaneya, B. Jiang, T. Tsumoto, (2005) RNAi-induced gene silencing by local electroporation in targeting brain region, Journal of neurophysiology 93 594-602.

[56] M. Golzio, L. Mazzolini, P. Moller, M.P. Rols, J. Teissie, (2005) Inhibition of gene expression in mice muscle by in vivo electrically mediated siRNA delivery, *Gene therapy* 12 246-251.

[57] T. Kishida, H. Asada, S. Gojo, S. Ohashi, M. Shin-Ya, K. Yasutomi, R. Terauchi, K.A. Takahashi, T. Kubo, J. Imanishi, O. Mazda, (2004) Sequence-specific gene silencing in murine muscle induced by electroporation-mediated transfer of short interfering RNA, *The journal of gene medicine* 6 105-110.

[58] B. Kim, Q. Tang, P.S. Biswas, J. Xu, R.M. Schiffelers, F.Y. Xie, A.M. Ansari, P.V. Scaria, M.C. Woodle, P. Lu, B.T. Rouse, (2004) Inhibition of ocular angiogenesis by siRNA targeting vascular endothelial growth factor pathway genes: therapeutic strategy for herpetic stromal keratitis, *The American journal of pathology* 165 2177-2185.

[59] Y. Takabatake, Y. Isaka, M. Mizui, H. Kawachi, F. Shimizu, T. Ito, M. Hori, E. Imai, (2005) Exploring RNA interference as a therapeutic strategy for renal disease, *Gene therapy* 12 965-973.

[60] Y. Takahashi, M. Nishikawa, N. Kobayashi, Y. Takakura, (2005) Gene silencing in primary and metastatic tumors by small interfering RNA delivery in mice: quantitative analysis using melanoma cells expressing firefly and sea pansy luciferases, *J Control Release* 105 332-343.

[61] J. Soutschek, A. Akinc, B. Bramlage, K. Charisse, R. Constien, M. Donoghue, S. Elbashir, A. Geick, P. Hadwiger, J. Harborth, M. John, V. Kesavan, G. Lavine, R.K.

Pandey, T. Racie, K.G. Rajeev, I. Rohl, I. Toudjarska, G. Wang, S. Wuschko, D. Bumcrot, V. Koteliansky, S. Limmer, M. Manoharan, H.P. Vornlocher, (2004) Therapeutic silencing of an endogenous gene by systemic administration of modified siRNAs, *Nature* 432 173-178.

[62] M. Sioud, D.R. Sorensen, (2003) Cationic liposome-mediated delivery of siRNAs in adult mice, *Biochemical and biophysical research communications* 312 1220-1225.

[63] D.R. Sorensen, M. Leirdal, M. Sioud, (2003) Gene silencing by systemic delivery of synthetic siRNAs in adult mice, *Journal of molecular biology* 327 761-766.

[64] J. Yano, K. Hirabayashi, S. Nakagawa, T. Yamaguchi, M. Nogawa, I. Kashimori, H. Naito, H. Kitagawa, K. Ishiyama, T. Ohgi, T. Irimura, (2004) Antitumor activity of small interfering RNA/cationic liposome complex in mouse models of cancer, *Clin Cancer Res* 10 7721-7726.

[65] M. Thomas, J.J. Lu, Q. Ge, C. Zhang, J. Chen, A.M. Klibanov, (2005) Full deacylation of polyethylenimine dramatically boosts its gene delivery efficiency and specificity to mouse lung, *Proceedings of the National Academy of Sciences of the United States of America* 102 5679-5684.

[66] A.L. Cardoso, P. Costa, L.P. de Almeida, S. Simoes, N. Plesnila, C. Culmsee, E. Wagner, M.C. Pedroso de Lima, (2009) Tf-lipoplex-mediated c-Jun silencing improves neuronal survival following excitotoxic damage in vivo, *J Control Release.*

[67] A.L. Cardoso, S. Simoes, L.P. de Almeida, J. Pelisek, C. Culmsee, E. Wagner, M.C. Pedroso de Lima, (2007) siRNA delivery by a transferrin-associated lipid-based vector: a non-viral strategy to mediate gene silencing, *The journal of gene medicine* 9 170-183.

[68] A.L. Cardoso, S. Simoes, L.P. de Almeida, N. Plesnila, M.C. Pedroso de Lima, E. Wagner, C. Culmsee, (2008) Tf-lipoplexes for neuronal siRNA delivery: a promising system to mediate gene silencing in the CNS, *J Control Release* 132 113-123.

[69] R.W. Malone, P.L. Felgner, I.M. Verma, (1989) Cationic liposome-mediated RNA transfection, *Proceedings of the National Academy of Sciences of the United States of America* 86 6077-6081.

[70] S. Simoes, A. Filipe, H. Faneca, M. Mano, N. Penacho, N. Duzgunes, M.P. de Lima, (2005) Cationic liposomes for gene delivery, *Expert Opin Drug Deliv* 2 237-254.

[71] A. Bonoiu, S.D. Mahajan, L. Ye, R. Kumar, H. Ding, K.T. Yong, I. Roy, R. Aalinkeel, B. Nair, J.L. Reynolds, D.E. Sykes, M.A. Imperiale, E.J. Bergey, S.A. Schwartz, P.N. Prasad, (2009) MMP-9 gene silencing by a quantum dot-siRNA nanoplex delivery to maintain the integrity of the blood brain barrier, *Brain Res* 1282 142-155.

[72] A.C. Bonoiu, S.D. Mahajan, H. Ding, I. Roy, K.T. Yong, R. Kumar, R. Hu, E.J. Bergey, S.A. Schwartz, P.N. Prasad, (2009) Nanotechnology approach for drug addiction therapy: gene silencing using delivery of gold nanorod-siRNA nanoplex in dopaminergic neurons, *Proceedings of the National Academy of Sciences of the United States of America* 106 5546-5550.

[73] Y. Zhang, Y.F. Zhang, J. Bryant, A. Charles, R.J. Boado, W.M. Pardridge, (2004) Intravenous RNA interference gene therapy targeting the human epidermal growth factor receptor prolongs survival in intracranial brain cancer, *Clin Cancer Res* 10 3667-3677.

[74] P. Kumar, H. Wu, J.L. McBride, K.E. Jung, M.H. Kim, B.L. Davidson, S.K. Lee, P. Shankar, N. Manjunath, (2007) Transvascular delivery of small interfering RNA to the central nervous system, *Nature* 448 39-43.

[75] Y. Zhang, R.J. Boado, W.M. Pardridge, (2003) In vivo knockdown of gene expression in brain cancer with intravenous RNAi in adult rats, *The journal of gene medicine 5* 1039-1045.

[76] C. Van den Haute, K. Eggermont, B. Nuttin, Z. Debyser, V. Baekelandt, (2003) Lentiviral vector-mediated delivery of short hairpin RNA results in persistent knockdown of gene expression in mouse brain, *Human gene therapy 14* 1799-1807.

[77] K.D. Foust, E. Nurre, C.L. Montgomery, A. Hernandez, C.M. Chan, B.K. Kaspar, (2009) Intravascular AAV9 preferentially targets neonatal neurons and adult astrocytes, *Nature biotechnology 27* 59-65.

[78] S.Q. Harper, P.D. Staber, X. He, S.L. Eliason, I.H. Martins, Q. Mao, L. Yang, R.M. Kotin, H.L. Paulson, B.L. Davidson, (2005) RNA interference improves motor and neuropathological abnormalities in a Huntington's disease mouse model, *Proceedings of the National Academy of Sciences of the United States of America* 102 5820-5825.

[79] V.M. Miller, H. Xia, G.L. Marrs, C.M. Gouvion, G. Lee, B.L. Davidson, H.L. Paulson, (2003) Allele-specific silencing of dominant disease genes, *Proceedings of the National Academy of Sciences of the United States of America* 100 7195-7200.

[80] Y.L. Wang, W. Liu, E. Wada, M. Murata, K. Wada, I. Kanazawa, (2005) Clinico-pathological rescue of a model mouse of Huntington's disease by siRNA, *Neuroscience research 53 241-249.*

[81] G.S. Ralph, P.A. Radcliffe, D.M. Day, J.M. Carthy, M.A. Leroux, D.C. Lee, L.F. Wong, L.G. Bilsland, L. Greensmith, S.M. Kingsman, K.A. Mitrophanous, N.D. Mazarakis, M. Azzouz, (2005) Silencing mutant SOD1 using RNAi protects against neurodegeneration and extends survival in an ALS model, *Nature medicine* 11 429-433.

[82] C. Raoul, T. Abbas-Terki, J.C. Bensadoun, S. Guillot, G. Haase, J. Szulc, C.E. Henderson, P. Aebischer, (2005) Lentiviral-mediated silencing of SOD1 through RNA interference retards disease onset and progression in a mouse model of ALS, *Nature medicine* 11 423-428.

[83] P. Gonzalez-Alegre, N. Bode, B.L. Davidson, H.L. Paulson, (2005) Silencing primary dystonia: lentiviral-mediated RNA interference therapy for DYT1 dystonia, *J Neurosci* 25 10502-10509.

[84] V.M. Miller, C.M. Gouvion, B.L. Davidson, H.L. Paulson, (2004) Targeting Alzheimer's disease genes with RNA interference: an efficient strategy for silencing mutant alleles, *Nucleic Acids Res* 32 661-668.

[85] M.K. Sapru, J.W. Yates, S. Hogan, L. Jiang, J. Halter, M.C. Bohn, (2006) Silencing of human alpha-synuclein in vitro and in rat brain using lentiviral-mediated RNAi, *Experimental neurology* 198 382-390.

[86] T.M. Fountaine, R. Wade-Martins, (2007) RNA interference-mediated knockdown of alpha-synuclein protects human dopaminergic neuroblastoma cells from MPP(+) toxicity and reduces dopamine transport, *Journal of neuroscience research* 85 351-363.

[87] A. Pfeifer, S. Eigenbrod, S. Al-Khadra, A. Hofmann, G. Mitteregger, M. Moser, U. Bertsch, H. Kretzschmar, (2006) Lentivector-mediated RNAi efficiently suppresses prion protein and prolongs survival of scrapie-infected mice, *The Journal of clinical investigation* 116 3204-3210.

[88] P. Lingor, P. Koeberle, S. Kugler, M. Bahr, (2005) Down-regulation of apoptosis mediators by RNAi inhibits axotomy-induced retinal ganglion cell death in vivo, *Brain* 128 550-558.

[89] P.H. Tan, L.C. Yang, H.C. Shih, K.C. Lan, J.T. Cheng, (2005) Gene knockdown with intrathecal siRNA of NMDA receptor NR2B subunit reduces formalin-induced nociception in the rat, *Gene therapy* 12 59-66.

[90] M.C. Luo, D.Q. Zhang, S.W. Ma, Y.Y. Huang, S.J. Shuster, F. Porreca, J. Lai, (2005) An efficient intrathecal delivery of small interfering RNA to the spinal cord and peripheral neurons, *Molecular pain* 1 29.

[91] D.R. Thakker, F. Natt, D. Husken, H. van der Putten, R. Maier, D. Hoyer, J.F. Cryan, (2005) siRNA-mediated knockdown of the serotonin transporter in the adult mouse brain, *Molecular psychiatry* 10 782-789, 714.

[92] U. Ulbricht, C. Eckerich, R. Fillbrandt, M. Westphal, K. Lamszus, (2006) RNA interference targeting protein tyrosine phosphatase zeta/receptor-type protein tyrosine phosphatase beta suppresses glioblastoma growth in vitro and in vivo, *Journal of neurochemistry* 98 1497-1506.

[93] F. Bai, T. Wang, U. Pal, F. Bao, L.H. Gould, E. Fikrig, (2005) Use of RNA interference to prevent lethal murine west nile virus infection, *The Journal of infectious diseases* 191 1148-1154.

[94] M. Murakami, T. Ota, S. Nukuzuma, T. Takegami, (2005) Inhibitory effect of RNAi on Japanese encephalitis virus replication in vitro and in vivo, *Microbiology and immunology* 49 1047-1056.

In: Gene Silencing: Theory, Techniques and Applications
Editor: Anthony J.Catalano

ISBN: 978-1-61728-276-8
© 2010 Nova Science Publishers, Inc.

Chapter IV

Polymeric Nanoparticles as Efficient siRNA Delivery Systems

Surendra Nimesh[1,*], Nidhi Gupta[1] and Ramesh Chandra[2,3]

[1]Laboratory of Biochemical Neuroendocrinology, Clinical Research Institute of Montreal, Montreal, Quebec, Canada

[2]Dr. B. R. Ambedkar Center for Biomedical Research, University of Delhi, Delhi -110007, India

[3]Department of Chemistry, University of Delhi, Delhi -110007, India

Abstract

Small interfering RNAs (siRNA) are emerging as promising therapeutic agents for the treatment of inherited and acquired diseases, as well as research tools for the elucidation of gene function. Since the molecules undergo rapid enzymatic degradation and have poor cellular uptake, there is a need to design a delivery system which can protect and efficiently transport siRNA to the target cells. Polymeric nanoparticles have emerged as systems of choice with reduced cytotoxicity and enhanced efficacy. These systems not only protect siRNA from enzymatic degradation by forming condensed complexes but also leads to tissue and cellular targeting, improve cellular penetration, release the siRNA in the right intracellular compartment. Nanoparticles prepared from polycationic polymers like polyethylenimine, chitosan have been widely investigated due to ease of manipulatibility, stability, low immunogenicity, low cost and high flexibility regarding the size of transgene delivered. This chapter presents an overview of siRNA delivery strategies employing polymeric nanoparticles, with emphasis on self-assembled polymeric nanoparticles with promising potential to evolve as therapeutic tool in gene therapy.

[*] Corresponding author. Laboratory of Biochemical Neuroendocrinology, Clinical Research Institute of Montreal, 110 West Pine Ave, Montreal, Quebec, Canada H2W 1R7. Tel: 514-987-5559. Fax: 514-987-5542. E-mail address: surendranimesh@gmail.com

1. Introduction

With the advent of genomic era new drug targets have been identified, posing a challenging task ahead for the development of newer drugs. The rapid progress in biotechnology has led to the development of antibodies, low molecular weight pharmacological drugs and gene-targeting techniques as newer alternate to the existing drugs for curing several deadly diseases such as cancer. Over the past few decades, the development of newer drugs has included the use of antisense oligonucleotides, ribozyme and most recently, RNA interference (RNAi). The antisense oligonucleotides (ODN) were first introduced in 1978 for specific inhibition of gene expression [1,2] and further carried forward since the mid 1980s when certain backbone modifications introduced in the ODN to increase the activity and stability. Later, the techniques for condensing and electro-neutralizing the net negative charge of ODN compounds and delivery to cells in culture were developed. The cationic and anionic lipid formulations have been widely used to condense and deliver ODN, followed by the development of cationic polymers and later nanoparticles. Although failure of some of the recent randomized trials have led to the reduction in interest for antisense ODN based therapeutics, several antisense ODN compounds are still in late-stage preclinical or clinical development [3].

In the early 1980s, it was discovered that ribozymes or small catalytically active RNAs not only have role in transfer of genetic information but can also act as enzymes independent of the presence of proteins [4-6]. The mechanism of action of ribozymes consists of breakdown of covalent bonds in RNA molecules with high sequence specificity it was considered as one of the most important step towards the development of nucleic acid based gene targeting. However, the therapeutic use of ribozymes, suffers from several problems such as poor delivery into the target tissue/target cell as well as their pronounced instability. It has been well established that the RNA ribozymes are stable for less than 6s in human serum due to their rapid degradation [7]. In order to improve upon these disadvantages several chemical modifications into the ribozyme backbone or the delivery of plasmid or viral vectors with subsequent intracellular expression of ribozymes have been carried out. Though ribozymes have been used for a number of *in vitro* studies for target specific downregulation of gene expression, they have very limited *in vivo* and therapeutic applications.

Recently, RNA interference (RNAi) has been proposed as one of the most potent, naturally occurring biological mechanism for gene silencing. The short stretches of 21 to 23 nucleotides double stranded RNAs, which bring about the silencing of genes, are referred to as siRNAs (small interfering RNAs). These siRNAs are capable of degrading mRNAs that are complementary to one of the siRNA strands.In late 1980s, RNAi was first observed by plant biologists [8], but its molecular mechanism was not established until the late 1990s, when studies on the nematode *C. elegans* showed that RNAi is an evolutionary conserved gene-silencing mechanism [9-11], serving as a natural defence mechanism against viral pathogens or uncontrolled transposons mobilization [12]. Later, the sequence-specific posttranscriptional gene silencing by double-stranded RNA was observed and experimentally demonstrated in several organisms: plants, Neurospora, Drosophila, *C. elegans*, and mammals [13]. In 1998, it was Fire *et al.* who identified the double-stranded RNA molecules (dsRNA) as the mediator of gene silencing in *C. elegans*. These observations also elaborated the earlier studies on the same organism on gene silencing employing antisense as well as sense ODNs [14]. On comparing the two strategies i.e. antisense and RNAi it was observed that small interfering RNA (siRNA) was quantitatively more efficient and more stable in cell culture [15,16]. One of the major advantages of RNAi above antisense oligonucleotides is the well observed fact that the siRNA is based on a catalytical mechanism. Thus, in case of siRNA only the catalytical amounts were needed rather than the stoichiometrical amounts. Finally, it was observed that RNAi is not restricted to any specific organism but rather, was initially described under different names (post-transcriptional gene silencing i.e. PTGS, quelling, co-suppression), appears to be present in almost all eukaryotic organisms.

Nanotechnology is one of the newer branch of science which deals with the design, preparation and application of functional structures in the nanometer range (usually 100 nm or smaller). When the bulk materials are engineered into nanometer scale particles, the properties of particles often differ from those of the corresponding bulk materials. The fundamental characteristics of a given material such as melting point, magnetic properties, or even color can be precisely controlled by nanotechnology without changing its chemical composition. Till date nanotechnology has found numerous applications, of which, diagnosis, treatment, monitoring and control of biological systems is of prime importance. Nanotechnology may consist of a number of different types of nanodevices, including nanoparticles, nanomachines, nanofibers, nanosensors, and other nanoscale microfabrication based entities [17].

Nanoparticles have evolved as one of the most promising candidate of nanotechnology, which has made rapid progress as systems for targeted drug and gene delivery. Nanoparticles have been successful in delivering drugs to specific tissues within the body with greater potential to achieve better drug penetration into cells and also to improve drug efficacy. It is a well know fact that the surface machinery and intracellular organelles of the cell operate at the nano level: they regulate the actions of messenger molecules, maintain ionic stability, and manufacture a wide variety of crucial building blocks for the body. Biologically potent molecules such as sugars, nucleic acids, and peptides (1-10 nm) 'dock' into larger nano sized structures (10-100 nm) to mediate specific functions, or are processed further through the active sites of receptors and enzymes. Hence, nanoparticles can be considered as ideal entities for interacting on the same scale and enabling effective and selective therapeutics.

Nanoparticles can be broadly divided into two categories, namely, metallic and polymeric nanoparticles. Till date various therapeutically important metallic nanoparticles developed includes gold nanoshells, iron oxide nanocrystals, and quantum dots [18-20]. During past few decades, polymeric nanoparticles have emerged as systems of choice with reduced cytotoxicity and enhanced efficacy. These systems not only target the drug to its site of action but also maintain the drug concentrations at therapeutically relevant levels for a sustained period of time. Polymeric nanoparticles can be designed to encapsulate various therapeutic agents such as low molecular weight drugs, and macromolecules such as proteins or plasmid DNA. They can be categorized as (1) nanospheres are spherical nanometer range particles where the desired molecules can be either entrapped inside the sphere or adsorbed on the outer surface or both (Fig 1), (2) nanocapsules have a solid polymeric shell and an inner liquid core and the desired molecules can be entrapped (Fig 1). The polymeric matrix not only prevents the degradation of the encapsulated drug, but also leads to sustained release of the drug from nanoparticles. Moreover, the time and amount of drug released from the nanoparticles can be monitored by varying the different formulation parameters such as drug to polymer ratio, or polymer molecular weight and composition. As the nanoparticles directly interacts with the cell membranes, the surface properties of nanoparticles play a vital role in determining the internalization mechanism and intracellular disposition of nanoparticles. The drug molecules are either entrapped within the core of the nanoparticles or entangled into the polymer network or maybe adsorbed onto the surface. The encapsulated drugs are gradually released from the polymer matrix by the swelling of the polymeric nanoparticles by hydration followed by the release through diffusion, or by an enzymatic reaction leading to the degradation of the polymeric network. In case of drug adsorbed onto nanoparticles surface, release occurs by cleavage of the drug from the swelled nanoparticles or bulk degradation of polymeric matrix. The fate of the nanoparticles within the body is governed by its size, shape, surface charge and nature (hydrophobic or hydrophilic) of the polymer. Several natural and synthetic water-soluble polymers and their derivatives have been investigated for their potential use in polymeric nanoparticles based therapeutics, including polyethylene glycol , N-(2-hydroxypropyl) methacrylamide copolymers, poly(vinyl pyrrolidone), polyethylenimine, hyaluronic acid, chitosan, dextran and poly(aspartic acid) [(21].

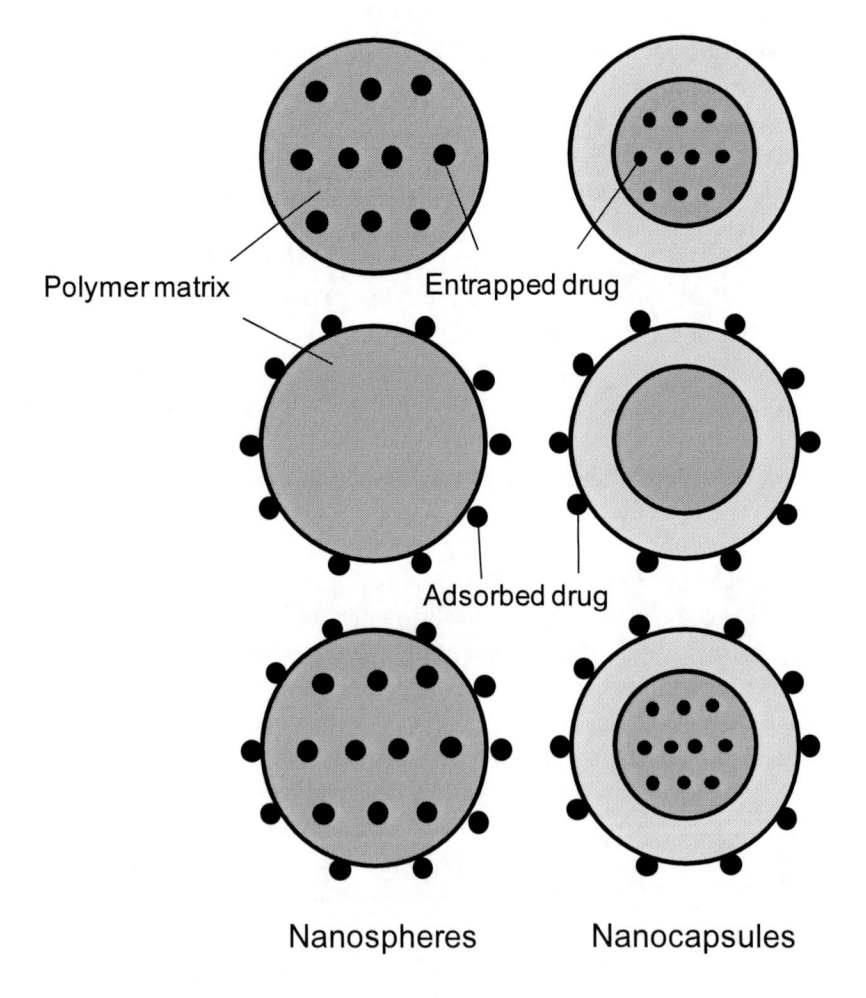

Figure 1. Different types of drug loaded nanoparticles (A) Nanospheres where drug is either embedded in the polymer matrix or adsorbed onto surface or both. (B) Nanocapsules where drug is either entrapped inside the hollow capsule or adsorbed onto surface or both.

Recently, polymeric nanoparticles have been investigated for the delivery of nucleic acids as therapeutic agents. Nucleic acids are expected to selectively regulate the genetic makeover that could be the cause of pathophysiological process in a disease. As compared to conventional therapeutics, the effect of nucleic acid therapeutics is upstream, at the level of gene expression producing those proteins. Transcription followed by translation leads to the amplification of nucleic acid activity on gene expression with prolonged effects. Also, nucleic acids offer unique prospects for therapeutic activity on previously intractable targets, such as transcription factors switching on or off the expression of entire groups of genes. Importantly, gene activity of nucleic acid agents depends on their sequence, so their pharmacological activity is largely independent of their physical chemistry.

Nucleic acids can be used to incorporate some desired functions in the system, i.e. gene therapy, or to stop some actions as in case of traditional therapeutics [22]. Gene therapy depends on the delivery of large nucleic acids consisting of sequences coding for desired proteins, while for inhibition of gene action only small nucleic acid sequences unique in the gene are required. The nucleic acid based systems for blocking the gene expression has been studied for many years through development of 'antisense' and 'ribozyme' oligonucleotides containing a sequence designed to bind a specific messenger RNA (mRNA) and then inhibit expression of the encoded protein [22]. Therapeutic use of these agents has been hampered by challenges such as limited stability, poor cellular internalization. However recently, a mechanism that cells themselves use for selective inhibition of gene expression, called RNA interference (RNAi), has become apparent [23-25]. This system has aroused numerous possibilities in the field of nucleic acid based therapeutic and has become centre of research for both academia and industry.

2. Basic Concept and Mechanism of RNA Interference

In 1998, Fire *et al.* reported that a few molecules of double-stranded RNA (dsRNA) injected into *C. elegans*, specifically knocked out the function of an endogenous gene and therefore protein [9]. These dsRNA molecules which are several hundred bases in length not only caused significant gene silencing, but were also more active than the corresponding single stranded antisense molecules. As these dsRNA molecules interfere with the function of the targeted gene, the process was named as 'RNA interference' (RNAi). RNAi regulates gene expression in mammalian cells through siRNA, which is a double-stranded RNA molecule having 21-23 base pairs. In this process the degradation of homologous mRNA by double-stranded RNA (dsRNA) is highly sequence-specific. The process of RNAi has been found to be very useful for genetic analysis and is rapidly evolving as a potent therapeutic approach for gene silencing [26,27]. The process of RNAi takes place in the cytoplasm of the target cells (Fig 2). The mechanism of RNAi was first deciphered from experiments using *Drosophila* extracts. The process initiates with the binding of dsRNA-specific endonuclease (a cytoplasmic ribonuclease III (RNase III)-like protein) called Dicer, which is capable of cleaving long dsRNAs [28]. Dicer cleaves the long strands of dsRNA into duplexes with 19 paired nucleotides and two nucleotide overhangs at both 3'-ends and these small dsRNAs are called small interfering RNAs (siRNAs) [29]. This siRNA molecule associates with a nuclease-containing multi-protein complex called RNA-induced silencing complex (RISC)[30]. Within RISC the siRNA unbinds by an RNA helicase activity and the sense strand removed for degradation by cellular nucleases [31]. Another RNA strand (the antisense strand) remains bound to the RISC complex and directs it to the target mRNA sequence, where it anneals complementarily by Watson-Crick base pairing. Finally, the RISC complex degrades the target mRNA by endonucleolytic activity thereby preventing its translation into protein [32]. Later, the cleavage products are released and degraded, leaving the disengaged RISC complex to further search for more target mRNAs [13].

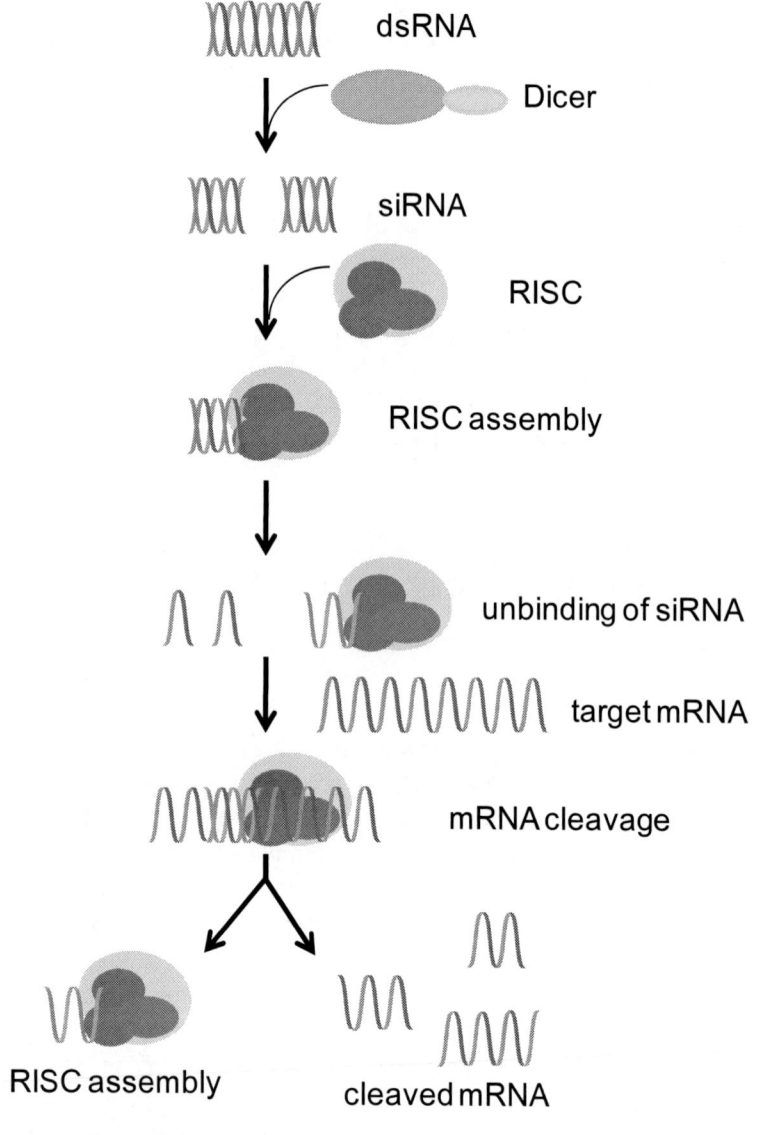

Figure 2. Mechanism of RNAi. Long dsRNA precursor molecules are cleaved by Dicer yielding siRNA. siRNA incorporates into RISC assembly, followed by unwinding of the double-stranded molecule by the helicase activity of RISC. The sense strand of siRNA is removed and the antisense strand binds targeted mRNA, which is then cleaved by RISC and subsequently degraded by cellular nucleases.

The first evidence for siRNAs capability to mediate sequence-specific gene silencing in mammalian cells and that the dicing step can be bypassed by the transfection of siRNA molecules into cells was shown by Tuschl et al in 2001 [29]. Afterwards, several studies have emphasized on the optimization of siRNAs with regard to *in vitro* and *in vivo* delivery, targeting efficiency and target specificity. RNAi has shown enormous potential for modulation of gene expression and to develop highly specific RNA based gene-silencing therapeutics. Since, RNAi takes advantage of the physiological gene-silencing machinery, it has higher efficacy than other nucleic acids based strategies. Moreover, siRNAs are more stable against nuclease degradation as compared to antisense oligonucleotides and, henceforth, have prolonged therapeutic effects than antisense therapy [33,34]. Although some non-specific effects have been observed, yet RNAi holds great promise as a new therapeutic tool for the manipulation and knock down of defective gene expression.

3. Barriers to siRNA delivery

During the *in vitro* transfection process, polymer/siRNA complexes face several barriers that diminish gene silencing efficiency. The process involves six key steps: the association of polymer/siRNA complexes to the cells, internalization, escape from degradation vesicles, intracellular movement or 'trafficking', the cytoplasmic translocation and finally the inhibition of gene expression (Fig 3). This last step requires separation of the complexes, at least partial, in order to allow nuclear machinery to mask the gene expression. The presence of anionic charge and chemical degradation of siRNA under physiological conditions poses major challenge with relevance to its delivery. Since, it is chemically similar to DNA and RNA, the barriers to their delivery are likely similar and may be grouped into intracellular and extracellular barriers [35,36]. For siRNA mechanism to have an effect, it is required to cross the cell membrane and also successfully escape lysosomal degradation. Due to presence of the negative charge and large size of siRNA it is obvious that it cannot bind to the cell surface or cross the cell membrane by passive diffusion. However, siRNA needs to enter the cytosol to function and exhibit the therapeutic effect. To improve upon the *in vivo* stability of siRNAs several chemical modifications have been carried out in the RNA backbone and various strategies are now available for improved *in vivo* stability of the system, which may eventually reduce the amount of dose required [37]. The *in vivo* gene silencing efficacy is governed by some important parameters such as bioavailability, cell-type-specific delivery, and *in vivo* pharmacokinetics (depending on the routes of administration).

The route of administration (e.g. intravenous (i.v.), intranasal (i.n.), intratracheal (i.t.), subcutaneous, intratumor, intramuscular, or oral) determines the extracellular barriers to siRNA delivery, which in turn, depends upon the targeted disease. For example in the case of i.v. delivery, the first major extracellular obstacle is its degradation by serum nucleases, which can be overcome by complexation with a suitable delivery system. The complexation system is selected in such a way that it not only facilitate cellular uptake but will also provide endosomal escape[38]. The delivery system should avoid both interactions with plasma proteins and uptake by the macrophages of the reticuloendothelial system (RES). It should be efficient enough to cross permeable endothelium such as in neovascularised tumors or inflammation. In this regard, complex size plays an important role; since only small particles can pass through the fenestrated barriers know as EPR for enhanced permeability and retention. However, before cellular entry, the delivery system should be able to cross the plasma membrane barrier composed of a variety of polysaccharides and proteins over the surface of cells that produce them [39]. Hence, all these barriers should be taken into consideration to design and develop new delivery carriers.

It is worth mentioning that the barriers encountered by native oligonucleotides i.e. siRNA, antisense molecules, or aptamers will be quite different from those encountered by oligonucleotides associated with various polymeric nanoparticles. One of the most significant biological barriers encountered by administered siRNA is the nuclease activity in plasma and tissues. The major enzymatic activity that occurs in the plasma is the 3′ exonuclease; although, cleavage of internucleotide bonds can also take place. However, the siRNA associated with polymeric nanoparticles easily bypass this barrier as the condensed siRNA is no more available for 3′ nuclease binding followed by cleavage. Another major obstacle is the renal clearance by the reticuloendothelial system (RES). Phagocytic cells of the RES, more specifically the Kupffer cells in the liver and splenic macrophages, can endocytose siRNA oligonucleotides, as well as carriers used to deliver them[40]. Foreign particles such as siRNA molecules in circulation are bound by opsonins, which consists of immunoglobulins, complement system proteins, and other serum proteins. These opsonized particles are recognized by a variety of receptors present on the cell surface of macrophages. Immunoglobulin G-opsonized particles are recognized by Fc receptors and complement-opsonized particles are internalized through complement receptors thereby leading to their degradation. However, surfaces of nanoparticles have been modified with hydrophilic polymers such as polyethylene glycol (PEG) [158], which reduces the adsorption of opsonins followed by reduced clearance by phagocytosis[41]. The covalent coupling of polymers with PEG results in shielding the surface charge of polycations. PEG is a preferred candidate for such steric stabilization due to its charge neutrality and water solubility. In principle, various factors should be considered both at the cellular and whole organism levels in the design of efficient *in vivo* delivery strategies for therapeutic oligonucleotides.

4. Polymeric Nanoparticles for the Delivery of siRNA

Nanoparticles made from polymers consisting of polyethyleneimine, chitosan and PEG based polymers have been employed to deliver siRNA. These are either polycation or polycation containing block copolymers that can efficiently form condensed complexes with siRNA in the nanometer size range. Strong electrostatic interactions between oppositely charged polyelectrolytes allow for "self-assembly", which can substantially hinder or protect the incorporated polynucleotide from enzymatic degradation in the bloodstream [1, 2]. Usually these nanoparticles possess a net positive surface charge, to readily attach to the cell surface via charge-charge interactions with the negatively charged membrane glycoproteins, thereby facilitating internalization by different endocytic mechanisms. Furthermore, polymers with derivable functional groups can be modified with surface ligands to achieve targeted delivery via receptor-mediated endocytosis.

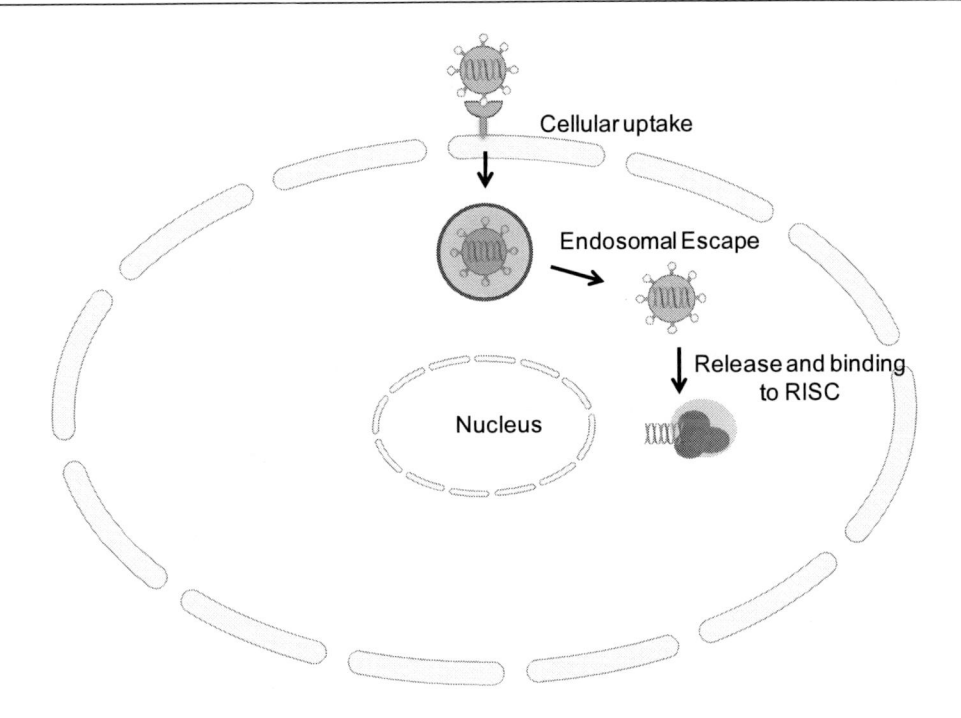

Figure 3. Schematic representation of siRNA loaded nanoparticles uptake by mammalian cells. siRNA is either compacted in the presence of polycationic polymers or adsorbed onto preformed nanoparticles. On reaching the target cell, they internalize by either receptor mediated endocytosis or macropinocytosis. The nanoparticles inhibit the degradation of siRNA by lysosomal enzymes and lead to escape into cytoplasm. The siRNA then binds to RISC assembly to inhibit the targeted gene expression.

4.1. Complexation of siRNA with Cationic Polymers

A plenty of nanoparticle systems based on polymers have been proposed to enhance the intracellular delivery of siRNA. At physiological pH siRNA possess negative charge and can readily interact electrostatically with cationic polymers to form complexes simply by mixing the constituents at optimal charge ratios. The positive charges of the complexes arises due to a high ratio of amine (in polycationic polymers) to low phosphate (in siRNA) known as N/P ratio. Being positively charged complexes attach to negatively charged cell surface followed by subsequent internalization.

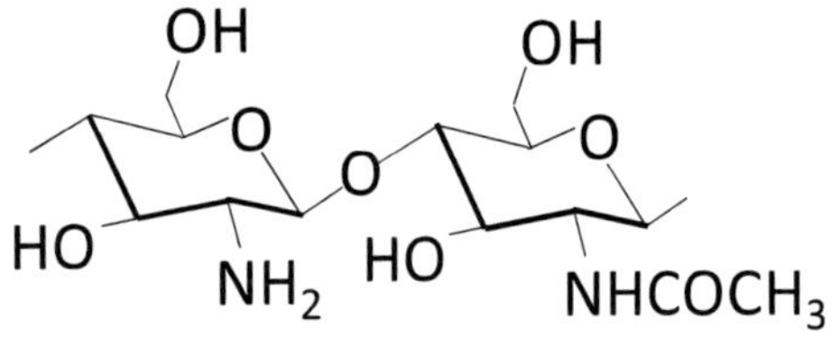

Figure 4. Chemical structure of chitosan.

4.1.1. Chitosan

Chitosan is one of the most widely investigated non-viral, naturally derived polymeric gene delivery vector. It is a linear cationic polysaccharide comprising of N-acetyl-D-glucosamine and β (1, 4)-linked D-glucosamine units; and it has been widely investigated as a DNA carrier since it is biodegradable, biocompatible and non immunogenic (Fig 4) [42-47].The transfection efficiency of chitosan/DNA systems depends on several factors such as the degree of deacetylation (DDA) and molecular weight (MW) of the chitosan, pH, protein interactions, charge ratio of chitosan to DNA (N/P ratio), cell type, nanoparticle size and interactions with cells [48]. The DNA binding affinity and transfection efficiency have been found to increase with increase in DDA or MW while maximum protein expression levels are achieved by obtaining an intermediate stability through control of MW and DDA [49,50]. Sato *et al.* compared transfection efficiency of chitosan/DNA complexes in A549 cells and also found it higher at pH 6.9 than at pH 7.6 [51]. Additionally, Zhao *et al.* investigated the effect of transfection medium pH on the transfection efficiency of chondrocytes using chitosan/DNA complexes and reported higher expression levels at pH 6.8 or 7.0 than at pH 7.4 [52].

A large number of reports have been published showing transfection with chitosan/DNA complexes both in the presence and absence of serum in the transfection medium [51,53,54]. Sato *et al.* demonstrated 2-3 times increase in gene expression level in the presence of serum compared to without serum and ascribed the effect as due to increased cell function [51]. However, upon addition of 50% serum, a reduction in transfection efficiency was observed and ascribed to cell damage induced by the high content of serum. Erbacher *et al.* reported higher transfection efficiency in HeLa cells in the presence of 10% serum than in the absence of serum [55]. Katas *et al.* seems to be the first group to report the use of chitosan to deliver siRNA *in vitro*[56]. The study involved two types of cell lines, CHO K1 and HEK 293 to reveal that the preparation method of siRNA association to the chitosan played an important role on the silencing effect. Nanoparticles of Chitosan-TPP entrapping siRNA were shown to be better vectors as siRNA delivery vehicles compared to chitosan-siRNA complexes possibly due to their high binding capacity and loading efficiency. Howard *et al.* engineered a novel chitosan-based siRNA nanoparticle delivery system for RNA interference *in vitro* and *in vivo*. The investigation revealed nanoparticle mediated knockdown of endogenous enhanced green fluorescent protein (EGFP) in both H1299 human lung carcinoma cells and murine peritoneal macrophages (77.9% and 89.3% reduction in EGFP fluorescence, respectively). Efficient *in vivo* RNA interference was achieved in bronchiole epithelial cells of transgenic EGFP mice after nasal administration of chitosan/siRNA formulations (37% and 43% reduction compared to mismatch and untreated control, respectively). These findings highlight the potential application of this novel chitosan based system in RNA-mediated therapy of systemic and mucosal disease.

A detailed study showed that the physicochemical properties (size, zeta potential, morphology and complex stability) as well as *in vitro* gene silencing of chitosan/siRNA nanoparticles are strongly dependent on chitosan MW and DDA [57]. High MW and DDA chitosan resulted in the formation of discrete stable nanoparticles ~ 200nm in size. Studies revealed that chitosan/siRNA formulations (N/P 50) prepared with low MW (~10 kDa) showed almost no knockdown of endogenous enhanced green fluorescent protein (EGFP) in H1299 human lung carcinoma cells, whereas those prepared from higher MW (64.8-170 kDa) and DD (~80%) showed greater gene silencing ranging between 45% and 65%. However, the highest gene silencing efficiency (80%) was achieved using chitosan/siRNA nanoparticles at N/P 150 using higher MW (114 and 170 kDa) and DDA (84%) that correlated with formation of stable nanoparticles of ~200 nm. Also, Rojanarata *et al.* reported that the chitosan-TPP mediated siRNA silencing of the endogenous EGFP gene occurred maximally with 70–73% efficiency with the lowest MW of chitosan (20 kDa) at a weight ratio of 80 [58]. Such high efficiency was associated with the increased siRNA binding ability and improved water solubility of the vector due to the addition of extra amine groups (especially those at N of thiazolium) from TPP, and salt formation between the phosphate group of TPP and the amine group of chitosan[58].

Anderson *et al.* described specific and efficient knockdown of EGFP in H1299 human lung carcinoma cells transfected in plates pre-coated with a chitosan/siRNA formulation containing sucrose as lyoprotectant (approximately 70%)[59]. This methodology alleviates the necessity for siRNA complexation immediately prior to use and addition onto cells. Moreover, chitosan/siRNA formulation displayed silencing activity over the period studied (approximately 2 months) when stored at room temperature. Delivery of shRNA employing chitosan nanoparticle have also been reported which leads to potent knockdown of TGFB1 by shRNA resulting in a decrease in rhabdomyosarcoma (RD) cell growth *in vitro* and tumorigenicity in nude mice[60]. The results suggest that chitosan nanoparticles mediated delivery of a shRNA produces efficient TGFB1 knockdown in RD cells. High levels of siRNA accumulation are observed due to delivery via chitosan/siRNA formulation within the kidney 24 hours post administration[61]. This comparative study between naked and chitosan/siRNA formulation, highlighted improvements to siRNA stability and pharmacokinetics[61].

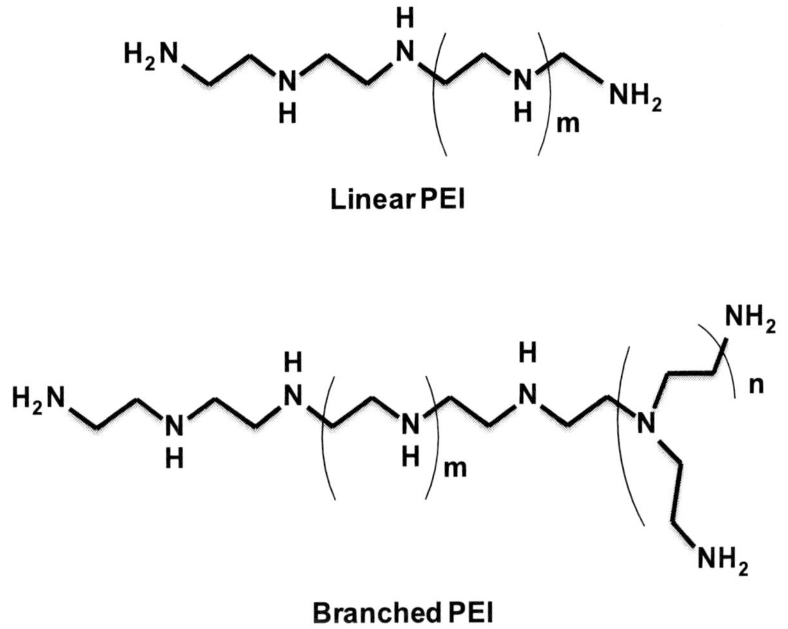

Linear PEI

Branched PEI

Figure 5. Chemical structure of Polyethylenimine (PEI). It is available in two forms i.e. linear and branched.

In order to control the size, chitosan nanoparticles were prepared using a coacervation method in the presence of polyguluronate (PG) encapsulating siRNA [62]. The mean diameter of siRNA-loaded chitosan nanoparticles was found from 110 to 430 nm, depending on the weight ratio between chitosan and siRNA. These nanoparticles were not only efficient in delivering siRNA to HEK 293FT and HeLa cells but also showed low cytotoxicity. In another study, chitosan/siRNA nanoparticles were observed as irregular, lamellar and dendritic structures with a hydrodynamic radius size of about 148 nm and net positive charges with zeta-potential value of 58.5 mV[63]. The knockdown effect of the chitosan/siRNA nanoparticles showed that FHL2 siRNA formulated within chitosan nanoparticles could knock down about 69.6% FHL2 gene expression, which is very similar to the 68.8% reduced gene expression when siRNA was transfected with liposome Lipofectamine. It was also observed that blocking FHL2 expression by siRNA could also inhibit the growth and proliferation of human colorectal cancer Lovo cells. These results proposed that chitosan-based siRNA nanoparticles were a very efficient delivery system for siRNA *in vivo*.

4.1.2. Polyethylenimine (PEI)

PEIs, ranging from low molecular weight to higher ones, have extensively been exploited as effective gene delivery vehicle [38,64-67]. It has also been shown to be an efficient versatile agent for *in vivo* gene delivery via a number of routes [68-70]. The transfection efficiency of PEI is directly related to size of the polymer and the charge-associated cytotoxicity [38,71-73]. High molecular weight PEI reported to exhibit superior transfection efficiency compared to low molecular weight PEI. However, the cytotoxicity limits its applications. It has been earlier reported that of linear and branched PEIs, the latter ones are more toxic and less efficient for transfection, particularly at higher polycation nitrogen to DNA phosphate ratio (Fig 5) [74]. PEI uses the "proton sponge" mechanism to promote the release of endocytosed polyplexes from the endosomes (Fig 6) [75-78]. According to this mechanism, the unprotonated amines with different pKa values confer a buffering effect over a wide range of pH. This buffering may protect the DNA from degradation in the endosomal compartment during the maturation of the early endosomes to late endosomes and their subsequent fusion with the lysosomes. The buffering property also allows the polycation polyethylenimine (PEI) to escape from the endosome: at lower pH, the buffering by PEI causes an influx of chloride ions and water into the endosomes, which eventually burst due to increased osmotic pressure, thus facilitating intracellular release of PEI-DNA polyplexes. High cation density of PEI (a potential positive charge per 43 Da, which is the monomer's molecular weight) also contributes to the formation of highly condensed particles by interacting with nucleic acids. However, it also results in significant cytotoxicity, particularly for large PEIs [76,79-82].

To further enhance the transfection efficiency of PEI mediated DNA-polymer complexes, the amine structures are modified[83]. One of the most common approaches is the attachment of a pendant group onto the primary amines of the cationic polymer. Various other modifications incorporated into PEI include coating with human serum albumin [77,84,85], dextran [86-88], PEGylation [89-92] and acylation [93]. Another strategy is to conjugate a ligand onto the primary amines of the cationic polymers to enhance the transgene expression by targeted delivery[94-96], such as the use of RGD peptide [97], mannose [98], and transferrin [99] for targeting endothelial cells, dendritic cells, and hepatocytes, respectively.

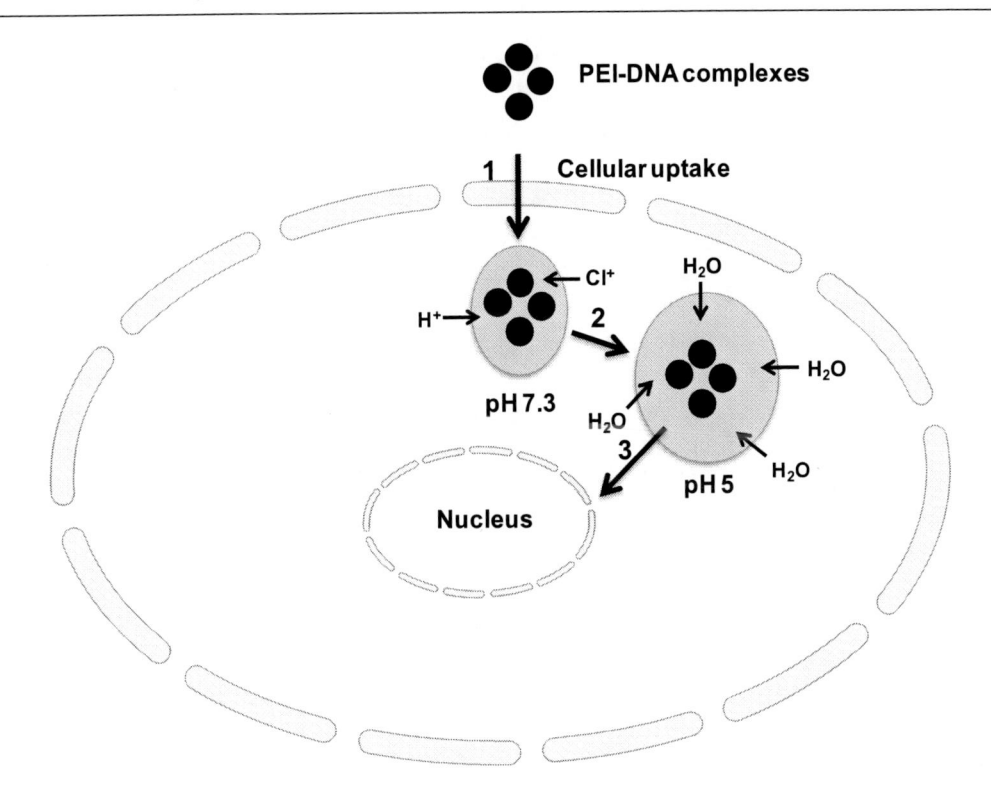

Figure 6. Schematic representation of the 'proton sponge effect': the initial step is endocytosis of the cationic complexes (1), followed by acidic endosome buffering (2) which leads to increased osmotic pressure and finally to lysis (3).

More recently, the application of PEIs as a siRNA delivery vector is gaining momentum. The atomic force microscopy (AFM) revealed that, complexation of PEI with siRNA molecules, compared to DNA, leads to the formation of more stable and uniformly sized particles [100]. These complexes not only completely covers the siRNAs but also capable of undergoing efficient endocytotic internalization. The non-covalent complexation of synthetic siRNAs with low molecular weight PEI efficiently stabilizes siRNAs and delivers siRNAs into cells where they display full bioactivity at completely non toxic concentrations [101]. Moreover, in case of subcutaneous administration in mouse tumor model, administration of complexed, but not of naked siRNAs, leads to the delivery of the intact siRNAs into the tumors. Also, the intraperitoneal injection of PEI complexed, but not naked siRNAs targeting the c-erbB2/neu (HER-2) receptor results in a marked reduction of tumor growth through siRNA-mediated HER-2 downregulation. Targeted delivery of PEI nanoparticles has been achieved by attachment of various targeting ligands to PEI. Self-assembling nanoparticles of siRNA with PEI that is PEGylated with an Arg-Gly-Asp (RGD) peptide ligand attached at the distal end of the PEG, acts as a means to target tumor neovasculature expressing integrins and used to deliver siRNA inhibiting vascular endothelial growth factor receptor-2 (VEGF R2) expression and thereby tumor angiogenesis[102]. Intravenous administration into tumor-bearing mice gave selective uptake at the tumor site, siRNA sequence-specific inhibition of protein expression within the tumor and inhibition of both tumor angiogenesis and growth rate.

Stable complexes of siRNA with PEI were prepared by incorporation of PEG into PEI better known as PEGylation. However, the high degree of PEGylation increased the size of the complexes to 300-400 nm, and condensation of siRNA only occurred at high N/P ratios [103]. Stability of siRNA complexes against heparin displacement and RNase digestion was found to increase with PEGylation. PEG chain length showed significant influence on biological activity of siRNA in knockdown experiments using NIH/3T3 fibroblasts stably expressing beta-galactosidase. PEGylated complexes with siRNA yielded knockdown efficiencies of around 70% higher than with PEI alone. This could be due to the fact that the high PEG content in PEI leads to formation of looser complexes which facilitates the efficient release of siRNA in cytoplasm [103]. The influence of size restriction of PEI to nanometer range by preparing nanoparticles before complexing with siRNA was investigated[104]. Nanoparticles prepared from high molecular weight PEI (750 kDa) by acylation with propionic anhydride followed by ionic cross-linking with derivatized PEG were found to be ~ 110 nm. Gene silencing experiment carried out by taking siRNA for reporter gene GFP in HEK 293 cells resulted in upto 83% inhibition of gene expression. Further, blending of PEI with a biocompatible polysaccharide (alginic acid) in appropriate ratio was investigated as it reduces the charge associated cytotoxicity of PEI and improves the transfection plausibly by aiding in the rupture of endosomes[105]. PEI-alginate nanocomposites can efficiently deliver siRNAs into HEK 293 cells leading to more than 80% knockdown of GFP reporter gene. In one of the *in vivo* studies, the effect of Fas-silencing siRNA (Fas siRNA) on diabetes development was evaluated in a cyclophosphamide (CY)-accelerated diabetes animal model after intravenous administration using PEI[106]. The systemic non-viral delivery of Fas siRNA showed significant delay in diabetes incidence up to 40days, while the control mice treated with naked Fas siRNA, scrambled dsRNA, or PBS were afflicted with diabetes within 20days.

4.1.3. Dendrimers

Dendrimers are highly branched synthetic polymers consisting of a central core molecule with branched surface, controllable nanometer size and highly monodisperse (Fig 7). They have a high density of positive surface charge and are quite efficient in condensing nucleic acids, thus providing protection against nuclease degradation. The positive surface charge enables them to adsorb onto cell surfaces leading to internalization followed by efficient escape from endosomal compartmentalization[107]. Preparation of stable and homogeneous nanoparticles using dendrimers has been found to depend on the size-to charge ratio and the dendrimer generation [108,109]. Dendrimers of ethylendiamine (EDA) core, polyamidoamine (PAMAM) have been used to deliver antisense oligonucleotides (AS-ODN) to clones of D5 mouse melanoma and Rat2 embryonal fibroblasts expressing luciferase cDNA [110]. Specific and dose dependent inhibition of luciferase expression about 25-50% was observed[111]. The ability of dendrimers to deliver AS-ODN to HeLa Luc/705 cells was evaluated in the absence or presence of serum. The results revealed that PAMAM dendrimers were more effective in delivering AS-ODN into the nucleus of cells in the presence of serum proteins as compared to other types of delivery agents. The intracellular distribution of a fluorescein isothiocyanate (FITC) labelled ODN delivered by PAMAM dendrimers showed that intracellular AS-ODN distribution was dependent on the phase of the cell cycle, with a nuclear localization predominantly in the G2/M phase[112].

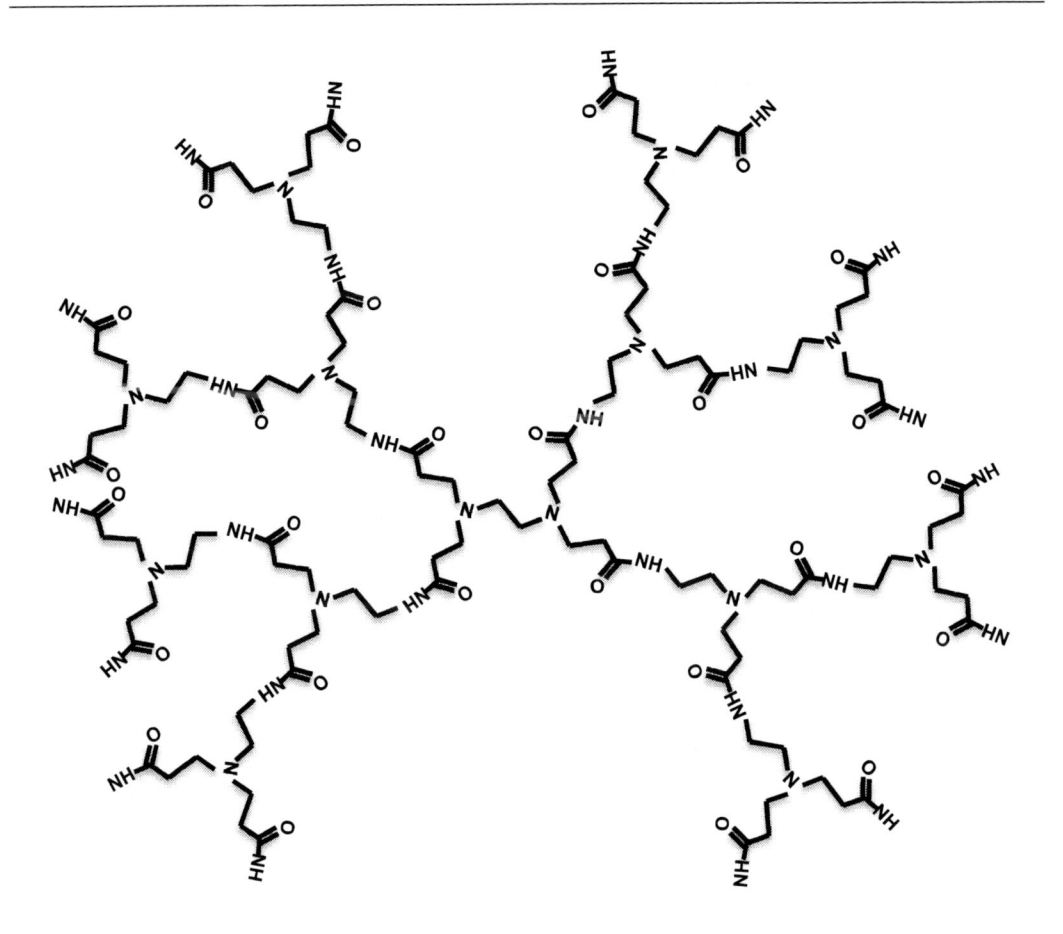

Figure 7. Chemical structure of Polyamidoamine (PAMAM) dendrimer.

Recently, dendrimers have been investigated for siRNA delivery both *in vitro* and *in vivo*. Nanometer size particles were formed on complexation with siRNA and stability found to depend on the dendrimer generation i.e. the higher the generation of the dendrimers, the stronger the interactions between the dendrimer and the RNA, and therefore the more stable the complex formed[108]. The siRNA-dendrimer complexes showed remarkable resistance to RNase degradation and further lead to efficient gene silencing in A549Luc cells stably expressing the GL3 luciferase gene. To fight against HIV infection amino-terminated carbosilane dendrimers (CBS) complexed with siRNA have been proposed [113]. CBS-siRNA complexes were shown to silence GAPDH expression and reduce HIV replication in lymphocytic cell line SupT1 and peripheral blood mononuclear cells (PBMC). In an attempt to develop tumor targeted delivery system poly(Propyleneimine) (PPI) dendrimers were surface modified with PEG with synthetic analog of Luteinizing Hormone-Releasing Hormone (LHRH) at distal end [114]. It was demonstrated that the layer-by-layer modification and targeting approach confers the siRNA-dendrimer nanoparticles stability in plasma and intracellular bioavailability, provides for their specific uptake by tumor cells, accumulation of siRNA in the cytoplasm of cancer cells, and efficient gene silencing. The nanoparticles were successfully internalized by LHRH-positive A549 and A2780 cancer cells that have over expressed LHRH receptors but not by LHRH-negative SKOV-3 cells which do not show a detectable level of LHRH receptors. *In vivo* studies in mice showed that non-targeted dendrimer-siRNA nanoparticles were found mainly in the liver and kidney 72 h after the injection, in contrast, targeted dendrimer-siRNA nanoparticles were mainly found in the tumors[114]. In one of the recent studies, Perez et al observed that the size of the siRNA-ethylendiamine (EDA) core PAMAM dendrimers was dependent on the ionic strength of the media[115]. The percentage inhibition of EGFP expression both in phagocytic (T98G-EGFP) and non-phagocytic (J-774-EGFP) cells was found to be 35 and 45% with these particles. It can be speculated from the available reports that dendrimers with the several advantages such as serum resistance, increased stability in biological liquids, tumor-specific targeting, effective penetration into cancer cells, along with nanoscale size can be used for the *in vivo* systemic delivery of siRNA for efficient cancer therapy.

4.2. Encapsulation of siRNA within Polymeric Nanoparticles

For optimal efficacy and reduced toxicity arising due to the cationic charges, siRNAs were entrapped within nanocapsules instead of being adsorbed or complexed with polycationic polymers. Polymeric nanocapsules have been well exploited to maintain a sufficient local concentration of siRNA. RhoA is one of the proteins over expressed in cancer cells. Studies on intravenously administering chitosan coated polyisohexylcyanoacrylate (PIHCA) nanoparticles encapsulated anti-RhoA siRNA in nude mice with aggressive breast cancer demonstrated that tumor growth can be inhibited by 90% without toxic effect [116]. Working towards *in vivo* experimental model, Toub *et al.* demonstrated the utility of polyisobutylcyanoacrylate (PIBCA) nanocapsules to deliver siRNAs into tumor[117]. Confocal microscopy data revealed that when a fluorescently labelled naked siRNA was added to the culture medium, only a small amount of fluorescence localized in the extracellular matrix was found. However, in case of nanocapsules encapsulating siRNA, punctuate fluorescence was observed within the cells, which suggest of an endosomal localization of the siRNA. A significant inhibition in tumor growth was observed, on administering a higher cumulative dose to a mice xenografted EWS-Fli1-expressing tumor. Nanocapsules of PIBCA, poly(ethylene)glycol-poly(D,L-lactide-co-glycolide) (PEG-PLGA) and PEG-epsilon-caprolactone-malic acid (PEG-PCL/MA) were synthesized encapsulating specific ERalpha-siRNAs [118]. A significant inhibition of ERalpha (90% over 5 days) was observed in MCF-7 human breast cancer cells that were exposed to PEG-PCL/MA NCs loaded with ERalpha-siRNA. Additionally, inhibition of tumor growth was observed with intravenous injection of these nanocapsules into estradiol-stimulated MCF-7 cell xenografts and also a decrease in ERalpha expression in tumor cells.

Nanoparticles of poly(D,L-lactide-co-glycolide)-polyethylenimine (PLGA-PEI) formulated using double emulsion-solvent evaporation technique encapsulating siRNA for a model gene, fire-fly luciferase, was investigated in MDA-Kb2 cells[119]. PLGA-PEI nanoparticles exhibited efficient silencing of the gene in cells stably expressing luciferase as well as in cells that could be induced to over express the gene. The presence of polycationic PEI in nanoparticles resulted in 2-fold higher cellular uptake of nanoparticles while fluorescence microscopy studies showed that PLGA-PEI nanoparticles delivered the encapsulated siRNA in the cellular cytoplasm; both higher uptake and greater cytosolic delivery could have contributed to the gene silencing effectiveness of PLGA-PEI nanoparticles. One of recent published study has used PLGA nanospheres containing a GFP gene silencing siRNA in GFP-expressing 293T cells[120]. A significant inhibition of GFP expression was observed after 1 day incubation and was still persistent 1 week later.

5. Advantages and Limits of Polymeric Nanoparticles

Polymeric carriers provide several advantages such as ease of manipulation with scope to change the molecular weight, geometry (linear and branched) and attachment of various targeting ligands by covalent linkage, stability, safety, low cost and high flexibility regarding the size of transgene delivered. They bear several advantages over the lipid based vectors as polymeric nanoparticles are relatively small in size with narrow distribution, provides better protection against degrading nucleases and ease of modification. Though polymeric nanoparticles have been successfully employed in numerous studies, they also have some limitations. The polycationic polymers undergo strong electrostatic interaction with plasma membrane proteins, which can lead to destabilization and ultimately rupture of the cell membrane. Fischer *et al.* demonstrated that the cytotoxicity of different types of polycationic polymers depend on the number and arrangement of the cationic charges which determines the degree of interaction with the cell membranes and the cells exposed to cationic polymers first show membrane leakage followed by a decrease in the metabolic activity [121]. A comparative study between polycationic, neutral and polyanionic polymers revealed that the polycationic polymers have the highest toxicity followed by neutral and anionic ones[122]. The transfection efficiency of PEI is directly related to size of the polymer and the charge-associated cytotoxicity [38,71-73]. Of linear and branched PEIs, the latter ones are more toxic and less efficient for transfection, particularly at higher polycation nitrogen to DNA phosphate ratio [74]. Additionally, PEI is a non-biodegradable polymer that can accumulate within the cells interfering with vital intracellular process. Two types of cytotoxicities have been correlated to PEI-mediated transfection i.e. an immediate toxicity due to free PEI (which occurs within 2h) and delayed toxicity (which occurs 7-9 h after transfection) associated due to cellular processing of PEI/DNA complexes [123,124]. Free PEI interacts with negatively charged serum proteins (such as albumin) and red blood cells, forms large clusters that adhere to the cells surface and destabilize the plasma membrane leading to immediate toxicity. This second type of cell death is possibly linked to PEI's uncomplexation from DNA and probable interaction with chromosomal DNA in the nucleus[124].

It has been shown that PEI/DNA complexes can activate complement system if the ratio of cation to anion is high, but the extent of complement activation is lowered as the PEI/DNA complexes approach neutrality [125,126]. It has also been asserted that circulating proteins can bind to PEI/DNA and inactivate the complexes if the complexes possess an excess of positive charge[125]. Natural polymers e.g. cationic chitosan, are largely explored in gene delivery applications as they are biocompatible, biodegradable and nontoxic. Chitosan-DNA nanoparticle systems are intensively investigated and are reported to possess effective nuclease resistance, endosomolytic activity and comparable transfection levels [45]. However, the optimum conditions for transfection efficiency of chitosan-DNA have not been clear, and must be elucidated for their application as gene carriers *in vitro* and *in vivo*.

Moreover, chitosan has also been found to be an efficient *in vitro* and *in vivo* siRNA delivery vector capable of mediating gene silencing with no toxicity [116,127].

6. Conclusions and Future Prospects

Since the discovery of RNAi, this approach has proven itself an indispensable tool for elucidating molecular pathways and phenotype/genotype relationships, and has many potential advantages over existing technologies. The success of RNAi depends on several parameters including (i) siRNA protection, (ii) high transfection efficacy, (iii) reduced toxicity and absence of non-specific effects, (iv) high potency even at small amounts of siRNAs, (v) adaptation to various treatment regimens and in various diseases as well as (vi) efficient vector bypass intracellular and extracellular barriers to reach their target tissue/organ. Though a plethora of reports have appeared detailing different types of vectors for siRNA delivery, safe and efficient delivery of siRNA into target cells or organs still remains a big challenge for therapeutic application. Polymeric nanoparticles are rapidly emerging as siRNA delivery systems both *in vitro* and *in vivo*. These systems can efficiently protect siRNA from degradation by nucleases even when they are in the *in vivo* environment. Numerous evidences have appeared in the literature that polymeric nanocarriers can, indeed, be used to deliver functional siRNA to their target cells after intravenous administration. Nanoparticles have been found successful in delivery of siRNA for antitumor treatments in different experimental tumor models. Despite excitement from a large number of animal model studies, including systemic delivery to nonhuman primates [128], there are a number of hurdles and concerns that must be overcome before RNAi will be harnessed as a new therapeutic modality. The ideal nanoparticles for siRNA carrier system can achieve long circulation time, low immunogenicity, good biocompatibility, selective targeting, and the efficient penetration of barriers such as the vascular endothelium and the blood brain barrier, self-regulating release without clinical side effects. Ample amount of work has to be done both in preparations of better polymers and in the development of better ways of encapsulating or complexing siRNA with them. There is a need for the collaborative work by academicians and industry groups in order to develop methods for preparation of stable polymeric nanoparticles and robust analytical methods to characterize formulations during formation and storage.

References

[1]   Stephenson, M.L. and Zamecnik, P.C. (1978) Inhibition of Rous sarcoma viral RNA translation by a specific oligodeoxyribonucleotide. *Proceedings of the National Academy of Sciences of the United States of America*, 75, 285-288.

[2]   Zamecnik, P.C. and Stephenson, M.L. (1978) Inhibition of Rous sarcoma virus replication and cell transformation by a specific oligodeoxynucleotide. *Proceedings of the National Academy of Sciences of the United States of America*, 75, 280-284.

[3]   Gleave, M.E. and Monia, B.P. (2005) Antisense therapy for cancer. *Nature Reviews Cancer*, 5, 468-479.

[4]   Cech, T.R., Zaug, A.J. and Grabowski, P.J. (1981) In vitro splicing of the ribosomal RNA precursor of tetrahymena: Involvement of a guanosine nucleotide in the excision of the intervening sequence. 27, 487-496.

[5]   Kruger, K., Grabowski, P.J., Zaug, A.J., Sands, J., Gottschling, D.E. and Cech, T.R. (1982) Self-splicing RNA: Autoexcision and autocyclization of the ribosomal RNA intervening sequence of tetrahymena. 31, 147-157.

[6]   Guerrier-Takada, C., Gardiner, K., Marsh, T., Pace, N. and Altman, S. (1983) The RNA moiety of ribonuclease P is the catalytic subunit of the enzyme. 35, 849-857.

[7]   Jarvis, T.C., Wincott, F.E., Alby, L.J., McSwiggen, J.A., Beigelman, L., Gustofson, J., DiRenzo, A., Levy, K., Arthur, M., MatulicAdamic, J. *et al.* (1996) Optimizing the cell efficacy of synthetic ribozymes - Site selection and chemical modifications of ribozymes targeting the proto-oncogene c-myb. *Journal of Biological Chemistry*, 271, 29107-29112.

[8]   Napoli, C., Lemieux, C. and Jorgensen, R. (1990) Introduction of a Chimeric Chalcone Synthase Gene into Petunia Results in Reversible Co-Suppression of Homologous Genes in Trans. *Plant Cell*, 2, 279-289.

[9]   Fire, A., Xu, S.Q., Montgomery, M.K., Kostas, S.A., Driver, S.E. and Mello, C.C. (1998) Potent and specific genetic interference by double-stranded RNA in Caenorhabditis elegans. *Nature*, 391, 806-811.

[10]   Ketzinel-Gilad, M., Shaul, Y. and Galun, E. (2006) RNA interference for antiviral therapy. *Journal of Gene Medicine*, 8, 933-950.

[11]   Reinhart, B.J., Slack, F.J., Basson, M., Pasquinelli, A.E., Bettinger, J.C., Rougvie, A.E., Horvitz, H.R. and Ruvkun, G. (2000) The 21-nucleotide let-7 RNA regulates developmental timing in Caenorhabditis elegans. *Nature*, 403, 901-906.

[12]   Ahlquist, P. (2002) RNA-dependent RNA polymerases, viruses, and RNA silencing. *Science (New York, N.Y*, 296, 1270-1273.

[13]   Leung, R.K.M. and Whittaker, P.A. (2005) RNA interference: From gene silencing to gene-specific therapeutics. *Pharmacology & Therapeutics*, 107, 222-239.

[14]   Guo, S. and Kemphues, K.J. (1995) par-1, a gene required for establishing polarity in C. elegans embryos, encodes a putative Ser/Thr kinase that is asymmetrically distributed. *Cell*, 81, 611-620.

[15]   Hough, S.R., Wiederholt, K.A., Burrier, A.C., Woolf, T.M. and Taylor, M.F. (2003) Why RNAi makes sense. *Nat Biotechnol*, 21, 731-732.

[16]   Lewis, D.L., Hagstrom, J.E., Loomis, A.G., Wolff, J.A. and Herweijer, H. (2002) Efficient delivery of siRNA for inhibition of gene expression in postnatal mice. *Nat Genet*, 32, 107-108.

[17]   Moghimi, S.M., Hunter, A.C. and Murray, J.C. (2005) Nanomedicine: current status and future prospects. *FASEB J*, 19, 311-330.

[18]   Loo, C., Lowery, A., Halas, N., West, J. and Drezek, R. (2005) Immunotargeted nanoshells for integrated cancer imaging and therapy. *Nano Lett*, 5, 709-711.

[19]   Kim, Y.K., Kwak, H.S., Kim, C.S., Chung, G.H., Han, Y.M. and Lee, J.M. (2006) Hepatocellular carcinoma in patients with chronic liver disease: comparison of SPIO-enhanced MR imaging and 16-detector row CT. *Radiology*, 238, 531-541.

[20]   Portney, N.G. and Ozkan, M. (2006) Nano-oncology: drug delivery, imaging, and sensing. *Anal Bioanal Chem*, 384, 620-630.

[21] Ogris, M. and Wagner, E. (2002) Targeting tumors with non-viral gene delivery systems. *Drug discovery today*, 7, 479-485.

[22] Scanlon, K.J. (2004) Anti-genes: siRNA, ribozymes and antisense. *Curr Pharm Biotechnol*, 5, 415-420.

[23] Matzke, M.A. and Birchler, J.A. (2005) RNAi-mediated pathways in the nucleus. *Nat Rev Genet*, 6, 24-35.

[24] Dykxhoorn, D.M., Novina, C.D. and Sharp, P.A. (2003) Killing the messenger: short RNAs that silence gene expression. *Nat Rev Mol Cell Biol*, 4, 457-467.

[25] Bantounas, I., Phylactou, L.A. and Uney, J.B. (2004) RNA interference and the use of small interfering RNA to study gene function in mammalian systems. *J Mol Endocrinol*, 33, 545-557.

[26] McManus, M.T. and Sharp, P.A. (2002) Gene silencing in mammals by small interfering RNAs. *Nat Rev Genet*, 3, 737-747.

[27] Hammond, S.M., Caudy, A.A. and Hannon, G.J. (2001) Post-transcriptional gene silencing by double-stranded RNA. *Nat Rev Genet*, 2, 110-119.

[28] Bernstein, E., Caudy, A.A., Hammond, S.M. and Hannon, G.J. (2001) Role for a bidentate ribonuclease in the initiation step of RNA interference. *Nature*, 409, 363-366.

[29] Elbashir, S.M., Harborth, J., Lendeckel, W., Yalcin, A., Weber, K. and Tuschl, T. (2001) Duplexes of 21-nucleotide RNAs mediate RNA interference in cultured mammalian cells. *Nature*, 411, 494-498.

[30] Hammond, S.M., Bernstein, E., Beach, D. and Hannon, G.J. (2000) An RNA-directed nuclease mediates post-transcriptional gene silencing in Drosophila cells. *Nature*, 404, 293-296.

[31] Nykanen, A., Haley, B. and Zamore, P.D. (2001) ATP requirements and small interfering RNA structure in the RNA interference pathway. *Cell*, 107, 309-321.

[32] Martinez, J. and Tuschl, T. (2004) RISC is a 5' phosphomonoester-producing RNA endonuclease. *Genes Dev*, 18, 975-980.

[33] Jana, S., Chakraborty, C., Nandi, S. and Deb, J.K. (2004) RNA interference: potential therapeutic targets. *Appl Microbiol Biotechnol*, 65, 649-657.

[34] Li, C.X., Parker, A., Menocal, E., Xiang, S., Borodyansky, L. and Fruehauf, J.H. (2006) Delivery of RNA interference. *Cell Cycle*, 5, 2103-2109.

[35] Thomas, M. and Klibanov, A.M. (2003) Non-viral gene therapy: polycation-mediated DNA delivery. *Applied Microbiology and Biotechnology*, 62, 27-34.

[36] Varga, C.M., Tedford, N.C., Thomas, M., Klibanov, A.M., Griffith, L.G. and Lauffenburger, D.A. (2005) Quantitative comparison of polyethylenimine formulations and adenoviral vectors in terms of intracellular gene delivery processes. *Gene therapy*, 12, 1023-1032.

[37] Behlke, M.A. (2006) Progress towards in vivo use of siRNAs. *Molecular Therapy*, 13, 644-670.

[38] Boussif, O., Lezoualch, F., Zanta, M.A., Mergny, M.D., Scherman, D., Demeneix, B. and Behr, J.P. (1995) A Versatile Vector for Gene and Oligonucleotide Transfer into Cells in Culture and in-Vivo - Polyethylenimine. *Proceedings of the National Academy of Sciences of the United States of America*, 92, 7297-7301.

[39] Li, W.J. and Szoka, F.C. (2007) Lipid-based nanoparticles for nucleic acid delivery. *Pharmaceutical research*, 24, 438-449.

[40] Alexis, F., Pridgen, E., Molnar, L.K. and Farokhzad, O.C. (2008) Factors affecting the clearance and biodistribution of polymeric nanoparticles. *Mol Pharm*, 5, 505-515.

[41] van Vlerken, L.E., Vyas, T.K. and Amiji, M.M. (2007) Poly(ethylene glycol)-modified nanocarriers for tumor-targeted and intracellular delivery. *Pharmaceutical research*, 24, 1405-1414.

[42] Özgel, G. and Akbuga, J. (2006) In vitro characterization and transfection of IL-2 gene complexes. *International journal of pharmaceutics*, 315, 44-51.

[43] MacLaughlin, F.C., Mumper, R.J., Wang, J., Tagliaferri, J.M., Gill, I., Hinchcliffe, M. and Rolland, A.P. (1998) Chitosan and depolymerized chitosan oligomers as condensing carriers for in vivo plasmid delivery. *J Control Release*, 56, 259-272.

[44] Richardson, S.C., Kolbe, H.V. and Duncan, R. (1999) Potential of low molecular mass chitosan as a DNA delivery system: biocompatibility, body distribution and ability to complex and protect DNA. *International journal of pharmaceutics*, 178, 231-243.

[45] Borchard, G. (2001) Chitosans for gene delivery. *Adv. Drug. Deliv. Rev.*, 52, 145-150.

[46] Guliyeva, Ü., Öner, F., Özsoy, S. and Haziroglu, R. (2006) Chitosan microparticles containing plasmid DNA as potential oral gene delivery system. *European Journal of Pharmaceutics and Biopharmaceutics*, 62, 17-25.

[47] Mansouri, S., Lavigne, P., Corsi, K., Benderdour, M., Beaumont, E. and Fernandes, J.C. (2004) Chitosan-DNA nanoparticles as non-viral vectors in gene therapy: strategies to improve transfection efficacy. *European Journal of Pharmaceutics and Biopharmaceutics*, 57, 1-8.

[48] Mansouri, S., Cuie, Y., Winnik, F., Shi, Q., Lavigne, P., Benderdour, M., Beaumont, E. and Fernandes, J.C. (2006) Characterization of folate-chitosan-DNA nanoparticles for gene therapy. *Biomaterials*, 27, 2060-2065.

[49] Lavertu, M., Methot, S., Tran-Khanh, N. and Buschmann, M.D. (2006) High efficiency gene transfer using chitosan/DNA nanoparticles with specific combinations of molecular weight and degree of deacetylation. *Biomaterials*, 27, 4815-4824.

[50] Ma, P.L., Lavertu, M., Winnik, F.M. and Buschmann, M.D. (2009) New Insights into Chitosan-DNA Interactions Using Isothermal Titration Microcalorimetry. *Biomacromolecules*.

[51] Sato, T., Ishii, T. and Okahata, Y. (2001) In vitro gene delivery mediated by chitosan. effect of pH, serum, and molecular mass of chitosan on the transfection efficiency. *Biomaterials*, 22, 2075-2080.

[52] Zhao, X., Yu, S.B., Wu, F.L., Mao, Z.B. and Yu, C.L. (2006) Transfection of primary chondrocytes using chitosan-pEGFP nanoparticles. *J Control Release*, 112, 223-228.

[53] Peng, S.F., Yang, M.J., Su, C.J., Chen, H.L., Lee, P.W., Wei, M.C. and Sung, H.W. (2009) Effects of incorporation of poly(gamma-glutamic acid) in chitosan/DNA complex nanoparticles on cellular uptake and transfection efficiency. *Biomaterials*, 30, 1797-1808.

[54] Bozkir, A. and Saka, O.M. (2004) Chitosan-DNA nanoparticles: effect on DNA integrity, bacterial transformation and transfection efficiency. *Journal of drug targeting*, 12, 281-288.

[55] Erbacher, P., Zou, S., Bettinger, T., Steffan, A.M. and Remy, J.S. (1998) Chitosan-based vector/DNA complexes for gene delivery: biophysical characteristics and transfection ability. *Pharmaceutical research*, 15, 1332-1339.

[56] Katas, H. and Alpar, H.O. (2006) Development and characterisation of chitosan nanoparticles for siRNA delivery. *Journal of Controlled Release*, 115, 216-225.

[57] Liu, X.D., Howard, K.A., Dong, M.D., Andersen, M.O., Rahbek, U.L., Johnsen, M.G., Hansen, O.C., Besenbacher, F. and Kjems, J. (2007) The influence of polymeric properties on chitosan/siRNA nanoparticle formulation and gene silencing. *Biomaterials*, 28, 1280-1288.

[58] Rojanarata, T., Opanasopit, P., Techaarpornkul, S., Ngawhirunpat, T. and Ruktanonchai, U. (2008) Chitosan-Thiamine Pyrophosphate as a Novel Carrier for siRNA Delivery. *Pharmaceutical research*, 25, 2807-2814.

[59] Andersen, M.Ø., Howard, K.A., Paludan, S.R., Besenbacher, F. and Kjems, J. (2008) Delivery of siRNA from lyophilized polymeric surfaces. *Biomaterials*, 29, 506-512.

[60] Wang, S.L., Yao, H.H., Guo, L.L., Dong, L., Li, S.G., Gu, Y.P. and Qin, Z.H. (2009) Selection of optimal sites for TGFB1 gene silencing by chitosan-TPP nanoparticle-mediated delivery of shRNA. *Cancer Genetics and Cytogenetics*, 190, 8-14.

[61] Gao, S., Dagnaes-Hansen, F., Nielsen, E.J.B., Wengel, J., Besenbacher, F., Howard, K.A. and Kjems, J. (2009) The Effect of Chemical Modification and Nanoparticle Formulation on Stability and Biodistribution of siRNA in Mice. *Mol Ther*, 17, 1225-1233.

[62] Lee, D.W., Yun, K.S., Ban, H.S., Choe, W., Lee, S.K. and Lee, K.Y. (2009) Preparation and characterization of chitosan/polyguluronate nanoparticles for siRNA delivery. *Journal of Controlled Release*, 139, 146-152.

[63] Ji, A.M., Su, D., Che, O., Li, W.S., Sun, L., Zhang, Z.Y., Yang, B. and Xu, F. (2009) Functional gene silencing mediated by chitosan/siRNA nanocomplexes. *Nanotechnology*, 20.

[64] Goula, D., Benoist, C., Mantero, S., Merlo, G., Levi, G. and Demeneix, B.A. (1998) Polyethylenimine-based intravenous delivery of transgenes to mouse lung. *Gene therapy*, 5, 1291-1295.

[65] Abdallah, B., Hassan, A., Benoist, C., Goula, D., Behr, J.P. and Demeneix, B.A. (1996) A powerful nonviral vector for in vivo gene transfer into the adult mammalian brain: polyethylenimine. *Human gene therapy*, 7, 1947-1954.

[66] Erbacher, P., Bettinger, T., Brion, E., Coll, J.L., Plank, C., Behr, J.P. and Remy, J.S. (2004) Genuine DNA/polyethylenimine (PEI) complexes improve transfection properties and cell survival. *Journal of drug targeting*, 12, 223-236.

[67] Kunath, K., von Harpe, A., Fischer, D., Petersen, H., Bickel, U., Voigt, K. and Kissel, T. (2003) Low-molecular-weight polyethylenimine as a non-viral vector for DNA delivery: comparison of physicochemical properties, transfection efficiency and in vivo distribution with high-molecular-weight polyethylenimine. *J Control Release*, 89, 113-125.

[68] Hong, J.W., Park, J.H., Huh, K.M., Chung, H., Kwon, I.C. and Jeong, S.Y. (2004) PEGylated polyethylenimine for in vivo local gene delivery based on lipiodolized emulsion system. *J Control Release*, 99, 167-176.

[69] Kichler, A., Chillon, M., Leborgne, C., Danos, O. and Frisch, B. (2002) Intranasal gene delivery with a polyethylenimine-PEG conjugate. *J Control Release*, 81, 379-388.

[70] Kichler, A., Leborgne, C., Coeytaux, E. and Danos, O. (2001) Polyethylenimine-mediated gene delivery: a mechanistic study. *The journal of gene medicine*, 3, 135-144.

[71] Fischer, D., Bieber, T., Li, Y., Elsasser, H.P. and Kissel, T. (1999) A novel non-viral vector for DNA delivery based on low molecular weight, branched polyethylenimine: effect of molecular weight on transfection efficiency and cytotoxicity. *Pharmaceutical research*, 16, 1273-1279.

[72] Godbey, W.T., Wu, K.K. and Mikos, A.G. (1999) Size matters: molecular weight affects the efficiency of poly(ethylenimine) as a gene delivery vehicle. *Journal of biomedical materials research*, 45, 268-275.

[73] Ogris, M., Brunner, S., Schuller, S., Kircheis, R. and Wagner, E. (1999) PEGylated DNA/transferrin-PEI complexes: reduced interaction with blood components, extended circulation in blood and potential for systemic gene delivery. *Gene therapy*, 6, 595-605.

[74] Wightman, L., Kircheis, R., Rossler, V., Carotta, S., Ruzicka, R., Kursa, M. and Wagner, E. (2001) Different behavior of branched and linear polyethylenimine for gene delivery in vitro and in vivo. *The journal of gene medicine*, 3, 362-372.

[75] Boussif, O., Lezoualc'h, F., Zanta, M.A., Mergny, M.D., Scherman, D., Demeneix, B. and Behr, J.P. (1995) A versatile vector for gene and oligonucleotide transfer into cells in culture and in vivo: polyethylenimine. *Proceedings of the National Academy of Sciences of the United States of America*, 92, 7297-7301.

[76] Thomas, M., Lu, J.J., Ge, Q., Zhang, C., Chen, J. and Klibanov, A.M. (2005) Full deacylation of polyethylenimine dramatically boosts its gene delivery efficiency and specificity to mouse lung. *Proceedings of the National Academy of Sciences of the United States of America*, 102, 5679-5684.

[77] Thomas, M. and Klibanov, A.M. (2002) Enhancing polyethylenimine's delivery of plasmid DNA into mammalian cells. *Proceedings of the National Academy of Sciences of the United States of America*, 99, 14640-14645.

[78] Akinc, A., Thomas, M., Klibanov, A.M. and Langer, R. (2005) Exploring polyethylenimine-mediated DNA transfection and the proton sponge hypothesis. *The journal of gene medicine*, 7, 657-663.

[79] Grayson, A.C., Doody, A.M. and Putnam, D. (2006) Biophysical and structural characterization of polyethylenimine-mediated siRNA delivery in vitro. *Pharmaceutical research*, 23, 1868-1876.

[80] Sonawane, N.D., Szoka, F.C., Jr. and Verkman, A.S. (2003) Chloride accumulation and swelling in endosomes enhances DNA transfer by polyamine-DNA polyplexes. *The Journal of biological chemistry*, 278, 44826-44831.

[81] Thomas, M. and Klibanov, A.M. (2003) Conjugation to gold nanoparticles enhances polyethylenimine's transfer of plasmid DNA into mammalian cells. *Proceedings of the National Academy of Sciences of the United States of America*, 100, 9138-9143.

[82] Thomas, M., Ge, Q., Lu, J.J., Chen, J. and Klibanov, A. (2005) Cross-linked Small Polyethylenimines: While Still Nontoxic, Deliver DNA Efficiently to Mammalian Cells in Vitro and in Vivo. *Pharmaceutical research*, 22, 373-380.

[83] Godbey, W.T., Wu, K.K. and Mikos, A.G. (1999) Tracking the intracellular path of poly(ethylenimine)/DNA complexes for gene delivery. *Proceedings of the National Academy of Sciences of the United States of America*, 96, 5177-5181.

[84]   Brisson, M., Tseng, W.C., Almonte, C., Watkins, S. and Huang, L. (1999) Subcellular trafficking of the cytoplasmic expression system. *Human gene therapy*, 10, 2601-2613.

[85]   Lin, W., Coombes, A.G., Davies, M.C., Davis, S.S. and Illum, L. (1993) Preparation of sub-100 nm human serum albumin nanospheres using a pH-coacervation method. *Journal of drug targeting*, 1, 237-243.

[86]   Rubino, O.P., Kowalsky, R. and Swarbrick, J. (1993) Albumin microspheres as a drug delivery system: relation among turbidity ratio, degree of cross-linking, and drug release. *Pharmaceutical research*, 10, 1059-1065.

[87]   Chen, C.Q., Lin, W., Coombes, A.G., Davis, S.S. and Illum, L. (1994) Preparation of human serum albumin microspheres by a novel acetone-heat denaturation method. *Journal of microencapsulation*, 11, 395-407.

[88]   Tiyaboonchai, W. and Limpeanchob, N. (2007) Formulation and characterization of amphotericin B-chitosan-dextran sulfate nanoparticles. *International journal of pharmaceutics*, 329, 142-149.

[89]   Tseng, W.C. and Jong, C.M. (2003) Improved stability of polycationic vector by dextran-grafted branched polyethylenimine. *Biomacromolecules*, 4, 1277-1284.

[90]   Vinogradov, S.V., Bronich, T.K. and Kabanov, A.V. (1998) Self-assembly of polyamine-poly(ethylene glycol) copolymers with phosphorothioate oligonucleotides. *Bioconjugate chemistry*, 9, 805-812.

[91]   Erbacher, P., Bettinger, T., Belguise-Valladier, P., Zou, S., Coll, J.L., Behr, J.P. and Remy, J.S. (1999) Transfection and physical properties of various saccharide, poly(ethylene glycol), and antibody-derivatized polyethylenimines (PEI). *The journal of gene medicine*, 1, 210-222.

[92]   Kataoka, K., Harada, A. and Nagasaki, Y. (2001) Block copolymer micelles for drug delivery: design, characterization and biological significance. *Adv Drug Deliv Rev*, 47, 113-131.

[93]   Sung, S.J., Min, S.H., Cho, K.Y., Lee, S., Min, Y.J., Yeom, Y.I. and Park, J.K. (2003) Effect of polyethylene glycol on gene delivery of polyethylenimine. *Biol Pharm Bull*, 26, 492-500.

[94]   Tang, G.P., Zeng, J.M., Gao, S.J., Ma, Y.X., Shi, L., Li, Y., Too, H.P. and Wang, S. (2003) Polyethylene glycol modified polyethylenimine for improved CNS gene transfer: effects of PEGylation extent. *Biomaterials*, 24, 2351-2362.

[95]   Forrest, M.L., Meister, G.E., Koerber, J.T. and Pack, D.W. (2004) Partial acetylation of polyethylenimine enhances in vitro gene delivery. *Pharmaceutical research*, 21, 365-371.

[96]   Kircheis, R., Kichler, A., Wallner, G., Kursa, M., Ogris, M., Felzmann, T., Buchberger, M. and Wagner, E. (1997) Coupling of cell-binding ligands to polyethylenimine for targeted gene delivery. *Gene therapy*, 4, 409-418.

[97]   Li, S., Tan, Y., Viroonchatapan, E., Pitt, B.R. and Huang, L. (2000) Targeted gene delivery to pulmonary endothelium by anti-PECAM antibody. *Am J Physiol Lung Cell Mol Physiol*, 278, L504-511.

[98]   O'Neill, M.M., Kennedy, C.A., Barton, R.W. and Tatake, R.J. (2001) Receptor-mediated gene delivery to human peripheral blood mononuclear cells using anti-CD3 antibody coupled to polyethylenimine. *Gene therapy*, 8, 362-368.

[99] Harbottle, R.P., Cooper, R.G., Hart, S.L., Ladhoff, A., McKay, T., Knight, A.M., Wagner, E., Miller, A.D. and Coutelle, C. (1998) An RGD-oligolysine peptide: a prototype construct for integrin-mediated gene delivery. *Human gene therapy*, 9, 1037-1047.

[100] Grzelinski, M., Urban-Klein, B., Martens, T., Lamszus, K., Bakowsky, U., Hobel, S., Czubayko, F. and Aigner, A. (2006) RNA interference-mediated gene silencing of pleiotrophin through polyethylenimine-complexed small interfering RNAs in vivo exerts antitumoral effects in glioblastoma xenografts. *Human gene therapy*, 17, 751-766.

[101] Urban-Klein, B., Werth, S., Abuharbeid, S., Czubayko, F. and Aigner, A. (2005) RNAi-mediated gene-targeting through systemic application of polyethylenimine (PEI)-complexed siRNA in vivo. *Gene therapy*, 12, 461-466.

[102] Schiffelers, R.M., Ansari, A., Xu, J., Zhou, Q., Tang, Q., Storm, G., Molema, G., Lu, P.Y., Scaria, P.V. and Woodle, M.C. (2004) Cancer siRNA therapy by tumor selective delivery with ligand-targeted sterically stabilized nanoparticle. *Nucleic acids research*, 32, e149.

[103] Mao, S., Neu, M., Germershaus, O., Merkel, O., Sitterberg, J., Bakowsky, U. and Kissel, T. (2006) Influence of polyethylene glycol chain length on the physicochemical and biological properties of poly(ethylene imine)-graft-poly(ethylene glycol) block copolymer/SiRNA polyplexes. *Bioconjugate chemistry*, 17, 1209-1218.

[104] Nimesh, S. and Chandra, R. (2009) Polyethylenimine nanoparticles as an efficient in vitro siRNA delivery system. *Eur J Pharm Biopharm*.

[105] Patnaik, S., Aggarwal, A., Nimesh, S., Goel, A., Ganguli, M., Saini, N., Singh, Y. and Gupta, K.C. (2006) PEI-alginate nanocomposites as efficient in vitro gene transfection agents. *Journal of Controlled Release*, 114, 398-409.

[106] Jeong, J.H., Kim, S.H., Lee, M., Kim, W.J., Park, T.G., Ko, K.S. and Kim, S.W. Non-viral systemic delivery of Fas siRNA suppresses cyclophosphamide-induced diabetes in NOD mice. *Journal of Controlled Release*, In Press, Corrected Proof.

[107] Svenson, S. and Tomalia, D.A. (2005) Dendrimers in biomedical applications--reflections on the field. *Adv Drug Deliv Rev*, 57, 2106-2129.

[108] Zhou, J., Wu, J., Hafdi, N., Behr, J.P., Erbacher, P. and Peng, L. (2006) PAMAM dendrimers for efficient siRNA delivery and potent gene silencing. *Chem Commun (Camb)*, 2362-2364.

[109] Shen, X.C., Zhou, J., Liu, X., Wu, J., Qu, F., Zhang, Z.L., Pang, D.W., Quelever, G., Zhang, C.C. and Peng, L. (2007) Importance of size-to-charge ratio in construction of stable and uniform nanoscale RNA/dendrimer complexes. *Org Biomol Chem*, 5, 3674-3681.

[110] Bielinska, A., Kukowska-Latallo, J.F., Johnson, J., Tomalia, D.A. and Baker, J.R., Jr. (1996) Regulation of in vitro gene expression using antisense oligonucleotides or antisense expression plasmids transfected using starburst PAMAM dendrimers. *Nucleic acids research*, 24, 2176-2182.

[111] Yoo, H., Sazani, P. and Juliano, R.L. (1999) PAMAM Dendrimers as Delivery Agents for Antisense Oligonucleotides. *Pharmaceutical research*, 16, 1799-1804.

[112] Helin, V., Gottikh, M., Mishal, Z., Subra, F., Malvy, C. and Lavignon, M. (1999) Cell cycle-dependent distribution and specific inhibitory effect of vectorized antisense oligonucleotides in cell culture. *Biochem Pharmacol*, 58, 95-107.

[113] Weber, N., Ortega, P., Clemente, M.I., Shcharbin, D., Bryszewska, M., de la Mata, F.J., Gomez, R. and Munoz-Fernandez, M.A. (2008) Characterization of carbosilane dendrimers as effective carriers of siRNA to HIV-infected lymphocytes. *J Control Release*, 132, 55-64.

[114] Taratula, O., Garbuzenko, O.B., Kirkpatrick, P., Pandya, I., Savla, R., Pozharov, V.P., He, H. and Minko, T. (2009) Surface-engineered targeted PPI dendrimer for efficient intracellular and intratumoral siRNA delivery. *J Control Release*, 140, 284-293.

[115] Perez, A.P., Romero, E.L. and Morilla, M.J. (2009) Ethylendiamine core PAMAM dendrimers/siRNA complexes as in vitro silencing agents. *International journal of pharmaceutics*, 380, 189-200.

[116] Pille, J.Y., Li, H., Blot, E., Bertrand, J.R., Pritchard, L.L., Opolon, P., Maksimenko, A., Lu, H., Vannier, J.P., Soria, J. *et al.* (2006) Intravenous delivery of anti-RhoA small interfering RNA loaded in nanoparticles of chitosan in mice: Safety and efficacy in xenografted aggressive breast cancer. *Human gene therapy*, 17, 1019-1026.

[117] Toub, N., Bertrand, J.R., Tamaddon, A., Elhamess, H., Hillaireau, H., Maksimenko, A., Maccario, J., Malvy, C., Fattal, E. and Couvreur, P. (2006) Efficacy of siRNA nanocapsules targeted against the EWS-Fli1 oncogene in Ewing sarcoma. *Pharmaceutical research*, 23, 892-900.

[118] Bouclier, C.l., Moine, L., Hillaireau, H., Marsaud, V.r., Connault, E., Opolon, P., Couvreur, P., Fattal, E. and Renoir, J.-M. (2008) Physicochemical Characteristics and Preliminary in Vivo Biological Evaluation of Nanocapsules Loaded with siRNA Targeting Estrogen Receptor Alpha. *Biomacromolecules*, 9, 2881-2890.

[119] Patil, Y. and Panyam, J. (2009) Polymeric nanoparticles for siRNA delivery and gene silencing. *International journal of pharmaceutics*, 367, 195-203.

[120] Yuan, X., Li, L., Rathinavelu, A., Hao, J., Narasimhan, M., He, M., Heitlage, V., Tam, L., Viqar, S. and Salehi, M. (2006) SiRNA drug delivery by biodegradable polymeric nanoparticles. *Journal of nanoscience and nanotechnology*, 6, 2821-2828.

[121] Fischer, D., Li, Y., Ahlemeyer, B., Krieglstein, J. and Kissel, T. (2003) In vitro cytotoxicity testing of polycations: influence of polymer structure on cell viability and hemolysis. *Biomaterials*, 24, 1121-1131.

[122] Jevprasesphant, R., Penny, J., Jalal, R., Attwood, D., McKeown, N.B. and D'Emanuele, A. (2003) The influence of surface modification on the cytotoxicity of PAMAM dendrimers. *International journal of pharmaceutics*, 252, 263-266.

[123] Godbey, W.T., Wu, K.K., Hirasaki, G.J. and Mikos, A.G. (1999) Improved packing of poly(ethylenimine)/DNA complexes increases transfection efficiency. *Gene therapy*, 6, 1380-1388.

[124] Godbey, W.T., Wu, K.K. and Mikos, A.G. (2001) Poly(ethylenimine)-mediated gene delivery affects endothelial cell function and viability. *Biomaterials*, 22, 471-480.

[125] Ferrari, S., Moro, E., Pettenazzo, A., Behr, J.P., Zacchello, F. and Scarpa, M. (1997) ExGen 500 is an efficient vector for gene delivery to lung epithelial cells in vitro and in vivo. *Gene therapy*, 4, 1100-1106.

[126] Plank, C., Mechtler, K., Szoka, F.C., Jr. and Wagner, E. (1996) Activation of the complement system by synthetic DNA complexes: a potential barrier for intravenous gene delivery. *Human gene therapy*, 7, 1437-1446.

[127] Howard, K.A., Rahbek, U.L., Liu, X.D., Damgaard, C.K., Glud, S.Z., Andersen, M.O., Hovgaard, M.B., Schmitz, A., Nyengaard, J.R., Besenbacher, F. *et al.* (2006) RNA interference in vitro and in vivo using a chitosan/siRNA nanoparticle system. *Molecular Therapy*, 14, 476-484.

[128] Zimmermann, T.S., Lee, A.C., Akinc, A., Bramlage, B., Bumcrot, D., Fedoruk, M.N., Harborth, J., Heyes, J.A., Jeffs, L.B., John, M. *et al.* (2006) RNAi-mediated gene silencing in non-human primates. *Nature*, 441, 111-114.

In: Gene Silencing: Theory, Techniques and Applications
Editor: Anthony J.Catalano
ISBN: 978-1-61728-276-8

Chapter V

Application of Negatively Charged Hydroxyproline-based DNA Mimics for Gene Silencing

Vladimir A. Efimov and Oksana G. Chakhmakhcheva

Shemyakin-Ovchinnikov Institute of Bioorganic Chemistry, ul. Miklukho-Maklaya
16/10, Moscow 117997, Russia

Abstract

With the aim in view to improve physicochemical and biological properties of natural oligonucleotides, several types of DNA analogues and mimics were designed, particularly negatively charged PNA-like mimics. Among them, two types of DNA mimics representing hetero-oligomers constructed from alternating monomers of phosphono peptide nucleic acids and monomers on the base of *trans*-1-acetyl-4-hydroxy-L-proline (HypNA-pPNAs) as well as oligomers constructed from chiral analogues of peptide nucleic acids with a constrained trans-4-hydroxy-N-acetylpyrrolidine-2-phosphonic acid backbone (pHypNAs) were developed. Their physico-chemical and biological properties were evaluated in the comparison with natural oligonucleotides, classical peptide nucleic acids and morpholino phosphorodiamidate oligonucleotide analogues. The results obtained in a set of experiments revealed a high potential of these phosphonate-containing PNA derivatives for a number of biological applications, such as diagnostic, nucleic acids analysis and inhibition of gene expression. HypNA-pPNA and pHypNA mimics combine high hybridization and mismatch discrimination characteristics with good water solubility and biological stability as well as the ability to penetrate cell membranes. Their effectiveness to provide the specific knockdown of a target protein production was demonstrated in research involving *in vitro* systems, living cells and intact organisms. As their effect lasts over a period of several days, due to their high stability in living cells, it represents a very potent technology for administrating antisense- or antigene-based drugs for future therapeutic applications.

Introduction

In the last years, the search for modified oligonucleotides and analogues with improved hybridization ability and selectivity toward nucleic acid targets attracted increased attention in connection with their application in molecular biology and medicine. A number of nucleic acid mimics have been developed, including successful examples such as charge-neutral peptide nucleic acids (PNAs)[1-3] and phosphorodiamidate morpholino oligomers (MOs) [4] (Figure 1). Thus, PNAs are attractive candidates for the discovery and development of antisense gene therapeutics and antigene drugs [3, 5]. These oligonucleotide mimics are unaffected by cellular degradative enzymes, especially by nucleases, and show strong nucleic acids binding. They demonstrated high affinity to DNA and RNA and stringent mismatch discrimination. However, rapid progress in biological applications of classical PNAs has been limited by their low water solubility, tendency to self-aggregation, poor propensity to cross cell membranes and, in part, their inappropriate cellular localization [6, 7]. Transfection reagents, such as cationic lipids and polymers, are positively charged and rely on self-assembly via complexation with the negatively charged DNA. Unmodified classical PNAs are charge neutral molecules, and they do not interact with cationic lipids and consequently are not effectively delivered to cells by such reagents. However, the improved delivery of PNAs into cells can be achieved by the use of carrier systems, such as their conjugates with cell-penetrating peptides and some other compounds as well as by the addition of special reagents to the culture medium [8, 9]. One of the attractive approaches to improve properties of PNAs is their supplying with negative charges in order to improve their water solubility and interaction with cellular delivery agents. Recently, PNAs conjugated to oligophosphonates via phosphonate glutamine and bisphosphonate lysine amino acid derivatives (pcPNAs) were reported by T.Shiraishi and. et al. [10], and their effectiveness as antisense agents was demonstrated. Up to twelve phosphonate moieties were introduced into a PNA oligomer by solid phase synthesis. It was shown that T_m values of the pcPNA/DNA and pcPNA/RNA complexes with phosphonate and bis-phosphonate conjugated PNAs were virtually identical to that of the unmodified PNA. However, the stability of the phosphonate conjugated PNA/DNA duplex decreased with decreasing ionic strength as opposed to the behavior of duplexes between classical PNAs and complementary DNA templates [11]. It was shown that such phosphonate conjugates may be delivered by a large range of systems developed for DNA and siRNA delivery including cationic polymers and dendrimers as well as nanoparticles, which all rely on electrostatic interactions between the antisense agent and the delivery vehicle.

At their turn, MOs exhibit good hybridization properties and the resistance to a broad range of cellular degradative enzymes. Also, they demonstrate high solubility in water despite their lack of charge, due to strong polarity. However, they exhibit lower affinity to nucleic acids and mismatch discrimination than PNAs [4, 12, 13].

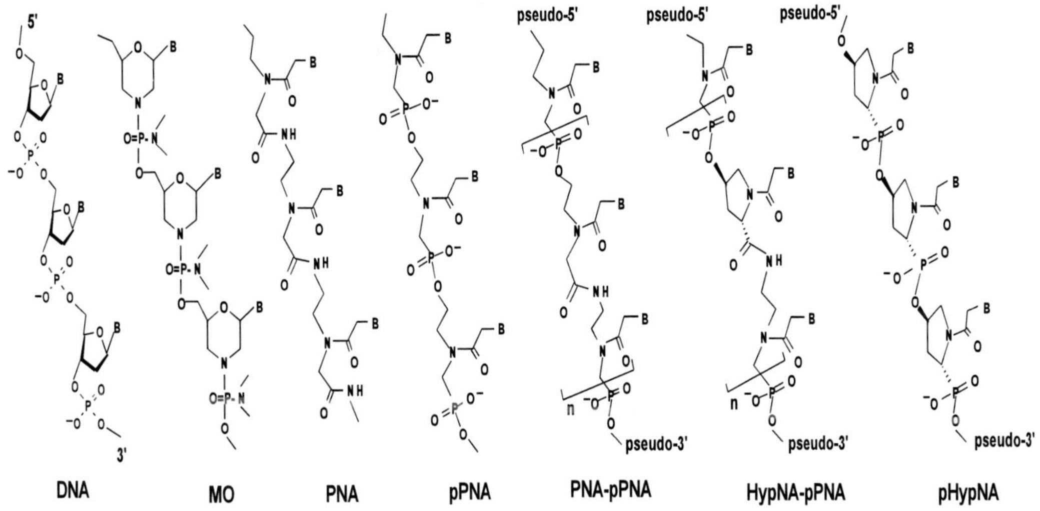

Figure 1. General chemical structures of DNA mimics.

Since the publication of classical PNAs, a range of other PNA analogues have been prepared. In this review we decided to focus on PNA-like negatively charged DNA mimics containing phosphono-ester bonds between monomer units and their use as antisense inhibitors of gene expression in cell cultures and in experimental animals.

Phophono-PNAs and their Hetero-oligomer Derivatives

In recent years, a class of DNA mimics representing phosphonate PNA analogues (pPNAs) was designed, and the synthesis of pPNA oligomers containing N-(2-hydroxyethyl)-phosphono glycine (Figure 1), or N-(2-aminoethyl)-phosphono glycine backbone has been accomplished in solution and on solid phase [14-16]. The most effective synthesis of pPNA oligomers was achieved using solid phase technique, which was similar to the phosphotriester oligonucleotide synthesis with the O-nucleophilic intra-molecular catalysis [14]. To construct pPNA oligomers, the procedures to obtain corresponding monomers of types (1) and (2) were developed (Figure 2). These building units have a combination of blocking groups compatible with the technique of the automated oligonucleotide synthesis, which was successfully applied with some modifications to the construction of pPNA oligomers. Also, a set of optically active and aromatic pPNA analogues was obtained using this technique [17]. The examination of properties of pPNA oligomers revealed that they are fully stable to the action of nucleases, and the introduction of negative charges into the PNA backbone led to the excellent solubility characteristics. However, the thermal stability of pPNA complexes with complementary DNA (or RNA) targets was lower than the stability of corresponding complexes formed by classical PNAs.

As both PNA and pPNA mimics are isosteric compounds, a set of chimeras composed of PNA and pPNA monomers (PNA-pPNAs) was obtained [18, 19]. The synthesis of these hetero-oligomers was carried out using specially synthesized dimer PNA-pPNA building units (3) (Figure 2) containing the internal amide bond between the monomer residues. The PNA-pPNA chimeras showed the improved hybridization characteristics in comparison with pPNAs and DNA oligomers, which were combined with good water solubility [18].

Later, the solid-phase synthesis of HypNA-pPNA hetero-oligomers, in which PNA monomers were replaced by monomers on the basis of *trans*-4-hydroxy-L-proline (HypNA), has been accomplished (Figure 1) [20, 21]. A HypNA monomer represents conformationally constrained chiral PNA analogue, in which the β-C atom of the hydroxyethyl group and the α-C atom of a glycine unit of the backbone are bridged with methylene group. Convenient schemes for the synthesis of the HypNA monomers and HypNA-pPNA dimers (4) (Figure 2) were developed [22]. The evaluation of properties of HypNA-pPNA hetero-oligomers synthesized from these dimers revealed that they demonstrate very strong binding to complementary DNA and RNA strands with melting temperatures very close to those of classical PNA/DNA, or PNA/RNA, complexes [21].

Locked Phosphono-PNA Analogues on the Basis of Hydroxyproline

Recently, DNA mimics totally constructed from the conformationally constrained chiral pPNA analogues on the basis of 1-acetyl-4-hydroxypyrrolidine-2-phosphonic acid were designed, and the corresponding monomers (5 a-d) were synthesized [23, 24]. Preliminary investigations on the binding properties of chiral homo-Thy oligo-mimics have shown that the oligomers totally constructed from the monomers with *cis*-L (5b) or *cis*-D (5d) configuration (Figure 2) were not able to form stable complexes with the complementary homo-A DNA or RNA targets. The homo-Thy oligomers constructed from *trans*-L (5a) or *trans*-D (5c) monomers were able to give stable complexes with complementary targets, particularly the oligomer composed of *trans*-L monomers. The latter type of oligomers exhibited strong binding to the complementary DNA/RNA target with high melting temperatures of complexes. Then, the procedures for construction of *trans*-L pHypNA (pHypNA) monomers (Figure 1) containing the four natural nucleobases were developed, and the corresponding oligomers were synthesized. Like for pPNA mimics, the solid–phase synthesis of pHypNA oligomers was carried out by the phosphorotriester method with O-nucleophilic intra-molecular catalysis. The evaluation of properties of pHypNA oligomers revealed that they have excellent solubility in water, and this type of mimics fully preserved all properties of HypNA-pPNAs including high affinity to DNA and RNA targets.

In an attempt to further improve the hybridization efficiency of DNA mimics, another type of monomers *trans*-L pHypNA-Me (6) was obtained (Figure 2). In these compounds, the residue of thymin-1-yl-acetic acid was replaced by the residue of (R)-2-(thymin-1-yl)propanoic or (S)-2-(thymin-1-yl)propanoic acid [25]. This type of molecules had an additional chiral methyl group in the position close to the heterocycle. The monomers of type (6) were used for the automated synthesis of the corresponding homo-Thy pHypNA-Me oligomers. The experiments on the hybridization of these stereo-specific oligomers with complementary DNA and RNA fragments revealed dramatic loss of the hybridization ability of *trans*-L pHypNA-Me oligomers in comparison with *trans*-L pHypNA oligomers in both cases. Earlier, it was shown that the distance between the nucleobases in PNA oligomers is of great importance for the stability of their complexes with the complementary nucleic acid targets [26]. Thus, the extension of a linker to the nucleobase from methylenecarbonyl to ethylenecarbonyl in the classical PNA molecule had a negative influence on hybridization properties of PNAs, and the thermal stability of the complexes formed from these modified PNA oligomers and the complementary DNA was significantly lower than that of the corresponding complexes involving unmodified PNA. In the case of *trans*-L pHypNA-Me derivatives, the length of this linker was not changed; however, the replacement of H atom in position 2 of (thymin-1-yl)acetic acid with methyl group also had a negative influence on hybridization properties of mimic oligomers.

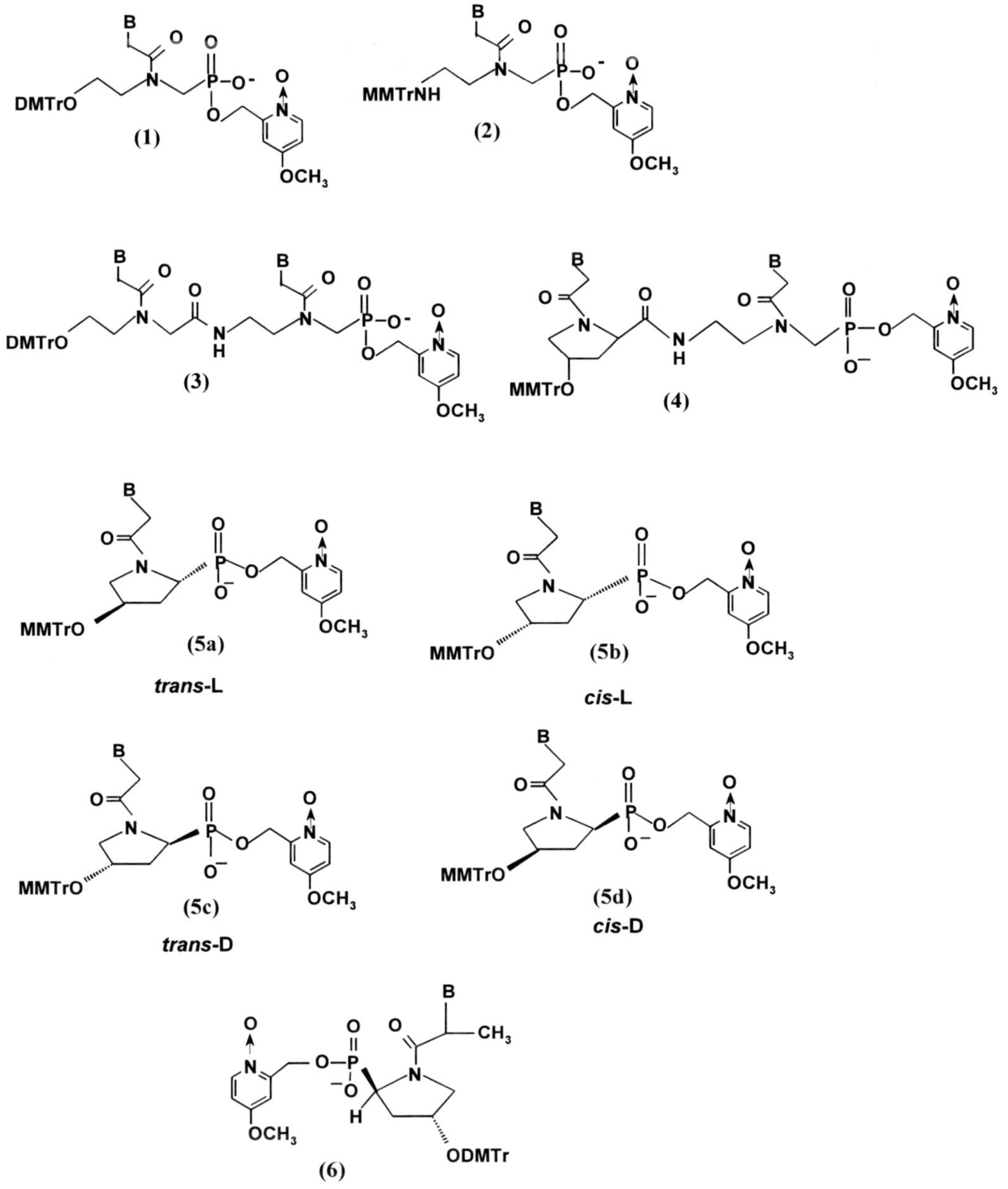

Figure 2. Building units for the solid phase synthesis of negatively charged PNA analogues.

Pyrrolidine-based Oligonucleotide Analogues

Some time ago, several research groups were interested in hydroxy-N-acetylprolinol derivatives as a sugar substitute in oligonucleotides. Thus, modified oligonucleotides incorporating *trans*-4-hydroxy-*N*-acetyl-L-prolinol (*trans*-4-OH-L-NAP-NA) (7), or its D-analogue, as sugar substitute were synthesized from the corresponding phosphoramidite monomers containing adenine and thymine as nucleobases [27] (Figure 3). This type of compounds represented an analogue of 2′-5′-connected 3′-deoxyribooligonucleotide with the same number of atoms between their repeating units, and this distance was not respected to the 6-atom spacing found in natural nucleic acids. It was shown that these oligomers display interesting hybridization features with complementary fragments of natural nucleic acids, which preferably hybridize with *trans*-L-hydroxyprolinol-containing oligomers over the oligomers of *trans*-D type. Also, the prolinol-based oligomers, 3-hydroxy-*N*-acetylprolinol derived nucleotide analogues (*trans*-3-OH-L-NAP-NA) (8) were prepared, where phosphorus spacing was as in DNA molecule [28]. Fully modified oligomers were synthesized from the O-phosphoramidites of the properly protected *trans*- and *cis*-3-hydroxy-*N*-acetylprolinol monomers. The *trans*-L as well as the *trans*-D homo-Ade oligomers of this type were capable of hybridization with complementary DNA and RNA, whereas no complexation was observed between homo-Thy oligomers (or the oligomers with the mixed Ade, Thy sequence) and complementary nucleic acid targets. In general, the stability of complexes formed by these oligonucleotide analogues was considerably lower than those detected for 4-hydroxy-*N*-acetylprolinol-based structures. Starting from *trans*-3-hydroxy-L-proline, the synthesis of the pyrrolidino-C-nucleotide analogues (*pyrro*-C-NA) (9) was accomplished to obtain oligomers bearing positive charges in the backbone [29, 30] (Figure 3). The synthesis of these oligomers was performed by the standard phosphoramidite method, and the thermal stability of complexes formed by the modified oligomers with complementary targets was measured. It was shown that the stability of these complexes was in general lower than those of the duplexes formed by unmodified oligonucleotides.

Recently, a series of prolinol-based modified monomers with phosphonomethyl (10), or phosphonoformyl (11), group attached to the pyrrolidine nitrogen (*pro*-N-NA) was prepared for the use in solid phase synthesis of oligonucleotide mimics [31] (Figure 3). In these monomers, the 3′- or 4′-carbon atom of ribose was replaced with the nitrogen atom of pyrrolidine, and the oxygen atom of ribose was replaced with methylene group. These protected nucleotide analogues give rise to two types of internucleotide linkages when incorporated into natural oligonucleotides: *N-CH₂-P-O-C* and *N-CO-P-O-C*. Each individual pair consisted of two streoisomers capable of forming the same type of the isosteric internucleotide linkages. Also, several types of nucleoside analogues representing phosphonate derivatives of 3-pyrrolidinol (*pyrro*-N-NA) (12) were obtained and incorporated into oligonucleotide chains. These three structurally diverse types of the protected pyrrolidine monomers were introduced into oligonucleotides using the solid phase synthesis. A series of nonamers containing two or three modified units, as well as the fully modified adenine 15-mer, were synthesized. The measurements of thermal characteristics of the complexes of modified oligonucleotides with the complementary DNA revealed a destabilizing effect of the introduced modification. So, oligonucleotides modified with such units exhibited lower affinity to complementary nucleic acids than natural ones [32].

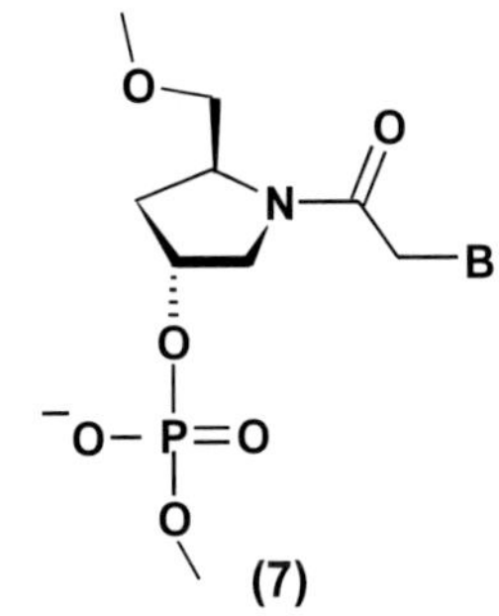

(7)
***trans*-4-HO-L-NAP-NA**

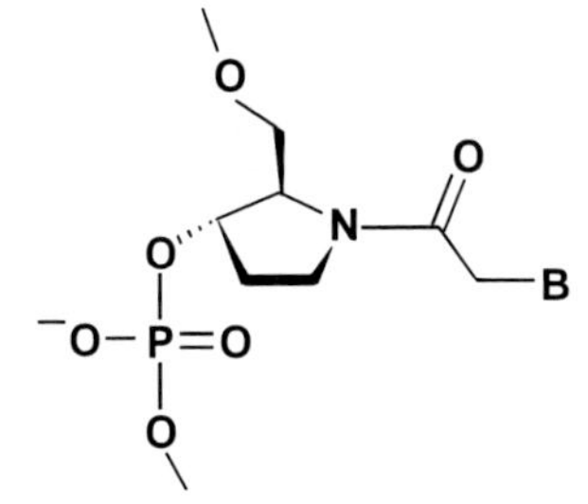

(8)
***trans*-3-HO-L-NAP-NA**

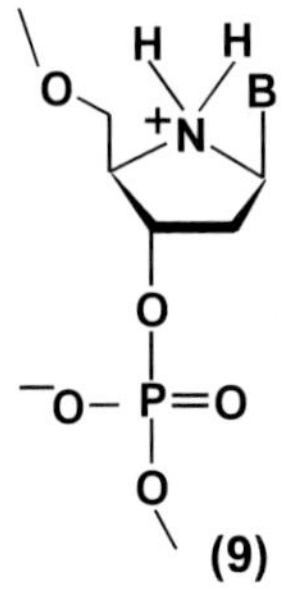

(9)
***Pyrro*-C-NA**

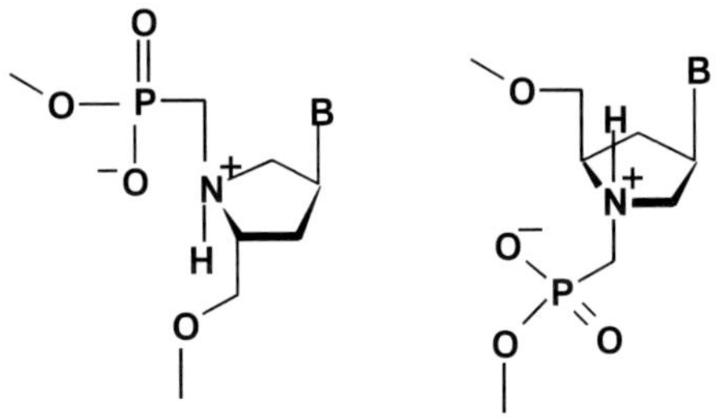

(10)

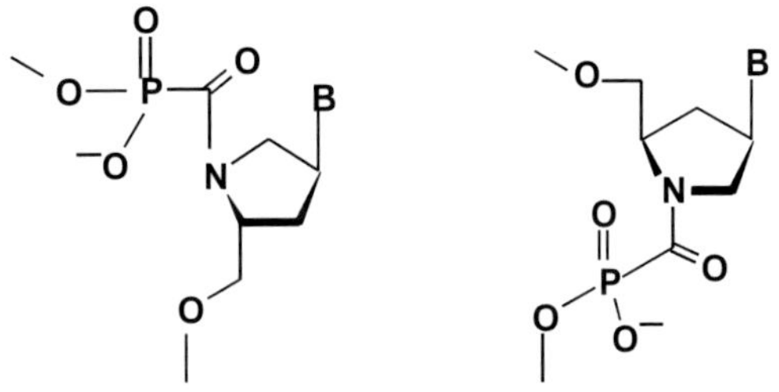

(11)

***Pro*-N-NA**

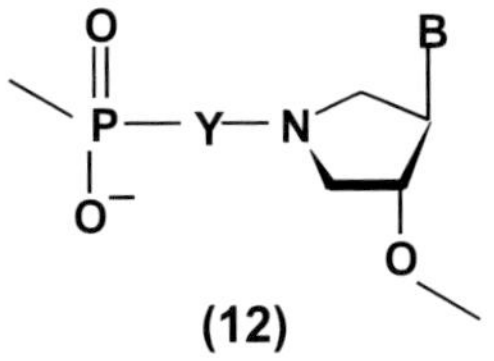

Y = CH_2, CH_2CH_2, CH_2CO

(12)
***Pyrro*-N-NA**

Figure 3. Some pyrrolidine-based analogues of nucleic acids.

Several other types of monomers on the base of pyrrolidine (PyrrNAs) (13-17), in which nucleobase, hydroxyl group and phosphonic acid residue were attached to different carbon atoms of the pyrrolidine ring were obtained (Figure 4) and the synthesis of the corresponding PyrrNA oligomers was accomplished by the modified solid-phase phosphotriester method similar to the synthesis of pHypNA and HypNA-pPNA oligomers [33]. Preliminary results obtained in the study of the hybridization properties of these mimics revealed that oligomers constructed from monomers with *cis*-L configuration are not able to form stable complexes with the complementary oligonucleotides, whereas oligomers constructed from *trans*-L monomers are able to form complexes with the complementary targets, but the stability of these complexes is lower than the stability of natural DNA/DNA and DNA/RNA duplexes and significantly lower than the stability of complexes between *trans*-L pHypNA oligomers and DNA (or RNA) targets.

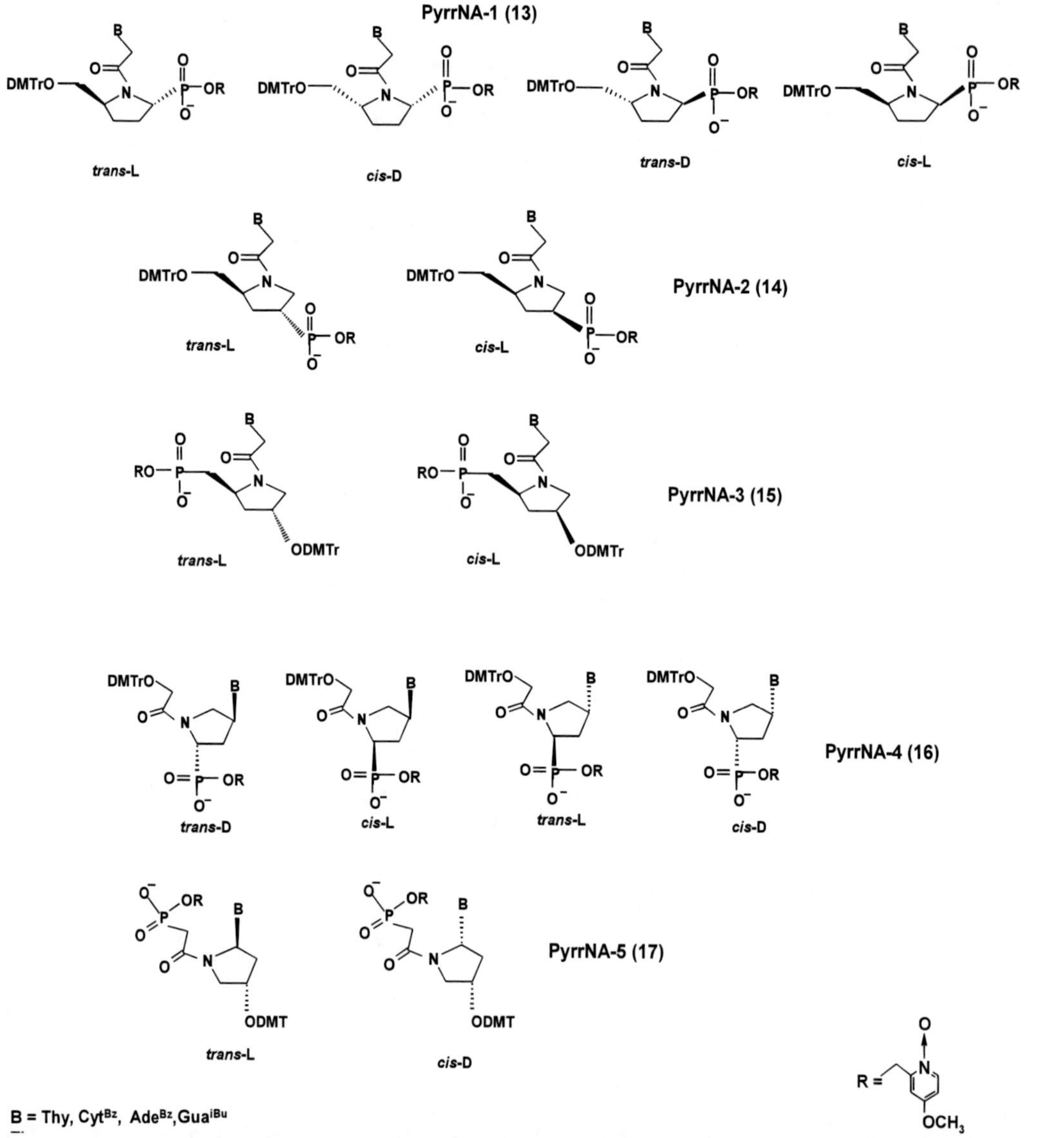

Figure 4. General chemical structures of chiral monomer units for the construction of PyrrNA mimic oligomers.

Biological Stability and Binding Selectivity of Negatively Charged PNA-like Mimics

The ideal antisense molecule should not only effectively recognize a target, but also be stable in biological fluids. Unlike the biological instabilities of close analogs of RNA and DNA, negatively charged pPNA. PNA-pPNA, HypNA-pPNA and pHypNA DNA mimic show stability to the action of nucleases and proteases [34]. Their half-live in serum was estimated as ~72 h. The similar stability of these mimics was observed in cytoplasmic and nuclear cell extracts. So, the high stability allow them to preserve their action in cells for a prolonged period of time after a single administration, and this stability to enzymatic degradation provides an advantage in applications requiring long-term activity in biological systems, such as studies in embryos and in therapeutics.

In contrast to natural oligonucleotides and their phosphorothioate analogues, these mimics as well as classical PNAs do not activate RNase H upon binding to complementary RNA [35]. Such blocker-type oligomers must tightly bind to their targeted RNA sequences in order to prevent RNA processing (e.g., splicing), readout (translation), or other functions (e.g., extension of telomers) of their targeted RNA transcripts. Negatively charged HypNA-pPNA and pHypNA mimics demonstrate excellent hybridization properties and the binding selectivity (Table 1). It was shown that the hybridization of these mimic oligomers with DNA and RNA targets occurred in a sequence specific manner [21, 23] and the formation of complexes between mimic probes and non-complementary targets was not detected. The introduction of negative charges into the backbone provides excellent solubility without hindering the hybridization properties of PNAs. From the titration data and the electrophoretic behavior, it was concluded that, similar to PNAs, homo-pyrimidine sequences of all phosphonate containing mimics formed with complementary DNA (or RNA) targets triple helixes, whereas oligomers with mixed nucleobase sequences formed duplexes with nucleic acid targets [18, 20, 24]. Similarly to PNAs, HypNA-pPNA and pHypNA probes with a chain length of 16-18-nt can effectively discriminate between single base mismatches or deletions in the target DNA (or RNA) sequence. The introduction of one mismatch in the center of a sequence gives a drop in the melting temperature 12-23°C depending on the length of oligomer, base mismatch position and the sequence. Oligomers with two separately situated mismatches were not able to form stable complexes with the targets.

Several other types of monomers on the base of pyrrolidine (PyrrNAs) (13-17), in which nucleobase, hydroxyl group and phosphonic acid residue were attached to different carbon atoms of the pyrrolidine ring were obtained (Figure 4) and the synthesis of the corresponding PyrrNA oligomers was accomplished by the modified solid-phase phosphotriester method similar to the synthesis of pHypNA and HypNA-pPNA oligomers [33]. Preliminary results obtained in the study of the hybridization properties of these mimics revealed that oligomers constructed from monomers with *cis*-L configuration are not able to form stable complexes with the complementary oligonucleotides, whereas oligomers constructed from *trans*-L monomers are able to form complexes with the complementary targets, but the stability of these complexes is lower than the stability of natural DNA/DNA and DNA/RNA duplexes and significantly lower than the stability of complexes between *trans*-L pHypNA oligomers and DNA (or RNA) targets.

In contrast to natural oligonucleotides and similarly to PNAs, negatively charged HypNA-pPNA and pHypNA oligomers can effectively hybridize with complementary targets at low salt concentrations, and T_m values of their complexes with DNA (or RNA) targets are not dependent on ionic strength [23, 24]. Recently, derivatives of classical PNAs representing their conjugates with phosphonate glutamine and bisphosphonate lysine amino acids via the N-terminus end (pcPNAs) were reported [10]. However, opposed to the behavior of HypNA-pPNAs and pHypNAs, pcPNAs bearing terminal oligophosphonates demonstrated decreasing in the stability of their duplexes with complementary DNA with decreasing ionic strength. It should be noted that the independence between T_m and salt concentration provides a dramatic advantage over classical anionic DNA and RNA oligonucleotides in probe diagnostic applications and some molecular biology applications.

Table 1. Stability of complexes formed by DNA mimics and natural oligonucleotides with complementary DNA, or RNA, targets*

Probe sequence**	Target	Oligonu-cleotide T_m (°C)	HypNA-pPNA T_m (°C)	pHypNA T_m (°C)
TTTTTTTTTTTTTT	d(AAAAAAAAAAAAAA)	37	78	76
TTTTgTTTTTcTTT	d(AAAAAAAAAAAAAA)	<10	20	29
TTTTTTTaTTTTTT	d(AAAAAAAAAAAAAA)	23	55	48
TTTTTTTTTTTTTT	r(AAAAAAAAAAAAAA)	35	62	64
AAAAAAAAAAAAAA	d(TTTTTTTTTTTTTT)	37	53	54
TGGTCTCAAGTCAGTG	d(CACTGACTTGAGACCA)	64	58	61
TGGTCTCAAGTCAGTG	d(CACTGAgTTGAGACCA)	57	41	45
TGGTCTCAAGTCAGTG	d(CACTGACTTGAGtCCA)	54	47	49
TCACTCAACACTCAC	d(GTGAGTGTTGAGTGA)	57	53	55
TCACTCAACACTCAC	d(GTGAGTGgTGAGTGA)	48	35	38
CTGCAAAGGACACCATGA	d(TCATGGTGTCCTTTGCAG)	54	69	67
CTGCAAAGcACACCATGA	d(TCATGGTGTCCTTTGCAG)	42	49	52
ATCATGGTCATAGCTGTT	d(AACAGCTATGACCATGAT)	61	58	63
ACACTTACACTTACAC	d(GTGTAAGTGTAAGTGT)	57	42	45
ACACTTACACTTACAC	r(GUGUAAGUGUAAGUGU)	52	57	62
CCCTATAGTGAGTGTCGT	d(ACGACACTCACTATAGGG)	61	60	63
CCCTATAGTGAGTGTCGT	r(ACGACACUCACUAUAGGG)	65	72	69
GCTCTCGTCGCTCTCCAT	d(CATGGAGAGCGACGAGAGC)	76	65	63
GCTCTCGTCGCTCTCCAT	r(CAUGGAGAGCGACGAGAGC)	70	78	77
GCTCTCGTCaCTCTCCAT	d(CATGGAGAGCGACGAGAGC)	65	45	42
GCTCTCGTCaCTCTCCAT	r(CAUGGAGAGCGACGAGAGC)	58	61	58

* Melting experiments were performed in 10 mM Tris-HCl (pH 7.5)/ 0.5 M NaCl/ 10 mM $MgCl_2$.
** Lowercase base in the sequence represents a mismatch position.

HypNA-pPNA mimics, as well as PNA-pPNAs, were tested as capture and detection probes for the construction of arrays for nucleic acid hybridization analysis, and their high potential was demonstrated [21, 38, 39]. Thus, the examination of mimics as capture probes in solid-phase hybridization assays revealed that PNA-pPNA and HypNA-pPNA probes worked well in the same way but giving increased signal intensities compared to natural oligonucleotides due to their higher binding affinity. This property was fully preserved by pHypNA oligomers. In solid-phase experiments, mimic capture probes showed better mismatch discrimination properties than oligonucleotide probes [21]. The application of negatively charged single-stranded PNA-relative mimics as capture probes in nucleic acids analysis is stimulated by their unique property do not interact with some intercalating dyes, particularly ethidium bromide, ethidium homodimer, homodimeric oxazole yellow (YOYO) and thiazole orange (TOTO) dyes. At the same time, the duplexes and triplexes of these mimics with captured DNA (RNA) targets exhibit fluorescence under UV-light after the staining [39]. The application of such visualization method in conjunction with these mimics is promising for the analysis of long nucleic acid targets, because it does not require a preliminary radioactive or fluorescent target labeling.

The phosphonate-containing PNA-like mimics were shown to be very useful for the isolation of intact RNAs, particularly polyadenylated mRNA, from cells and tissues, which is an essential step for many functional genomic applications. A highly efficient procedure for isolation of intact 3′ poly(A)-tailed mRNA from cells and tissues using homo-Thy affinity ligands based on triplex forming mimic oligomers conjugated with biotin residue was developed [40]. Homo-Thy pPNA, PNA-pPNA and HypNA-pPNA oligomers of "linear" and "clamping" types conjugated with biotin residue were tested. The "clamping" molecules represented bis-oligomers composed of homo-Thy strands, which were connected with a neutral flexible linker (Figure 4a). It was demonstrated that mimic oligomers hybridized to poly-A mRNA tails with a high degree of specificity and give very low background binding of the probe to unwanted rRNA. An exceptionally high affinity for DNA and RNA targets, good water solubility and nuclease resistance of these mimics improves the mRNA isolation procedure giving rise to a several-fold increase of mRNA yield and allowed for the isolation of a representative mRNA population including mRNA with short poly-A tracks [40]. The application of mimic probes enables efficient isolation of RNA from extracted total RNA samples in buffers with low salt concentrations, and this decreases the variety of non-polyadenylated RNA co-purification due to the enhanced destabilization of mRNA secondary structure in low salt concentration. Another of their advantages over natural oligonucleotides is reduced DNA contamination of RNA samples that is a consequence of possibility of DNase treatment during RNA preparation.

The properties of molecular beacons containing HypNA-pPNA and pHypNA segments were also tested. Molecular beacons are used for RNA and DNA monitoring in biosensor applications and for gene monitoring in living systems [41, 42]. The construction of novel types of chimeric molecular beacons including the beacons with classical PNA segments [43] as well as chimeric beacons composed of DNA, RNA and negatively charged PNA analogues were reported [44]. The evaluation of properties in model assays with the complementary and mismatched synthetic DNA and RNA targets has shown that the chimeric beacons are about 10 times more sensitive in the detection of nucleic acid targets in solution than oligodeoxyribonucleotide beacons of the same sequence [25]. As the rapid and sensitive detection of RNA in living cells using modified molecular beacons that possess self-delivery, targeting and reporting functions was demonstrated [45], the mimic beacons can be very useful for *in vivo* applications in view of their higher biological stability, an improved discriminatory power and increased affinity for nucleic acid targets in the comparison with natural oligonucleotide probes. Together with the insensitivity of PNA-related beacons to the presence of a salt and the DNA-binding/processing proteins, their potential as robust tools for *in vivo* recognition of specific sequences can be estimated as very high.

The sequence-specific recognition of double-stranded DNA (dsDNA) is a topic of considerable interest in the development of tools for molecular biology, diagnostics, therapeutics and bionanotechnology. Oligomers that bind to duplex DNA would have the potential to directly inactivate gene expression at the chromosome level. A remarkable feature of thymine-rich classical PNAs is their ability to recognize complementary sequences within duplex DNA by strand invasion that results in local strand displacement (formation of a D-loop) [46]. It was shown that at high ionic strength and low concentrations of homopyrimidine PNA oligomer, conventional PNA–dsDNA Hoogsteen type triplexes are formed predominately in a parallel orientation (PNA amino terminal facing the 5'-end of the purine DNA strand). At low ionic strength, high homopyrimidine PNA oligomer concentrations, or long reaction times, triplex invasion complexes dominate. These complexes are extremely stable and contain an internal PNA–DNA–PNA triplex involving combined Hoogsteen (parallel orientation) and Watson–Crick (antiparallel orientation) base pairing and an unbound DNA strand displaced in a P-loop structure [3]. Negatively charged PNA analogues, similarly to PNAs, can invade ds DNA (Figure 4b) [25]. The melting experiments and the gel mobility shift assays for the interaction of a poly-purine/poly-pyrimidine DNA duplex with the corresponding poly-pyrimidine pHypNA dodecamers have shown that under the addition of a mimic oligomer in various molar ratios to the duplex, the DNA strand containing a polypyrimidine sequence was displaced from the duplex by the competing homopyrimidine mimic chain giving the invasion complex of the triplex type. Also, it was shown that pHypNA oligomer with the sequence (TCACTCAACACTCAC) was involved in the sequence specific duplex formation with the complementary DNA strand from dsDNA. The simplicity and generality of their recognition make pHypNAs attractive candidates for application as antigene reagents [25].

Gene Silencing in a Cell-Free Translation System

In first experiments, the evaluation of antisense properties of HypNA-pPNA and pHypNA mimics against their targets was performed in an *in vitro* translation assay with purified mRNA transcripts containing the coding sequences for firefly (FLuc) and *Renilla* (RLuc) luciferases [34]. The pHypNA and HypNA-pPNA probes complementary to the start codon region of the firefly luciferase coding sequence were synthesized, and the relative efficiency of these oligomers against the same transcript was tested. After the incubation in a cell-free translation system containing *FLuc* and *RLuc* mRNAs, the relative effect of antisense oligomers was evaluated by measuring the ratio between firefly and *Renilla* luciferase production in the presence and in the absence of oligomers. It was found that both types of oligomers exhibited strong antisense effect at 50-200 nM concentrations with up to 85% reduction of the FLuc production (Figure 5). The inhibition was dose-dependent and still observed at 5 nM of oligomers. Practically no reduction of RLuc activity was seen in any case [34]. The control oligomers with two mismatches and oligomers with scrambled sequences had no inhibitory effect under the same conditions.

Cell Delivery of HypNA-pPNA and pHypNA Mimics

The design of potent systems for the delivery of antisense and antigene agents that target genes of interest remains a major challenge in therapeutics [47, 48]. Phosphonate containing PNA analogues may be delivered by a large range of systems developed for nucleic acids delivery, particularly by cationic polymers and dendrimers as well as by nanoparticles, which all rely on electrostatic interactions between the antisense agent and the delivery vehicle. Investigations on the cell delivery of negatively charged HypNA-pPNAs and HypNAs revealed that free oligomers can slowly penetrate into bacterial and eukaryotic living cells and be distributed in the cytoplasm [49], but the addition of Lipofectamine-2000 can significantly improve their delivery into eukaryotic cells with localization mostly in nucleus (Figure 6) [34]. The other type of delivery vehicles is a non-covalent peptide delivery system. Peptide-based nanoparticle devices representing short amphipathic peptides are able to form stable nanoparticles with nucleic acids and their mimics [50-52]. These carriers mainly enter cells independently of the endosomal pathway and efficiently deliver cargoes in a large variety of challenging cell lines as well as in animal models. Thus, it was shown that the addition of a non-covalent peptide delivery system Pep-2, which is based on a peptide having the sequence (K E T W F E T W F T E W S Q P K K K R K V), considerably speeds up the delivery of HypNA-pPNAs into more than 90 % of living cells with localization in the cytoplasm and nucleus, and the maximal uptake can be achieved in 2 h [35]. Recently, a new peptide carrier, a short amphipathic peptide Pep-3, with the sequence (K W F E T W F T E W P K K R K) was developed [53].

Pep-3 combines a tryptophan/phenylalanine domain with a lysine/arginine-rich hydrophilic motif and has a tendency to adopt a helical structure within membranes. It forms stable nano-size complexes with both uncharged PNAs and negatively charged PNA analogues (a dissociation constant of ~10–20 nM) and promotes their efficient cellular uptake in different cell lines, including primary (*HUVECs* and *Jurkat T* cells) and suspension cell lines without any associated cytotoxicity [53]. Also, it was shown that Pep-3-mediated nanoparticle systems efficiently delivered negatively charged PNA analogues into animal tumour models, and the delivery of antisense-cyclin B1- oligomers blocks tumour growth upon intratumoral and intravenous injection. PEGylation of Pep-3 significantly improves complex stability *in vivo* and consequently the efficiency of antisense oligomer administered intravenously. It allows reducing the dose required to induce a specific and robust biological response, thereby limiting non-specific cytotoxic effects at high concentrations. Both HypNA-pPNA and pHypNA oligomers show no toxic effect on cell growth in the concentration up to 5-7 µM.

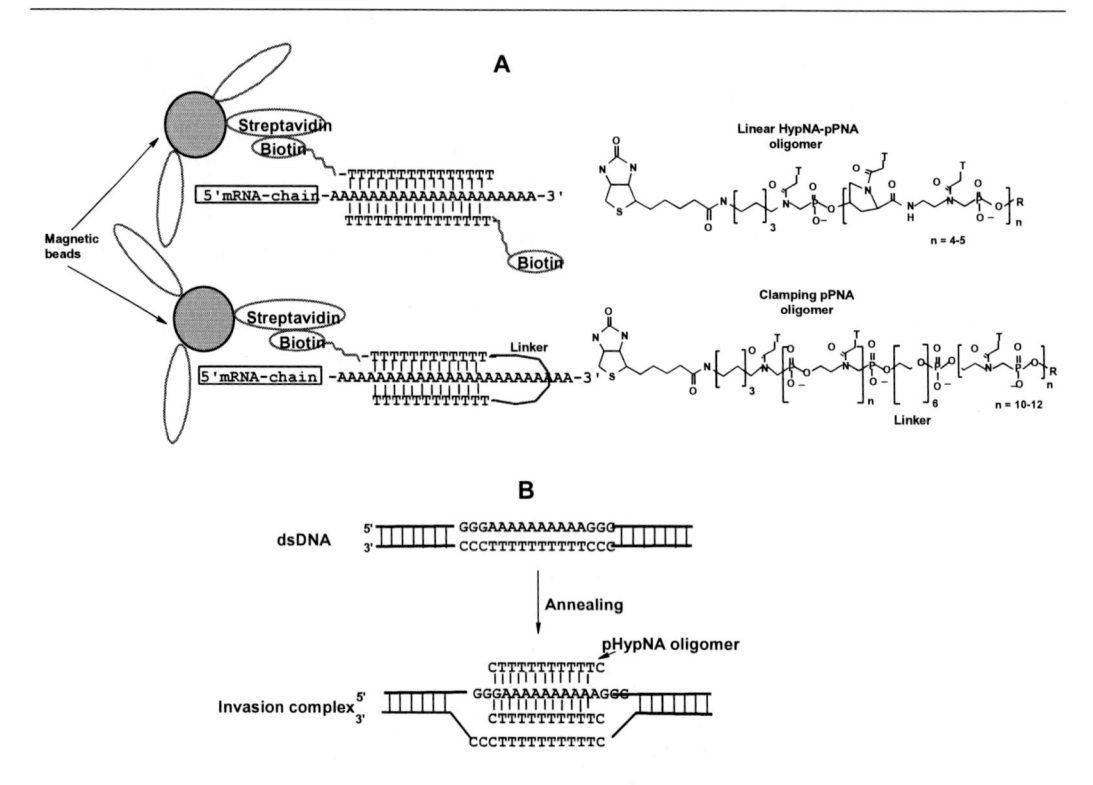

Figure 5. Schematic representation of the formation of complexes of poly-A$^+$-mRNA with biotinylated oligo-Thy mimics and streptavidine magnetic beads during the process of mRNA isolation (A) and the formation of a triplex invasion complex between dsDNA and a pHypNA oligomer (CTTTTTTTTTTC) (B).

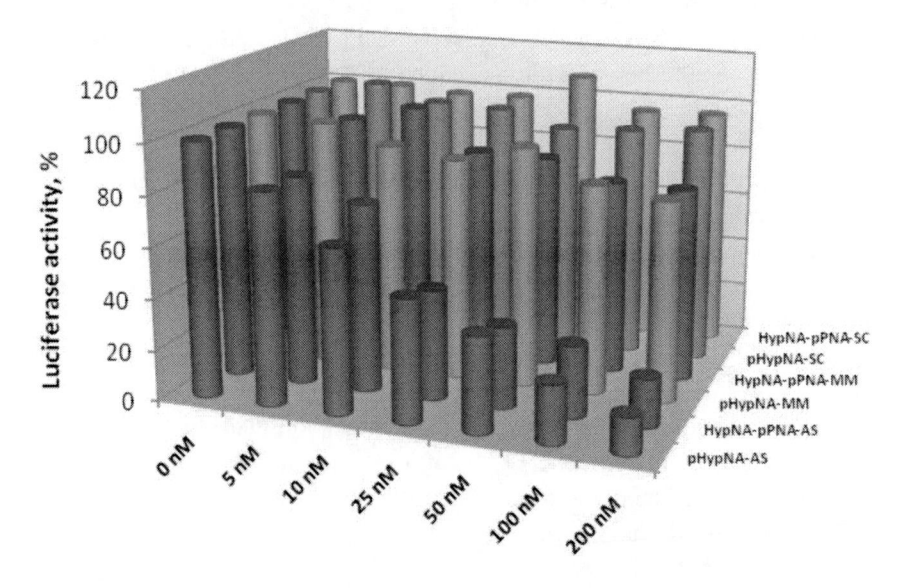

Figure 6. Analysis of firefly luciferase translation inhibition *in vitro* by HypNA-pPNA and pHypNA antisense oligomers (TGGCGTCGGTGACCAT) (AS) targeted against the translational start site of FLuc mRNA. As controls, the oligomers containing two mismatches (MM) and scrambled sequences were used. Firefly luciferase activity was calculated relative to the activity in the absence of any oligomer and normalized respect to *Renilla* luciferase production.

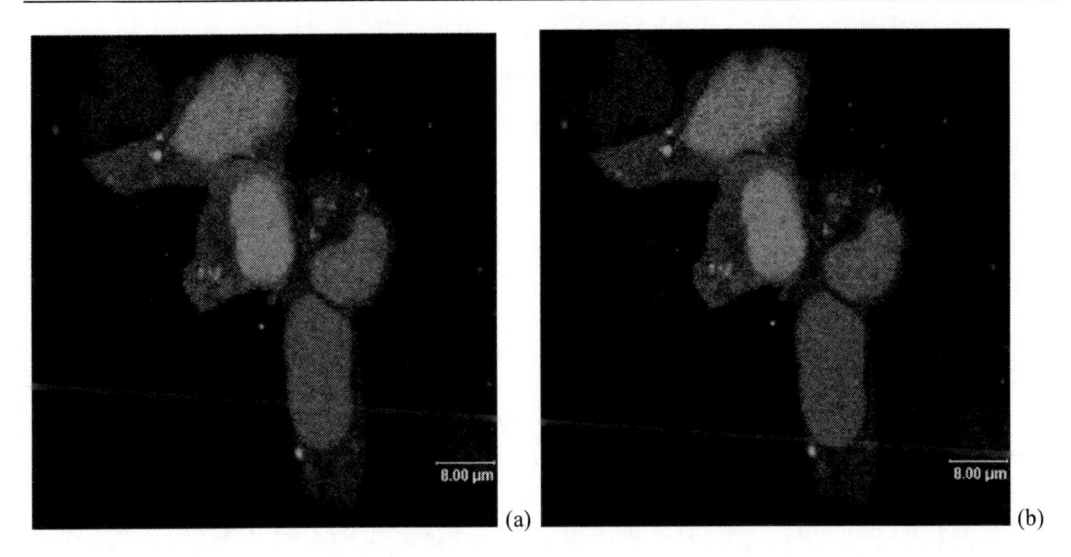

(a)   (b)

Figure 7. Intracellular distribution of the fluorescein-labeled 15-mer mimic oligomers of HypNA (a) and HypNA-pPNA (b) types. Confocal microscopy image was taken after 20 h incubation with unfixed Phoenix Eco cells in the presence of Lipofectamine-2000.

Down-regulation of Protein Biosynthesis in Cell Cultures

In first experiments on the down-regulation of protein production, the ability of negatively charged PNA-like mimic oligomers to suppress expression of green fluorescent protein in *E. coli* cells was tested [24, 49]. An antisense HypNA-pPNA oligomer demonstrated ~70% inhibition of GFP production, whereas the control oligomers with scrambled and mismatched sequences had no effect. Also, HypNA-pPNAs were tested in the suppression of AChE-R production in rat brain cell culture in comparison with a phosphorothioate oligonucleotide of the same sequence. It was found that the HypNA-pPNA antisense oligomer specifically inhibited the production of AChE-R, when supplemented to the cell culture medium in 0.5 µM concentration and its effect was comparable with that of the control phosphorothioate oligonucleotide [24, 49].

The down-regulation of the levels of cyclin B1 production was demonstrated for antisense HypNA-pPNA octadecamer in 50-100 nM concentrations [35]. This effect was analyzed in several cell lines, particularly *HeLa* cells, human fibroblasts (*HS 68*) and *293* cells in comparison with the effect of classical PNAs and antisense phosphorothioate oligonucleotides with the same sequence. In these experiments, an antisense effect of HypNA-pPNA oligomer was estimated after the quantification of cyclin B1 protein levels 8.5-fold and 25-fold higher than that of PNA oligomer and phosphorothioate oligonucleotide, respectively. Also, it was shown that with the use of the peptide delivery system Pep-2, antisense cyclin B1 HypNA-pPNA oligomer was able to reduce proliferation of human breast cancer cells *MCF-7* in the 0.5-1 µM concentration by 70-92 %, whereas classical PNA oligomer reduced cell proliferation only by 35 % for 1 µM and by 64% for 2 µM under the same conditions [35].

Later, complexes of HypNA-pPNA oligomers with sequences against the open ready frame of the cyclin B1 gene with the delivery system Pep-3 at a molar ratio of 1/20 were obtained [53]. At first, their impact on cyclin B1 protein levels was evaluated on *HeLa* cells. A concentration of 50 nM of antisense Cyc-B1 HypNA-pPNA altered cyclin B1 protein levels by up to 85 % in a specific fashion. Dose response analysis of the antisense oligomers revealed that, when complexed to Pep-3, they efficiently downregulated cyclin B1protein levels in *HeLa* cells, and one of oligomers, Cyc-B1a, was able to reduce cyclin B1 levels by more than 90%. In contrast, no effect on cyclin B1 protein levels was observed in the presence of a control HypNA-pPNA sequence bearing two mutations in the same conditions.

Also, HypNA-pPNA oligomers were successfully used for the investigation on the mechanism of Mallory body (MB) formation in the liver cells of chronic liver diseases [54, 55]. Gene-specific mimic antisense oligomers designed to inhibit the expression of p62 (a scaffolding protein that binds to polyubiquitin) and valosin-containing protein (VCP) were added to the medium of the primary mouse cell cultures. Pep-2 was used as vehicle for transfecting the oligomers into the cells. The transfection efficiency was nearly 100%. The results obtained with HypNA-pPNAs indicate that p62 is involved in the mechanism of MB formation, while VCP is located in MBs in mouse and human livers and plays an important role in inducing MB formation.

In recent reports, the HypNA-pPNA transient gene-silencing technique was shown to be effective in various studies to inhibit gene expression in cell cultures. Thus, HypNA-pPNA oligomers were used for inhibition of *lrg-47* gene expression during the investigation on the role of the IFN-induced GTPase LRG-47 in host resistance to infection with *T. cruzi*. [56]. HypNA-pPNA oligomers were used for sarcoendoplasmic reticulum calcium ATPase (SERCA1) knockdown experiments in the investigation of a role of SERCA1 truncated isoform in the proapoptotic calcium transfer from endoplasmic reticulum (ER) to mitochondria during ER stress [57]. Also, they were applied for silencing myeloid differentiation factor 88 (MyD88) in human monocyte–derived dendric cells to investigate if blockade of MyD88 human would inhibit eosinophil-derived neurotoxin induced IL-6 production [58].

Recently, we have reported that pHypNA oligomers are also highly effective in gene silencing experiments in eukaryotic cells [34]. To illustrate this, a dual-color vector, p2FP-RNAi, was used to test functionality of pHypNAs as antisense agents. This plasmid is a mammalian expression vector encoding two fluorescent reporters: red fluorescent protein, JRed, which functions as a positive transfection marker, and green fluorescent protein, TurboGFP, which stands as an indicator of siRNA or antisense oligomer efficiency. TurboGFP is an improved variant of the green fluorescent protein CopGFP cloned from copepoda *Pontellina plumata* [59]. It reveals bright green fluorescence and fast maturation when expressed in eukaryotic cells and does not form aggregates in long-term cultures. Synthetic pHypNA and HypNA-pPNA octadecamers directed against TurboGFP sequence as well as the corresponding short double-stranded RNA (dsRNA) were synthesized. *Phoenix Eco* cells were transfected with the p2FP- RNAi vector mixed with one of these types of oligomers in the presence of Lipofectamine-2000 reagent. The microscopic fluorescent analysis of cells performed 24 h after delivery clearly indicates that TurboGFP has been more efficiently silenced by pHypNA antisense oligomer in comparison with the corresponding HypNA-pPNA oligomer, and its efficiency was close to that of dsRNA targeted on the same sequence (Figure 7) [34]. In control experiments, pHypNA and HypNA-pPNA oligomers with scrambled or mismatched sequences practically did not influence the production of the green fluorescent protein.

Gene Knockdown in Living Model Organisms

Recent advances in sequence-based approaches to "knockdown" gene function have opened a door to an array of approaches to uncover functions for genes of interest in living organisms. One practical consequence of the excellent aqueous solubility of negatively charged PNA-analogues (up to 30 mM) is that a minimal volume of a highly concentrated oligomer solution can be injected into quite small eggs, or early-stage embryos of model otganizms, as is required for developmental studies. This high water solubility, combined with exquisite sequence specificity, low toxicity, lack of non-antisense effects and stability in biological systems, makes HypNA-pPNAs and pHypNAs the useful tools for selective gene knockdown studies in developmental biology.

Zebrafish (*Danio rerio*) is one of the most appropriate species for this analysis. Zebrafish embryos develop rapidly and are optically clear, permitting direct observation of the developing organs. Zebrafish also contain orthologs for almost all human genes [60]. Vertebrate knockdown strategies using morpholinos (MOs) have been demonstrated to be effective, rapid, and cost-efficient reverse genetic approaches for studying gene function in zebrafish. Several years ago, negatively charged PNA-like DNA mimics were evaluated as an alternative to MOs for oligonucleotide inhibition of gene expression in zebrafish embryos.

Thus, gene knockdown experiments targeting the zebrafish *Egfl7* were performed using two different sets of antisense oligomers, which represent both MOs and HypNA-pPNAs. One set (AS$_{-47}$) hybridized to the 5'-untranslated region (UTR) and blocks translation, and the other (AS$_{195}$) hybridized with an exon-intron junction, resulting in intron retention and hence premature translation termination. Both types of oligomers gave identical results, and their microinjections in the amount of 4 ng per embryo caused specific vascular defects with minimal non-specific effects [61].

In the study of zebrafish *chordin, notail, uroD* and *dharma* knockdown effects, it was also shown that HypNA-pPNA octadecamers displayed potency comparable with MO pentacosamers [62]. It was observed that HypNA-pPNAs enabled specific mRNA knockdown with single mismatch specificity, rather than the three to four mismatches necessary for MO negative controls (Figure 8). Thus, a single-base mismatch in the *ntl* HypNA–pPNA oligomer induced the *ntl*$^{/-}$ phenotype in only 30 % of the injected larvae, while the fully complementary HypNA–pPNA induced the *ntl*$^{/-}$ phenotype in virtually all embryos injected. In contrast, a pentadecamer MO with the same single-base mismatch produced the *ntl* null phenotype in 98 % of injected embryos. In addition, a *dharma* mutant, that had not been found susceptible to MO knockdown, was successfully phenocopied with HypNA–pPNAs. In general, these results demonstrate that HypNA–pPNAs display higher sequence specificity in comparison with MOs [37, 63]. Another key advantage of HypNA–pPNAs for gene knockdown studies is the lack of nonspecific effects; their doses, which are high enough to produce strong loss-of-function phenotypes, produced few or no nonspecific effects. Also, these compounds offer the possibility of delivery at time points later than the one-cell stage and thus might be useful for analysis of late stage phenotypes in zebrafish [48].

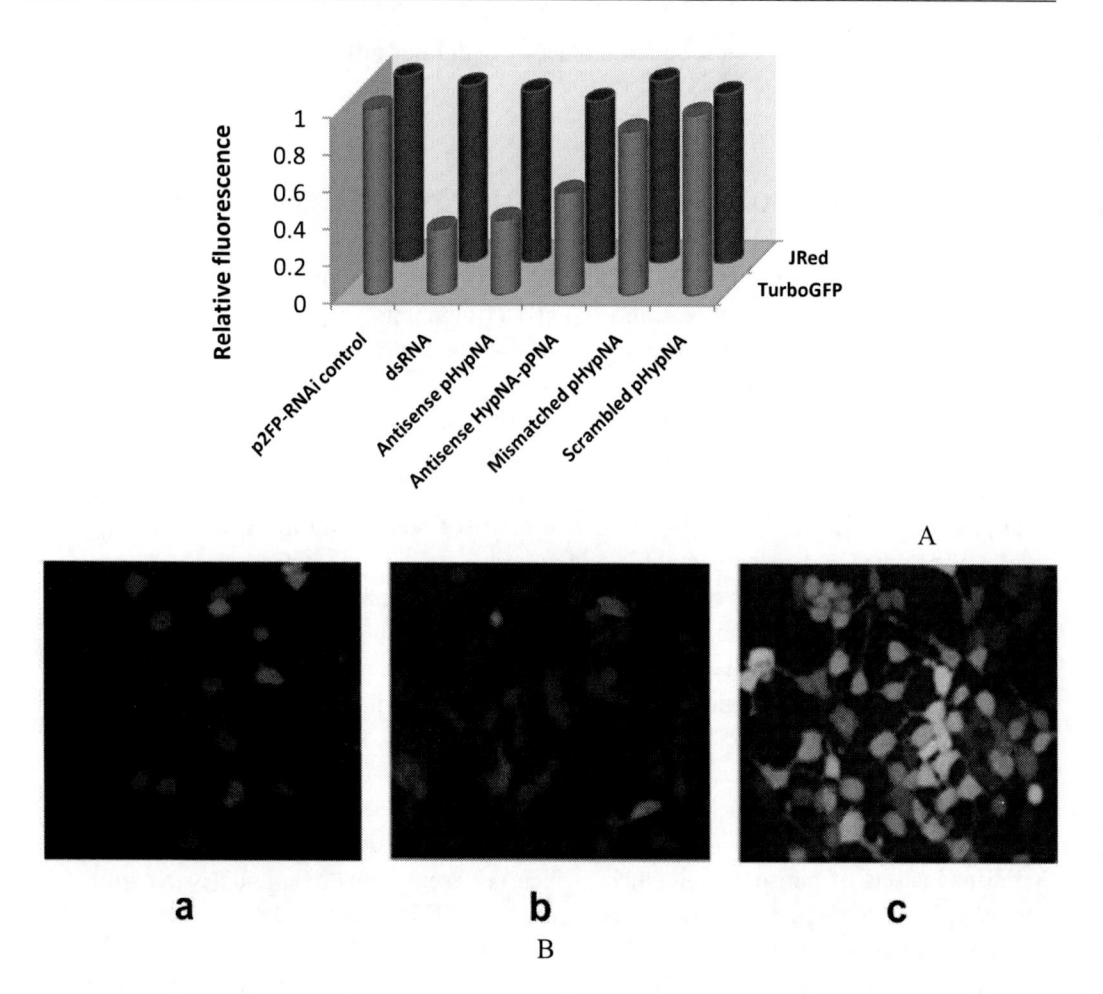

Figure 8. Suppression of TurboGFP production in eukaryotic cells. A. - The cells transfected with p2FP-RNAi dual-color vector were treated in the growth medium for 24 h at 37 °C with 0.5 μM solutions of the pHypNA or HypNA-pPNA antisense 18-mers (GCTCTCGTCGCTCTCCAT) designed to target the start codon region of TurboGFP mRNA. As controls, the cells treated with the corresponding dsRNA targeting the same sequence, cells treated with pHypNA oligomers with mismatched (GCTCTCGTCaCTCTCCAT) and scrambled (ACTTACACTTACACTTAC) sequences, and cells untreated with antisense oligomers (p2FP-RNAi control) were analyzed. The average fluorescence data for three separate experiments are shown. B. - Fluorescent microscopy images of cells transfected with p2FP-RNAi and treated with 0.5 μM mimic oligomers: (a) HypNA-pPNA, (b) pHypNA, and (c) untreated cells.

Also, zebrafish embryos were injected with MO and HypNApPNA antisense oligomers against a tumor suppressor gene, whose mechanism of action remains unclear. First pass bioinformatics analysis of the transcription profiles from RNA samples extracted from treated and control embryos identified up- and downregulated genes, of which one clear example was p53. Both MOs and HypNA-pPNAs against a tumor suppressor gene induced comparable upregulation of p53, illustrating similar effects on transcription profiles. The mimics were also used for the investigation of lipid methabolism and the development of vertebrate central nervous system in zebrafish [64, 65]. Later, microinjection of zebrafish embryos with HypNA-pPNAs to knock down *tp53 mRNA* was reported [66]. It was found that interfering with p53 expression and function using these oligomers protected developing zebrafish embryos against radiation damage. So, knockdown of *tp53* mRNA could alter the efficacy of radiotherapy against developing zebrafish embryos as a model for tumors.

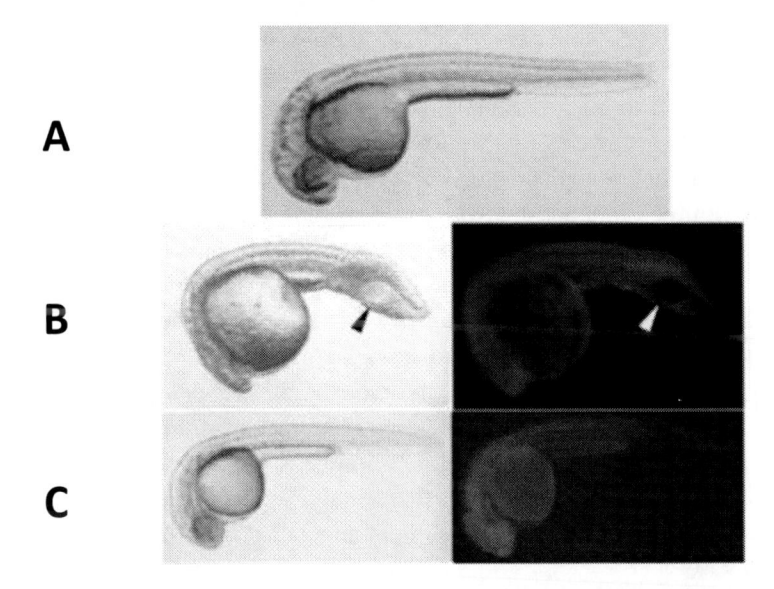

Figure 9. Fluorescein labeled antisense HypNA-pPNA *chordin* oligomer (GCAGCCCCTCCATCATCC) (0.2 mM) inhibits chordin function interdicting development of zebrafish embryos. A. – Wild-type morphology at 24 h postfertilization. B. - Embryos injected with *chordin* HypNA-pPNA oligomer phenocopy the null *chordin*$^{-/-}$ mutation. C. - Embryos injected with HypNA-pPNA oligomer with 2 mismatches appear wild-type. Images obtained in UV-light show the distribution of labeled mimic oligomer inside zebrafish bodies.

Zebrafish was used as a simple model to study cell cycle regulation during vertebrate embryogenesis. To investigate the regulation of this process, the translation of *ccnd1* mRNA, encoding cyclin D1 potein, was impaired by microinjection of ccnd1-specific MO and HypNA-pPNA antisense oligomers. Both, MOs and HypNA-pPNAs were compared for their ability to knock down *ccnd1* mRNA translation to cyclin D1. It was found that downregulation of cyclin D1 by antisense oligomers with different backbones resulted in icroophthalmia and microcephaly, but not lethality. This phenotype restricted to the organ sites characterized by highest expression of cyclin D1 during gastrulation, i.e. the developing eye and the head region. Elimination of knockdown effects by a single mismatch in the HypNA-pPNA 16mer and 3-mismatches in the MO 20mer correlated with the observation of nonactivity with one and two mismatches in HypNA-pPNA 18mers, versus the need for four mismatches in an MO 25mer to abrogate activity [63]. In more recent reports, only HypNA-pPNA oligomers were used for downregulation of cyclin D1 by the same authors [67].

The transparency of the embryonic body of the zebrafish (*Danio rerio*) makes gene manipulation by photo-illumination possible. Nowadays, the interest has increased in applying this technology to the temporally and spatially controlled repression of specific genes by the photoreactivation of antisense oligonucleotides. Thus, the application of light-activated "caging" strategies to negatively charged PNA-like mimics can provide an "on - off" switch for controlling gene expression. Recently, the synthesis, characterization, and *in vivo* application of light-activated antisense HypNA-pPNA, which hybridization to mRNA is conditionally blocked by a complementary relatively short 2`-OMe RNA strand, has been reported [68]. An amine-terminated, 18-mer antisense HypNA-pPNA targeting the mRNA was attached to a thiol-terminated 8-12-mer 2`-OMe-RNA sense strand (sRNA) via a 1-(5-(N-maleimidomethyl)-2-nitrophenyl)ethanol N-hydroxysuccinimide ester photocleavable linker (PL) (Figure 9). Cleavage of caged HypNA-pPNA (T_m ~80-90°C) by irradiation with 365-nm UV light yielded the less stable HypNA-pPNA /sRNA heteroduplex (ΔT_m 20-40°C). The drastic reduction in the stability of the heteroduplex after photocleavage of PL enables HypNA-pPNA oligomer to bind to complementary mRNA and block translation. The proper design of relatively short sense 2`-OMe-RNA, which significantly reduces the T_m of the HypNA-pPNA/sRNA heteroduplex after uncaging, appears to be essential to successfully activate antisense mimic oligomer *in vivo*. Using caged negatively charged mimic oligomers, Tang *et al.* succeeded in regulating two developmentally important genes in zebrafish embryos: *chordin* and *bozozok* genes. Caged HypNA-pPNA oligomers blocked expression in a light- and concentration-dependent manner. These data highlight the power of caged negatively charged PNA analogues to control complex gene networks. Biological problems of great complexity motivate the development of quantitative methods for controlling gene activity with high spatial and temporal resolution, using light as an external trigger. By this strategy of conditional repression, it could be possible to vary protein levels quantitatively throughout the embryo by UV laser. Also, the negative effects of uncaged mimics on gene expression could be made positive by targeting them against mRNAs encoding the repressor of gene expression [69].

Several other examples of downregulation of gene expression by injection of antisense HypNA-pPNA oligomers into zebrafish embryos include the downregulation of nicalin, nomo and progranulin production [70, 71]. Also, they were used for *spp* and *sppls* knockdowns [72] and for the downregulation of *wnt5* [73]. Thus, it was found that HypNA-pPNA-based transient gene silencing is feasible in zebrafish embryos and provides a valuable reverse genetic screening strategy. Preliminary results from our collaborators at the Thomas Jefferson University (Pennsylvania, U.S.A.) suggest that HypNAs are also effective antisense agents for gene knockdown experiments in zebrafish. (E. Wickstrom, H. Kaji unpublished data).

Also, antisense HypNA-pPNAs were applied for downregulation of gene expression in embrios of other model fishes, particularly medaka (*Oryzias latipes*) and rainbow trout (*Oncorhynchus mykiss*). One of the examples is the knockdown of *kal1.1 and kal1.2* to characterize their influence in forebrain GnRH neuronal development [74]. The other example is the investigation of the plausible role of DMY, the second vertebrate sex-determining gene identified during the early gonadal sex differentiation in the XY medaka, by using the way of sex reversal by DMY knock-down with antisense HypNA-pPNA oligomer [75]. Loss-of-function experiments performed with the help of HypNA-pPNA antisense knockdown technique were used to elucidate the role of Müllerian inhibiting substance (MIS) signaling during early sex differentiation in medaka [76], and to elucidate the molecular mechanisms of primordial germ cell (PGC) proliferation in rainbow trout [77].

The investigation on pHypNA and HypNA-pPNA effectiveness as antisense agents in the downregulation experiments in *Xenopus laevis* has been performed, also. Earlier, it was reported that the MO downregulation of the gene for a protein Ras-like GTPase Ras-dva, which is expressed during *Xenopus laevis* neurulation in a very restricted area surrounding the anterior margin of the neural plate, resulted in head development abnormalities, which include reduction of the forebrain, olfactory pits, otic vesicles, branchial arches and malformations of the head cartilages [78]. Similar abnormalities were observed on 60-85 % of developing tadpoles after microinjecting antisense pHypNA or HypNA-pPNA oligomers

(Figure 10). These results clearly demonstrated that negatively charged mimics can act as specific antisense inhibitors of gene function in *Xenopus* embryos [25, 34].

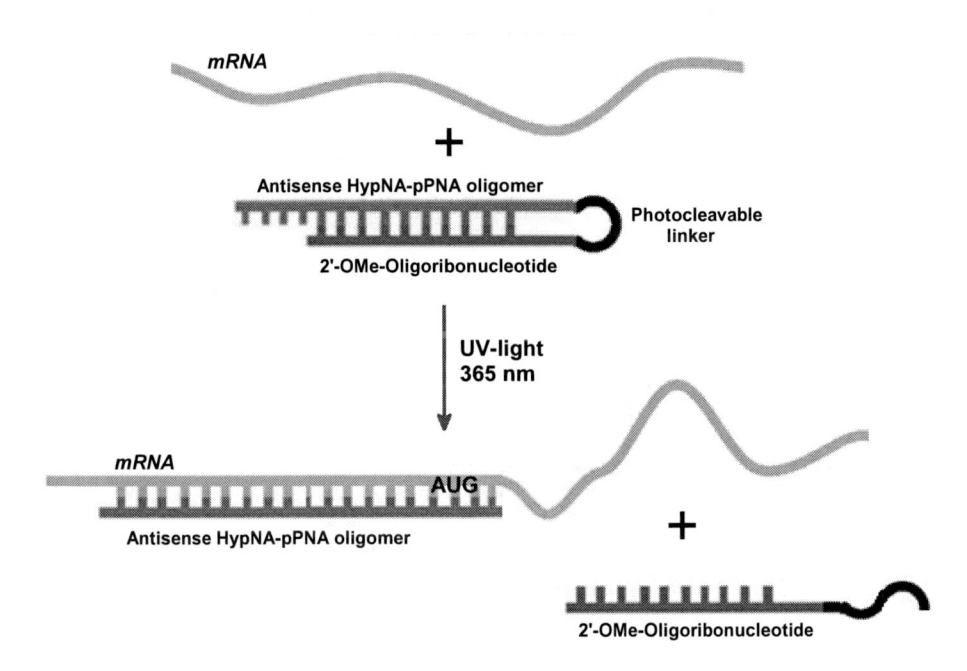

Figure 10. Scheme for controlling gene expression using HypNA-pPNA oligomer attached to complementary 2'-OMe oligoribonucleotide via a photo-cleavable linker.

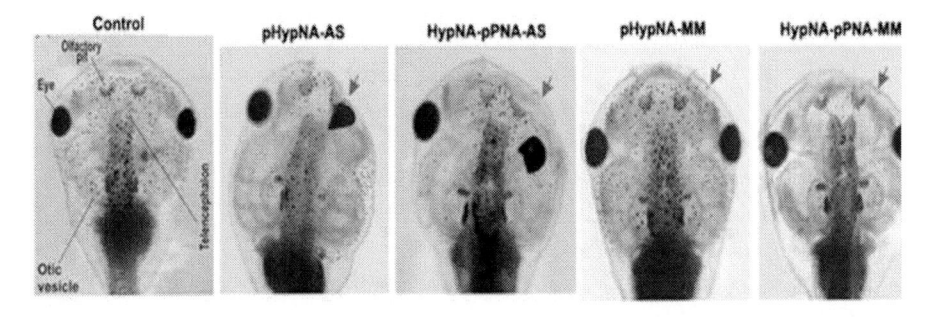

Figure 11. Images of heads of *Xenopus* tadpoles developed from the embryos microinjected with the *Ras-dva* targeted antisense pHypNA and HypNA-pPNA oligomers (TGCGCTTTCTTTTGTCGT) and mismatched oligomers (TGCtCTTTCTTcTGTCGT). The injection of antisense oligomers results in severe abnormalities of the head structure. The side of injection is shown by red arrow.

A potential of antisense cyclin B1 HypNA-pPNA in conjunction with Pep-3 delivering system was evaluated on human prostate carcinoma cell (PC3) xenografted mice [53]. This mouse tumour model was used to test systemic administration of Pep-3/antisense on the growth inhibition of established subcutaneous tumours. Formulations of Pep-3/HypNA-pPNA complexes at a 20/1 molar ratio, were administered intravenously or directly into the tumour every three days, and changes in tumour size were monitored over a course of two weeks following injection. As a control, the administration of 50 µg (intratumorally) and 100 µg (intravenously) of free HyPNA-pPNA or Pep-3 do not produce any significant effect on tumour growth. Intratumoral injection of Pep-3/HyPNA-pPNA potently inhibits tumour growth inconcentration-dependent manner, with 50% inhibition for 1 µg and more than 92% for 5 µg of antisense HyPNA-pPNA. In contrast, intravenous administration of Pep-3/HyPNA-pPNA (10 µg of antisense) only reduces tumour growth by ~20%. PEGylation improved Pep-3 intravenous efficiency and inhibited tumour growth by more than 90%, which is 4–5-fold more efficient than unPEGylated Pep-3. Moreover, the effect on tumour growth was sequence specific as an antisense HyPNA-pPNA bearing two mutations complexed to PEG-Pep-3 was unable to inhibit tumour growth.

Conclusion

PNA-relative DNA mimics bearing negative charges are relatively new types of compounds, which are promising tools for diagnostic and therapeutic applications. These compounds combine high hybridization and mismatch discrimination characteristics with good water solubility and biological stability as well as the ability to penetrate cell membranes. The binding affinity of HypNA-pPNA and pHypNA mimics for complementary nucleic acids is practically insensitive to the ionic strength of the medium, and this property provides a dramatic advantage over anionic oligonucleotides and their analogues in several molecular biology applications. The analysis of biological properties of HypNA-pPNAs and pHypNAs, allows to conclude that these DNA mimics are very promising tools for the application as specific antisense agents for reverse genetic studies as well as potential antisense and antigene therapeutics. Their ability to turn of individual genes at will in cells provides a powerful tool for elucidating the role of a particular gene. The antisense effect of HypNA-pPNAs and pHypNAs lasts over a period of several days, due to their high stability, and this represents a very potent technology for administering antisense-based drugs. Similarly to MOs, these compounds can act as effective antisense inhibitors in living organisms over a range of developmental stages, and the results obtained in gene knockdown experiments *in vivo* validate these negatively charged mimics as a valuable alternative to MOs for reverse genetic studies. The stronger hybridization and greater specificity of HypNA-pPNAs enable knockdown of mRNAs unaffected by MO oligomers. Moreover, the ability of these mimics efficiently blocks cancer cell proliferation at very low concentrations suggests that these compounds are good candidates for the development of a potential anticancer drug. In general, the examination of the properties of negatively charged PNA-related mimics clearly demonstrated their high potential for further evaluation as potential compounds for diagnostic and therapeutic applications.

References

[1]   Nielsen, P., Egholm, M., Berg, R. & Buchardt, O. (1991). Sequence-selective recognition of DNA by strand displacement with a thymine-substituted polyamide. *Science, 254*, 1497-1500.

[2]   *Peptide Nucleic Acids, Protocols and Applications*, eds.: Nielsen, P. and Egholm, M., Horizon Scientific Press, Norfolk, England, 1999.

[3]   P. E. Nielsen, P. E. (2006). The Many Faces of PNA. In C.G. Janson, & M.J. During (Eds.) Medical Intelligence Unit: Peptide Nucleic Acids, Morpholinos and Related Antisense Biomolecules (pp. 3-17). Eurekah and Kluwer Academic / Plenum Publishers.

[4]   Summerton, J. & Weller, D. (1997). Morpholino antisense oligomers: design, preparation, and properties. *Antisense Nucleic Acid Drug Dev., 7*, 187-195.

[5]   Nielsen, P.E. (2000). Antisense peptide nucleic acids. *Curr. Opin. Mol. Ther., 2*, 282-287.

[6]   Weiler, J., Gausephohl, H., Hauser, N., Jensen, O. & Hoheisel, J. (1997). Hybridisation based DNA screening on peptide nucleic acid (PNA) oligomer arrays. *Nucleic Acids Res., 25,* 2792-2799.

[7]   Bergman, F., Bannwarth, W. & Tam, S. (1995). Solid phase synthesis of directly linked PNA-DNA-hybrids. *Tetrahedron Lett., 36*, 6823-6826.

[8]   Turner, J., Ivanova, G., Verbeure, B., Williams, D., Arzumanov, A., Abes, S., Lebleu, B., Gait, M. (2005). Cell-penetrating peptide conjugates of peptide nucleic acids (PNA) as inhibitors of HIV-1 Tat-dependent *trans*-activation in cells. *Nucleic Acids Res., 33*, 6837-6849.

[9]   Shiraishi, T. & Nielsen, P. (2004). Down-regulation of MDM2 and activation of p53 in human cancer cells by antisense 9-aminoacridine–PNA (peptide nucleic acid) conjugates. *Nucleic Acids Res., 32*, 4893-4902.

[10]   Shiraishi, T., Hamzavi, R. & Nielsen, P.E. (2008). Subnanomolar antisense activity of phosphonate-peptide nucleic acid (PNA) conjugates delivered by cationic lipids to *HeLa* cells. *Nucleic Acids Res.,36*, 4424-4432.

[11]   Egholm, M., Buchardt, O., Christensen, L., Behrens, C., Freier, S.M., Driver, D.A., Berg, R.H., Kim, S.K., Norden, B. & Nielsen, P.E. (1993) PNA hybridizes to complementary oligonucleotidesobeying the Watson-Crick hydrogen-bonding rules. *Nature, 365*, 566–568.

[12]   Stein, D., Foster, E., Huang, S., Weller, D. & Summerton, J. (1997). A specificity comparison of four antisense types: morpholino, 2'-O-methyl RNA, DNA, and phosphorothioate DNA. *Antisense Nucleic Acid Drug Dev., 7*, 151–157.

[13]   Summerton, J., Stein, D., Huang, S., Matthews, P., Weller. D. & Partridge, M. (1997). Morpholino and phosphorothioate antisense oligomers compared in cellfree and in-cell systems. *Antisense Nucleic Acid Drug Dev., 7*, 63–70.

[14]   Van der Laan, A., Stromberg, R., van Boom, J., Kuyl-Yeheskiely, E., Efimov, V. & Chakhmakhcheva, O. (1996). An approach towards the synthesis of oligomers containing a N-2-hydroxyethyl-aminomethyl phosphonate backbone: a novel PNA analogue. *Tetrahedron Lett., 37,* 7857-7860.

[15] Peyman, A., Uhlmann, E., Wagner, K., Augustin, S., Breipohl, G.,Will, D., Schafer, A. & Wallmeier, H. (1996). Phosphonic ester nucleic acids (PHONAs): oligonucleotide analogues with an achiral phosphonic acid ester backbone. *Angew. Chem. Int. Ed. Engl.*, *35*, 2636-2638.

[16] Kehler, J., Henriksen, U., Vejbjerg, H. & Dahl, O. (1998). Synthesis and hybridization properties of anacyclic achiral phosphonate DNA analogue. *Bioorg. Med. Chem.*, *6*, 315-322.

[17] Efimov, V.A. , Buryakova, A.A., Choob, M.V. & Chakhmakhcheva, O.G. (1999) Phosphonate analogues of peptide nucleic acids and related compounds: synthesis and hybridization properties. *Nucleosides & Nucleotides, 18*, 1393-1396.

[18] Efimov, V., Choob, M., Buryakova, A., Kalinkina, A. & Chakhmakhcheva, O. (1998). Synthesis and evaluation of some properties of chimeric oligomers containing PNA and phosphono-PNA residues. *Nucleic Acids Res.*, *26, 566-575.*

[19] Peyman, A., Uhlmann, E., Wagner, K., Augustin, S., Weiser, C., Will, D. & Breipohl, G. (1997). PHONA - PNA co-oligomers: nucleic acid mimetics with interesting properties. *Angew. Chem. Int. Ed. Engl.*, *36*, 2809-2812.

[20] Efimov, A., Buryakova, A., Choob, M. & Chakhmakhcheva, O. (1999). Phosphonate analogues of peptide nucleic acids and related compounds: synthesis and hybridization properties. *Nucleosides & Nucleotides*, *18*, 1393-1396.

[21] Efimov, V., Buryakova, A. & Chakhmakhcheva, O. (1999). Synthesis of polyacrylamides N-substituted with PNA-like oligonucleotide mimics for molecular diagnostic applications. *Nucleic Acids Res.*, *27, 4416-4426.*

[22] Efimov, V. & Chakhmakhcheva, O. (2004). Synthesis of DNA mimics representing HypNA-pPNA hetero-oligomers. *Meth. Mol. Biol.*, *288*, 147-163.

[23] Efimov, V. & Chakhmakhcheva, O. (2002). Phosphono-PNAs: synthesis, properties and applications. *Collection Symp. Series*, *5*, 136-144.

[24] Efimov, V., Klykov, V. & Chakhmakhcheva, O. (2003). Phosphono peptide nucleic acids with a constrained hydroxyproline-based backbone. *Nucleosides, Nucleotides & Nucleic Acids*, *22*, 593-599.

[25] Efimov, V.A. & Chakhmakhcheva, O.G. (2006) Hydroxyproline-based DNA mimics: a review on synthesis and properties. *Collect. Czech. Chem. Commun., 71, 929-955.*

[26] Hyrup, B., Egholm, M., Nielsen, P., Witting, P., Norden, B. & Buchardt, O. (1994) Structure-activity studies of the binding of modified peptide nucleic acids (PNAs) to DNA. *J. Am.Chem. Soc.*, *116*, 7964-7970.

[27] Ceulemans, G., van Aerschot, A., Wroblowski, B., Rozenski, J., Hendrix, C. & Herdewijn, P. (1997) Oligonucleotide Analogues with 4-Hydroxy-*N*-Acetylprolinol as Sugar Substitute. *Chem. Eur. J.*, *3*, 1997-2010.

[28] Ceulemans, G., van Aerschot, A., Rozenski, J. & Herdewijn, P. (1997) Oligonucleotides with 3-hydroxy-*N*-acetylprolinol as sugar substitute. *Tetrahedron*, *53*, 14957-14974.

[29] Häberli, A. & Leumann C. (2002) DNA binding properties of oligodeoxynucleotides containing pyrrolidino *C*-nucleosides. *Org. Lett.*, *4*, 3275-3278.

[30] Häberli, A., Mayer, A. & Leumann, C. (2003) Pyrrolidino-DNA. *Nucleosides, Nucleotides & Nucleic Acids*, *22*, 1187-1189.

[31] Vaněk, V., Buděšinský, M., Rinnová, M. & Rosenberg, I. (2009) Prolinol-based nucleoside phosphonic acids: new isosteric conformationally flexible nucleotide analogues. *Tetrahedron, 65,* 862–876.

[32] Rejman, D., Kočalka, P., Pohl, R., Točik, Z. & Rosenberg, I. (2009) Synthesis and hybridization of oligonucleotides modified at AMP sites with adenine pyrrolidine phosphonate nucleotides. *Collect. Czech. Chem. Commun., 74, 935-955.*

[33] Efimov, V.A., Klykov, V.N., & Chakhmakhcheva, O.G. (2007) Synthesis and properties of pyrrolidine-based negatively charged DNA mimics. *Nucleosides, Nucleotides & Nucleic Acids, 26,* 1595 -1599.

[34] Efimov, V.A., Birikh, K.R., Staroverov, D.B., Lukyanov, S.A., Tereshina, M.B., Zaraisky, A.G. & Chakhmakhcheva, O.G. (2006). Hydroxyproline-based DNA mimics provide an efficient gene silencing *in vitro* and *in vivo. Nucleic Acids Res., 34,* 2247–2257

[35] Morris, M., Chaloin, L., Choob, M., Archdeacon, J., Heitz, F. & Divita, G. (2004). Combination of a new generation of PNAs with a peptide-based carrier enables efficient targeting of cell cycle progression. *Gene Therapy., 11,* 757–764

[36] Urtishak, K., Choob, M., Tian, X., Sternheim, N., Talbot, W., Wickstrom, E. & Farber, S. (2003). Targeted gene knockdown in zebrafish using negatively charged peptide nucleic acid mimics. *Developmental Dynamics, 228,* 405–413.

[37] Wickstrom, E., Choob, M., Urtishak, K., Tian, X., Sternheim, N., Talbot, S., Archdeacon, J., Efimov, V. & Farber, S. (2004). Sequence specificity of alternating hydroxyprolyl/phosphono peptide nucleic acids against zebrafish embryo mRNAs. *Journal of Drug Targeting, 12,* 363–372.

[38] Efimov, V., Choob, M., Buryakova, A., Phelan, D. & Chakhmakhcheva, O. (2000). PNA-related oligonucleotide mimics and their evaluation for nucleic acid hybridization studies and analysis. *Nucleosides, Nucleotides & Nucleic Acids, 20,* 419-428.

[39] Efimov, V. & Chakhmakhcheva, O. (2001). Solid phase synthesis of PNA-like oligonucleotide mimics and their use for polyacrylamide-based molecular diagnostic assays. In: R. Empton (Ed.). *Solid phase synthesis and combinatorial libraries. (*pp.75-78), Mayflower Worldwide, UK.

[40] Phelan, D., Hondorp, K., Choob, M., Efimov, V. & Fernandez, J. (2001). Messenger RNA isolation using novel PNA analogues. *Nucleosides, Nucleotides & Nucleic Acids, 20,* 1107-1111.

[41] Tyagi, S. & Kramer, F. (1996). Molecular beacons: probes that fluoresce upon hybridization. *Nature Biotechnology, 14,* 303-308.

[42] Tan, V., Wang, K. & Drake, J. (2004). Molecular beacons. *Current Opinion in Chemical Biology, 8,* 547-553.

[43] Ortiz, E., Estrada, G. & Lizardi, P. (1998). PNA molecular beacons for rapid detection of PCR amplicons. *Molecular and Cellular Probes, 12,* 219-226.

[44] Efimov, V., Chakhmakhcheva, O., Choob, M. & Archdeacon, D. (2002). Using chimeric DNA/RNA molecular beacons for target-specific signal amplification. *Collection Symp. Series, 5,* 308-311.

[45] Nitin, N., Santangelo, P., Kim, G., Nie, S & Bao, G. (2004). Peptide-linked molecular beacons for efficient delivery and rapid mRNA detection in living cells. *Nucleic Acids Res., 32,* e58.

[46] Peffer N., Hanvey J., Bisi J., Thomson S., Hassman F., Noble S. & Babiss L. (1993) Strand-invasion of duplex DNA by peptide nucleic acid oligomers. *Proc. Natl. Acad. Sci. U.S.A.*, *90*, 10648-10652.

[47] Wyrozumska, P., Stebelska, K., Grzybek, M. & Sikorski, A.F. (2006) Synthetic vextors for genetic drug delivery. In: M.R. Mozafari (Ed.) *Nanocarrier technologies: frontiers of nanotherapy* (pp. 139-174) Springer.

[48] Doan, T.N., Eilertson, C.D. & Rubinstein, A.L. (2004) High-throughput target validation in model organisms. *DDT:TARGETS*, *3*, 191-197

[49] Efimov, V., Chakhmakhcheva, O. & Wickstrom, E. (2005). Synthesis and application of negatively charged PNA analogues. *Nucleosides, Nucleotide & Nucleic Acids*, *24*, 1853-1874.

[50] Herbig, M.E., Weller, K.M. & Merkle, H.P. (2007). Reviewing biophysical and cell biological methodologies in cell-penetrating peptide (CPP) research. *Critical reviews in therapeutic drug carrier systems, 24*, 203-255.

[51] Crombez, L., Morris, M.C., Deshayes, S., Heitz, F. & Divita, G. (2008). Peptide-based nanoparticle for *ex vivo* and *in vivo* drug delivery. *Current Pharmaceutical Design, 14*, 3656-3665.

[52] Morris, M.C., Deshayes, S., Heitz, F. & Divita, G. (2008). Cell-penetrating peptides: from molecular mechanisms to therapeutics. *Bio. Cell, 100*, 201-217.

[53] Morris, M.C., Gros, E., Aldrian-Herrada, G., Choob, M., Archdeacon, J., Heitz, F. & Divita, G. (2007) A non-covalent peptide-based carrier for in vivo delivery of DNA mimics. *Nucleic Acids Res., 35*, e49.

[54] Nan, L., Wu, Y., Bardag-Gorce, F., Li, J., French, B.A., Fu, A.N., Francis, T., Vu, J. & French S.W. (2004). p62 is involved in the mechanism of Mallory body formation. *Exp. Mol. Pathol.*, *77*, 168-175.

[55] Nan, L., Wu, Y., Bardag-Gorce, F., Li, J., French, B.A., Wilson, L.T., Nguyen, S.K. & French S.W. (2005). RNA interference of VCP/p97 increases Mallory body formation. *Exp. Mol. Pathol., 78*, 1-9.

[56] Santiago, H.C. , Feng, C.G., Bafica, A., Roffe, E., Arantes, R.M., Cheever, A., Taylor, G., Vierira, L.Q., Aliberti, J., Gazzinelli, R.T. & Sher, A. (2005) Mice Deficient in LRG-47 Display Enhanced usceptibility to *Trypanosoma cruzi* Infection Associated with Defective Hemopoiesis and Intracellular Control of Parasite Growth. *The Journal of Immunology, 175*, 8165–8172.

[57] Chami, M., Oulès, B., Szabadkai, G., Tacine, R., Rizzuto, R. & Paterlini-Bréchot, P. (2008) Role of SERCA1 Truncated Isoform in the Proapoptotic Calcium Transfer from ER to Mitochondria during ER Stress. *Molecular Cell, 32*, 641–651, 2008.

[58] Yang , D., Chen , Q., Bo Su , S., Zhang, P., Kurosaka, K., Caspi , R.R., Michalek, S.M., Rosenberg, H.F., Zhang, N. & Oppenheim, J.J. (2008) Eosinophil-derived neurotoxin acts as an alarmin to activate the TLR2 – MyD88 signal pathway in dendritic cells and enhances Th2 immune responses. *J. Exp. Med., 205*, 79–90.

[59] Shagin, D., Barsova, E., Yanushevich, Y., Fradkov, A., Lukyanov, K., Labas, Y., Ugalde, J., Meyer, A., Nunes, J., Widder, E., Lukyanov, S. & Matz, M. (2004) GFP-like proteins as ubiquitous Metazoan superfamily: evolution of functional features and structural complexity. *Mol. Biol. Evol,. 21*, 841-850.

[60] Deiters, A. & Yoder, J.A. (2006) Conditional transgene and gene targeting methodologies in zebrafish. *Zebrafish, 3*, 415-429.

[61] Parker, L., Schmidt, M., Jin, S.-W., Gray, A., Beis, D., Pham, T., Frantz, G., Palmieri, S., Hillan, K., Stainier, Y., de Sauvage, F. & Ye, W. (2004). The endothelial-cell-derived secreted factor *Egfl7* regulates vascular tube formation. *Nature*, *428*, 754-758.

[62] Wickstrom, E., Urtishak, K.A., Choob, M., Tian, X., Sternheim, N., Cross, L.M., Rubinstein, A. & Farber, S.A. (2004). Downregulation of gene expression with negatively charged peptide nucleic acids (PNAs) in zebrafish embryos. *Methods Cell Biol.*, *77*, 137-158.

[63] Duffy, K., McAleer, M., Davidson, W., Kari, L., Kari, C., Liu, C., Farber, S., Cheng, K., Mest, J., Wickstrom, E., Dicker, A. & Rodeck, U. (2005) Coordinate control of cell cycle regulatory genes in zebrafish development tested by cyclin D1 knockdown with morpholino phosphorodiamidates and hydroxyprolyl-phosphono peptide nucleic acids. *Nucleic Acids Res.*, *33*, 4914–4921.

[64] Ho, S.Y., Thorpe, J.L., Deng, Y., Santana, E., DeRose, R.A. and Farber, S.A. 2004. Lipid methabolism in zebrafish. *Methods Cell Biol.*, *76*, 87-108.

[65] Ninkovic, J., Tallafuss, A., Leucht, C., Topczewski, J., Tannhauser, B., Solnica-Krezel, L. & Bally-Cuif, L. (2005). Inhibition of neurogenesis at the zebrafish midbrain-hindbrain boundary by the combined and dose-dependent activity of a new hairylE (spl) gene pair. *Development*, *132*, 75-88.

[66] Duffy, K.T. & Wickstrom, E. (2007). Zebrafish *tp53* knockdown extends the survival of irradiated zebrafish embryos more effectively than the p53 Inhibitor Pifithrin. *Cancer Biology & Therapy, 6*, 675-678.

[67] Mcaleer, M.F., Duffy, K. T., Davidson, W.R., Kari, G., Dicker, A.P., Rodeck, U. & Wickstrom, E. (2006) Antisense inhibition of cyclin D1 expression is equivalent to flavopiridol for radiosensitization of zebrafish embryos. *Int. J. Radiation Oncology Biol. Phys., 66*, 546–551.

[68] Tang, X.J. , Maegawa, S., Weinberg, E.S. & Dmochowski, I. J. (2007). Regulating gene expression in zebrafish embryos using light-activated, negatively charged peptide nucleic acids. *J. Am. Chem. Soc., 129*, 11000-11001.

[69] Okamoto, H. (2007). Yin–Yang Ways of Controlling Gene Expression Are Now in Our Hands. *ACS Chemical Biology*, *2*, 646–648.

[70] Haffner, C., Frauli, M., Topp, S., Irmler, M., Hofmann, K., Regula, J. T., Bally-Cuif, L. & Haass, C. (2004). Nicalin and its binding partner Nomo are novel Nodal signaling antagonists. *The EMBO Journal*, *23*, 3041–3050.

[71] Shankaran, S.S. , Capell, A., Hruscha, A.T., Fellerer, K., Neumann, M., Schmid, B. & Haass, C. (2008) Missense mutations in the progranulin gene linked to frontotemporal lobar degeneration with ubiquitin-immunoreactive inclusions reduce progranulin production and secretion. *J. Biol. Chem.*, *283*, 1744-1753.

[72] Krawitz, P., Haffner, C., Fluhrer, R., Steiner, H., Schmid, B. & Haass, C. (2005) Differential localization and identification of a critical aspartate suggest non-redundant proteolytic functions of the presenilin homologues sppl2b and sppl3. *J. Biol. Chem.*, *280*, 39515–39523.

[73] Robu, M.E., Larson, J.D., Nasevicius, A., Beiraghi, S., Brenner, C., Farber, S.A. & Ekker, S.C. (2007) p53 Activation by knockdown technologies. *PLoS Genetics, 3,* e78.

[74] Okubo, K., Sakai, F., Lau, E.L., Yoshizaki, G., Takeuchi, Y., Naruse, K., Aida, K. & Nagahama, Y. (2006) Forebrain gonadotropin-releasing hormone neuronal

development: insights from transgenic medaka and the relevance to X-linked Kallmann syndrome. *Endocrinology, 147*, 1076-1084.

[75] Paul-Prasanth, B., Matsuda, M., Lau, E.-L., Suzuki, A., Sakai, F., Kobayashi, T. & Nagahama, Y. (2006) Knock-down of DMY initiates female pathway in the genetic male medaka, *Oryzias latipes*. *Biochem. Biophys. Res. Commun., 351,* 815–819.

[76] Shiraishi, E., Yoshinaga, N., Miura, T., Yokoi, H., Wakamatsu, Y., Abe, S.-I. & Kitano, T. (2008) Müllerian inhibiting substance is required for germ cell proliferation during early gonadal differentiation in medaka (*Oryzias latipes*). *Endocrinology, 149,* 1813–1819.

[77] Sawatari , E., Shikina, S., Takeuchi, T. & Yoshizaki, G. (2007) A novel transforming growth factor-β superfamily member expressed in gonadal somatic cells enhances primordial germ cell and spermatogonial proliferation in rainbow trout (*Oncorhynchus mykiss*). *Dev. Biol., 301*, 266–275.

[78] Tereshina, M., Zaraisky, A. & Novoselov, V. (2006) Ras-dva, a member of novel family of small GTPases, is required for the anterior ectoderm patterning in the *Xenopus laevis* embryo. *Development, 133*, 485-494.

In: Gene Silencing: Theory, Techniques and Applications ISBN: 978-1-61728-276-8
Editor: Anthony J.Catalano © 2010 Nova Science Publishers, Inc.

Chapter VI

Gene Silencing and the Analysis of Immune Response in Model Insects

Davide Malagoli and Mauro Mandrioli[*]

Department of Biology, University of Modena and Reggio Emilia, Via Campi 213/D,
41125 Modena, Italy

Abstract

Insects are organisms of considerable interest for comparative biology and medicine, therefore it is not surprising that several publications referred to them as model organisms. Insect and vertebrate evolution diverged more than 500 million years ago, but the molecular bases of several fundamental biological functions, including innate immune response, were already established in their common progenitor and have been conserved. Consequently, starting from information collected in insects, new insights into human biology and pathology were gained. Gene silencing includes several powerful methods, such as the production of loss-of-function mutants and RNA interference. These procedures, in particularly when performed in models for which molecular databases are already available, allow the genetic dissection of several immune-related processes and pathways. In the present review, we will concentrate our attention on the information derived from gene silencing techniques on insect immune signalling with particular attention for *Drosophila melanogaster* and *Anopheles gambiae*.

Introduction

In the last decade the literature in the field of the immunology of invertebrates has been widely increasing and important contributions have provided a better understanding of the

---

[*] Correspondence to: Mauro Mandrioli, Department of Biology, University of Modena and Reggio Emilia, Modena, Via Campi 213/D, 41125, Italy, E-mail: mauro.mandrioli@unimore.it

processes and the strategies adopted by their immune systems (Ballarin et al., 2008). In view of their large number and considerable functional diversity (Novotny et al., 2007), insects have long been used as experimental organisms in several disciplines, including comparative biology, medicine and parasitology. However, despite such interest, much of the attention of scientific community has been focused on a small number of species that are either genetically well characterized, e.g., *Drosophila melanogaster* (Arias, 2008), or represent vectors of primary importance diseases, e.g., *Anopheles gambiae* (Beier, 1998; Breman et al., 2004).

Even though more than 500 million years of evolution separate insects and humans (Hedges et al., 2004), the fundaments of basic biological functions, including immunity, have been conserved. As a consequence, information derived from studies of immune responses in *D. melanogaster* and *A. gambiae* host-parasite interactions have provided a better comprehension of human immune system functioning and pathology (Lemaitre and Hoffmann, 2007), and host-parasite interactions, including those with the malarial parasite *Plasmodium falciparum* (Vizioli et al., 2001).

As all the other insects, *D. melanogaster* and *A. gambiae* present an immune system endowed with only innate immune components, comprised of cellular and humoral factors (Nappi et al., 2004). Cell-mediated immunity includes phagocytosis and encapsulation, exerted by specific cell types (Ballarin et al., 2008) while humoral mediators comprise several factors among which the antimicrobial peptides (AMPs) and the components of the pro-phenoloxidase (pro-PO) cascade have been the most elucidated (Bulet and Stöcklin, 2005). It is important to observe that cellular and humoral components have not to be considered as separate elements, because several findings have indicated that secreted factors are fundamental for clotting, pathogen recognition and engulfment. However, we will concentrate on the two components separately, and will not further consider clotting (Dushai, 2009) and pathogen recognition (Lemaitre and Hoffmann, 2007) that recently have been reviewed in considerable detail.

The sequencing of the genomes of *D. melanogaster* (Adams, 2000) and *A. gambia*e (Holt et al., 2002) opened the possibility to disentangle and dissect immune-related pathways that regulate cell-mediated and humoral immunity, especially through the observation of the immune response in gain-of-function or loss-of-function mutants (Lemaitre and Hoffmann, 2007). Even more precise knowledge has been obtained by the associated application of RNA interference (RNAi) protocols, that allow the transient silencing of gene expression. The present review will summarize recent findings derived from the application of gene silencing techniques on some insect immune-related functions, with special reference to the major models, *D. melanogaster* and *A. gambiae*.

Cell-mediated Immunity

Phagocytosis of invading pathogens is a critical component of metazoan innate immune systems, and therefore it is an highly conserved aspect of cell-mediated immunity. In invertebrates, the morphological changes and cytoskeletal rearrangements that occur during phagocytosis have been repeatedly described, but it was only after functional wide screening studies (Rämet et al., 2002) that it was possible to identify genes whose expressions were essential for phagocytosis. In these respects, *D. melanogaster* has represented an excellent model, because beside the information gained with genome sequencing (Adams, 2000) it is also available the *Drosophila* S2 cell line (Rämet et al., 2001, 2002), derived from embryonic plasmatocytes and sharing many characteristics of larval and adult plasmatocytes, including an active phagocytic activity.

Gene silencing protocols indicated that the most important genes for phagocytosis in *Drosophila* are not transcriptionally regulated (Kallio et al., 2005). However, this does not mean that gene silencing protocols could not provide fundamental insights into phagocytic process. RNAi experiments confirmed the importance of cytoskeletal components and molecules involved in vesicle trafficking and allowed the identification in *Drosophila* hemocytes of elements essentials for the uptake of different species of bacteria (Philips et al., 2005) and fungi (Stroschein-Stevenson et al., 2006). The processes by which the binding of a potential pathogen to the phagocytic receptors leads to its engulfment depends on the products of several genes (Rämet et al., 2002; Pearson et al., 2003), whose regulation is poorly understood (Kallio et al., 2005). Numerous researches have addressed this subject, especially in the attempt to elucidate how intracellular bacteria can escape or even take advantage by phagocytic processes. Using a genome-wide RNAi screen, Agaisse et al. (2005) identified host factors that are involved in intracellular pathogenesis or that specifically affect access to the cytosol by intracellular pathogens, such as *Lysteria monocytogenes*, but many aspects in the elimination of engulfed particles remain to be identified (Lemaitre and Hoffmann, 2007).

RNAi has been used to dissect the phagocytosis pathways also in *A. gambiae* and in particular functional gene analysis became possible in mosquitoes with the development of RNAi silencing, which allowed targeted inactivation of gene expression either by transfection of cultured cells with double-stranded RNA (dsRNA) (Levashina et al., 2001) or by direct injection of dsRNA into adult mosquitoes (Blandin et al., 2002). Using these approaches the role of more of than 70 candidate genes in phagocytosis has been tested showing that the inactivation of 26 of them affects the phagocytic activity by more than 45% (Moita et al., 2005). Interestingly, the pathway used in *A. gambiae* for phagocytosis of microorganisms is similar to that mediating apoptotic cell removal in *Caenorhabditis elegans* (Moita et al., 2005).

Silencing of genes that are associated with phagocytosis pathways leads to bacterial accumulation by 8 hr after infection (Moita et al., 2005). However, the inactivation did not significantly affected mosquito survival after *E. coli* infection. Similar results were obtained in bacterial-infected *D. melanogaster* (Elrod-Erickson et al., 2000), suggesting that insects use phagocytosis in the early antibacterial response, and rely on other effector mechanisms (mainly antimicrobial peptides, AMPs) to fight bacterial infections in the longer term (Blandin and Levashina, 2007)

Very recently, the complex frame of cellular processes related to the phagocytic processes has been enriched by another major component, i.e., macroautophagy. Macroautophagy, hereafter referred to as autophagy, is an evolutionary conserved mechanism in which cells engulf cytoplasmic material into double membrane vacuoles that then fuse with lysosomes, forming an autophagosome (Xie and Klionsky, 2007). At the end of the process, the engulfed material is degraded and can possibly be reutilized by the cells. Even though autophagy is mainly a housekeeping process activated in cells by starvation and other insults, some evidence suggests the involvement of autophagy in innate immunity of vertebrates and invertebrates (Hussey et al., 2009). In adult *D. melanogaster* the intracellular peptidoglycan-recognition protein (PGRP)-LE, upon binding the Gram negative and intracellular bacterium *L. monocytogenes* promotes autophagy that, in its turn, inhibits intracellular growth of the bacteria. *PGRP-LE* null mutant flies are more sensitive versus the injection of *L. monocytogenes* with respects wild-type flies whereas selective expression of PGRP-LE in mutant hemocytes rescue the phenotype. Further RNAi experiments in primary cultured hemocytes and S2 cells indicated that *autophagy related gene* (*atg*) 5 plays a key role in the elimination of *L. monocytogenes*, indicating that autophagy is fundamental *in vivo* for host survival after the infection with an intracellular pathogen (Yano et al., 2008). Interestingly, it has been observed that the autophagic process promoting the clearance of intracellular bacteria is activated independently from the two major branches that regulate the humoral response in *D. melanogaster*, namely the Toll and IMD pathways. Since the largest part of antimicrobial response appears to be controlled by Toll and IMD pathways (Silverman et al., 2009), immune-related autophagy could be promoted by a cytoplasmic sensor or a distinct innate immune pathway (Yano and Kurata, 2008). Beside its role in antimicrobial defense, autophagy has recently been observed to display also antiviral features in fruit flies and cultured hemocytes infected with the mammalian viral pathogen, vesicular stomatitis virus (VSV) (Shelly et al., 2009). Adult flies and cultured cells in which the expression of several *atg* genes was silenced by mean of RNAi were more susceptible to the VSV infection, even though phagocytic activity remains unaffected. Importantly, VSV-induced autophagy appears to rely on a signalling pathway that is related with starvation (the PI3K-Akt pathway) and not with immune response (Shelly et al., 2009).

Encapsulation is a peculiar and intriguing immune reaction that requires the communication between distant organs and involves, among others, a particular hemocyte lineage, the lamellocyte (Lemaitre and Hoffmann, 2007). Lamellocytes are circulating hemocytes observed essentially in *Drosophila* larvae. They are barely observable in the body cavity of healthy larvae, but significantly increase in number after parasitization, especially in presence of eggs of parasitic wasps (Eslin et al., 2009). Parasitic wasps may attack the larval stage, and feed externally on the pupae inside the puparium. Usually, the wasp lays one egg into the host and this may stimulate *Drosophila* immune response. After the parasitoid's egg has been injected into the larva, the few circulating lamellocytes adhere to the egg surface and induce, through unknown signaling molecules, a strong lamellocyte proliferation associated with the massive differentiation of lamellocytes from either prohemocytes of the lymph gland or the sessile hemocytes (Nappi et al., 2009). Encapsulation is usually followed by melanization, which seems to be connected with the killing of the parasite (Nappi et al., 2009). Genome-wide studies of *Drosophila* immunity are often focused on the immune response versus microbial pathogens, and much less information is available on the genes that are up- or down-regulated against parasitoids. A microarray study of the transcriptional response of *D. melanogaster* to the parasitoid wasp *Asobara tabida* attack demonstrated that numerous genes that responded to the parasitization had not previously been associated with innate immunity, confirming the substantial differences in immunologic responses against microbial pathogens and macro-parasites (Wertheim et al., 2005). Even though the signalling underpinning the whole process of encapsulation is far from being elucidated (Lemaitre and Hoffmann, 2007), genome-wide analysis and RNAi assays demonstrated it involves cell adhesion molecules (Irving et al., 2005), Rho-GTPases family members (Williams et al., 2005, 2006) and at least one glycophorin-like proteins (Kurucz et al., 2003).

The responses to parasites, in particular involving melanization, have been extensively studied in mosquitoes, and the use of gene silencing techniques has lead to the identification of several molecules limiting *Plasmodium* ookinete loads (Blandin and Levashina, 2007). The first identified molecule has been the complement-like protein thioester-containing protein (TEP) 1, which binds to the surface of ookinetes and triggers their killing once they have crossed the midgut epithelium (Blandin et al., 2004). *tep1* gene silencing increases the number of *Plasmodium berghei* oocysts developing in the midgut. Similarly, depletion of two proteins with leucine-rich repeats, LRIM1 and *Anopheles Plasmodium* responsive leucine-rich repeat 1 (APL1), leads to increased parasite numbers (Osta et al., 2004a; Riehle et al., 2006). Conversely, silencing of two C-type lectins, CTLMA2 and CTL4, causes partial parasite melanization in the susceptible strain (Osta et al., 2004a).

Proteolytic cascades mediate several immune responses in insects, including wound healing, blood clotting, and melanization, and are tightly regulated by serine protease inhibitors to avoid potentially detrimental overreactions (Levashina et al., 1999). The knockdown of *serpin2* (*srpn2*) induces the formation of melanotic masses, increases parasite killing, and triggers melanization in strains that normally did not melanize killed parasites (Levashina et al., 1999). Silencing of *serpin6* (*srpn6*), whose expression is strongly upregulated upon parasite infection, does not alter the number of developing parasites in *A. gambiae* but rather influences the balance between lysis and melanization during clearance of killed parasites (Volz et al., 2005). Consistent with these results, several clip-domain serine proteases have also been shown to be involved in limiting parasite numbers and/or affecting the regulation of melanization (Volz et al., 2005).

In contrast to insects, phagocytosis defects in vertebrates cause profound reduction in survival of the animal after bacterial infection suggesting that this contrasting effect could result not only from the direct pathogen elimination but also from a second distinct role of phagocytosis unique to vertebrates consisting in the antigen presentation and initiation of the adaptive response (Moita et al., 2005; Underhill and Ozinsky, 2002). The comparison of the repertoire of molecules used in phagocytosis in evolutionarily very distinct groups of metazoans such as insects (mainly *A. gambiae* and *D. melanogaster*) and vertebrates (such as *Mus musculus* and *Homo sapiens*) may reveal altogether novel defense strategies as well as diversification of common pathways by the recruitment of other effectors or molecular mediators of defense reactions.

Humoral Immunity

In holometabolous insects, a tissue injury or foreign particles (e.g., bacteria or parasites) that contaminate the body cavity stimulate proteolytic cascades that lead to blood clotting, melanization and secretion of several humoral factors such as AMPs and cytokines.

In wounded or parasitized insects, the deposition of melanin is usually a rapid phenomenon that occurs at the injury site. In *Drosophila*, the crystal cell is the principal type of larval hemocyte implicated in melanotic encapsulation of parasites (Nappi and Christensen, 2005). Melanin formation is the consequence of the activation of the inactive precursor prophenoloxidase (proPO) to the active form phenoloxidase (PO). After this initial step, a complex series of reactions occurs (see for review Nappi et al., 2009) which finally leads to formation of the pigment at wound sites and around large intruding organisms. Melanin synthesis is associated with the production of intermediate cytotoxic compounds that are considered to be important immune effectors against pathogens, even though this assumption is currently controversial (Leclerc et al., 2006; Nappi et al., 2009). Gene silencing protocols or experiments on mutants have been principally directed towards the elements that regulates the cascade, namely the proPO activating enzymes (PAEs) and serpins, serine protease inhibitor family members usually considered as negative regulators of PO-triggered cascade. Experiments on mutants larvae and adult flies deficient for the serpin Spn27A indicated that in wild-type *Drosophila* this factor acts as a negative regulator of a PAE and is quickly removed from the hemolymph by a *de novo* synthesized (and so far uncharacterized) co-factor whose synthesis is under the control of the Toll pathway alone (Ligoxygakis et al., 2002). In accordance with these observations, the majority of *spn27A* mutant larvae present spontaneous melanization and high degree of lethality at the mid-pupal stages (Ligoxygakis et al., 2002). More recently, experiments on a second serpin, Spn28D, refined our knowledge on the regulation of PO activity. Through loss-of-function mutants and *in vivo* RNAi it has been possible to observe that the knockdown of *spn28D* causes an over-reactive PO activity after pricking in larvae, leading to extensive melanization of tissues in contact with air and hemolymph PO exhaustion (Scherfer et al., 2008). The different responses raised by *spn27A* mutants and *spn28D*-RNAi in pricking experiments lead to the hypothesis that the former limits the melanization reaction to the wound site whereas the latter regulates PO availability by controlling its initial release from crystal cells (Scherfer et al., 2008). Despite this information, the definitive role of PO in fly innate immunity is still to be established. Indeed *PAE-1* null mutant flies do not present any PO activity in the hemolymph but display a survival rate comparable to that of wild-type flies after natural infection with the entomopathogenic fungus *Beauveria bassiana*, or micro-injection of yeast, or Gram positive/Gram negative bacteria (Leclerc et al., 2006).

The mosquito genome encodes 14 serpins, 10 of which are inhibitory (Christophides et al., 2002). Interestingly gene expansions/losses result in species-specific diversification and only one orthologous pair and four orthologous groups are evident (Christophides et al., 2002). The *Drosophila* Serpin encoded by the *nec* locus, which is a partner of the Persephone Clip-domain protease in the Toll-mediated antifungal response (Levashina et al., 1999), also has no orthologue in the mosquitoes. The functions corresponding to Persephone and Nec must be served by independently evolved *Anopheles* CLIPs and Serpins. The *Drosophila* Spn27A, involved in control of melanization, forms an orthologous group with three mosquito serpins, which constitute interesting potential modulators of prophenoloxidases (Christophides et al., 2002).

Intriguing elements emerged from comparative studies of the role of various serpins in different mosquitoes. Indeed, knockdown of *srpn6* expression by RNAi in susceptible *A. stephensi* leads to substantially increased parasite numbers, whereas depletion in susceptible *A. gambiae* delayed the progression of parasite lysis without affecting the number of developing parasites. However, the *A. gambiae srpn6* knockdown increases the number of melanized parasites in the L3-5 refractory strain and in susceptible G3 mosquitoes depleted of the C-type lectin, CTL4 (Abraham et al., 2005). These results indicate that *As*SRPN6 is involved in the parasite-killing process, whereas *Ag*SRPN6 acts on parasite clearance by inhibiting melanization and/or promoting parasite lysis suggesting that these observed phenotypic differences are due to changed roles of the respective target serine proteases in the two mosquito species (Abraham et al., 2005).

AMPs are secreted into the hemolymph mainly by the fat body (a major immune tissue of mesodermal origin) and directly kill invading pathogens (De Gregorio et al., 2001). To date, seven structural classes of AMPs with a different spectrum of activity have been identified in *D. melanogaster* (Lemaitre and Hoffmann, 2007) and *A. gambiae* (Christophides et al., 2002; Osta et al., 2004b) and considerable efforts have been made to characterize these fundamental molecules that are conserved among different metazoan taxa (Hangock and Scott, 2000). At present, there are no not mutants available for the AMPs (Lemaitre and Hoffmann, 2007), and the information derived from gene silencing approach concerns the two principal immune-related pathways that control the inducibility of AMPs, Toll and Imd pathways (De Gregorio et al., 2002; Aggarwal et al., 2008; Malagoli et al., 2008; Silverman et al., 2009). *In vivo* studies performed in flies deficient for both the Imd and Toll pathways, but constitutively expressing different AMPs under the control of a non-inducible promoter, showed that in some cases the expression of a single AMP is sufficient to rescue the mutant susceptibility to microbial infection. For example, mutant flies over-expressing *defensin* show a wild-type survival after infection by the Gram positive *Micrococcus luteus*, and the co-expression of *drosocin* with *attacin A* enhances fly resistance against the Gram negative *Agrobacterium tumefaciens* (Tzou et al., 2002). Recently, *in vivo* experiments on mutant flies lead to the isolation of a gene called *psidin*, whose peculiarity resides in being a molecule with two distinct immune related functions. By one side, psidin is a lysosomial protein necessary in phagocytizing hemocytes for the degradation of engulfed bacteria. On the other side, hemocyte-derived Psidin is necessary for the induction of *defensin* in the fat-body, even though the mediator produced by the hemocytes and able to modulate the fat body response remains unknown (Brennan et al., 2007).

At present, defensin, cecropin and gambicin, which are produced by the fat body and hemocytes and secreted into hemolymph upon immune challenge (Vizioli *et al.*, 2000, 2001), represent the best characterized AMPs in mosquitoes. These peptides exhibit bactericidal and/or fungicidal activities *in vitro* and are thought to constitute the first line of defense against microbial infections (see for review Dimopoulos et al., 2001; Hoffmann and Reichhart, 2002). The RNAi on genes coding AMPs has been a valuable tool to determine the function of these mosquito AMPs. In this regard, Blandin et al. (2002) injected *defensin* dsRNA into adults and observed a silencing of the *defensin* gene assessing that this peptide is required for the mosquito antimicrobial defense against Gram-positive bacteria. In contrast, the injection of *defensin* dsRNA in mosquitoes infected by *P. berghei* did not result in a loss of mosquito viability neither in significant effects on the development and morphology of the parasite midgut stages suggesting that defensin is not a major antiparasitic factor in *A. gambiae in vivo* (Blandin et al., 2002).

Using RNAi experiments Meister et al. (2005) showed that in *A. gambiae*, the expression of *cecropin 1* is regulated by REL2, an NF-κB-like transcription factor orthologous to *Drosophila* Relish. However, in contrast to Relish (which responds principally to Gram-negative bacteria), the *Anopheles* REL2-F and REL2-S isoforms are involved in defense against the Gram positive *Staphylococcus aureus* and the Gram negative *Escherichia coli* bacteria, respectively. REL2-F also regulates the intensity of mosquito infection with the malaria parasite, *Plasmodium berghei*. A further element of interest is that in *Anopheles* there is an alternative splicing of the *rel2* gene that gives rise to two isoforms, *REL2-F* and *REL2-S*, which are differentially involved in defense against Gram positive and Gram negative bacteria, respectively. In this way, the single mosquito gene *rel2* mediates alternative immune responses, which, in the fly, require two genes: *rel2*'s orthologue *relish* and *dif* so that the post-transcriptional processing of *rel2* appears to compensate the absence of a *dif* gene in mosquitoes (Meister et al., 2005).

Recently, Pinto et al. (2009) performed an RNAi screening to identify *Plasmodium* immune modulators based on gene knockdown experiments on a set of 63 putative immune modulators to be tested for their effect on *P. berghei* early sporogonic development. Few silenced genes affected the presence of melanized ookinetes in the susceptible G3 strain of mosquitoes used for these experiments without affecting oocyst load. Silenced genes acting as modulators of *Plasmodium* consist of different genes coding for proteins with functional domains typically found within proteins that have been shown previously to affect parasite development, e.g., leucine-rich repeats, Ig-like domains and Fibrinogen C domains (Pinto et al., 2009). The contribution of these functional domains to antiparasite immunity is diverse, and each is likely to affect several pathways. In addition to constitutively producing factors that affect parasite development, circulating hemocytes change their transcriptional profiles upon ookinete traversing the midgut epithelium. Of the 119 transcripts with increased hemocyte expression levels identified by Pinto et al. (2009), 16 encode for known immunity-related proteins (Waterhouse et al., 2007) and six of those have already been shown to have direct or indirect antiparasite activity (such as AMPs) (Kim et al., 2004). Through gene silencing, Pinto and collaborators provided an extensive catalogue of hemocyte-enriched molecules that could be involved in the mosquitoes immune responses against malaria parasites (Pinto et al., 2009).

Among humoral factors, AMPs are probably the molecules whose structures and functions have been better elucidated, whereas *Drosophila* cytokines, i.e. Spätzle, Unpaired (Upd)-3 and the putative helical cytokine, *Drosophila* helical factor (DHF) still await further characterization. The cytokine Spätzle is the first described and best characterized *Drosophila* cytokine (Morisato and Anderson, 1994). Spätzle activates the Toll pathway, as a consequence of a proteolytic cascade triggered mainly by Gram positive bacteria and fungi (Silverman et al., 2009). The conformation of Spätzle resembles that of vertebrate NGF and the horseshoe crab coagulogen (Mizuguchi et al., 1998). In terms of function, the gene *spätzle* encodes for a secreted protein, involved in development and immunity, that requires proteolytic processing for activity (Morisato and Anderson, 1994) and acts immediately upstream of the receptor Toll (Ferrandon et al., 2004). After the knock-down of *spätzle* expression, mutant flies can recover the inducibility *drosomycin* in the fat body after injection of either recombinant full-length Spätzle or hemolymph from wild-type flies. However, the recovery of *drosomycin* induction is always subsequent to an immune challenge with Gram positive bacteria or fungi. These observations lead to the conclusion that Spätzle is a cytokine present in the hemolymph as an inactive precursor converted to the active form after an immune challenge (Ferrandon et al., 2004). Very recently, experiments on mutant flies indicated that Spätzle behaves in a way more similar to a mammalian cytokine than previously thought (Shia et al., 2009). In hemocyte-ablated and infected larvae, the fat body produces AMP with a very low efficiency when compared to wild-type larvae. Specific hemocyte ablation has been obtained using the GAL4-UAS system (Brand and Perrimon, 1993) and selecting the driver *peroxidasin*-GAL4 (*pxn*-GAL4) due to its minimal developmental interference (Shia et al., 2009). Moreover, the silencing of *spätzle* specifically in hemocytes results in a low production of Drosomycin by the fat body after infection with *M. luteus*, supporting the hypothesis that hemocyte-derived Spätzle is crucial for regulating AMP production by the fat body during septic injury (Shia et al., 2009).

The second cytokine observed in *D. melanogaster*, Upd-3, has been characterized as a member of the *upd* family that activates the JAK/STAT pathway during the embryogenesis of *Drosophila* (Harrison et al., 1998). Even though Upd-3 has no homology or similarity with known mammalian cytokines, it is secreted by hemocytes after septic injury and promotes the expression of *totA* (a protein playing probably a general role in stress response) by the fat body (Agaisse and Perrimon, 2004). RNAi data collected in S2 cells have suggested that *upd-3* is not induced by activation of the Imd-pathway, or at least not by the branch of the pathway controlled by the kinase dTAK1 (Malagoli et al., 2008).

Using a bioinformatics approach, the putative helical cytokine, *Drosophila* helical factor (DHF), displays a predicted structure highly similar to that of mammalian helical cytokines (Malagoli and Ottaviani, 2007; Malagoli et al., 2007). Its role as a full title cytokine is still under investigation (Malagoli et al., 2010), but several evidences indicate that DHF is a good candidate as the third *Drosophila* cytokine. *dhf* expression increases in hemocytes and larvae after immune challenge (Malagoli et al., 2007), but the observed induction is impaired after blocking the Imd pathway by mean of RNAi versus the kinase dTAK1 (Malagoli et al., 2008). DHF is secreted by S2 hemocytes during the incubation with Gram-negative bacteria, the administration of the recombinant form of DHF dose-dependently stimulate the transcription of AMPs by S2 cells and the silencing of *dhf* through RNAi results in a relevant reduction of AMP transcriptional levels during an immune challenge (Malagoli et al., unpublished results).

RNAi as a Natural Anti-viral Process in *D. melanogaster* and *A. gambiae*

Even though RNAi is usually considered a useful technique to specifically and transiently silence the expression of one or more specific genes, it also represents a fundamental immune-related process among all organisms especially directed against viruses. The antiviral functions of RNAi relies on genes presenting and elevated rate of adaptive evolution and have been utilized for studies finalized at the quantification of the rate of evolution of immunity-related genes (Obbard et al., 2009). A fundamental and conserved trait of RNA silencing aimed at limiting viral infection and observed in all organisms is the production of small RNAs by the endoribonuclease Dicer.

Differently from other metazoans, in *D. melanogaster* the dicing of the imperfectly base-paired precursor into microRNAs (miRNAs) is realized by Dicer-1, whilst the perfectly base-paired double-stranded RNA is cut into small-interfering RNAs (siRNAs) by Dicer-2 (Li and Ding 2005). siRNAs and miRNAs act as components of RNA-induced silencing complexes (RISCs) whose core protein component is represented by a member of the Argonaute (Ago) family of small RNA-guided RNA-binding proteins. The *D. melanogaster* genome encodes for at least five Ago family members, among which Ago1 and Ago2 have been reported to bind miRNAs and siRNAs, respectively (Förstermann et al., 2007). The functional specialization of Ago1 and Ago2 observed in *Drosophila* is still under investigation, but it has been proposed that the partitioning of virus-derived siRNAs into Ago2-RISC would reduce the probability of off-target silencing of host genes during an anti-viral RNAi response (Förstermann et al., 2007). Recent screening of RNAi library leads to the discovery of Ars-2, another important element of anti-viral immunity of the fruit fly. Ars-2 interacts with Dicer-2 influencing *in vitro* the activity of the latter. Moreover, S2 cells or adult flies depleted of Ars-2 present a highly reduced anti-viral immune-response (Sabin et al., 2009).

RNAi has also been reported as a natural antagonist to virus replication in mosquitoes (Hoa et al., 2003; Keene et al., 2004). In *A. gambiae* the components of RNAi machinery has been identified by Hoa et al. (2003) in a cell line using a pre-treatment with dsRNA of putative genes, followed by monitoring the effect of RNAi against luciferase expression driven by the promoter of the *cecA* gene (Zheng and Zheng, 2002). These experiments showed that *Ag*Dcr2, *Ag*Ago2 and *Ag*Ago3 play important roles in RNAi in mosquitoes. Surprisingly, no significant result was found with *Ag*Ago1, whereas mutations in *Dm*Ago1 protein resulted in late embryonic/early larval lethality (Williams and Rubin, 2002) suggesting that different Ago proteins were acquired by *Drosophila* and *Anopheles* during evolution of the RNAi machineries. This hypothesis is strengthened by the observation that also *Ag*Dicer-1 seems to be differently used in respect to *Drosophila* Dicer-1 that is necessary for degradation of dsRNA into siRNA (Bernstein et al., 2001). Similar results have been obtained by Keene and colleagues (2004) working with whole mosquitoes and showing that injection of dsRNA cognate to *Agago2*, which is a gene known to function in the *A. gambiae* RNAi pathway (Hoa et al., 2003), silenced the RNAi machinery, thereby permitting the O'nyong-nyong virus (ONNV) to replicate and disseminate quickly in mosquitoes. Similarly, the injection of dsRNA derived from *Agago3* also made mosquitoes more permissive to ONNV replication. In contrast, dsRNAs derived from *Agago1*, *Agago4*, and *Agago5* did not alter virus replication significantly suggesting a regulatory role for RNAi in controlling arbovirus infections in mosquitoes, with *Ag*Ago2 and *Ag*Ago3 proteins playing as critical components of the mosquito RNAi pathway involved in the inhibition of alphavirus replication. Interestingly, Ago2 has been reported as an important RISC component in *Drosophila* (Hammond, et al., 2001) making *Ag*Ago2 the most important component of the mosquito RNAi pathway.

These data support the idea that RNAi is a mechanism to protect mosquitoes from viral infection, even if the efficacy of such mechanisms could be specific for each virus (Keene et al., 2004; Sanchez-Vargas et al., 2004). Indeed, data demonstrating dsRNA effectiveness against a myriad of targets in *A. gambiae* (Blandin et al., 2004; Osta et al., 2004b) clearly indicate that mosquitoes generally have a robust RNAi response that could explain why anopholine mosquitoes are such poor vectors for arboviruses. Interestingly, *Aedes aegypti* readily transmit both alphaviruses (such as chikungunya and sindbis) and flaviviruses (including yellow fever and dengue) indicating that these mosquitoes have a weaker RNAi response. In support of this hypothesis, Keene et al. (2004) observed that ONNV readily disseminates in *A. aegypti* tissues, whereas the Sindbis virus MRE16 strain (Foy et al., 2004), which disseminates very efficiently in all of the culicines, could not replicate in *A. gambiae* tissues (Keene et al., 2004).

Conclusion

The insects *D. melanogaster* and *A. gambiae* represent two of the best-characterized host defense systems of invertebrates. The success of *D. melanogaster* relies especially on its tractability for genetic and genomic approaches, together with a relevant cost-effectiveness in comparison with vertebrate models. Studies in *D. melanogaster* gave a significant contribution to the discovery of the relevant similarities between *D. melanogaster* and vertebrate innate immunity basis, while *A. gambiae* is a model adopted essentially for being a vector of one of the most important human disease worldwide. In particular, the recent discovery of RNAi and its adaptation to mosquitoes is now providing new tools for the dissection of *Plasmodium*-mosquito interactions and for the analysis of aspects influencing the mosquito vectorial capacity that will improve our ability to interfere with the *Plasmodium* life cycle reducing malaria diffusion.

The diverse perspectives leading the experiments on *D. melanogaster* and *A. gambiae* are also reflected by the different level of knowledge acquired so far on the immune systems of these two models. Even though limited by the high rate of evolution that the immune-related genes present, the integration of information acquired from studies of innate immune responses in *D. melanogaster* and *A. gambiae* will likely help in understanding several currently unresolved issues in the immune system of the most represented group of the animal kingdom and in exploring the new intriguing applicative perspectives opened by the development of easy-to-apply methods for gene silencing in insects.

Acknowledgments

We gratefully acknowledge Prof. E Ottaviani (University of Modena and Reggio Emilia, Modena, Italy) and Prof. AJ Nappi (Loyola University Chicago, Chicago, IL, USA) for helpful discussion on this manuscript. This work was supported by FAR 2009 grants of the University of Modena and Reggio Emilia to DM and MM.

References

Abraham EG, Pinto SB, Ghosh A, Vanlandingham DL, Budd A, Higgs S, Kafatos FC, Jacobs-Lorena M, Michel K. An immune-responsive serpin, SRPN6, mediates mosquito defense against malaria parasites. *Proc Natl Acad Sci U S A,* 2005, 102, 16327-16332.

Adams, MD. The genome sequence of *Drosophila melanogaster. Science,* 2000, 287, 2185–2195.

Agaisse H, Perrimon, N. The roles of JAK/STAT signalling in *Drosophila* immune responses. *Immunol Rev,* 2004, 198, 72-82.

Agaisse H, Burrack LS, Philips JA, Rubin EJ, Perrimon N, Higgins DE. Genome-wide RNAi screen for host factors required for intracellular bacterial infection. *Science,* 2005, 309, 1248-1251.

Aggarwal K, Rus F, Vriesema-Magnuson C, Ertürk-Hasdemir D, Paquette N, Silverman N. Rudra interrupts receptor signalling complexes to negatively regulate the IMD pathway. *PLoS Pathog,* 2008, 4, e1000120.

Arias AM. *Drosophila melanogaster* and the development of biology in the 20th century. *Methods Mol Biol,* 2008, 420, 1-25.

Ballarin L, Cammarata M, Cima F, Grimaldi A, Lorenzon S, Malagoli D, Ottaviani E. Immune-neuroendocrine biology of invertebrates: a collection of methods. *Inv Surv J,* 2008, 5, 192-215.

Beier JC. Malaria parasite development in mosquitoes. *Ann Rev Entomol,* 1998, 43, 519-543.

Bernstein E, Caudy AA, Hammond SM, Hannon GJ. Role for a bidentate ribonuclease in the initiation step of RNA interference. *Nature,* 2001, 409, 363–366.

Blandin S, Moita LF, Kocher T, Wilm M, Kafatos FC, Levashina EA. Reverse genetics in the mosquito *Anopheles gambiae*: targeted disruption of the defensin gene. *EMBO Report,* 2002, 3, 852–856.

Blandin S, Shiao SH, Moita LF, Janse CJ, Waters AP, Kafatos FC, Levashina EA. Complement-like protein TEP1 is a determinant of vectorial capacity in the malaria vector *Anopheles gambiae. Cell,* 2004,116, 661–670.

Blandin SA, Levashina EA, Phagocytosis in mosquito immune responses. *Immunol Reviews,* 2007, 219, 8–16.

Brand, AH, Perrimon, N. Targeted gene expression as a means of altering cell fates and generating dominant phenotypes. *Development,* 1993, 118, 401-415.

Breman JG, Alilio MS, Mills A. Conquering the intolerable burden of malaria: what's new, what's needed: a summary. *Am J Trop Med Hyg,* 2004, 71, 1-15

Brennan CA, Delaney JR, Schneider DS, Anderson KV. Psidin is required in *Drosophila* blood cells for both phagocytic degradation and immune activation of the fat body. *Curr Biol,* 2007, 17, 67-72.

Bulet P, Stöcklin, R. Insect antimicrobial peptides: structures, properties and gene regulation. *Protein Pept Lett,* 2005, 12, 3-11.

Christophides GK, Zdobnov E, Barillas-Mury C, Birney E, Blandin S, Blass C, Brey PT, Collins FH, Danielli A, Dimopoulos G, Hetru C, Hoa NT, Hoffmann JA, Kanzok SM, Letunic I, Levashina EA, Loukeris TG, Lycett G, Meister S, Michel K, Moita LF, Müller HM, Osta MA, Paskewitz SM, Reichhart JM, Rzhetsky A, Troxler L, Vernick KD, Vlachou D, Volz J, von Mering C, Xu J, Zheng L, Bork P, Kafatos FC. Immunity-related genes and gene families in *Anopheles gambiae. Science,* 2002, 298, 159-165.

De Gregorio E, Spellman PT, Rubin GM, Lemaitre B. Genome-wide analysis of the *Drosophila* immune response by using oligonucleotide microarrays. *Proc Natl Acad Sci USA,* 2001, 98, 12590-12595.

De Gregorio E, Spellman PT, Tzou P, Rubin GM, Lemaitre B. The Toll and Imd pathways are the major regulators of the immune response in *Drosophila. EMBO J,* 2002, 21, 2568-2579.

Dimopoulos G, Müller HM, Levashina EA, Kafatos FC. Innate immune defense against malaria infection in the mosquito. *Curr Opin Immunol,* 2001, 13, 79–88.

Dushay MS. Insect hemolymph clotting. Cell Mol Life Sci, 2009, 66, 2643-2650.

Elrod-Erickson M, Mishra S, Schneider D. Interactions between the cellular and humoral immune responses in *Drosophila. Curr Biol,* 2000, 10, 781–784.

Eslin P, Prévost G, Havard S, Doury G. Immune resistance of *Drosophila* hosts against *Asobara parasitoids*: cellular aspects. *Adv Parasitol* 2009, 70, 189-215.

Ferrandon D, Imler JL, Hoffmann JA. Sensing infection in *Drosophila*: *Toll and beyond*. *Semin Immunol*, 2004, 16, 43-53.

Förstermann K, Horwich MD, Wee L, Tomari Y, Zamore PD. *Drosophila* microRNAs are sorted into functionally distinct Argonaute complexes after production by Dicer-1. *Cell*, 2007, 130, 287-297.

Foy BD, Myles KM, Pierro DJ, Sanchez-Vargas I, Uhlirova M, Jindra M, Beaty BJ, Olson KE. Development of a new Sindbis virus transducing system and its characterization in three culicine mosquitoes and two lepidopteran species. *Insect Mol Biol*, 2004, 13, 89–100.

Hammond SM, Boettcher S, Caudy AA, Kobayashi R, Hannon G. Argonaute2, a link between genetic and biochemical analyses of RNAi. *Science*, 2001, 293, 1146–1150.

Harrison DA, McCoon PE, Binari R, Gilman M, Perrimon N. *Drosophila* unpaired encodes a secreted protein that activates the JAK signaling pathway. *Genes Dev*, 1998, 12, 3252–3263.

Hedges SB, Blair JE, Venturi ML, Shoe JL. A molecular timescale of eukaryote evolution and the rise of complex multicellular life. *BMC Evol Biol*, 2004, 4, 2.

Hoa NT, Keene KM, Olson KE, Zheng L. Characterization of RNA interference in an *Anopheles gambiae* cell line. *Insect Biochem Mol Biol*, 2003, 33, 949-957.

Hoffmann JA, Reichhart JM. *Drosophila* innate immunity: an evolutionary perspective. *Nat Immunol*, 2002, 3, 121–126.

Holt RG, Subramanian G, Halpern M et al. The genome sequence of the malaria mosquito *Anopheles gambiae*. *Science*, 2002, 298, 129 – 149.

Hussey S, Travassos LH, Jones NL. Autophagy as an emerging dimension to adaptive and innate immunity. *Semin Immunol*, 2009, 21, 233-241.

Irving P, Ubeda JM, Doucet D, Troxler L, Lagueux M, Zachary D, Hoffmann JA, Hetru C, Meister M. New insights into *Drosophila* larval haemocyte functions through genome-wide analysis. *Cell Microbiol*, 2005, 7, 335-350.

Kallio J, Leinonen A, Ulvila J, Valanne S, Ezekowitz RA, Rämet M. Functional analysis of immune response genes in *Drosophila* identifies JNK pathway as a regulator of antimicrobial peptide gene expression in S2 cells. *Microbes Infect*, 2005, 7, 811-819.

Keene KM, Foy BD, Sanchez-Vargas I, Beaty BJ, Blair CD, Olson KE. RNA interference acts as a natural antiviral response to O'nyong-nyong virus (Alphavirus; Togaviridae) infection of *Anopheles gambiae*. *Proc Natl Acad Sci USA*, 2004, 101, 17240-17245.

Kim W, Koo H, Richman AM, Seeley D, Vizioli J, Klocko AD, O'Brochta DA. Ectopic expression of a cecropin transgene in the human malaria vector mosquito *Anopheles gambiae* (Diptera: Culicidae): effects on susceptibility to *Plasmodium*. *J Med Entomol*, 2004, 41, 447–455.

Kurucz E, Zettervall CJ, Sinka R, Vilmos P, Pivarcsi A, Ekengren S, Hegedüs Z, Ando I, Hultmark D. Hemese, a hemocyte-specific transmembrane protein, affects the cellular immune response in *Drosophila*. *Proc Natl Acad Sci USA*, 2003, 100, 2622-2627.

Leclerc V, Pelte N, El Chamy L, Martinelli C, Ligoxygakis P, Hoffmann JA, Reichhart JM. Prophenoloxidase activation is not required for survival to microbial infections in *Drosophila*. *EMBO Rep*, 2006, 7, 231-235.

Lemaitre, B, Hoffmann, JA. The host defense of *Drosophila melanogaster*. *Annu Rev Immunol,* 2007, 25, 697-743.

Levashina EA, Langley E, Green C, Gubb D, Ashburner M, Hoffmann JA, Reichhart JM Constitutive activation of toll-mediated antifungal defense in serpin-deficient *Drosophila*. *Science,* 1999, 285, 1917–1919.

Levashina EA, Moita LF, Blandin S, Vriend G, Lagueux M, Kafatos FC. Conserved role of a complement-like protein in phagocytosis revealed by dsRNA knockout in cultured cells of the mosquito, *Anopheles gambiae*. *Cell,* 2001,104, 709–718.

Li HW, Ding SW. Antiviral silencing in animals. *FEBS Lett,* 2005, 579, 5965-5973.

Ligoxygakis P, Pelte N, Ji C, Leclerc V, Duvic B, Belvin M, Jiang H, Hoffmann JA, Reichhart JM. A serpin mutant links Toll activation to melanization in the host defence of *Drosophila*. *EMBO J,* 2002, 21, 6330-6337.

Malagoli D, Ottaviani E. Helical cytokines and invertebrate immunity: a new field of research. *Scand J Immunol,* 2007, 66, 484-485.

Malagoli D, Sacchi S, Ottaviani E. unpaired (upd)-3 expression and other immune-related functions are stimulated by interleukin-8 in *Drosophila melanogaster* SL2 cell line. *Cytokine,* 2008, 44, 269-274.

Malagoli D, Sacchi S, Ottaviani E. Lectins and cytokines in celomatic invertebrates: two tales with the same end. *Inv Surv J,* 2010, 7, 1-10.

Meister S, Kanzok SM, Zheng XL, Luna C, Li TR, Hoa NT, Clayton JR, White KP, Kafatos FC, Christophides GK, Zheng L. Immune signaling pathways regulating bacterial and malaria parasite infection of the mosquito *Anopheles gambiae*. *Proc Natl Acad Sci U S A,* 2005, 102, 11420-11425.

Mizuguchi K, Parker JS, Blundell TL, Gay NJ. Getting knotted: a model for the structure and activation of Spätzle. *Trends Biochem Sci,* 23, 239-242, 1998.

Moita LF, Wang-Sattler R, Michel K, Zimmermann T, Blandin S, Levashina EA, Kafatos FC. *In vivo* identification of novel regulators and conserved pathways of phagocytosis in *A. gambiae*. *Immunity,* 2005, 23, 65–73.

Morisato D, Anderson KV. The *spätzle* gene encodes a component of the extracellular signalling pathway establishing the dorsal-ventral pattern of the *Drosophila* embryo. *Cell,* 1994, 76, 677-688.

Nappi AJ, Christensen BM. Melanogenesis and associated cytotoxic reactions: applications to insect innate immunity. *Insect Biochem Mol Biol,* 2005, 35, 443-459.

Nappi, AJ, Kohler, L, Mastore, M. Signalling pathways implicated in the cellular innate immune responses of *Drosophila*. *Inv Surv J,* 2004, 1, 5-33.

Nappi A, Poirié M, Carton Y. The role of melanization and cytotoxic by-products in the cellular immune responses of *Drosophila* against parasitic wasps. *Adv Parasitol* 2009, 70, 99-121.

Novotny V, Miller SE, Hulcr J, Drew RAI, et al. Low beta diversity of herbivorous insects in tropical forests. *Nature,* 2007, 448, 692-695.

Obbard DJ, Welch JJ, Kim KW, Jiggins FM. Quantifying adaptive evolution in the *Drosophila* immune system. *PLoS Genet,* 2009, 5, e1000698.

Osta MA, Christophides GK, Kafatos FC. Effects of mosquito genes on *Plasmodium* development. *Science,* 2004a, 303, 2030–2032.

Osta MA, Christophides GK, Vlachou D, Kafatos FC. Innate immunity in the malaria vector *Anopheles gambiae*: comparative and functional genomics. *J Exp Biol,* 2004b, 207, 2551-2563.

Pearson AM, Baksa K, Rämet M, Protas P, McKee M, Brown D, Ezekowitz RAB. Identification of cytoskeletal regulatory proteins required for efficient phagocytosis in *Drosophila. Microbes Infect,* 2003, 5, 815–824.

Philips JA, Rubin EJ, Perrimon N. *Drosophila* RNAi screen reveals CD36 family member required for mycobacterial infection. *Science,* 2005, 309, 1251-1253.

Pinto SB, Lombardo F, Koutsos AC, Waterhouse RM, McKay K, An C, Ramakrishnan C, Kafatos FC, Michel K. Discovery of *Plasmodium* modulators by genome-wide analysis of circulating hemocytes in *Anopheles gambiae. Proc Natl Acad Sci U S A,* 2009, 106, 21270-21275.

Rämet M, Pearson A, Manfruelli P, Li X, Koziel H, Göbel V, Chung E, Krieger M, Ezekowitz RA. *Drosophila* scavenger receptor CI is a pattern recognition receptor for bacteria. *Immunity,* 2001, 15, 1027-1038.

Rämet M, Manfruelli P, Pearson A, Mathey-Prevot B, Ezekowitz RA. Functional genomic analysis of phagocytosis and identification of a *Drosophila* receptor for *E. coli. Nature,* 2002, 416, 644-648.

Riehle MM, Markianos K, Niaré O, Xu J, Li J, Touré AM, Podiougou B, Oduol F, Diawara S, Diallo M, Coulibaly B, Ouatara A, Kruglyak L, Traoré SF, Vernick KD. Natural malaria infection in *Anopheles gambiae* is regulated by a single genomic control region. *Science,* 2006, 312, 577–579.

Sabin LR, Zhou R, Gruber JJ, Lukinova N, Bambina S, Berman A, Lau CK, Thompson CB, Cherry S. Ars2 regulates both miRNA- and siRNA- dependent silencing and suppresses RNA virus infection in *Drosophila. Cell,* 2009, 138, 340-351.

Sanchez-Vargas I, Travanty EA, Keene KM, Beaty BJ, Blair CD, Olson KE. RNA interference, arthropod-borne viruses, and mosquitoes. *Virus Res,* 2004, 102, 65–74.

Scherfer C, Tang H, Kambris Z, Lhocine N, Hashimoto C, Lemaitre B. *Drosophila* Serpin-28D regulates hemolymph phenoloxidase activity and adult pigmentation. *Dev Biol,* 2008, 323, 189-196.

Shelly S, Lukinova N, Bambina S, Berman A, Cherry S. Autophagy is an essential component of *Drosophila* immunity against vesicular stomatitis virus. *Immunity,* 2009, 30, 588-598.

Shia AK, Glittenberg M, Thompson G, Weber AN, Reichhart JM, Ligoxygakis P. Toll-dependent antimicrobial responses in *Drosophila* larval fat body require Spätzle secreted by haemocytes. *J Cell Sci,* 2009, 122, 4505-4515.

Silverman N, Paquette N, Aggarwal K. Specificity and signaling in the *Drosophila* immune response. *Inv Surv J,* 6, 163-174, 2009

Stroschein-Stevenson SL, Foley E, O'Farrell PH, Johnson AD. Identification of *Drosophila* gene products required for phagocytosis of *Candida albicans. PLoS Biol,* 2006, 4: e4.

Tzou P, Reichhart JM, Lemaitre B. Constitutive expression of a single antimicrobial peptide can restore wild-type resistance to infection in immunodeficient *Drosophila* mutants. *Proc Natl Acad Sci USA,* 2002, 99, 2152-2157.

Underhill DM, Ozinsky A. Phagocytosis of microbes: complexity in action. *Annu Rev Immunol* 2002, 20, 825–852.

Vizioli J, Bulet P, Charlet M, Lowenberger C, Blass C, Müller HM, Dimopoulos G, Hoffmann J, Kafatos FC, Richman A. Cloning and analysis of a cecropin gene from the malaria vector mosquito, *Anopheles gambiae. Insect Mol Biol,* 2000, 9, 75–84.

Vizioli J, Bulet P, Hoffman JA, Kafatos FC, Muller HM, Dimopoulos G. Gambicin: a novel immune responsive antimicrobial peptide from the malaria vector *Anopheles gambiae. Proc Natl Acad Sci USA,* 2001, 98, 12630-12635.

Volz J, Osta MA, Kafatos FC, Muller HM. The roles of two clip domain serine proteases in innate immune responses of the malaria vector *Anopheles gambiae. J Biol Chem,* 2005, 280, 40161–40168.

Waterhouse RM, Kriventseva EV, Meister S, Xi Z, Alvarez KS, Bartholomay LC, Barillas-Mury C, Bian G, Blandin S, Christensen BM, Dong Y, Jiang H, Kanost MR, Koutsos AC, Levashina EA, Li J, Ligoxygakis P, Maccallum RM, Mayhew GF, Mendes A, Michel K, Osta MA, Paskewitz S, Shin SW, Vlachou D, Wang L, Wei W, Zheng L, Zou Z, Severson DW, Raikhel AS, Kafatos FC, Dimopoulos G, Zdobnov EM, Christophides GK. Evolutionary dynamics of immune-related genes and pathways in disease-vector mosquitoes. *Science,* 2007, 316, 1738–1743.

Wertheim B, Kraaijeveld AR, Schuster E, Blanc E, Hopkins M, Pletcher SD, Strand MR, Partridge L, Godfray HC. Genome-wide gene expression in response to parasitoid attack in *Drosophila. Genome Biol,* 2005, 6, R94.

Williams and Rubin, 2002. R.W. Williams and G.M. Rubin, Argonaute1 is required for efficient RNA interference in *Drosophila* embryos. *Proc Natl Acad Sci USA,* 2002, 99, 6889–6894.

Williams MJ, Ando I, Hultmark D. *Drosophila melanogaster* Rac2 is necessary for a proper cellular immune response. *Genes Cells,* 2005, 10, 813-823.

Xie Z, Klionsky DJ. Autophagosome formation: core machinery and adaptations. *Nat Cell Biol,* 2007, 9, 1102-1109.

Yano T, Kurata S. Induction of autophagy via innate bacterial recognition. *Autophagy,* 2008, 4, 958-960.

Yano T, Mita S, Ohmori H, Oshima Y, Fujimoto Y, Ueda R, Takada H, Goldman WE, Fukase K, Silverman N, Yoshimori T, Kurata S. Autophagic control of *Listeria* through intracellular innate immune recognition in *Drosophila. Nat Immunol,* 2008, 9, 908-916.

Zheng XL, Zheng, AL. Genomic organization and regulation of three cecropin genes in *Anopheles gambiae. Insect Mol Biol,* 2002, 11, 517–525.

In: Gene Silencing: Theory, Techniques and Applications   ISBN: 978-1-61728-276-8
Editor: Anthony J. Catalano   © 2010 Nova Science Publishers, Inc.

Chapter VII

RNAi in Plants: Recent Developments and Applications in Agriculture

Muthappa Senthil-Kumar and Kirankumar S. Mysore[*]

The Samuel Roberts Noble Foundation, 2510 Sam Noble Pky, Ardmore, OK 73402 USA

Abstract

Sequencing of plant genome and expressed sequence tag (EST) have provided abundant sequence information in several plant species. Elucidating function(s) of all of these genes is a huge undertaking. Even in well-studied plants like Arabidopsis, function is not known for majority of genes. Hence, a powerful tool that can be widely used to understand gene function is necessary. Several functional genomics tools were developed in the recent past to achieve this goal. RNA interference (RNAi) is one such tool widely used to analyze gene function. RNAi is also proved to be a tool for plant researchers to produce improved crop varieties.

First part of this review is focused on three RNAi based concepts that has potential applications in plant functional genomics and agriculture. These concepts are tissue specific silencing, inducible silencing and host delivered RNAi (hdRNAi) during plant-pest interaction. Tissue specific promoters driving RNAi constructs can induce gene silencing in a particular organelle or tissue. Also, RNAi constructs with stress or chemical inducible promoters can be used to induce gene silencing only when required. These two concepts together can be used to achieve temporal and spatial control of gene silencing in plants. In the hdRNAi, dsRNA generated in an RNAi transgenic plant is delivered to interacting target organism (pest), activating gene silencing in the target organism. A comprehensive review pertaining to these areas is presented.

---

[*] Corresponding author. Plant Biology Division, The Samuel Roberts Noble Foundation, 2510 Sam Noble Pky, Ardmore, OK 73402 USA. Email: ksmysore@noble.org Fax: 580-224-6692

Second part of the review deals with applications of RNAi in agriculture, animal husbandry and biofuel industry. As suppression of gene expression by RNAi is inheritable, this has been a tool for developing transgenic crop plants for resistance against disease, pests, drought and in other areas of agriculture. This review summarizes developments in these areas with major emphasis on application of RNAi for development of biotic stress tolerant crops. We also note limitations of RNAi technology and ways to overcome the same.

Introduction

RNAi interference (RNAi) has emerged as one of the major plant functional genomics tool. RNAi involves production of double-stranded RNA (dsRNA) homologous to the gene being targeted for post transcriptional gene silencing (PTGS). This dsRNA is processed into approximately 21-nucleotide RNAs, known as small interfering RNAs (siRNAs), by the enzyme Dicer. These siRNAs then provide specificity to endonuclease-containing, RNA-induced silencing complex (RISC), which targets homologous RNAs for degradation (reviewed in [1-3]). RNAi has rapidly gained importance as a reverse genetics tool to knock down expression of targeted genes in plants, lower animals and microorganisms. It has several inherent advantages over other functional genomics tools. The ability to target multiple gene family members with a single RNAi-inducing trigger gene sequence is one such advantage. Another advantage is the dominant nature of gene knockdowns due to RNAi. Dominance of RNAi allows one to save time by eliminating additional generations needed to identify individuals that are homozygous for recessive loss-of-function alleles [4]. Likewise, orthologs can be knocked down in F1 hybrids in which RNAi-inducing transgene is introduced through only one of the parents. Finally, RNAi allows knock down of genes in polyploid genomes that contain several orthologs and are thus not suitable to traditional mutagenesis.

RNAi based gene silencing mechanism is conserved across plant and animal kingdoms [5] thus holding promise not only for functional genomics but also in agricultural applications like engineering for stress tolerance. In addition, improvements in RNAi method have unleashed its enormous potential to tackle several challenges in agriculture. Here we review its wide range of applications in crop improvement. Prospects of RNAi in genetic improvement of crop plants, possible future problems and remedy are discussed.

RNAi Vectors and Methods for their Delivery into Plants

RNAi in plants has been achieved by expressing hairpin RNA (hpRNA) that fold back to create a dsRNA [6]. These hpRNAs are potent inducers of PTGS and give rise to siRNAs derived from the dsRNA. A list of RNAi vectors, many sharing overall design but differ in cloning strategies, selectable markers and other elements developed during recent years are reviewed [4, 7]. Commonly used methods for RNAi-mediated gene silencing in plants are transforming plants with hpRNA producing vectors by *Agrobacterium*-mediated plant transformation or particle bombardment to produce stable transgenic plants, and infiltration of *Agrobacterium* cultures harboring hpRNA producing construct for transient gene silencing [8-10]. Apart from these, direct siRNA delivery [11], artificial micro RNA (amiRNA) [12] based vectors and exogenous dsRNA spray [13] are also shown to be effective for gene silencing. For the past one decade several novel methods to achieve stable-, transient-, inducible-, specific-, differential-, comprehensive- RNAi have been developed, thus diversifying the application of RNAi. We describe below few of these methods and their applications.

Transient and Stable Gene Silencing Using RNAi

Agrobacterium-mediated transient gene silencing assay in intact tissues has emerged as a rapid and useful method to analyze gene function in plants. Infiltration of *Agrobacterium* cultures harboring hpRNA construct into intercellular spaces (referred as Agroinfiltration) induces silencing of corresponding endogenous gene in infiltrated cells and non-infiltrated cells adjacent to infiltration zone. Agroinfiltration of hpRNA producing vectors is a suitable method for gene suppression even in difficult to transform plants and this facilitate high-throughput functional genomics screening. RNAi vector (pANDA-mini) was Agroinfiltrated into leaves and transient suppression of phytoene desaturease gene has been demonstrated in rice [14]. Similarly, this method has been widely used in several other studies to identify gene function in plants [10, 15, 16]. Apart from Agroinfiltration, direct introduction of a plasmid producing hpRNA by particle bombardment is useful to analyze gene function even in transformation recalcitrant plants. Using this method transient gene silencing of glutathione synthetase (*GSHS*) in somatic embryos of *Camellia sinensis* L. has been shown [17]. Further, biolistic method has been used for transient gene silencing to study gene function in petals tissues of *Antirrhinum* flowers [9]. Transient RNAi method is also used to silence genes in protoplasts and suspension-cultured cells. This has been effectively used to silence genes in Arabidopsis [18] and *Coptis japonica* protoplasts [19]. The transient silencing has some advantages as a tool in functional genomics. Transient RNAi helps in rapid identification of function of plant genes whose stable genetic mutants are not available at present. It also complements existing methods for better functional genomics of crop plants. This method can be applied to silence any gene for a short period (7-10 days) with no limitation of plant species. Although transient gene silencing is effective for analyzing gene function at tissue level, this cannot be used for characterization of gene function in whole plants, because such study requires a stable and systemic gene silencing.

Hence, genetically stable RNAi plants are widely used for gene function analysis at whole plant level. Stable RNAi transgenics can be developed for all transformable plants. Studies showed that *Agrobacterium*-mediated transformation is preferred over biolistic method for developing stable RNAi plants, especially for large scale projects. Earlier paper by McGinnis and colleagues describe detailed protocol for developing stable RNAi plants for high-throughput screening in Arabidopsis and maize [20]. Literature information shows that RNAi is inherited for long duration in Arabidopsis [21] facilitating its application in agriculture. Another stable but not genetically inherited gene silencing method is RNAi via *Agrobacterium rhizogenes*-mediated hairy root transformation. This is a valuable tool to study genes involved in root development and root–microbe interactions. It is a very fast and efficient system to silence genes in roots and has been widely used to study genes involved in symbiosis in legumes like *Medicago truncatula* [22].

Temporal and Spatial Control of RNAi

Inducible RNAi

Inducible gene silencing is one of the effective ways of dealing with developmentally abnormal and lethal phenotypes associated with constitutive silencing of essential genes in plants. Inducible RNAi can be achieved by expressing hpRNA vector under the control of a specific inducible promoter. Presently, chemical (ethanol) and stress (heat and drought) inducible promoters are widely used [23-26]. Ethanol-inducible gene expression system (alc switch) [24, 25] has been successfully used in Arabidopsis, tobacco, potato, oilseed rape and other plant species. Further, a heat-shock gene promoter (HSP18.2) has been successfully used to trigger heat inducible RNAi in Arabidopsis [27] and a drought inducible promoter (RD-29) has been used for gene silencing exclusively during drought stress in canola [26]. Similarly, other stress specific promoters [e.g. ABRE] also have potential to be used to silence genes only during specific stress.

Tissue Specific RNAi

Tissue specific RNAi is achieved by silencing genes in specific tissue or organelle by expressing hpRNA construct under a tissue specific promoter. By silencing a particular gene in a specific tissue, function of that gene in that particular tissue can be studied without blocking its essential functions in other tissues. Tissue specific silencing in seed, root, shoot and other organelle has been demonstrated. RNAi with seed specific promoter has been used to disrupt gossypol biosynthesis exclusively in cotton seed tissue by silencing δ-cadinene synthase gene during seed development [28]. Similarly, endosperm-specific RNAi in maize to silence lysine-ketoglutarate reductase encoding gene has been shown [29]. Further, down-regulation of a farnesyl transferase gene specifically in shoot of canola plant has been achieved by using hydroxypyruvate reductase gene (*AtHPR1*) promoter driving the RNAi construct [26]. Although plant siRNAs were reported to induce systemic gene silencing [30], tissue specific silencing in stable transgenic plants can still be achieved through disruption of gene(s) responsible for systemic movement. Further, studies have also shown that tissue specific silencing can prevent transitive RNAi [29] an important cause of non specific silencing. By combining inducible and tissue specific RNAi, temporal and spatial control over gene silencing in plants can be achieved. Such controlled silencing has widened the use of RNAi technology both in functional genomics and agriculture.

Host Delivered RNAi

Host delivered RNAi (hdRNAi) is achieved by engineering crop plants with hpRNA vector to produce dsRNA against target (pest) organism so that this dsRNA will be ingested into pest organisms during feeding, leading to initiation of RNAi and silencing of essential genes in the pest. In this case crop plants are merely used as factories to produce dsRNA or siRNA, without silencing their endogenous genes because they do not have homology to silencing trigger. Further, delivery of dsRNA or siRNA from host plant to target organisms cause endogenous gene silencing in the target organism. Two major aspects facilitated the application of hdRNAi. First, dsRNA-mediated gene silencing in lower animals are similar to plants [1, 31, 32]. Hence, silencing signals produced in plants induces gene silencing in target organisms like insects. Second, dsRNA has been shown to be taken up by insects [33] or nematodes [34] through feeding tubes during artificial feeding. Based on this preliminary understanding, studies further proved that dsRNA molecules processed by plant RNAi machinery can be ingested into feeding nematodes [34], insects [32] and parasites [31]. This has potential application in pest and disease control. Apart from this, hdRNAi also has potential use for functional genomics of plant interacting target organisms where RNAi protocols are still primitive or difficult to establish. Tissue specific or stress inducible promoter driven vectors can add further use to such functional genomics studies. Nevertheless, hdRNAi has already found wide application to control agricultural pests that feed on plant tissues.

RNAi as a Tool for Plant Gene Function Analysis

Development of hpRNA producing vectors [6] have boosted the application of RNAi for effective plant gene silencing. RNAi has now become a routine lab test for gene function analysis and used in wide range of plant species for a variety of genes. RNAi as a reverse genetics approach has been successfully applied to identify function of genes involved in plant development [35], secondary metabolism [36], symbiosis [22], abiotic stresses [8] and several biotic stresses [37, 38]. To completely exploit the potential of RNAi in plant functional genomics, large scale projects using hpRNA technology are now aimed to generate RNAi knock down plants in several plant species (http://maizegdb.org/; http://chromdb.org/; https://mtrnai.msi.umn.edu/). Apart from this amiRNA based silencing method to generate silenced plants is also gaining importance (http://2010.cshl.edu/scripts/main2.pl). Out of several of these projects high-throughput generation of Medicago, Arabidopsis and maize RNAi lines have made significant progress [7, 39]. Yet another project, REGIA, that aimed to characterize transcription factors in Arabidopsis is also using RNAi as one of the tool to achieve its goals. Information generated over period using RNAi as functional genomic tool has been thoroughly reviewed in several recent literatures [3, 5, 20, 34]. Hence, this review is focused on discussing literature information emerging in commercial and applied aspects of RNAi technology.

RNAi Technology for Improved Agriculture

RNAi can be applied for crop improvement by two ways. First, silencing of certain plant genes are known to improve agronomic traits or increase plant fitness against stress [40-43]. To provide more clarity in subsequent part of this review, the method used for generation of plants with stable gene silencing is referred as host gene silencing -hair pin RNAi (HGS-hpRNAi). In this method the host plant is genetically engineered to alter its own gene expression for improving its agronomic superiority or commercial importance for improving plant fitness against stresses (biotic and abiotic). Second, by using hdRNAi method, the gene(s) are silenced only in target organism (e.g. insect or pathogen or plant parasite) that damages crop (referred here as hdRNAi). In this method, the plant is provided with a tool (i.e. ability to silence the gene of target organism) in order to defend against various biotic stresses. This hdRNAi approach can be sub divided in to two categories. One is silencing of genes in organisms that externally feed on the plant and cause damage (hdRNAi-1). This hdRNAi-1 can be used to silence genes in biotic stress causal agents like insect pests, nematodes and parasitic weeds. Second category is targeting gene silencing in viruses that enter into the host plant cell (hdRNAi-2). We discuss application of HGS-hpRNAi and hdRNAi for crop improvement in the following part of review.

Improving Disease Resistance

HGS-hpRNAi can be used to improve plant disease resistance against bacterial and fungal pathogens. In Arabidopsis, bacterial component, flagellin, induces expression of a specific miRNA, which in turn leads to down-regulation of signaling pathways that increases the plant's resistance [44]. This study clearly indicated the potential of gene down-regulation to control disease resistance in plants. Application of RNAi-mediated oncogene silencing to control crown gall disease has been shown [45]. Also tobacco plants silenced for glutathione S-transferase gene transcripts showed resistance to black shank disease [46]. In addition, studies have shown that lowering lignin content increases plant fitness to fungal pathogen. Lowered lignin content in soybean stem provided resistance to *Sclerotinia sclerotiorum* [47].

Fatty acids and their derivatives play important signaling roles by negatively regulating plant bacterial disease resistance [48, 49]. Suppressing *SACPD* gene which encodes a fatty acid desaturase enhanced resistance of Arabidopsis and soybean plants to multiple pathogens [48]. Further, RNAi-mediated knockdown of a rice homolog of this gene namely *OsSSI2* markedly enhanced resistance to leaf blight bacterium *Xanthomonas oryzae* and blast fungus *Magnaporthe grisea* [48]. Yet another study by Yara and colleagues demonstrated that suppression of two ω-3 fatty acid desaturases namely *OsFAD7* and *OsFAD8* genes enhanced disease resistance against *Magnaporthe grisea* in rice[38]. These studies clearly showed that fatty acid metabolism related genes can be key target for developing disease resistant RNAi transgenic crop plants. A yet unexplored aspect in bacteriology and mycology is the silencing of bacterial or fungal pathogen genes through hdRNAi-1 method. Potential target gene candidates for such hdRNAi-1 can come from one or all of three types of pathogen genes namely (i) essential genes involved in pathogen metabolism; (ii) genes which cause pathogen to be resistant to plant toxins; (iii) genes that encode effectors that are involved in pathogenicity.

Nevertheless, hdRNAi-2 based strategy has been used to engineer plants to impart virus resistance. In this case RNAi vector carrying viral target sequence in transgenic plants produce dsRNA that eventually silences virus multiplying in the cell. Transgenic tomato plants producing dsRNA against *Potato spindle tuber viroid* (PSTVd) sequences exhibited resistance to PSTVd infection [37]. Similarly, cassava plants were successfully engineered to resist *African cassava mosaic virus* (ACMV) [50]. Yet another method (transient) used to control virus infection in plants is dsRNA spray on the plants. Inoculation of plant leaves with dsRNA derived from viral sequences prevented virus infection [13]. Mass production of dsRNA can be done either through *in vitro*-transcription or *in vivo* expression in bacteria [13]. As viruses has high levels of variability, our ability to frequently change dsRNA that can suitably defend each new target variant of virus is necessary. Although feasibility of dsRNA spray method under field conditions are yet to be tested, this has potential for future applications not only for virus control but also other phytopathogens because of three reasons. First, dsRNA spray can prevent pathogen resistance, as researchers can predict and continuously modify dsRNA synthesis based on previous pattern of changes in pathogen resistance mechanism. It also allows options to target different pathogen genes each time. Second, this method can also be applied to control pathogens that already developed resistance against particular RNAi transgenic by suitably modifying dsRNA characteristics. Under this specific situation dsRNA spray will be a timely alternate to stable RNAi transgenic plants. Third, this method can be easily modified to control season specific- and multiple-pathogens. Alternatively, dsRNA spray is a potential tool to test efficacy of particular dsRNA (thereby RNAi construct) and as well as to select key target genes before developing stable RNAi plants. However, this method also has some inherent limitations namely lack of systemic pathogen control, absence of cost effective ways for mass producing dsRNA and non availability of methods for precise application of large quantities of dsRNA on target crop at an appropriate time.

Improving Insect and Nematode Resistance

RNAi is useful not only for functional genomic research in insects, but also for the control of insect pests. For this purpose, insect should be able to take up dsRNA through plant feeding [51]. Recent results have shown that dsRNA fed as a diet component to insect effectively down-regulated targeted genes in insect. More significantly, hdRNAi-1 method has been shown to induce silencing in plant feeding insects, opening the way for development of a new generation of insect-resistant crops [33]. Transgenic corn plants engineered to express dsRNAs for vacuolar ATPase or tubulin genes showed resistance to western corn rootworm (WCR) [32]. In this study, resistance was reflected as reduced plant damage, larval stunting and mortality. Cotton bollworm (*Helicoverpa armigera*) larvae feeding on plant material expressing dsRNA specific to cytochrome P450 gene (CYP6AE14) has showed retarded larvae growth [52]. Thus hdRNAi-1 technology is also called as species-specific insecticide and is a potential alternative to chemical pesticides [53]. Apart from stable RNAi transgenic plants, dsRNA sprayed onto plant leaves in conjunction with dsRNA enriched soil can minimize insect damage. Studies testing efficiency of dsRNA spray against insect pests are reviewed earlier [54].

However to realize complete potential of this technology, further understanding of RNAi mechanism in insects and their genomic information are necessary. Some important questions that need to be addressed are; can the dsRNA in insect be delivered back to plants via insect saliva?; can the RNAi mechanism in the pest be transmitted from feeding stage (larvae or adult) to next growth stage or generation (pupa or egg)?; what is the time taken for effective gene silencing and mortality of insect from the time of its first feeding of RNAi transgenic plant? Moreover, in vertebrates dsRNA can itself induce immune responses, thus possibly giving resistance to progeny insects. These aspects need to be researched for better understanding and suitable modifications in hd-RNAi method.

Another important aspect for crop improvement is engineering plants for nematode resistance. This can be done by two ways: (i) increasing plant resistance by silencing certain plant genes using HGS-hpRNAi; (ii) using hdRNAi-1 method. Among these two ways hdRNAi-1 is most effective. In this method dsRNA, or its siRNAs, is delivered from plant to nematode through ingestion of transgenic plant tissue [55]. Transgenic plants with hdRNAi-1 have been so far more successful with root-knot nematode (RKN) than cyst nematode (CN). Transgenic tobacco plants producing dsRNA targeting a RKN (*Meloidogyne javanica*) putative zinc finger transcription factor gene (*MjTis11*) has been shown to be effective in controlling this nematode [56]. In order to increase efficacy of CN nematode control, hdRNAi-1 was used to target up to four crucial CN (*Heterodera schachtii*) genes by expressing different hpRNA constructs in Arabidopsis. This led to a higher reduction in number of mature CN nematode females feeding on this plant [57]. Further understanding of RNAi machinery in plant parasitic nematodes will help to achieve higher efficiency of gene silencing in CN [34].

Improving Resistance against Parasitic Weeds

Management of plant parasitic weeds through genetic engineering has been difficult as no genetic loci in crop plants conferring resistance has been found yet. Hence, potential for application of RNAi in this area has been researched in recent years. In order for hdRNAi-1 to be effective, the delivery of signals from host plant to parasitic weeds is necessary. Hence, RNAi signal spread was examined. *Triphysaria versicolor* parasite expressing the GUS reporter gene was allowed to parasitize transgenic lettuce roots expressing hpRNA containing a fragment of the GUS gene. When stained for GUS activity, *Triphysaria* roots attached to non-transgenic lettuce showed full GUS activity, but those parasitizing RNAi transgenic lettuce lacked GUS activity [31]. This indicates that silencing signal can move from host to parasite and silence targeted gene. Transgenic RNAi tomato plants bearing mannose 6-phosphate reductase gene (*M6PR*) hpRNA construct from *Orobanche aegyptiaca* has been effective against this parasite by causing death to its tubercles attached to tomato plants [58]. Other studies also showed that parasites can take up the specific RNAi from host via haustorium [59]. These studies clearly demonstrated future potential for hdRNAi-1 technology in plant parasite control.

Improving Drought Tolerance

Gene silencing has been used to develop drought tolerant plants. Conditional and specific down-regulation of farnesyltransferase in canola using the *AtHPR1* promoter driving an RNAi construct showed yield protection under drought stress [26]. Down-regulation of canola farnesyltransferase for drought tolerance was a conditional and reversible process, which depended on the amount of available water in the soil. Hence, transgenic plants were more resistant to water deficit-induced seed abortion during flowering without affecting yield [26]. Earlier, Wang and colleagues have also demonstrated the use of farnesyltransferase as an effective target for engineering drought tolerance in few other plants under laboratory and field conditions [26]. In addition, information about potential candidate genes as negative regulator of drought tolerance is emerging. HGS-hpRNAi transgenic rice plants that cause silencing of the receptor for activated C-kinase 1 (RACK1) showed tolerance to drought stress [60]. Previously, our group found that *Nicotiana benthamiana* plants down-regulated with a proteinase (*APRO2*) and a transcription factor (*JMJC*) genes maintained higher membrane integrity under water deficit stress [61]. While RNAi technology is being improved, list of key candidate genes for HGS-hpRNAi is concomitantly growing. RNAi approach is expected to produce many more drought tolerant crop plants in near future.

Improving Nutritional Value

We discuss here the application of HGS-hpRNAi in improving nutritional qualities of wheat grain, oil seeds and cotton seeds. RNAi was successfully used to reduce γ-gliadins in bread wheat [62]. Similarly, transgenic RNAi wheat with reduced levels of two starch-branching enzyme encoding genes (*SBEIIa* and *SBEIIb*) has showed increased amylose content [63] that has health benefits. Further, RNAi of oleate desaturase (*FAD2*) gene resulted in 70% increase in oleic acid content in seeds of transgenic peanut plants [64]. Production of beneficial unsaturated fats with no trans-fat in oil seed crops using RNAi technology is yet another aspect of nutritional improvement. Apart from nutritional improvement in wheat and oilseed crops, RNAi has also been used to identify new protein sources by making cotton seed edible for animal consumption. Reducing gossypol content in cotton seeds has been shown to make this seeds a suitable protein source for animals and humans [28]. More strategies on applications of RNAi for the improvement of plant nutritional value [65] and list of potential candidate genes [66] have been reviewed elsewhere.

Metabolic Engineering

Instead of overexpressing desirable genes, potentially deleterious genes can be silenced (by HGS-hpRNAi) to manipulate metabolic pathways for better utilization of plants for commercial production of beneficial metabolites [67]. RNAi has been applied in genetic engineering of starches, oils and storage proteins. We describe here two such applications. First, a study by Kim and colleagues demonstrated successful metabolic engineering of lignans. Lignans have numerous applications in mammals, including antitumor and antioxidant activities. A transgenic cell line with RNAi construct of *pinoresinol-lariciresinol reductase* gene was developed in *Forsythia koreana* cell suspension for engineering of lignan production [68]. Second, peanut allergy contributing gene *Arah2* has been knocked down using RNAi approach and the results showed potential for removing allergenicity in peanut [69]. HGS-hpRNAi also has potential for engineering plants for pharmaceutical application.

Improving Forage Digestibility, Biomass and Grain Yield

Plant lignin composition can be altered using RNAi [70, 71] and the resultant product has been used to improve digestibility characteristics in forage crops for better livestock production [72]. Down-regulation of specific cytochrome P450 enzymes involved in lignin synthesis genes improved forage quality in alfalfa, by increasing digestibility of fodder and thus reducing bloating in cows foraging this crop [42]. Reduced lignin containing biomass also has application in paper pulp industry.

Engineering for agronomic traits requires a particular degree of gene suppression of target genes. Varying degrees of gene suppression is difficult to achieve using other available genetic modifications tools, except for gene silencing. Thus RNAi has been exploited to produce transgenics with improved agronomic traits. Initially RNAi was applied to manipulate plant architecture. Loss of a gene called as *OsDWARF4* in rice resulted in shorter plants with erect leaves. Erect leaf architecture increased photosynthesis in the lower leaves and thus such plants has potential for improved yields under dense planting conditions [41]. Similarly, RNAi rice knockout transgenic lines for *qSW5* (QTL for seed width on chromosome 5) showed increased seed weight [43]. Further, semi-dwarf plants were generated from a taller rice variety, QX1, by RNAi-mediated suppression of GA 20-oxidase (*OsGA20ox2*) gene expression. This RNAi transgenic plants had increased panicle length, higher number of seeds per panicle and higher test (1000 grain) weight [35]. RNAi trait can be introgressed with high yielding target genotypes in back-cross breeding program. Since RNAi approach can be easily integrated into plant breeding programs, it has enormous potential to facilitate new genetic studies in hybrids. Adding to this, tissue specificity and inducible silencing coupled to HGS-hpRNAi can be used to suppress gene expression 'at will' thus preventing possible adverse affects that may occur due to constitutive silencing.

Improved Bio-fuel Production

Recalcitrance of plant material to a process called saccharification is a major limitation for conversion of lignocellulosic biomass to ethanol. Genetic reduction of lignin content effectively overcame cell wall recalcitrance to bioconversion [40]. Stems of transgenic alfalfa lines independently down-regulated in each of six lignin biosynthetic enzymes, yielded nearly twice as much sugar from cell walls as wild-type plants [42]. Down-regulation of lignin genes like cinnamate 4-hydroxylase (C3H), shikimate hydroxycinnamoyl transferase (HCT) and 4-coumarate-coA ligase (4CL) in plants reduced total lignin content, increased dry matter degradability, and improved accessibility of cellulases for cellulose degradation [70]. This lignin modification also facilitated bypassing the need for acid pretreatment [40]. Thus, it is possible to reduce or eliminate the costly pretreatment step by using biomass from low-lignin transgenic plants, thereby greatly reducing the cost of biofuel production [70]. Since the biosynthetic pathways to lignin monomers are conserved across species, knowledge gained from above studies should be applicable to modify lignin composition in major biofuel important crops like switchgrass and Miscanthus. While manipulating lignin content, care must be taken not to severely impact crop productivity. Studies have shown that altering lignin composition does not compromise plant fitness and on the contrary in some cases resistance to certain phyto-pathogens was observed [47]. Recently HGS-hpRNAi has been effectively used to reduce lignin content in switch grass [70] and commercial products from these studies are expected in future.

Off-target Gene Silencing during RNAi and Remedy

Although initially gene silencing by RNAi was believed to be specific, several recent reports have showed that off-target gene silencing can widely occur during RNAi [73]. This effect is likely to have potential negative impact in RNAi transgenics. For example, RNAi aimed for intended insect control can also have un-intended impact on beneficial organisms. This warrants researchers to perform long-term research to completely understand the factors contributing for off-target silencing and prevent potential problems to the environment. In addition, the effect of off-target during hdRNAi can potentially silence plant genes itself (becoming HGS-hpRNAi). This is possible because the dsRNA is being constantly produced with high threshold levels in plant cell (not utilized in the absence of pest) and hence increase chances of access to off-target endogenous plant mRNA to RISC complex.

Only very limited information about genome and EST sequences are available for several of insect species, plant parasites, and many crop plants. In this scenario assessing whether a particular gene is off-targeted or not is difficult and cannot rely on computational tools that can design specific RNAi trigger fragments to prevent off-targeting in these organisms. If a hypothetical scenario is considered, at some point of time in future if a plant, insect, weed and nematodes (all are RNAi transgenic) share a same environment, the off-target effect will be magnified leading to alteration in balance of plant-pest interactions. This could potentially be a threat to ecology and evolution in future. Sequence identity between RNAi trigger and target mRNA, gene region selected for trigger, trigger size, and transitive RNAi influences off-target silencing. By using specific silencing trigger sequence in RNAi vector, by tissue specific and inducible silencing, some of off-target effects can be prevented [29]. Our group has developed a publicly available web-based computational tool called siRNA Scan (http://bioinfo2.noble.org/RNAiScan.htm) to identify potential off-targets during PTGS in plants [73]. This software can be effectively used to design appropriate RNAi construct to prevent off-target silencing.

Other Precautions

Before the exciting possibilities of RNAi applications can be realized, some significant challenges remain to be addressed. The questions to be answered include: will the successes seen in the lab translate to field? What about effect of dsRNA on non-target organisms that co-exist with pests in the same environment? What are the long term effect on environment due to cross pollination between RNAi transgenic plant and non-transgenic plant of same and different species? Further, a large gap exists in our understanding of RNAi mechanisms across different organisms. Hence, there is a need for long term trials to estimate influence of RNAi transgenic crop on environment considering all stages of ecological cycle or food chain. List of genetically engineered crops in USA using RNAi is reviewed elsewhere [74].

Future Perspectives of RNAi Technology

RNAi has been an incredible functional genomics tool in the past decade. This technology will remain as potential genomic tool for the next decades to come as large scale knock down lines are being developed in major crop plants. Application of RNA silencing in single cell organisms like *Chlamydomonas reinhardtii*, an algae close to plants, will enable high-throughput plant gene function identification (e.g. for abiotic stress tolerance). A stress specific cDNA library cloned in a vector that can induce gene silencing (e.g. sense strand overexpression to induce RNAi) could be a potential source for developing knock down *Chlamydomonas* lines for each gene. By implementing high-throughput stress screening in such gene silenced *Chlamydomonas*, the gene function identification can be hastened. Potential of RNAi to engineer crop plants tolerant to other abiotic stresses apart from drought has not yet exploited. Despite few limitations, application of RNAi in crop improvement especially pest and disease control is expected to be significant in the future. Since stable RNAi transgenic plants can be easily blended with plant breeding program, molecular breeders can potentially use this technology to incorporate insect or disease resistance into a high yielding commercial hybrid or variety. Alternatively, an agronomically superior cultivar can be engineered for additional plant fitness (e.g. stress tolerance) by using RNAi technology.

Acknowledgments

RNAi- and VIGS-related projects in KSM laboratory are supported by the Samuel Roberts Noble Foundation, National Science Foundation (grant no. 0445799), Oklahoma Center for the Advancement of Science and Technology (grant no. PSB09-020) and the U.S.-Israel Binational Agricultural Research Development Fund (BARD grant no. IS-3922-06).

References

[1]   Agrawal, N., Dasaradhi, P. V. N., Mohmmed, A., Malhotra, P., Bhatnagar, R. K. & Mukherjee, S. K. (2003). RNA interference: biology, mechanism, and applications. *Microbiology and Molecular Biology Reviews* 67, 657-685.

[2]   Qi, Y. & Hannon, G. J. (2005). Uncovering RNAi mechanisms in plants: Biochemistry enters the foray. *FEBS Letters* 579, 5899-5903.

[3]   Watson, J. M., Fusaro, A. F., Wang, M. & Waterhouse, P. M. (2005). RNA silencing platforms in plants. *FEBS Letters* 579, 5982-5987.

[4]   Preuss, S. & Pikaard, C. S. (2003) Targeted gene silencing in plants using RNA interference, in *RNA Interference (RNAi)~Nuts & Bolts of siRNA Technology* (Engelke, D., Ed.), pp 23-36, DNA Press, LLC.

[5]   Baulcombe, D. (2004). RNA silencing in plants. *Nature* 431, 356-363.

[6]   Wesley, V., Helliwell, C., Smith, N., Wang, M., Rouse, D., Liu, Q., Gooding, P., Singh, S., Abbott, D., Stoutjesdijk, P., Robinson, S., Gleave, A., Green, A. & Waterhouse, P.

(2001). Construct design for efficient, effective and high-throughput gene silencing in plants. *The Plant Journal* 27, 581-590.

[7]   Isshiki, M. & Kodama, H. (2010) Plant RNAi and crop improvement, in *Molecular Techniques in Crop Improvement*, pp 653-673.

[8]   Senthil-Kumar, M., Hema, R., Suryachandra, T. R., Ramegowda, H. V., Gopalakrishna, R., Rama, N., Udayakumar, M. & Mysore, K. S. (2010). Functional characterization of three water deficit stress-induced genes in tobacco and Arabidopsis: An approach based on gene down regulation. *Plant Physiology and Biochemistry* 48, 35-44.

[9]   Shang, Y., Schwinn, K., Bennett, M., Hunter, D., Waugh, T., Pathirana, N., Brummell, D., Jameson, P. & Davies, K. (2007). Methods for transient assay of gene function in floral tissues. *Plant Methods* 3, 1.

[10]   Silhavy, D. (2005) Agro-infiltration: A versatile tool for RNAi studies in plants, in *Gene silencing by RNA interference: Technology and application* (Sohail, M., Ed.), pp 357-363, CRC press, Boca Raton.

[11]   Tang, W., Weidner, D. A., Hu, B. Y., Newton, R. J. & Hu, X.-H. (2006). Efficient delivery of small interfering RNA to plant cells by a nanosecond pulsed laser-induced stress wave for posttranscriptional gene silencing. *Plant Science* 171, 375-381.

[12]   Warthmann, N., Chen, H., Ossowski, S., Weigel, D. & Herve, P. (2008). Highly specific gene silencing by artificial miRNAs in rice. *PLoS ONE* 3, e1829.

[13]   Tenllado, F., Llave, C. & Díaz-Ruíz, J. R. (2004). RNA interference as a new biotechnological tool for the control of virus diseases in plants. *Virus Research* 102, 85-96.

[14]   Miki, D. & Shimamoto, K. (2004). Simple RNAi vectors for stable and transient suppression of gene function in rice. *Plant Cell Physiology* 45, 490-495.

[15]   Bhaskar, P. B., Venkateshwaran, M., Wu, L., Ane, J.-M. & Jiang, J. (2009). Agrobacterium-mediated transient gene expression and silencing: A rapid tool for functional gene assay in potato. *PLoS ONE* 4, e5812.

[16]   Thomas, H., Gregor, K. & Wilfried, S. (2006). RNAi-induced silencing of gene expression in strawberry fruit by agroinfiltration: a rapid assay for gene function analysis. *The Plant Journal* 48, 818-826.

[17]   Mohanpuria, P., Rana, N. & Yadav, S. (2008). Transient RNAi based gene silencing of glutathione synthetase reduces glutathione content in *Camellia sinensis* (L.) O. Kuntze somatic embryos. *Biologia Plantarum* 52, 381-384.

[18]   An, C.-I., Sawada, A., Fukusaki, E.-i. & Kobayashi, A. (2003). A transient RNA interference assay system using Arabidopsis protoplasts. *Bioscience, Biotechnology, and Biochemistry* 67, 2674-2677.

[19]   Dubouzet, J. G., Morishige, T., Fujii, N., An, C.-I., Fukusaki, E.-i., Ifuku, K. & Sato, F. (2005). Transient RNA silencing of scoulerine 9-O-methyltransferase expression by double stranded RNA in *Coptis japonica* protoplasts. *Bioscience, Biotechnology, and Biochemistry* 69, 63-70.

[20]   McGinnis, K. M. (2010). RNAi for functional genomics in plants. *Brief Funct Genomic Proteomic*, elp052.

[21]   Stoutjesdijk, P. A., Singh, S. P., Liu, Q., Hurlstone, C. J., Waterhouse, P. A. & Green, A. G. (2002). hpRNA-mediated targeting of the Arabidopsis FAD2 gene gives highly efficient and stable silencing. *Plant Physiology* 129, 1723-1731.

[22] Limpens, E., Ramos, J., Franken, C., Raz, V., Compaan, B., Franssen, H., Bisseling, T. & Geurts, R. (2004). RNA interference in *Agrobacterium rhizogenes*-transformed roots of Arabidopsis and *Medicago truncatula*. *Journal of Experimental Botany* 55, 983-992.

[23] Anna, W., Helen, T., Ian, M., Peter, W. & Chris, H. (2005). A high-throughput inducible RNAi vector for plants. *Plant Biotechnology Journal* 3, 583-590.

[24] Li, R., Jia, X. & Mao, X. (2005). Ethanol-inducible gene expression system and its applications in plant functional genomics. *Plant Science* 169, 463-469.

[25] Shuai, C., Daniel, H., Uwe, S. & Frederik, B. (2003). Temporal and spatial control of gene silencing in transgenic plants by inducible expression of double-stranded RNA. *The Plant Journal* 36, 731-740.

[26] Wang, Y., Beaith, M., Chalifoux, M., Ying, J., Uchacz, T., Sarvas, C., Griffiths, R., Kuzma, M., Wan, J. & Huang, Y. (2009). Shoot-specific down-regulation of protein farnesyltransferase (a-subunit) for yield protection against drought in canola. *Molecular Plant* 2, 191-200.

[27] Masclaux, F., Charpenteau, M., Takahashi, T., Pont-Lezica, R. & Galaud, J.-P. (2004). Gene silencing using a heat-inducible RNAi system in Arabidopsis. *Biochemical and Biophysical Research Communications* 321, 364-369.

[28] Sunilkumar, G., Campbell, L. M., Puckhaber, L., Stipanovic, R. D. & Rathore, K. S. (2006). Engineering cottonseed for use in human nutrition by tissue-specific reduction of toxic gossypol. *Proceedings of the National Academy of Sciences* 103, 18054-18059.

[29] Nancy, M. H., Jonnelle, L. M., Christopher, P. B., Shihshieh, H., Michael, H. L. & Thomas, M. M. (2007). High-lysine corn generated by endosperm-specific suppression of lysine catabolism using RNAi. *Plant Biotechnology Journal* 5, 605-614.

[30] Klahre, U., Crete, P., Leuenberger, S. A., Iglesias, V. A. & Meins, F. (2002). High molecular weight RNAs and small interfering RNAs induce systemic posttranscriptional gene silencing in plants. *Proceedings of the National Academy of Sciences of the United States of America* 99, 11981-11986.

[31] Alexey, A. T., Natalia, B. T., Tadeusz, W., Richard, M. & John, I. Y. (2008). Trans-specific gene silencing between host and parasitic plants. *The Plant Journal* 56, 389-397.

[32] Baum, J. A., Bogaert, T., Clinton, W., Heck, G. R., Feldmann, P., Ilagan, O., Johnson, S., Plaetinck, G., Munyikwa, T., Pleau, M., Vaughn, T. & Roberts, J. (2007). Control of coleopteran insect pests through RNA interference. *Nature Biotechnology* 25, 1322-1326.

[33] Price, D. R. G. & Gatehouse, J. A. (2008). RNAi-mediated crop protection against insects. *Trends in Biotechnology* 26, 393-400.

[34] Rosso, M. N., Jones, J. T. & Abad, P. (2009). RNAi and functional genomics in plant parasitic nematodes. *Annual Review of Phytopathology* 47, 207-232.

[35] Qiao, F., Yang, Q., Wang, C.-L., Fan, Y.-L., Wu, X.-F. & Zhao, K.-J. (2007). Modification of plant height via RNAi suppression of OsGA20ox2 gene in rice. *Euphytica* 158, 35-45.

[36] Wagner, G. J. & Kroumova, A. B. (2008) The use of RNAi to elucidate and manipulate secondary metabolite synthesis in plants, in *Current Perspectives in microRNAs (miRNA)*, pp 431-459.

[37] Nora, S., MichÈLe, Z., Asuka, I., Biao, D., Ming-Bo, W., Gabi, K. & Michael, W. (2009). RNAi-mediated resistance to *Potato spindle tuber viroid* in transgenic tomato expressing a viroid hairpin RNA construct. *Molecular Plant Pathology* 10, 459-469.

[38] Yara, A., Yaeno, T., Hasegawa, M., Seto, H., Montillet, J.-L., Kusumi, K., Seo, S. & Iba, K. (2007). Disease resistance against *Magnaporthe grisea* is enhanced in transgenic rice with suppression of o-3 fatty acid desaturases. *Plant Cell Physiology* 48, 1263-1274.

[39] Ivashuta, S., Liu, J., Liu, J., Lohar, D. P., Haridas, S., Bucciarelli, B., VandenBosch, K. A., Vance, C. P., Harrison, M. J. & Gantt, J. S. (2005). RNA interference identifies a calcium-dependent protein kinase involved in *Medicago truncatula* root development. *Plant Cell* 17, 2911-2921.

[40] Chen, F. & Dixon, R. A. (2007). Lignin modification improves fermentable sugar yields for biofuel production. *Nature Biotechnology* 25, 759-761.

[41] Feldmann, K. A. (2006). Steroid regulation improves crop yield. *Nature Biotechnology* 24, 46-47.

[42] Reddy, M. S. S., Chen, F., Shadle, G., Jackson, L., Aljoe, H. & Dixon, R. A. (2005). Targeted down-regulation of cytochrome P450 enzymes for forage quality improvement in alfalfa (Medicago sativa L.). *Proceedings of the National Academy of Sciences of the United States of America* 102, 16573-16578.

[43] Shomura, A., Izawa, T., Ebana, K., Ebitani, T., Kanegae, H., Konishi, S. & Yano, M. (2008). Deletion in a gene associated with grain size increased yields during rice domestication. *Nature Genetics* 40, 1023-1028.

[44] Fritz, J. H., Girardin, S. E. & Philpott, D. J. (2006). Innate immune defense through RNA interference. *Sci. STKE* 2006, pe27-.

[45] Escobar, M. A., Civerolo, E. L., Summerfelt, K. R. & Dandekar, A. M. (2001). RNAi-mediated oncogene silencing confers resistance to crown gall tumorigenesis. *Proceedings of the National Academy of Sciences of the United States of America* 98, 13437-13442.

[46] Hernández, I., Chacón, O., Rodriguez, R., Portieles, R., López, Y., Pujol, M. & Borrás-Hidalgo, O. (2009). Black shank resistant tobacco by silencing of glutathione S-transferase. *Biochemical and Biophysical Research Communications* 387, 300-304.

[47] Peltier, A. J., Hatfield, R. D. & Grau, C. R. (2009). Soybean stem lignin concentration relates to resistance to *Sclerotinia sclerotiorum*. *Plant Disease* 93, 149-154.

[48] Jiang, C.-J., Shimono, M., Maeda, S., Inoue, H., Mori, M., Hasegawa, M., Sugano, S. & Takatsuji, H. (2009). Suppression of the rice fatty-acid desaturase gene OsSSI2 enhances resistance to blast and leaf blight diseases in rice. *Molecular Plant-Microbe Interactions* 22, 820-829.

[49] Li, K., Yuh-Shuh, W., Srinivasa Rao, U., Keri, W., Yuhong, T., Vatsala, V., Barney, J. V., Kent, D. C., Elison, B. B. & Kirankumar, S. M. (2008). Overexpression of a fatty acid amide hydrolase compromises innate immunity in Arabidopsis. *The Plant Journal* 56, 336-349.

[50] Vanderschuren, H., Alder, A., Zhang, P. & Gruissem, W. (2009). Dose-dependent RNAi-mediated geminivirus resistance in the tropical root crop cassava. *Plant Molecular Biology* 70, 265-272.

[51] Huvenne, H. & Smagghe, G. (2010). Mechanisms of dsRNA uptake in insects and potential of RNAi for pest control: A review. *Journal of Insect Physiology* 56, 227-235.

[52] Mao, Y.-B., Cai, W.-J., Wang, J.-W., Hong, G.-J., Tao, X.-Y., Wang, L.-J., Huang, Y.-P. & Chen, X.-Y. (2007). Silencing a cotton bollworm P450 monooxygenase gene by plant-mediated RNAi impairs larval tolerance of gossypol. *Nature Biotechnology* 25, 1307-1313.

[53] Whyard, S., Singh, A. D. & Wong, S. (2009). Ingested double-stranded RNAs can act as species-specific insecticides. *Insect Biochemistry and Molecular Biology* 39, 824-832.

[54] Katherine, A. (2009). Using RNA interference to increase crop yield and decrease pest damage. *MMG 445 Basic Biotechnology* 5, 7-12.

[55] Gheysen, G. & Vanholme, B. (2007). RNAi from plants to nematodes. *Trends in Biotechnology* 25, 89-92.

[56] Fairbairn, D., Cavallaro, A., Bernard, M., Mahalinga-Iyer, J., Graham, M. & Botella, J. (2007). Host-delivered RNAi: an effective strategy to silence genes in plant parasitic nematodes. *Planta* 226, 1525-1533.

[57] Sindhu, A. S., Maier, T. R., Mitchum, M. G., Hussey, R. S., Davis, E. L. & Baum, T. J. (2009). Effective and specific in planta RNAi in cyst nematodes: expression interference of four parasitism genes reduces parasitic success. *J. Exp. Bot.* 60, 315-324.

[58] Radi, A., Hila, C., Daniel, M. J., Diana, L., Benjamin, S., Aaron, Z., Anna, N., Oded, Y. & Amit, G.-O. (2009). Gene silencing of mannose 6-phosphate reductase in the parasitic weed *Orobanche aegyptiaca* through the production of homologous dsRNA sequences in the host plant. *Plant Biotechnology Journal* 7, 487-498.

[59] John, I. Y., Pradeepa, G., Biao, W., Natalya, T. & Alexey, A. T. (2009). Engineering host resistance against parasitic weeds with RNA interference. *Pest Management Science* 65, 460-466.

[60] Li, D.-h., Liu, H., Yang, Y.-l., Zhen, P.-p. & Liang, J.-s. (2009). Down-regulated expression of RACK1 gene by RNA interference enhances drought tolerance in rice. *Rice Science* 16, 14-20.

[61] Senthil-Kumar, M., Govind, G., Kang, L., Mysore, K. & Udayakumar, M. (2007). Functional characterization of *Nicotiana benthamiana* homologs of peanut water deficit-induced genes by virus-induced gene silencing. *Planta* 225, 523-539.

[62] Gil-Humanes, J., Pistón, F., Hernando, A., Alvarez, J. B., Shewry, P. R. & Barro, F. (2008). Silencing of g-gliadins by RNA interference (RNAi) in bread wheat. *Journal of Cereal Science* 48, 565-568.

[63] Regina, A., Bird, A., Topping, D., Bowden, S., Freeman, J., Barsby, T., Kosar-Hashemi, B., Li, Z., Rahman, S. & Morell, M. (2006). High-amylose wheat generated by RNA interference improves indices of large-bowel health in rats. *Proceedings of the National Academy of Sciences of the United States of America* 103, 3546-3551.

[64] Yin, D., Deng, S., Zhan, K. & Cui, D. (2007). High-oleic peanut oils produced by HpRNA-mediated gene silencing of oleate desaturase. *Plant Molecular Biology Reporter* 25, 154-163.

[65] Tang, G., Galili, G. & Zhuang, X. (2007). RNAi and microRNA: breakthrough technologies for the improvement of plant nutritional value and metabolic engineering. *Metabolomics* 3, 357-369.

[66] Mansoor, S., Amin, I., Hussain, M., Zafar, Y. & Briddon, R. W. (2006). Engineering novel traits in plants through RNA interference. *Trends in Plant Science* 11, 559-565.

[67]   Hebert, C. G., Valdes, J. J. & Bentley, W. E. (2008). Beyond silencing -- engineering applications of RNA interference and antisense technology for altering cellular phenotype. *Current Opinion in Biotechnology* 19, 500-505.

[68]   Kim, H. J., Ono, E., Morimoto, K., Yamagaki, T., Okazawa, A., Kobayashi, A. & Satake, H. (2009). Metabolic engineering of lignan biosynthesis in Forsythia cell culture. *Plant Cell Physiology* 50, 2200-2209.

[69]   Hortense, W. D., Koffi, N. K., Fur, C. C., Marceline, E. & Olga, M. V. (2008). Alleviating peanut allergy using genetic engineering: the silencing of the immunodominant allergen Ara h 2 leads to its significant reduction and a decrease in peanut allergenicity. *Plant Biotechnology Journal* 6, 135-145.

[70]   Hisano, H., Nandakumar, R. & Wang, Z.-Y. (2009). Genetic modification of lignin biosynthesis for improved biofuel production. *In Vitro Cellular & Developmental Biology - Plant* 45, 306-313.

[71]   Sticklen, M. B. E. L., MI, US). (2008) Altering regulation of maize lignin biosynthesis enzymes via RNAi technology, Michigan State University, United States.

[72]   Marino, M. (2008). Profile of Richard Dixon. *Proceedings of the National Academy of Sciences* 105, 2263-2265.

[73]   Xu, P., Zhang, Y., Kang, L., Roossinck, M. J. & Mysore, K. S. (2006). Computational estimation and experimental verification of off-target silencing during posttranscriptional gene silencing in plants. *Plant Physiology* 142, 429-440.

[74]   Auer, C. & Frederick, R. (2009). Crop improvement using small RNAs: applications and predictive ecological risk assessments. *Trends in Biotechnology* 27, 644-651.

In: Gene Silencing: Theory, Techniques and Applications ISBN: 978-1-61728-276-8
Editor: Anthony J. Catalano © 2010 Nova Science Publishers, Inc.

Chapter VIII

Silencing of Galectin-3 by Promoter Methylation during Prostate Cancer Progression: A Novel Transcriptional Regulation of Galectin-3 Expression

Hafiz Ahmed[*]

Department of Biochemistry and Molecular Biology, Program in Oncology, Greenebaum Cancer Center, University of Maryland School of Medicine, Baltimore, MD, USA

Abstract

Protein-carbohydrate interactions play significant role in modulating cell-cell and cell-extracellular matrix interactions, which, in turn, mediate various biological processes such as growth regulation, immune function, cancer metastasis, and apoptosis. Galectin-3, a member of the β-galactoside-binding protein family, is found multifunctional and is involved in normal growth development as well as cancer progression and metastasis, but the detailed mechanisms of its functions or its transcriptional regulations are not well understood. Besides, several regulatory elements such as GC box, CRE motif, AP-1 site, and NF-κB sites, the promoter of galectin-3 gene (*LGALS3*) contains several CpG islands that can be methylated during tumorigenesis leading to the gene silencing. This review discusses the galectin-3 epigenetics, which represents a novel regulatory mechanism of its transcription.

[*] Address for correspondence: Hafiz Ahmed, Ph.D. Columbus Center. 701 East Pratt Street, Suite 236, Baltimore, MD 21202, USA. E-mail: ahmed@umbi.umd.edu; hahmed@som.umaryland.edu

Introduction

Interactions between cells and between cells and the extracellular matrix (ECM) are pivotal for proper cellular function. In recent years, protein-carbohydrate interactions have been considered as very important for modulation cell-cell and cell-ECM interactions (Figure 1A), which, in turn, mediate various biological processes such as cell activation, growth regulation, cancer metastasis, and apoptosis. Thus, the identification of carbohydrate-binding proteins (lectins) and their partners (carbohydrate ligands), and the detailed understanding of the molecular mechanisms and downstream effects of these protein-carbohydrate interactions are subjects of current intense research. Galectins (gal), a family of β-galactoside-binding proteins, are involved in growth development as well as cancer progression and metastasis [1-5]. However, the detailed mechanisms of these functions remain largely unknown. Of the fifteen members of the galectin family identified so far, gal1, 2, 5, 7, 10, 11, 13, 14, and 15 are examples of the "proto" type galectins (one carbohydrate-recognition domain [CRD] per subunit), while Gal4, 6, 8, 9, and 12 are "tandem-repeat" type galectins, which contain two CRDs [6] (Figure 1B). Gal3 is the only representative of the "chimera" galectin type, containing three structurally distinct domains: a 12-amino acid short N-terminal domain (ND), proline and glycine rich long ND, and a C-terminal CRD [7]. Gal3 is a monomer, but can form multimer at certain circumstances such as at high concentration [8].

Gal3, previously known as Mac-2, L-29, L-31, L-34, IgE binding-protein, CBP35, and CBP30, is an interesting multifunctional molecule and probably the most studied member of the galectin family [2, 9-12]. The short ND is highly conserved in all mammalian gal3 [10] and may have at least two roles; its deletion blocks secretion of gal3 [7], while mutation of the conserved Ser6 affects gal3 anti-apoptotic signaling activity [13]. The long ND is responsible for multimerization of gal3 and shows positive cooperativity in carbohydrate binding [8]. For example, matrix metalloproteinases, MMP-2 and MMP-9 cleave gal3 at the position Ala62–Tyr63 resulting in 22 kDa fragment that fails to self associate [14]. The C-terminal domain of gal3 is composed of about 130 amino acids that form a globular structure like other galectins [10]. It accommodates whole carbohydrate-binding site, which is responsible for lectin activity [15, 16]. Within the CRD particularly interesting amino acid sequence is NWGR; this motif is highly conserved within the BH1 domain of the Bcl-2 family proteins, and it was shown to be responsible for the anti-apoptotic activity of both Bcl-2 and gal3 [17]. The NWGR motif is also involved in self-association of gal3 molecules through the CRDs in the absence of saccharide ligands [18]. The replacement of tryptophan with leucine (W181L) within the NWGR motif abolishes homodimerization through the CRDs of gal3. However, this mutant can still bind wild-type gal3 through the interactions of N-terminal domains.

Gal3 plays an important role in normal development and tumorigenesis through regulating cell proliferation, apoptosis, cell adhesion, invasion, angiogenesis and metastasis by binding to the cell surface β-galactose-containing glycoconjugates or glycolipids. Gal3 may exert its multiple biological roles intracellularly within the nucleus or the cytoplasm, or after its secretion, at the cell surface and/or the extracellular space, mediating interactions between cells and the extracellular matrix [3, 4, 6]. Its expression in normal and tumor tissues and regulation of its expression including the newly discovered epigenetic regulation in some tumor tissues have been discussed in this review.

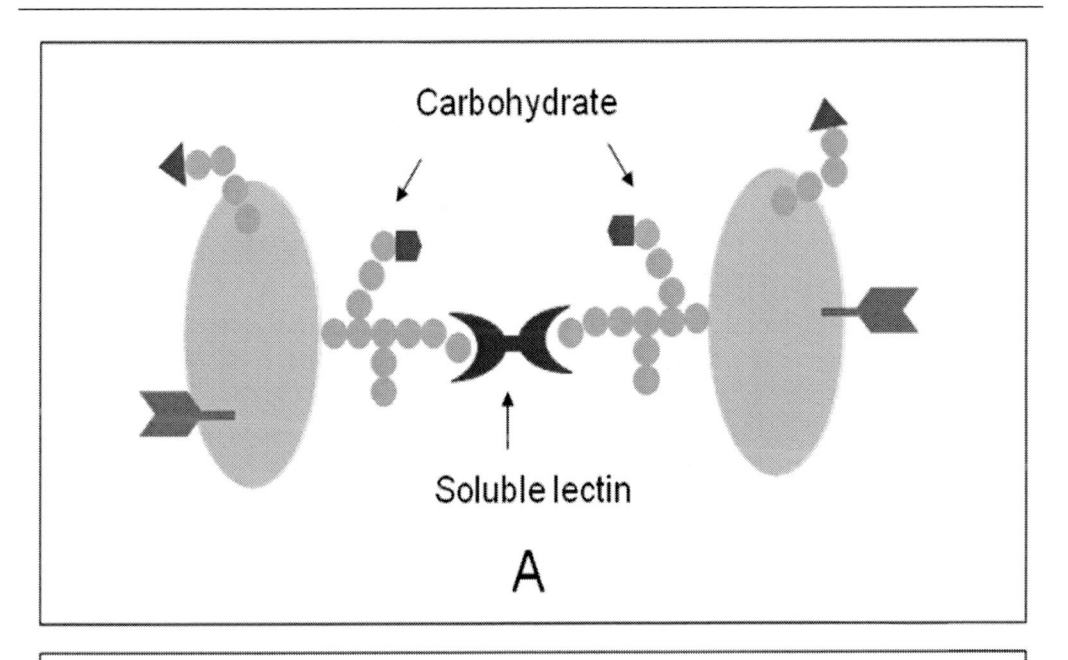

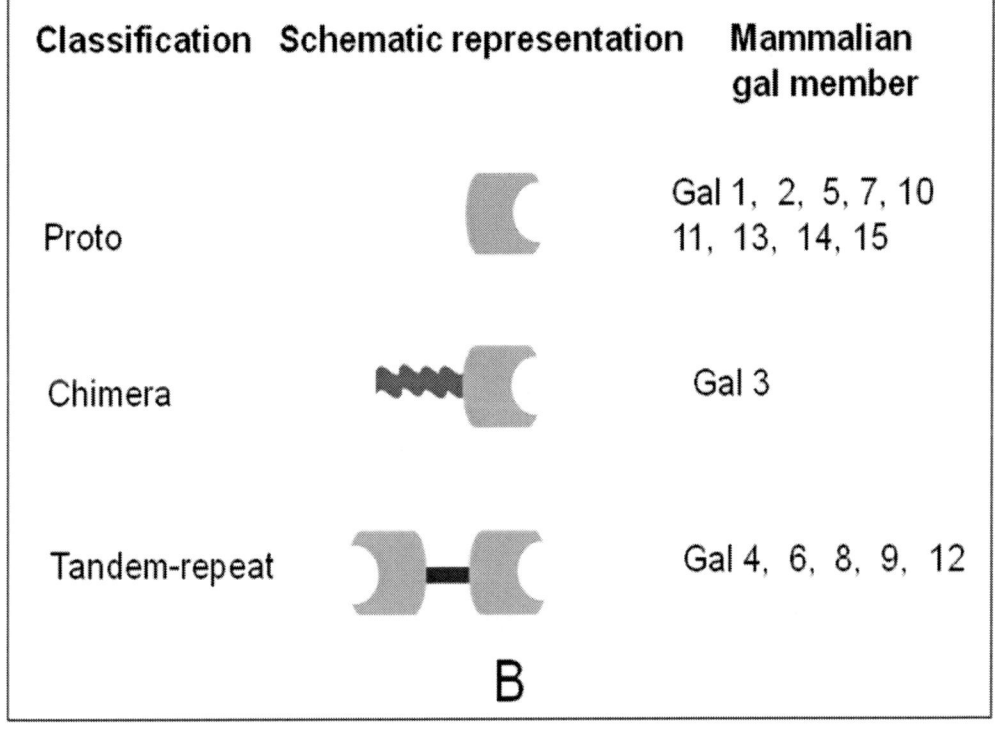

Figure 1. A. Schematic representation of lectin-carbohydrate-mediated cell-cell interaction. Lectins can be soluble or membrane-bound. B. Classification of galectins: Schematic representation of proto-, chimera, and tandem-repeat type galectins. They are numbered according to the order of their discovery.

Gal3 Expression in Normal Tissues-Role in Growth Development

Gal3 is developmentally regulated and expressed in many tissues of adults [19, 20]. During mouse embryogenesis, gal3 first appears at fourth day of gestation in the trophectoderm of blastocyst, followed by its expression in the notochord cells between 8.5 and 11.5 days of gestation [19]. In later stages of mouse development, gal3 is expressed in the cartilage, ribs, facial bones, suprabasal layer of epidermis, endodermal lining of the bladder, larynx and oesophagus [10]. In adult, gal3 is mainly expressed in the epithelial cells such as small intestine [21], colon [22], cornea [23, 24], kidney [25], lung [26], thymus [27], breast [28], and prostate [29]. The expression of gal3 is also detected in ductal cells of salivary glands [30], pancreas [31], kidney [32], and eye [33] and in intrahepatic bile ducts [34]. Regarding cell type, gal3 expression is observed in fibroblasts [35], chondrocytes and osteoblasts [36], osteoclasts [37], keratinocytes [38], Schwann cells [39] and gastric mucosa [40], endothelial cells [41], and also immune related cells such as neutrophils [42], eosinophils [43], basophils and mast cells [44], Langerhans cells [38, 45], dendritic cells [46], as well as monocytes [47] and macrophages from different tissues [3, 9, 48, 49].

Increased or Decreased Expression of Gal3 in Tumors- Role in Tumor Progression and Metastasis

Gal3 mediates homotypic and heterotypic aggregation and promotes interactions between tumor cells and endothelial cells, angiogenesis, and tumor metastasis [2, 4, 9]. It was shown that cell surface gal3 mediates homotypic cell adhesion by binding to soluble complementary glycoconjugates [50]. Interactions of metastatic cancer cells with vasculatory endothelium are critical during early stages of cancer metastasis. Recent studies demonstrate that the endothelium gal3 participates in docking of cancer cells including breast and prostate cancers on capillary endothelium by specifically interacting with cancer cells-associated Thomsen-Friedenreich disaccharide (Gal β1,3GalNAc) [51]. Intracelluar gal3 in cancer cells confers anti-apoptotic activity. For example, gal3 acts as a specific binding partner of activated K-Ras and this interaction promotes strong K-Ras activation of PI3K (phosphoinositide 3-kinase) [52]. Gal3 is the only member of the galectin family that contains the NWGR anti-death domain of the Bcl-2 family. Bcl-2 translocation to the mitochondrial membrane leads to anti-apoptosis activity resulting from blocking cytochrome c release [9]. In human breast carcinoma BT547 cells, gal3 can also inhibit cytochrome c release thereby protecting cells against nitric oxide-induced apoptosis [9]. Moreover, gal3 binds Bcl-2 protein *in vitro* to form heterodimers and thus behaves as a mitochondria-associated apoptotic regulator in the cytoplasm. Interestingly, synexin (annexin 7, a Ca^{2+}- and phospholipid-binding protein), is required for gal3 prevention of mitochondrial damage [9]. Extracellular gal3 secreted from tumor cells induces apoptosis of cancer-infiltrating T-cells suggesting its role in the immune escape mechanism during tumor progression [53].

Expression of gal3 is observed in many tumors. However, the intensity of the gal3 expression in tumors depends on the type of tumor, its invasiveness and metastatic potential [54, 55]. For example, increased expression of gal3 is observed in colon, head and neck, gastric, endometrial, thyroid, liver, bladder cancers and breast carcinomas [56-61]. Gal3 plays an important role in tumor progression and metastasis in many cancers such as breast cancer, colon cancer, and human brain tumors [2, 4, 9, 49, 54, 55]. Gal3 transfected human breast cancer cells BT549, which is gal3 null, after intrasplenic injection, formed metastatic colonies in the liver, while native (non-transfected) BT549 cells did not [62]. Change in cellular localization of gal3 is observed during progression of various cancers. Down-regulation of gal3 expression has been demonstrated in colorectal cancer, with increased cytoplasmic expression of gal3 at more advanced stages [54, 55, 63]. In tongue cancer, nuclear gal3 is decreased, but cytoplasmic gal3 is increased during progression from normal to cancer [54, 55]. The decreased expression of galectin-3 was also observed in prostate [64-66], kidney [67], and pituitary cancers [68]. Recent data by us and others indicated that decreased expression of gal3 in pituitary and prostate tumors is, in part, due to its promoter methylation [66, 68-70, please see described later].

Gal3 Gene (*LGALS3*): Structure and Transcriptional Regulation

The human gal3 gene (*LGALS3*), located on chromosome 14 and locus q21–q22 [71], is about 17 kb long and is composed of six exons and five introns [72]. Exon I contains the major part of the 5′ untranslated sequence of mRNA; while exon II encodes the remaining part of the 5′ untranslated region, the translation initiation site and codon sequence for the first six amino acids including the initial methionine. The N-terminal domain of the gal3 is located in exon III; while the CRD is in exon V. Interestingly, the intron II of *LGALS3* contains an internal promoter that drives production of alternative transcripts preferentially in peripheral blood leukocytes [73, 74]. These transcripts arise from an internal gene embedded within *LGALS3*, named *galig* (galectin-3 internal gene) [74]. Galig's CRD is incapable of binding carbohydrates as it contains two overlapping open-reading frames out-of-frame within the lectin coding sequence. However, the galig protein promotes cytochrome c release upon direct interaction with the mitochondria [75].

Although a large body of data about gal3 expression is available in the literature, the mechanisms of its transcriptional regulation are not well understood. However, the expression of gal3 depends on cell type, external stimuli and environmental conditions and involves numerous transcription factors and signaling pathways [10]. Gal3 expression may serve as differentiation marker for certain cell types. For example, the differentiation of the human monocytes or promyelocytic cell line HL-60 to macrophage-like cells induced by phorbol ester is accompanied by increased expression of gal3 [76]. Gal3 expression is up-regulated in phagocytic macrophages and thus considered as a "macrophage activation marker" [77]. Gal3 expression is also elevated in microglia and macrophages activated by phagocytosis of myelin or when exposed to granulocyte-macrophage colony-stimulating factor [78]. In contrast, activation of human monocytes by lipopolysaccharide and interferon-γ is accompanied by decrease of gal3 expression [47]. The reduced expression of gal3 was also observed in monocytic THP-1 cells treated with non-steroidal [79] or cortico-steroidal anti-inflammatory drugs [80]. Interestingly, gal3 expression is absent or barely detected in the resting lymphocytes [81, 82], but the activated B and T cells induce gal3 expression [82]. Gal3 could be considered also as a transformation marker since the gal3 expression is increased in fully *ras*-transformed fibroblasts, when cells have lost their anchorage-dependent growth [83].

In the promoter region of the gal3 gene, several regulatory elements such as five putative Sp1 binding sites (GC boxes), five cAMP-dependent response element (CRE) motifs, four AP-1- and one AP-4-like sites, two NF-κB-like sites, one sis-inducible element (SIE) and a consensus basic helix–loop–helix (bHLH) core sequence are found [72]. The presence of multiple GC box motifs for binding ubiquitous expressed Sp1 transcription factor is a characteristic of constitutively expressed "housekeeping" genes. The activation of the Sp1 binding transcription factor is responsible for gal3 induction by Tat protein of HIV [84]. The SIE that binds sis-inducible factors was suggested to be a possible candidate for the growth-induced activation of gal3 gene expression, caused by the addition of serum. The presence of CRE and NF-κB-like site in the gal3 promoter suggests that the activation of gal3 expression could be regulated through the signaling pathways involving the cAMP-response element-binding protein (CREB) or the NF-κB transcription factor. The CREB/ATF and the NF-κB/Rel transcription factors pathways may be involved in the regulation of gal3 expression by the Tax protein during HTLV-I infection of T cells [85]. The involvement of the NF-κB transcription factor in regulation of gal3 expression, as well as the Jun protein, a component of AP-1 transcription factor has recently been confirmed [86]. The regulation of gal3 expression through the NF-κB transcription factor was shown to be mediated by nucling, a novel apoptosis-associated protein, which interferes with NF-κB via the nuclear translocation process of NF-κB/p65, thus inhibiting gal3 expression on both protein and mRNA level [36, 87]. In skeletal tissues, the regulation of gal3 expression is mediated by the transcription factor Runx2 [9]. Very recently, gal3 expression is found to be regulated in pituitary and prostate tumors by methylation of CpG islands in promoter region [66, 68, 69], which is described later in a greater detail.

Epigenetic Events- DNA Methylation in the Promoter Results in Transcriptional Inactivation

Epigenetic mechanisms such as DNA methylation and histone modification play essential roles in many molecular and cellular alterations associated with the development and progression of cancer [88]. DNA methylation refers to the covalent binding of a methyl group specifically to the carbon-5 position of cytosine residues of the dinucleotide CpG. This is catalyzed by a family of enzymes, the DNA methyltransferases (DNMTs). Two types of DNA methylation alterations have been demonstrated in human cancers. The first refers to global hypomethylation in which the genomes of cancer cells show decreased methylation compared to normal cells [89-91]. The hypomethylation is primarily due to the loss of methylation in repetitive elements and other non-transcribed regions of the genome. This genome-wide hypomethylation potentially leads to loss of imprinting, chromosomal instability, cellular hyperproliferation, and activation of oncogenes such as K-ras and PU.1 [92-94].

The second type of methylation alteration in cancer cells is the hypermethylation of CpG islands in the promoter regions of tumor suppressor and other regulatory genes that are normally unmethylated. The promoter regions of these genes may be inactivated by methylation, which silences their expression. However, differential methylation is not a general mechanism for regulating gene expression, because most inactive promoters remained unmethylated [95]. It is thought that DNA methylation alters chromosome structure and defines regions for transcriptional regulation. Clusters of CpG sites are found dispersed around the genome and are referred to as CpG islands [96]. These islands are found in the promoter region of about 60% of genes, and in exons, introns, and repetitive elements of most genes. In normal cells, most CpG islands in the promoter regions are unmethylated whereas CpG islands in intronic regions and repetitive elements are heavily methylated, perhaps to help the cell identify regions for gene transcription.

Although the importance of CpG island methylation has been demonstrated in cancer, the mechanisms that lead to these changes in cancer are not yet understood. Of three members (DNMT1, DNMT3a, and DNMT3b) of the DNA methyltransferase family, DNMT1 is believed to be primarily involved in the maintenance of CpG methylation [97, 98]. However, other studies suggest that DNMT3b, independently or in cooperation with DNMT1, also contributes to hypermethylation [99-101]. The suppression of transcription by DNA methylation may occur by either direct inhibition [102] or indirect inhibition [103] of transcription factor binding. For the latter, a family of proteins known as methyl binding domain (MBD) proteins is believed to specifically bind DNA containing methylated CpG sites [103, 104]. At least three of the five known members of this family (MeCP2, MBD2 and MBD3) have been shown to be associated with large protein complexes containing histone deacetylase (HDAC1 and HDAC2) and chromatin-remodeling (Sin3a and mi-2) activities [105, 106]. Histone deacetylase (HDAC1 and 2) and chromatin remodeling activities (Mi-2 and Sin3a) produce alterations in chromatin structure that make it refractory to transcriptional activation [107]. In addition to the large protein complexes, the MBD proteins may associate with several other complexes involved in transcriptional repression. Recently, MeCP2 was shown to interact with at least two other proteins, c-ski and N-CoR, known to be involved in transcriptional repression [108]. However, Ohm et al. recently hypothesize that the stem cell-like chromatin pattern may predispose tumor suppressor genes to DNA hypermethylation and heritable gene silencing during tumor initiation and progression [109].

As mentioned above, cancer cells exhibit two apparently opposing changes in the DNA methylation pattern: a decrease of DNA methylation in the intronic CpG islands and an increase of DNA methylation in the promoter CpG islands. Recent studies suggest that both changes may play important roles in the tumorigenic process. However, the increased methylation at CpG islands has been by far the most studied and has a much clearer role in carcinogenesis. Increased CpG island methylation can result in inactivation of many well-characterized tumor suppressor genes (*e.g., BRCA1*, breast cancer 1 gene) as well as inactivation of DNA repair genes, resulting in increased levels of genetic damage. The most striking example is the pi isoform of glutathione S-transferase (*GSTP1*), which is involved in detoxification of potentially DNA-damaging electrophiles [110].

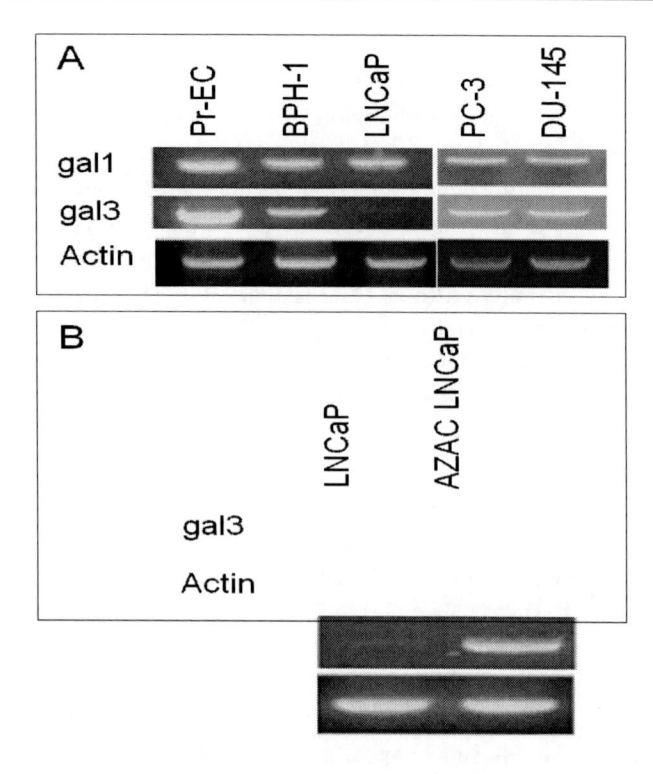

Figure 2. Expression of gal3 determined by reverse transcriptase polymerase chain reaction (RT-PCR). A. Expression of gal3 in normal prostate cells (Pr-EC), benign prostatic hyperplasia cells (BPH-1), androgen-dependent prostate cancer cells LNCaP, androgen-independent prostate cancer cells PC-3 and DU-145. B. Expression of gal3 in azacytidine (AZAC)-treated LNCaP.

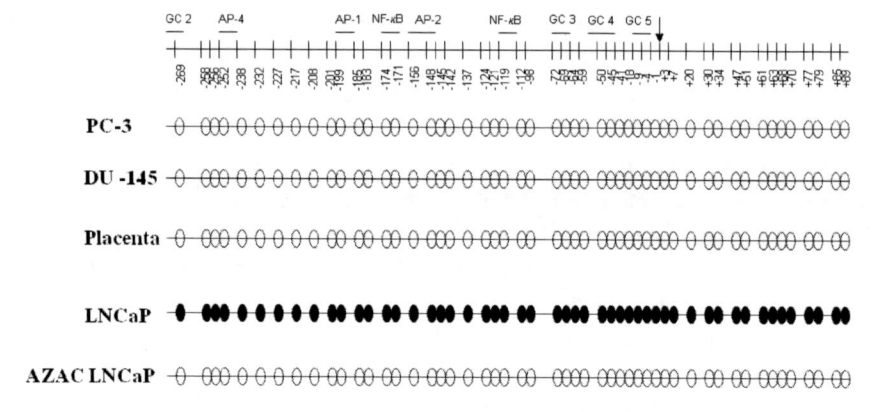

Figure 3. Methylation profile of gal3 promoter region (384 bp) from prostate cancer cell lines LNCaP (untreated and AZAC-treated), PC-3, and DU-145. Each row represents a single cloned allele, and each oval represents a single CpG site (open oval, unmethylated; closed oval, methylated). The numbering in the schematic diagram at the top represents the position relative to the published transcription site (+1, indicated by the arrow).

Expression of Gal3 Gene is Transcriptionally Regulated by DNA Methylation

Gal3 was shown highly expressed in androgen independent PC-3 and DU-145 cells, but weakly expressed in androgen dependent LNCaP cells [69] (Figure 2A). Treatment of LNCaP cells with azacytidine (DNA methyltransferase inhibitor) showed restored expression of gal3 indicating that the promoter methylation is responsible for gal3 gene silencing [69] (Figure 2B). We have also demonstrated DNA methylation on the gal3 promoter in LNCaP cells. For this purpose, DNA was treated with sodium bisulfite and subsequently amplified by PCR [69, 111] with primer pairs located outside the CpG sites. This method allows precise analysis of methylation in a selected region by converting all non-methylated cytosines (C) into uracil (U), while methylated cytosines remain unchanged. All cytosines of CpG sites (about 50) in the gal3 promoter from LNCaP DNA were methylated in the 384 bp PCR product (Figure 3). No gal3 promoter methylation was observed in azacytidine-treated LNCaP [66, 69]. Similarly, no methylation in the gal3 promoter from either PC-3 or DU-145 DNA was observed [66, 69]. Gal3 silencing by promoter methylation is not specific to prostate cancer cells, rather it is a common mechanism for gal3 gene silencing in cancer, as treatment of two gal3 null breast cancer cell lines BT549 [69] and SKBR3 [68] with azacytidine induced expression of gal3. Expression of gal1 was also shown to be regulated by promoter methylation [112]. A small genomic region of approximately 100 base pairs surrounding the transcriptional start site (-50/ +50) accounts for most transcriptional activity of gal1 [112].

Decreased Expression of Gal3 in Pituitary and Prostate Tumors is Associated with Increased Methylation in the Gal3 Promoter

Expression of gal3 in conjunction with its promoter methylation has been recently been investigated in pituitary [68] and prostate [66, 69] tumors. Among pituitary tumors, mainly in follicle-stimulating hormone/luteinizing hormone-producing (38%) and null cell (57%) adenomas, the gal3 promoter was found to be methylated and silenced [68]. Interestingly, the prolactin- and adrenocorticotropic hormone-producing tumors were unmethylated and expressed gal3 [68]. We investigated gal3 expression in normal, benign prostatic hyperplasia (BPH), and various stages of prostate adenocarcinoma [66]. Gal3 was found strongly expressed in normal, BPH, and high-grade prostatic intraepithelial neoplasia (HGPIN, a precursor lesion to development of invasive prostatic adenocarcinoma) tissues in both the nucleus and the cytoplasm. However, gal3 showed decreasing immunopositivity during stage evolution. Moreover, localization of gal3 is interesting during stage evolution. In particular, stage I tumors showed a strong immunopositivity both in nucleus and cytoplasm, while in more advanced stages immunostaining was less intense and localized mainly in cytoplasm, with rare, occasional nucleus positivity. However, two stage I specimens (out of 10) showed little or no gal3 immunopositivity [66]. Similar observation on gal3 expression in prostate tumor was made by others [64, 113]. The role of cytoplasmic and nuclear gal3 in cancer progression was examined by specifically expressing gal3 in either cytoplasm or nucleus of

LNCaP, a gal3-negative human prostate cancer cell line [55]. Cytoplasmic gal3 was anti-apoptotic and promoted tumor growth and angiogenesis. Interestingly, nuclear gal3 affected these parameters in an opposite fashion with an overall antitumoral activity [55]. The role of decreased expression of gal3 in early stage of prostate tumor is not known. However, in normal cells gal3 is believed to interact to the members of Nkx homeodomain family [55], especially to a prostate tumor suppressor, Nkx3.1 [114] and decrease the expression of the cancer phenotype [55]. Therefore, the silencing of gal3 gene during the development of prostate tumor could be necessary to suppress the influence of Nkx3.1 gene and thereby to help tumor cells to proliferate.

Since gal3 expression is decreased in stage evolution of prostate adenocarcinoma as evidenced from our studies [66] and others [64], we hypothesized that DNA isolated from early stages of prostate cancer would be methylated in the gal3 promoter region. Results revealed that the gal3 promoter from multiple specimens of stage II tumor is heavily methylated throughout its entire length, but that from multiple specimens of stage III and IV tumor is lightly methylated. Whereas gal3 promoter in stage III showed few methylation sites, mostly between -199 to -252 nt, the gal3 promoter from stage IV tumor specimens was methylated between -112 to -227 nt. In stage I prostate cancer, however, both light and heavy methylation is evident in the gal3 promoter. In multiple normal prostate and BPH samples, the gal3 promoter was almost unmethylated. Overall, results indicated that the decreased expression of gal3 in tumor prostate is associated with the hypermethylation of its promoter.

Conclusion

Recent studies by us and others have shown that the expression of gal3 in prostate and breast cancer cell lines is regulated by the DNA methylation in its promoter and is negatively correlated [66, 68, 69]. Moreover, decreased expression of gal3 in pituitary and prostate tumors was found associated with hypermethylation in the gal3 promoter [66, 68]. The gal3 promoter showed almost complete methylation in all CpG sites in early stages of prostate cancer, but is lightly methylated in later stages. The methylation pattern of the gal3 promoter at various stages of prostate cancer, and in particular its complete methylation in early stages, is unique, because in other genes such as *GSTP1*, CpG methylation correlates positively with tumor grade and stage (i.e. low methylation in early stages and high methylation in late stages) [115]. However, the degree of gal3 promoter methylation and its expression at various stages of prostate cancer cannot be correlated from this study [66]. This is because the tumor tissue is heterogeneous and so DNA population extracted from the tumor tissue may not necessarily represent single tumor stage. To answer this question, however, further studies are needed from the micro-dissected tissues such as those obtained from laser capture micro-dissection approach.

Acknowledgments

I apologize to the authors whose works are not cited due to the limited space. The work carried out in the author's laboratory has been supported by the UMBI Presidential Proof of Concept Award and the National Institute of Health Grant RO3 CA133935-01.

References

[1]   Ahmed H, Du SJ, O'Leary N, Vasta GR. Biochemical and molecular characterization of galectins from zebrafish (*Danio rerio*): notochord-specific expression of a prototype galectin during early embryogenesis. *Glycobiology.* 2004; 14:219-32.

[2]   Nakahara S, Raz A. Regulation of cancer-related gene expression by galectin-3 and the molecular mechanism of its nuclear import pathway. *Cancer Metastasis Rev.* 2007; 26:605-10.

[3]   Rabinovich GA, Liu FT, Hirashima M, Anderson A. An emerging role for galectins in tuning the immune response: lessons from experimental models of inflammatory disease, autoimmunity and cancer. *Scand. J. Immunol.* 2007; 66:143-58.

[4]   Nakahara S, Raz A. Biological modulation by lectins and their ligands in tumor progression and metastasis. *Anticancer Agents Med. Chem.* 2008; 8:22-36.

[5]   Vasta GR.Roles of galectins in infection. Nat Rev Microbiol. 2009; 7:424-38.

[6]   Elola MT, Wolfenstein-Todel C, Troncoso MF, Vasta GR, Rabinovich GA. Galectins: matricellular glycan-binding proteins linking cell adhesion, migration, and survival. *Cell Mol Life Sci.* 2007; 64:1679-700.

[7]   Gong HC, Honjo Y, Nangia-Makker P, Hogan V, Mazurak N, Bresalier RS, Raz A. The NH2 terminus of galectin-3 governs cellular compartmentalization and functions in cancer cells, *Cancer Res.* 1999; 59:6239-45.

[8]   Massa SM, Cooper DN, Leffler H, Barondes SH. L-29, an endogenous lectin, binds to glycoconjugate ligands with positive cooperativity. *Biochemistry.* 1993; 32:260-7.

[9]   Nakahara S, Oka N, Raz A. On the role of galectin-3 in cancer apoptosis. *Apoptosis.* 2005; 10: 267-75.

[10]   Dumic J, Dabelic S, Flögel M. Galectin-3: an open-ended story. *Biochim Biophys Acta.* 2006; 1760:616-35.

[11]   Henderson NC, Sethi T. The regulation of inflammation by galectin-3. *Immunol Rev.* 2009; 230:160-71.

[12]   Hsu DK, Chen HY, Liu FT. Galectin-3 regulates T-cell functions. *Immunol Rev.* 2009; 230:114-27.

[13]   Yoshii T, Fukumori T, Honjo Y, Inohara H, Kim HR, Raz A. Galectin-3 phosphorylation is required for its anti-apoptotic function and cell cycle arrest. *J Biol Chem.* 2002; 277:6852-7.

[14]   Ochieng J, Green B, Evans S, James O, Warfield P. Modulation of the biological functions of galectin-3 by matrix metalloproteinases. *Biochim Biophys Acta.* 1998; 1379:97-106.

[15] Hsu DK, Zuberi RI, Liu FT. Biochemical and biophysical characterization of human recombinant IgE-binding protein, an S-type animal lectin. *J Biol Chem.* 1992; 267:14167-74.

[16] Ochieng J, Platt D, Tait L, Hogan V, Raz T, Carmi P, Raz A. Structure-function relationship of a recombinant human galactoside-binding protein. *Biochemistry.* 1993; 32:4455-60.

[17] Yang RY, Hsu DK, Liu FT. Expression of galectin-3 modulates T-cell growth and apoptosis. *Proc Natl Acad Sci U S A.* 1996; 93:6737-42.

[18] Yang RY, Hill PN, Hsu DK, Liu FT. Role of the carboxyl-terminal lectin domain in self-association of galectin-3. *Biochemistry.* 1998; 37:4086-92.

[19] Colnot C, Ripoche MA, Scaerou F, Foulis D, Poirier F.Galectins in mouse embryogenesis. *Biochem Soc Trans.* 1996; 24:141-6.

[20] Poirier F. Roles of galectins in vivo. *Biochem Soc Symp.* 2002; 69:95-103.

[21] Mercer N, Guzman L, Cueto Rua E, Drut R, Ahmed H, Vasta GR, Toscano MA, Rabinovich GA, Docena GH. Duodenal intraepithelial lymphocytes of children with cow milk allergy preferentially bind the glycan-binding protein galectin-3. *Int J Immunopathol Pharmacol.* 2009; 22:207-17.

[22] Dumont P, Berton A, Nagy N, Sandras F, Tinton S, Demetter P, Mascart F, Allaoui A, Decaestecker C, Salmon I. Expression of galectin-3 in the tumor immune response in colon cancer. *Lab Invest.* 2008; 88:896-906.

[23] Z. Cao, N. Said, S. Amin, H.K. Wu, A. Bruce, M. Garate, D.K. Hsu, I. Kuwabara, F.-T. Liu, N. Panjwani, Galectins-3 and -7, but not galectin-1, play a role in re-epithelialization of wounds, *J. Biol. Chem.* 2002; 277:42299–305.

[24] Kaji Y, Amano S, Usui T, Oshika T, Yamashiro K, Ishida S, Suzuki K, Tanaka S, Adamis AP, Nagai R, Horiuchi S. Expression and function of receptors for advanced glycation end products in bovine corneal endothelial cells. *Invest Ophthalmol Vis Sci.* 2003; 44:521-8.

[25] Kang EH, Moon KC, Lee EY, Lee YJ, Lee EB, Ahn C, Song YW. Renal expression of galectin-3 in systemic lupus erythematosus patients with nephritis. *Lupus.* 2009; 18:22-8.

[26] Won YS, Jeong ES, Park HJ, Lee CH, Nam KH, Kim HC, Park JI, Choi YK. Upregulation of galectin-3 by Corynebacterium kutscheri infection in the rat lung. *Exp Anim.* 2007; 56:85-91.

[27] Silva-Monteiro E, Reis Lorenzato L, Kenji Nihei O, Junqueira M, Rabinovich GA, Hsu DK, Liu FT, Savino W, Chammas R, Villa-Verde DM. Altered expression of galectin-3 induces cortical thymocyte depletion and premature exit of immature thymocytes during Trypanosoma cruzi infection. *Am J Pathol.* 2007; 170:546-56.

[28] Shekhar MP, Nangia-Makker P, Tait L, Miller F, Raz A. Alterations in galectin-3 expression and distribution correlate with breast cancer progression: functional analysis of galectin-3 in breast epithelial-endothelial interactions. *Am J Pathol.* 2004; 165:1931-41.

[29] R.A. Pacis, M.J. Pilat, K.J. Pienta, K. Wojno, A. Raz, V. Hogan, C.R. Cooper, Decreased galectin-3 expression in prostate cancer. *Prostate* 2000; 44:118–23.

[30] X.C. Xu, J.J. Sola Gallego, R. Lotan, A.K. El-Naggar, Differential expression of galectin-1 and galectin-3 in benign and malignant salivary gland neoplasms. *Int. J. Oncol.* 2000; 17:271–6.

[31] L. Wang, H. Friess, Z. Zhu, L. Frigeri, A. Zimmermann, M. Korc, P.O. Berberat, M.W. Buchler, Galectin-1 and galectin-3 in chronic pancreatitis. *Lab. Invest.* 2000; 80:1233-41.

[32] S. Sasaki, Q. Bao, R.C. Hughes, Galectin-3 modulates rat mesangial cell proliferation and matrix synthesis during experimental glomerulonephritis induced by anti-Thy1.1 antibodies. *J. Pathol.* 1999; 187:481-9.

[33] M.P. Fautsch, A.O. Silva, D.H. Johnson, Carbohydrate binding proteins galectin-1 and galectin-3 in human trabecular meshwork. *Exp. Eye Res.* 2003; 77:11-6.

[34] T. Shimonishi, K. Miyazaki, N. Kono, H. Sabit, K. Tuneyama, K. Harada, J. Hirabayashi, K. Kasai, Y. Nakanuma, Expression of endogenous galectin-1 and galectin-3 in intrahepatic cholangiocarcinoma. *Hum. Pathol.* 2001; 32:302-10.

[35] I.K. Moutsatsos, M. Wade, M. Schindler, J.L. Wang, Endogenous lectins from cultured cells: nuclear localization of carbohydrate-binding protein 35 in proliferating 3T3 fibroblasts. *Proc. Natl. Acad. Sci. U. S. A.* 1987; 84:6452-6.

[36] M. Stock, H. Schafer, S. Stricker, G. Gross, S. Mundlos, F. Otto, Expression of galectin-3 in skeletal tissues is controlled by Runx2. *J. Biol. Chem.* 2003; 278:17360-7.

[37] S. Niida, N. Amizuka, F. Hara, H. Ozawa, H. Kodama, Expression of Mac-2 antigen in the preosteoclast and osteoclast identified in the op/op mouse injected with macrophage colony-stimulating factor. *J. Bone Miner. Res.* 1994; 9:873-81.

[38] Wollenberg, H. de la Salle, D. Hanau, F.-T. Liu, T. Bieber, Human keratinocytes release the endogenous beta-galactoside-binding soluble lectin immunoglobulin E (IgE-binding protein) which binds to Langerhans cells where it modulates their binding capacity for IgE glycoforms. *J. Exp. Med.* 1993; 178:777-85.

[39] F. Reichert, A. Saada, S. Rotshenker, Peripheral nerve injury induces Schwann cells to express two macrophage phenotypes: phagocytosis and the galactose-specific lectin MAC-2. *J. Neurosci.* 1994; 14:3231-45.

[40] R. Lotan, H. Ito, W. Yasui, H. Yokozaki, D. Lotan, E. Tahara, Expression of a 31-kDa lactoside-binding lectin in normal human gastric mucosa and in primary and metastatic gastric carcinomas. *Int. J. Cancer* 1994; 56:474-80.

[41] R. Lotan, P.N. Belloni, R.J. Tressler, D. Lotan, X.C. Xu, G.L. Nicolson, Expression of galectins on microvessel endothelial cells and their involvement in tumour cell adhesion. *Glycoconj. J.* 1994; 11:462-8.

[42] M.J. Truong, V. Gruart, J.P. Kusnierz, J.P. Papin, S. Loiseau, A. Capron, M. Capron, Human neutrophils express immunoglobulin E (IgE)-binding proteins (Mac-2/epsilon BP) of the S-type lectin family: role in IgE dependent activation. *J. Exp. Med.* 1993; 177:243-8.

[43] M.J. Truong, V. Gruart, F.-T. Liu, L. Prin, A. Capron, M. Capron, IgEbinding molecules (Mac-2/epsilon BP) expressed by human eosinophils. Implication in IgE-dependent eosinophil cytotoxicity. *Eur. J. Immunol.* 1993; 23:3230-5.

[44] S.S. Craig, P. Krishnaswamy, A.M. Irani, C.L. Kepley, F.-T. Liu, L.B. Schwartz, Immunoelectron microscopic localization of galectin-3, an IgE binding protein, in human mast cells and basophils. *Anat. Rec.* 1995; 242:211-9.

[45] K. Smetana, Z. Holikova, R. Klubal, N.V. Bovin, B. Dvorankova, J. Bartunkova, F.-T. Liu, H.J. Gabius, Coexpression of binding sites for A (B) histo-blood group trisaccharides with galectin-3 and Lag antigen in human Langerhans cells. *J. Leukoc. Biol.* 1999; 66:644-9.

[46] A.B. Dietz, P.A. Bulur, G.J. Knutson, R. Matasic, S. Vuk-Pavlovic, Maturation of human monocyte-derived dendritic cells studied by microarray hybridization. Biochem. Biophys. *Res. Commun.* 2000; 275:731-8.

[47] F.T. Liu, D.K. Hsu, R.I. Zuberi, I. Kuwabara, E.Y. Chi, W.R. Henderson Jr., Expression and function of galectin-3, a beta-galactoside-binding lectin, in human monocytes and macrophages. *Am. J. Pathol.* 1995; 147:1016-28.

[48] Saada, F. Reichert, S. Rotshenker, Granulocyte macrophage colony stimulating factor produced in lesioned peripheral nerves induces the upregulation of cell surface expression of MAC-2 by macrophages and Schwann cells. *J. Cell Biol.* 1996; 133:159–167.

[49] Liu FT, Rabinovich GA. Galectins as modulators of tumour progression. *Nat Rev Cancer.* 2005; 5:29-41.

[50] Inohara, H., et al., Interactions between galectin-3 and Mac-2-binding protein mediate cell-cell adhesion. *Cancer Res.* 1996; 56:4530-4.

[51] Glinsky VV, Glinsky GV, Rittenhouse-Olson K, Huflejt ME, Glinskii OV, Deutscher SL, Quinn TP. The role of Thomsen-Friedenreich antigen in adhesion of human breast and prostate cancer cells to the endothelium. *Cancer Res.* 2001; 61:4851-7.

[52] Elad-Sfadia G, Haklai R, Balan E, Kloog Y. Galectin-3 augments K-Ras activation and triggers a Ras signal that attenuates ERK but not phosphoinositide 3-kinase activity. *J Biol Chem.* 2004; 279:34922-30.

[53] Fukumori T, Takenaka Y, Yoshii T, Kim HR, Hogan V, Inohara H, Kagawa S, Raz A. CD29 and CD7 mediate galectin-3-induced type II T-cell apoptosis. *Cancer Res.* 2003; 63:8302-11.

[54] Danguy A, Camby I, Kiss R. Galectins and cancer. *Biochim Biophys Acta.* 2002; 1572:285-93.

[55] Califice S, Castronovo V, Van Den Brûle F. Galectin-3 and cancer. *In J Oncol.* 2004; 25:983-92.

[56] Hsu DK, Dowling CA, Jeng KC, Chen JT, Yang RY, Liu FT. Galectin-3 expression is induced in cirrhotic liver and hepatocellular carcinoma. *Int J Cancer* 1999; 81:519-26.

[57] Miyazaki J, Hokari R, Kato S, Tsuzuki Y, Kawaguchi A, Nagao S, Itoh K, Miura S. Increased expression of galectin-3 in primary gastric cancer and the metastatic lymph nodes. *Oncol Rep.* 2002; 9:1307-12.

[58] Yoshimura A, Gemma A, Hosoya Y, Komaki E, Hosomi Y, Okano T, Takenaka K, Matuda K, Seike M, Uematsu K, Hibino S, Shibuya M, Yamada T, Hirohashi S, Kudoh S. Increased expression of the LGALS3 (galectin 3) gene in human non-small-cell lung cancer. *Genes Chromosomes Cancer* 2003; 37:159-64.

[59] Sakaki M, Oka N, Nakanishi R, Yamaguchi K, Fukumori T, Kanayama HO. Serum level of galectin-3 in human bladder cancer. *J Med Invest.* 2008; 55:127-32.

[60] Saussez S, Glinoer D, Chantrain G, Pattou F, Carnaille B, André S, Gabius HJ, Laurent G. Serum galectin-1 and galectin-3 levels in benign and malignant nodular thyroid disease. *Thyroid* 2008; 18:705-12.

[61] Saussez S, Decaestecker C, Mahillon V, Cludts S, Capouillez A, Chevalier D, Vet HK, André S, Toubeau G, Leroy X, Gabius HJ. Galectin-3 upregulation during tumor progression in head and neck cancer. *Laryngoscope* 2008; 118:1583-90.

[62] Song YK, Billiar TR, Lee YJ. Role of galectin-3 in breast cancer metastasis: involvement of nitric oxide. *Am J Pathol.* 2002; 160:1069-75.

[63] Nakamura M, Inufusa H, Adachi T, Aga M, Kurimoto M, Nakatani Y, Wakano T, Nakajima A, Hida JI, Miyake M, Shindo K, Yasutomi M. Involvement of galectin-3 expression in colorectal cancer progression and metastasis. *Int J Oncol.* 1999;15:143-8.

[64] Pacis RA, Pilat MJ, Pienta KJ, Wojno K, Raz A, Hogan V, Cooper CR. Decreased galectin-3 expression in prostate cancer. *Prostate* 2000; 44:118-23.

[65] Merseburger AS, Kramer MW, Hennenlotter J, Simon P, Knapp J, Hartmann JT, Stenzl A, Serth J, Kuczyk MA. Involvement of decreased galectin-3 expression in the pathogenesis and progression of prostate cancer. *Prostate* 2008; 68:72-7.

[66] Ahmed, H., Cappello, F., Rodolico, V., and Vasta, G.R. Evidence of heavy methylation in the galectin-3 promoter in early stages of prostate adenocarcinoma: Development and validation of a methylated marker for early diagnosis of prostate cancer. *Trans. Oncol.* 2009; 2:146-56.

[67] Merseburger AS, Kramer MW, Hennenlotter J, Serth J, Kruck S, Gracia A, Stenzl A, Kuczyk MA. Loss of galectin-3 expression correlates with clear cell renal carcinoma progression and reduced survival. *World J Urol.* 2008; 26:637-42.

[68] Ruebel KH, Jin L, Qian X, Scheithauer BW, Kovacs K, Nakamura N, Zhang H, Raz A, Lloyd RV. Effects of DNA methylation on galectin-3 expression in pituitary tumors. *Cancer Res* 2005; 65:1136-40.

[69] Ahmed H, Banerjee PB, Vasta GR. Differential expression of galectins in normal, benign and malignant prostate epithelial cells: Silencing of galectin-3 expression in prostate cancer by its promoter methylation. *Biochem. Biophys. Res. Commun.* 2007; 358: 241-6.

[70] Ahmed H. Promoter methylation in prostate cancer and its application for the early detection of prostate cancer using serum and urine samples. *Biomarkers in Cancer* 2010; 2:17-31.

[71] J. Raimond, D.B. Zimonjic, C. Mignon, M. Mattei, N.C. Popescu, M. Monsigny, A. Legrand, Mapping of the galectin-3 gene (LGALS3) to human chromosome 14 at region 14q21–22. *Mamm. Genome* 1997; 8:706-7.

[72] Kadrofske MM, Openo KP, Wang JL. The human LGALS3 (galectin-3) gene: determination of the gene structure and functional characterization of the promoter. *Arch Biochem Biophys.* 1998; 349:7-20.

[73] J. Raimond, F. Rouleux, M. Monsigny, A. Legrand, The second intron of the human galectin-3 gene has a strong promoter activity down-regulated by p53, *FEBS Lett.* 1995; 363:165-9.

[74] M. Guittaut, S. Charpentier, T. Normand, M. Dubois, J. Raimond, A. Legrand, Identification of an internal gene to the human Galectin-3 gene with two different overlapping reading frames that do not encode Galectin-3, *J. Biol. Chem.* 2001; 276:2652-7.

[75] M. Duneau, M. Boyer-Guittaut, P. Gonzalez, S. Charpentier, T. Normand, M. Dubois, J. Raimond, A. Legrand, Galig, a novel cell death gene that encodes a mitochondrial protein promoting cytochrome c release, *Exp. Cell Res.* 2005; 302:194-205.

[76] Abedin MJ, Kashio Y, Seki M, Nakamura K, Hirashima M. Potential roles of galectins in myeloid differentiation into three different lineages. *J Leukoc Biol.* 2003; 73:650-6.

[77] M.J. Elliott, A. Strasser, D. Metcalf, Selective up-regulation of macrophage function in granulocyte-macrophage colony-stimulating factor transgenic mice. *J. Immunol.* 1991; 147:2957-63.

[78] F. Reichert, S. Rotshenker, Galectin-3/MAC-2 in experimental allergic encephalomyelitis. *Exp. Neurol.* 1999; 160:508-14.

[79] S. Dabelic, M. Flogel, J. Dumic, Effects of aspirin and indomethacin on galectin-3. *Croat. Chem. Acta* 2005; 178:433-40.

[80] S. Dabelic, M. Flogel, J. Dumic, Corticosteroids affect galectin-3 expression. *Period. Biol.* 2005; 107:175-81.

[81] F.-T. Liu, K. Albrandt, E. Mendel, A. Kulczycki Jr., N.K. Orida, Identification of an IgE-binding protein by molecular cloning. *Proc. Natl. Acad. Sci. U. S. A.* 1985; 82:4100-4.

[82] H.G. Joo, P.S. Goedegebuure, N. Sadanaga, M. Nagoshi, W. von Bernstorff, T.J. Eberlein, Expression and function of galectin-3, a beta- galactoside-binding protein in activated T lymphocytes. *J. Leukoc. Biol.* 2001; 69:555–564.

[83] E. Hebert, M. Monsigny, Galectin-3 mRNA level depends on transformation phenotype in ras-transformed NIH 3T3 cells. *Biol. Cell* 1994; 81:73-6.

[84] S. Fogel, M. Guittaut, A. Legrand, M. Monsigny, E. Hebert, The tat protein of HIV-1 induces galectin-3 expression. *Glycobiology* 1999; 9:383-7.

[85] D.K. Hsu, S.R. Hammes, I. Kuwabara,W.C. Greene, F.-T. Liu, Human T lymphotropic virus-I infection of human T lymphocytes induces expression of the beta-galactoside-binding lectin, galectin-3. *Am. J. Pathol.* 1996; 148:1661-70.

[86] J. Dumic, G. Lauc, M. Flogel, Expression of galectin-3 in cells exposed to stress-Roles of jun and NF-kappaB, *Cell. Physiol. Biochem.* 2000; 10:149-58.

[87] Costessi, A. Pines, P. D'Andrea, M. Romanello, G. Damante, L. Cesaratto, F. Quadrifoglio, L. Moro, G. Tell, Extracellular nucleotides activate Runx2 in the osteoblast-like HOBIT cell line: a possible molecular link between mechanical stress and osteoblasts' response. *Bone* 2005; 36:418-32.

[88] McKenna ES, Roberts CW. Epigenetics and cancer without genomic instability. *Cell Cycle* 2009; 8:23-6.

[89] Ehrlich M. DNA methylation in cancer: too much, but also too little. *Oncogene.* 2002; 21:5400-13.

[90] Vucic EA, Brown CJ, Lam WL. Epigenetics of cancer progression. *Pharmacogenomics* 2008; 9:215-34.

[91] Pogribny IP, Beland FA. DNA hypomethylation in the origin and pathogenesis of human diseases. *Cell Mol Life Sci.* 2009; 66:2249-61.

[92] Franco R, Schoneveld O, Georgakilas AG, Panayiotidis MI. Oxidative stress, DNA methylation and carcinogenesis. *Cancer Lett.* 2008; 266:6-11.

[93] Benbrahim-Tallaa L, Waterland RA, Styblo M, Achanzar WE, Webber MM, Waalkes MP. Molecular events associated with arsenic-induced malignant transformation of human prostatic epithelial cells: aberrant genomic DNA methylation and K-ras oncogene activation. *Toxicol Appl Pharmacol.* 2005; 206:288-98.

[94] Nagarajan RP, Costello JF. Molecular epigenetics and genetics in neuro-oncology. *Neurotherapeutics* 2009; 6:436-46.

[95] Weber M, Hellmann I, Stadler MB, Ramos L, Pääbo S, Rebhan M, Schübeler D. Distribution, silencing potential and evolutionary impact of promoter DNA methylation in the human genome. *Nat Genet.* 2007; 39:457-66.

[96] Illingworth RS, Bird AP. CpG islands--'a rough guide'. *FEBS Lett.* 2009; 583:1713-20.

[97] Bestor, T.H. Activation of mammalian DNA methyltransferase by cleavage of a Zn binding regulatory domain. *Embo J.* 1992; 11:2611- 7.

[98] Robert MF, Morin S, Beaulieu N, Gauthier F, Chute IC, Barsalou A, MacLeod AR. DNMT1 is required to maintain CpG methylation and aberrant gene silencing in human cancer cells. *Nat Genet.* 2003; 33:61-5.

[99] El-Osta A. DNMT cooperativity--the developing links between methylation, chromatin structure and cancer. *Bioessays* 2003; 25:1071-84.

[100] Benbrahim-Tallaa L, Waterland RA, Dill AL, Webber MM, Waalkes MP. Tumor suppressor gene inactivation during cadmium-induced malignant transformation of human prostate cells correlates with overexpression of de novo DNA methyltransferase. *Environ Health Perspect.* 2007;115:1454-9.

[101] Roll JD, Rivenbark AG, Jones WD, Coleman WB. DNMT3b overexpression contributes to a hypermethylator phenotype in human breast cancer cell lines. *Mol Cancer* 2008; 7:15.

[102] Tate PH, Bird AP. Effects of DNA methylation on DNA-binding proteins and gene expression. *Curr Opin Genet Dev.* 1993; 3:226-31.

[103] Hendrich, B. and Bird, A. Identification and characterization of a family of mammalian methyl-CpG binding proteins. *Mol Cell Biol.* 1998; 18:6538-47.

[104] Sansom OJ, Maddison K, Clarke AR. Mechanisms of disease: methyl-binding domain proteins as potential therapeutic targets in cancer. *Nat Clin Pract Oncol.* 2007; 4:305-15.

[105] Zhang, Y. et al. Analysis of the NuRD subunits reveals a histone deacetylase core complex and a connection with DNA methylation. *Genes Dev.* 1999; 13:1924-35.

[106] Nan, X. et al. Transcriptional repression by the methyl-CpG-binding protein MeCP2 involves a histone deacetylase complex. *Nature* 1998; 393:386-9.

[107] Tyler JK, Kadonaga JT. The "dark side" of chromatin remodeling: repressive effects on transcription. *Cell* 1999; 99:443-6.

[108] Kokura, K. et al. The Ski protein family is required for MeCP2-mediated transcriptional repression. *J Biol Chem.* 2001; 276:34115-21.

[109] Ohm JE, McGarvey KM, Yu X, Cheng L, Schuebel KE, Cope L, Mohammad HP, Chen W, Daniel VC, Yu W, Berman DM, Jenuwein T, Pruitt K, Sharkis SJ, Watkins DN, Herman JG, Baylin SB. A stem cell-like chromatin pattern may predispose tumor suppressor genes to DNA hypermethylation and heritable silencing. *Nat Genet.* 2007; 39:237-42.

[110] Lee, W.H. et al. Cytidine methylation of regulatory sequences near the pi-class glutathione S-transferase gene accompanies human prostatic carcinogenesis. *Proc Natl Acad Sci USA.* 1994; 91:11733-7.

[111] Frommer M, McDonald LE, Millar DS, Collis CM, Watt F, Grigg GW, Molloy PL, and Paul CL. A genomic sequencing protocol that yields a positive display of 5-methylcytosine residues in individual DNA strands. *Proc Natl Acad Sci US.* 1992; 89:1827-31.

[112] Salvatore P, Benvenuto G, Caporaso M, Bruni CB, Chiariotti L. High resolution methylation analysis of the galectin-1 gene promoter region in expressing and nonexpressing tissues. *FEBS Lett.* 1998; 421:152-8.

[113] van den Brûle FA, Waltregny D, Liu FT, Castronovo V. Alteration of the cytoplasmic/nuclear expression pattern of galectin-3 correlates with prostate carcinoma progression. *Int J Cancer.* 2000; 89:361-7.

[114] Kim MJ, Bhatia-Gaur R, Banach-Petrosky WA, Desai N, Wang Y, Hayward SW, Cunha GR, Cardiff RD, Shen MM, Abate-Shen C. Nkx3.1 mutant mice recapitulate early stages of prostate carcinogenesis. *Cancer Res.* 2002; 62:2999-3004.

[115] Jerónimo C, Henrique R, Hoque MO, Mambo E, Ribeiro FR, Varzim G, Oliveira J, Teixeira MR, Lopes C, and Sidransky D. A quantitative promoter methylation profile of prostate cancer. *Clin Cancer Res.* 2004; 10:8472-8.

In: Gene Silencing: Theory, Techniques and Applications
Editors: Anthony J. Catalano

ISBN: 978-1-61728-276-8
©2010 Nova Science Publishers, Inc.

Chapter IX

Silencing Hepatitis B Virus Replication with Antiviral Pri-Mir Shuttles Generated from Liver-Specific Pol II Promoters

Abdullah Ely and Patrick Arbuthnot[*]

Antiviral Gene Therapy Research Unit, Department of Molecular Medicine and Haematology, University of the Witwatersrand Medical School, WITS, SOUTH AFRICA.

Abstract

Chronic infection with hepatitis B virus (HBV) occurs in approximately 6% of the world's population and is often complicated by cirrhosis and hepatocellular carcinoma (HCC). Existing therapy rarely has durable effects and improving treatment to counter the infection remains an important medical priority. Although harnessing the RNA interference (RNAi) pathway to achieve therapeutic HBV gene silencing holds promise, precise regulation of the expression of silencing sequences is critically important for safe application of this approach. Earlier work from our laboratory demonstrated that pri-miR-31- and pri-miR-122-based anti HBV shuttles were capable of potent, safe antiviral activity and can be used in modular multimeric arrangements. To advance this approach, and limit the potential problems caused by extrahepatic expression of anti HBV RNAi activators, these sequences were placed under control of liver specific transcription control elements, viz. the human Factor VIII (FVIII), alpha-1-antitrypsin (A1AT), HBV preS2 and HBV basic core (BCP) promoters. Using a luciferase reporter gene assay optimal liver-specific transcription control was observed with A1AT and BCP regulating sequences. These elements were then incorporated into pri-miR-expression cassettes and

* Corresponding author: Tel: +27 (0)11 717 2365 ; Fax: +27 (0)11 717 2395 ; E-mail:
Patrick.Arbuthnot@wits.ac.za

were tested for antiviral efficacy in cell culture and a murine model of HBV replication. Results showed that silencing of HBV replication was achieved. Importantly there was no evidence for disruption of endogenous miR function, which is a significant advantage over use of stronger and constitutively active RNA polymerase (Pol) III promoter RNAi expression cassettes. The use of anti HBV pri-miR shuttles in the context of liver-specific Pol II promoters is likely to have usefulness for therapeutic HBV knockdown, and should also have general applicability to silencing of pathology causing genes in the liver.

Introduction

The hepatitis B virus (HBV) is a human pathogen that infects the liver and causes both acute and chronic disease. An estimated 2 billion people worldwide have been infected with HBV and 350 million individuals are chronic carriers of the virus [1]. Individuals persistently infected with HBV are at an increased risk of developing the severe complications of cirrhosis and hepatocellular carcinoma (HCC) [2]. HCC prognosis is very poor and annual incidence is almost equal to annual mortality [3]. An effective anti HBV vaccine is available and universal vaccination of infants has been implemented in parts of Asia and sub Saharan Africa, where the infection is endemic [4]. Immunization is however prophylactic and offers little therapeutic benefit to existing chronic carriers. Immunomodulators and nucleotide and nucleoside analogues remain the only licensed therapies for management of chronic infections [5]. However current treatment regimens are limited by expense, side effects, development of resistance and efficacy that is not durable. The advancement of novel treatment strategies to limit severe sequelae of chronic HBV infection therefore remains an important medical goal. HBV replication involves reverse transcription of a greater than genome length 3.5 kb pregenomic RNA (pgRNA), which is essential for the formation of partly double stranded relaxed circular DNA (rcDNA) that is typically found in the HB virion (Fig. 1A) [6, 7]. After infection of hepatocytes, the rcDNA is converted to a covalently closed circular DNA (cccDNA), which serves as template for transcription of sequences encoding viral proteins and pgRNA. The viral genome has a highly compact arrangement with overlapping open reading frames (ORFs), which limits its sequence plasticity. Methods that utilize RNA interference (RNAi) to interfere with the stability of pgRNA therefore have the potential to disrupt an essential step of viral replication.

An important consideration in development of RNAi-based therapy for HBV infection is the controlled expression of silencing sequences to avoid potentially harmful unintended effects. Achieving tissue-specific expression of RNAi effecter sequences also mitigates the need for strategies to deliver expression cassettes to target tissues exclusively. We have previously shown that pri-miR-31 or pri-miR-122 may be used as scaffolds for embedding anti HBV guide sequences. These shuttles could be incorporated into CMV RNA polymerase (Pol) II promoter cassettes and used for efficient knockdown of markers of HBV replication in cultured cells and in vivo [8-10]. Moreover, this approach was used to generate Pol II-expressed multimeric miR shuttles that simultaneously target multiple sites to overcome viral escape [10]. Demonstration that RNAi expression cassettes under the control of the CMV promoter are effective paves the way for investigating use of other Pol II-driven cassettes that have properties of tissue-specific expression.

Human Factor VIII (FVIII) and alpha-1-antitrypsin (A1AT) promoters and HBV preS2 and core promoters are well-characterized liver-specific promoters and are potentially useful for specific expression of anti HBV pri-miR shuttles in liver tissue. The promoter of FVIII contains sequence elements that are recognized by the liver-enriched transcription factor hepatocyte nuclear factor 1 (HNF-1) [11]. Similarly liver-specific expression of the A1AT promoter is regulated by the liver transcription factors HNF-1α and HNF-4 [12]. Employing HBV regulatory elements for the liver-specific expression of pri-miR shuttle sequences is also appealing as the liver tropism exhibited by HBV is conferred in part by its transcription regulatory elements (reviewed in [6]). Use of a pri-miR scaffold that is derived from the pri-miR-122 sequence that is abundantly expressed liver cells only [13] is also potentially useful for conferring liver specificity on antiviral RNAi activators. In this study, we report on the generation of liver-specific pri-miR-122-derived anti HBV shuttle expression cassettes that are capable of efficient viral gene silencing without evidence for disrupting the endogenous miR pathway.

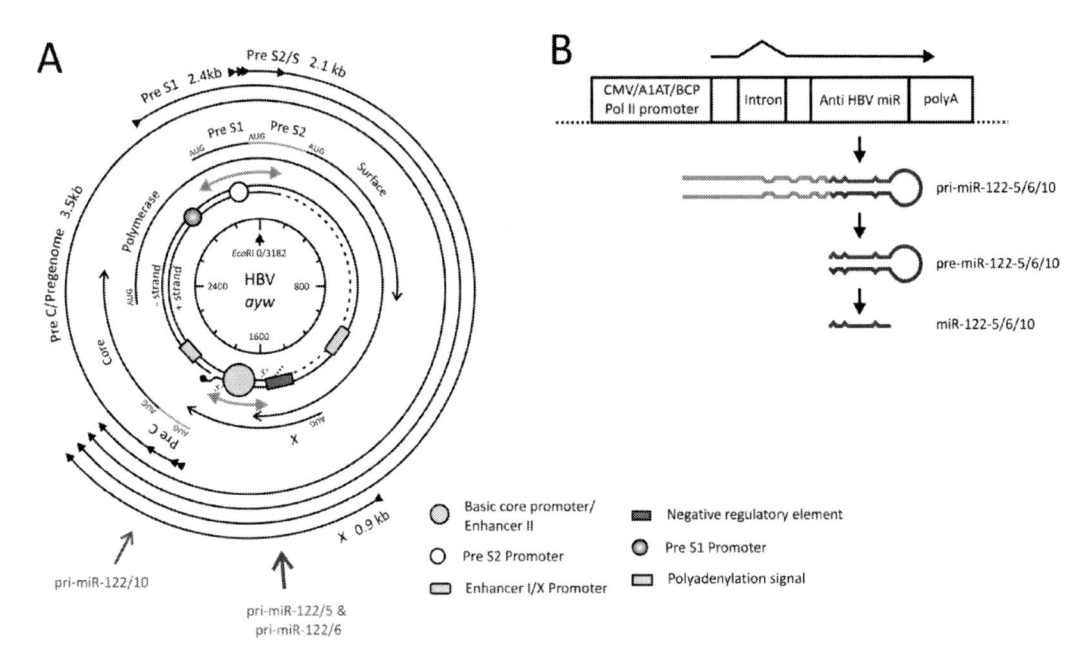

Figure 1. Organization HBV genome and Pol II antiviral expression cassettes. A. The partially double stranded circular genome with cis elements and coordinates calculated relative to the unique EcoRI restriction site are indicated at the center. Viral open reading frames are shown as arrows immediately surrounding the genome. The outermost arrows represent the four major viral transcripts that terminate at a single polyadenylation site. Sites within the *HBx* ORF that are targeted by the miR shuttles are indicated. The BCP and PreS2 promoter sequences that were incorporated into Pol II expression cassettes are shown as double headed arrows. B. Schematic illustration of pri-miR-122 shuttle expression cassettes showing upstream Pol II promoter (CMV, A1AT, FVIII, BCP or PreS2), intron and pri-miR mimic sequences with downstream transcription termination signal (poly A). Pre-miR shuttles were inserted in an exon and were flanked by 51 nt of pri-miR-122-derived sequences. The putative pre-miRs generated after Drosha/DGCR8 processing are indicated in color (purple and red) and the mature processed guide sequences that are selected after Dicer processing and strand selection by RISC are indicated in red only.

Materials and Methods

Plasmids encoding HBV target sequences, constitutively active reporters and miR-16 sponges. The pCH-9/3091 [14], pTZ-FLuc and pCH-FLuc [9], psiCHECK-miR-16T×7 and pU6 miR-16S×7 [10] and pCI-neo eGFP [15] vectors have been described before. To generate pCMV FLuc, a vector that encodes Firefly luciferase under control of the CMV immediate early promoter, the Firefly luciferase-encoding DNA fragment was removed from pTZ-FLuc after XhoI and SpeI restriction and inserted into XhoI and XbaI sites of pCI-neo (Promega, WI, USA).

CMV Pol-II promoter driven pri-miR-122 shuttle expression cassettes. Anti HBV pri-miR-122 shuttle sequences were designed to target sites within the *HBx* ORF (accession number J02203.1). pri-miR-122/5 encodes a guide that is complementary to co-ordinates 1575-1597, pri-miR-122/6 targets co-ordinates 1580-1602 and pri-miR-122/10 targets co-ordinates 1863-1885 of HBV (Fig. 1A). Construction of the CMV pri-miR-122/5 shuttle expression cassette has been described previously [8, 9]. Propagation of Pol II-driven pri-miR-122/6 and pri-miR-122/10 shuttle cassettes was carried out using a similar approach, which initially involved formation of pre-miR-encoding sequences by extension of partially complementary forward (F) and reverse (R) oligonucleotide primers. The complete pri-miR-encoding sequences were then constituted by PCR amplification of the pre-miR template and subsequently ligated to liver-specific Pol II promoters. All oligonucleotides were synthesized using standard phosphoramidite chemistry (Inqaba Biotech, South Africa). The oligonucleotide sequences were as follows: pre-miR-122/6 F: 5'- GAG TTT CCT TAG CAG AGC TGG TGC AGA GGT GAA GCG AAG TGC AGT CTA AAC AAT GCA CTT CG - 3', pre-miR-122/6 R: 5'- GGA TTG CCT AGC AGT AGC TAG GTC AGA GGT TAA GCG AAG TGC ATT GTT TAG ACT GCA CTT CG -3', pre-miR-122/10 F: 5'- GAG TTT CCT TAG CAG AGC TGA GGC ACA GCT GGG AGG CTT GAA CGT CTA AAC TAT TTC AAG CC -3' and pre-miR-122/10 R: 5'- GGA TTG CCT AGC AGT AGC TAA TTC ACA GCT GGG AGG CTT GAA ATA GTT TAG ACG TTC AAG CC -3'. Purified products from primer extension were used as template in PCR using universal pri-miR-122 F (5'- GAC TGC TAG CTG GAG GTG AAG TTA ACA CCT TCG TGG CTA CAG AGT TTC TTT AGC AGA GCT G -3') and pri-miR-122 R (5'- GAT CAC TAG TAA AAA AGC AAA CGA TGC CAA GAC ATT TAT CGA GGG AAG GAT TGC CTA GCA GTA GCT A-3') primers. The pri-miR-122 shuttle fragments were ligated to the PCR cloning vector pTZ57R/T (InsTAclone™ PCR cloning Kit, Fermentas, MD, USA) according to the manufacturer's instructions to generate the pTZ-pri-miR-122 vectors. To generate CMV-driven pri-miR-122/6 and pri-miR-122/10 expression cassettes, the pri-miR-122/6 and pri-miR-122/10 shuttle sequences were excised from their respective pTZ vectors using NheI and XbaI and inserted into equivalent sites of pCI-neo (Promega, WI, USA). This procedure was similar to that previously described for propagating pCMV pri-miR-122/5 [9].

Liver-specific reporter and pri-miR-122 shuttle expression vectors. The human A1AT (Genbank accession number D38257.1) and FVIII (Genbank accession number NT_167198.1) promoter sequences were amplified from total human genomic DNA extracted from Huh7 cells using standard PCR procedures. Primers A1AT F (5'- GAT CTG ATC ATT CCC TGG TCT GAA TGT GTG -3') and A1AT R (5'- GAT CAA GCT TAC TGT CCC AGG TCA GTG GTG -3') were used to amplify the A1AT promoter. Similarly FVIII F (5'-

GAT CAG ATC TGA GCT CAC CAT GGC TAC ATT -3') and FVIII R (5'- GAT CAA GCT TGA CTT ATT GCT ACA AAT GTT CAA C -3') generated a FVIII promoter sequence. The PCR amplicons extended from nucleotides 75856176 - 75854902 of chromosome 14 for the A1AT promoter sequence and nucleotides 5169940 - 5168766 of the X chromosome for the FVIII promoter sequence. The HBV Basic Core Promoter (BCP) and preS2 promoter sequences (Genbank accession number J02203.1) were amplified from the plasmid pCH-9/3091 using the following primer sets: BCP F (5'- GAT CAG ATC TGC ATG GAG ACC ACC GTG AAC -3'), BCP R (5'- GAT CAA GCT TCA CCC AAG GCA CAG CTT GGA -3'), PreS2 F (5'- GAT CAG ATC TGC CTT CAG AGC AAA CAC CGC -3') and PreS2 R (5'- GAT CAA GCT TAC AGG CCT CTC ACT CTG GGA -3'). Amplicons extended from viral nucleotide coordinates 1606 to 1890 for BCP and from 2911 - 50 for the PreS2 promoter (Fig. 1A). All oligonucleotides were synthesized by standard phosphoramidite chemistry (Inqaba Biotech, South Africa). Primers were designed such that amplification introduced a BglII (BclI in the case of A1AT) site at the 5' end and a HindIII site at the 3' end of the amplicons. PCR was carried out using the Expand High Fidelity PCR^{PLUS} System (Roche Diagnostics GmbH, Germany) according to manufacturer's instructions. Purified amplicons were ligated to the linearized PCR cloning vector pTZ57R/T to generate pTZ-A1AT, pTZ-FVIII, pTZ-BCP and pTZ-PreS2. Positive plasmid clones were identified using restriction enzyme mapping and sequences were verified using standard automated dideoxy chain termination procedures (Inqaba Biotech, South Africa).

To generate liver-specific Pol II expression plasmids, the CMV immediate early enhancer promoter transcriptional regulatory element within pCI-neo (Promega, WI, USA) was substituted with the sequences of the liver-specific promoters. The sequences encoding the FVIII, BCP and PreS2 promoter elements were removed from pTZ-FVIII, pTZ-BCP and pTZ-PreS2 using BglII and HindIII restriction and the A1AT sequence was excised from pTZ-A1AT using BclI and HindIII. Since BclI is sensitive to methylation, pTZ-A1AT was propagated in the dcm- and dam-methylase deficient strain of *E. coli*, GM2929. pCI-neo was digested with HindIII and EcoRI to yield 3815 bp, 1317 bp and 340 bp fragments. Separately, pCI-neo was digested with EcoRI and BglII to yield 4371 and 1101 bp fragments. Liver-specific promoter sequences were ligated with the 340 bp HindIII-EcoRI and the 4371 bp EcoRI-BglII fragments in a three-way ligation reaction to generate backbone liver-specific expression vectors (pA1AT, pFVIII, pBCP and pPreS2). BclI and BglII generated complementary overhangs thus allowing the A1AT promoter sequence to be ligated to the pCI-neo backbone. These plasmids contained the liver-specific promoter sequences followed by a chimeric intron and an SV40 late polyadenylation signal downstream of the insert (Fig. 1B).

To generate Pol II reporter gene expression cassettes, the Firefly luciferase sequence was removed from pCMV FLuc [9] after digestion with NheI and SmaI then ligated to equivalent sites of the liver-specific expression vectors to generate pA1AT FLuc, pFVIII FLuc, pBCP FLuc and pPreS2 FLuc. Plasmids containing liver-specific pri-miR-122-derived cassettes were propagated after ligation of pri-miR-122 shuttles, excised from pCMV pri-miR-122/5, pCMV pri-miR-122/6 and pCMV pri-miR-122/10 using NheI and SmaI, to equivalent sites of pA1AT and pBCP.

Cell Culture. Functionality of liver-specific expression cassettes was assessed in Huh7 cells and the HEK293-derived 116 cells by co-transfecting 100 ng of each of the different Pol II Firefly luciferase expression vectors, 100 ng of phRL-CMV (Promega, WI, USA) and 100 ng of pCI-neo eGFP. Transfection was achieved using Lipofectamine™ 2000 (Invitrogen, CA, USA) according to manufacturer's instructions. Firefly and *Renilla* luciferase activities were assayed using the Dual-Luciferase® Assay System (Promega, WI, USA) and measured on a Veritas Dual Injection Luminometer (Turner BioSystems, CA, USA). Firefly luciferase activity was normalized to *Renilla* luciferase activity. Detection of Firefly luciferase activity that was exclusive to the liver-derived cell line was interpreted as liver-specific expression. To assess the tissue-specificity of silencing by the liver-specific pri-miR shuttles, Huh7 and 116 cells were transfected with 80 ng of target plasmid (pCH-FLuc). Together with this plasmid, 800 ng of the A1AT- and BCP-driven pri-miR shuttle vectors, 50 ng of phRL-CMV and 50 ng of pCI-neo eGFP were also transfected. Plasmid DNA was made up to a total of 1 μg with pCI-neo (Promega, WI, USA) and transfected as described above. Forty eight hours post-transfection cells were lysed and measurement of Firefly and *Renilla* luciferase activities was carried out.

Possible disruption of the miR biogenesis pathway by the A1AT and BCP Pri-miR-122 shuttle expression cassettes was assessed in Huh7 cells after co-transfecting RNAi effecter expression cassettes and the miR-16 target vector, psiCHECK-miR-16T×7. This assay, an adaptation of the use of miR sponges to derepress miR function [16], has been described previously [10]. Cells were co-transfected with 80 ng of psiCHECK-miR-16T×7, 800 ng of RNAi effecter or sponge (pU6-miR-16S×7) vectors and 120 ng of pCI-neo eGFP using Lipofectamine™ 2000.

Assessing efficacy in vivo of liver-specific pri-miR shuttle sequences. In vivo efficacy of the A1AT- and BCP-driven pri-miR-122 shuttle expression cassettes was further assessed by employing the hydrodynamic injection model of HBV replication. Five micrograms of liver-specific anti HBV expression cassettes were co-injected with 5 μg each of pCH-9/3091 and pCI-neo eGFP. Blood samples were collected from mice 3 and 5 days post-injection and HBsAg concentrations determined using the MONOLISA® HBs Ag Assay kit (Bio-Rad, CA, USA). All procedures were carried out in accordance with protocols approved by the Animal Ethics Screening Committee of the University of the Witwatersrand.

Statistical Analysis. The student's t-test was carried out to determine statistical significance and this was determined using the GraphPad Prism Software version 4.03 (GraphPad Software, CA, USA).

Results

Selection of liver-specific Pol II promoters for expression of pri-miR shuttle cassettes. Initially, to establish whether the selected promoter sequences are functional and exhibit tissue-specificity, the ability of these transcriptional elements to express the Firefly luciferase reporter gene was assessed. This was determined by quantitation of Firefly luciferase activity in human liver-derived (Huh7) and non hepatic human kidney-derived (116) cells. Sequences spanning nucleotides -1175 to -9 upstream of the *FVIII* [17] gene and nucleotides -1200 to -32 upstream of the *A1AT* [18] gene have been shown to drive transcription of reporter genes in liver-derived cell lines and were included in the expression cassettes used here. Plasmids expressing the reporter gene from a CMV immediate early promoter enhancer element (pCMV FLuc) and each from the panel of putative liver-specific Pol II promoters (pA1AT

FLuc, pFVIII FLuc, pBCP FLuc and pPreS2 FLuc) were used to transfect cells in culture. Firefly luciferase was strongly expressed in both Huh7 and 116 cells after transfection with pCMV FLuc (Fig. 2). This is not unexpected as the CMV immediate early promoter enhancer is known to be ubiquitously active and powerful [19]. Measurement of Firefly luciferase activity in Huh7 cells that were transfected with the liver-specific expression plasmids indicated that these promoter sequences were functional. In contrast to the expression from the CMV immediate early promoter enhancer Firefly luciferase activity was not detected in kidney-derived cells when expressed from any of the liver-specific promoters, which indicated that these cassettes exhibit tissue-specific expression. However, reporter gene activity derived from expression of the CMV promoter sequence was significantly stronger than that of the other liver-specific promoters. FVIII and PreS2 promoters showed very low activity, which was unlikely to achieve sufficient pri-miR shuttle expression, and were excluded from additional investigation. To advance the analysis of expression of anti HBV shuttles from liver specific Pol II promoters, CMV, A1AT and BCP transcriptional regulatory elements were selected for further study. A total of nine pri-miR-122 shuttle Pol II expression cassettes were then tested for silencing of HBV replication in cell culture and in vivo. These were the pri-miR-122/5, pri-miR-122/6 and pri-miR-122/10 shuttle sequences, which were each placed under transcriptional control of the CMV, A1AT or BCP promoters (Fig. 1B).

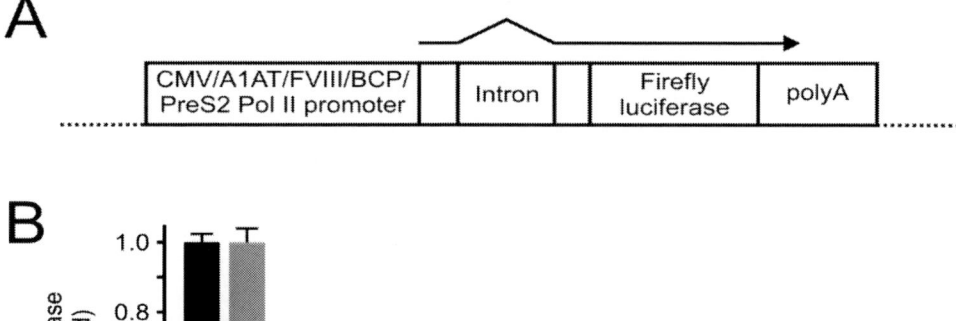

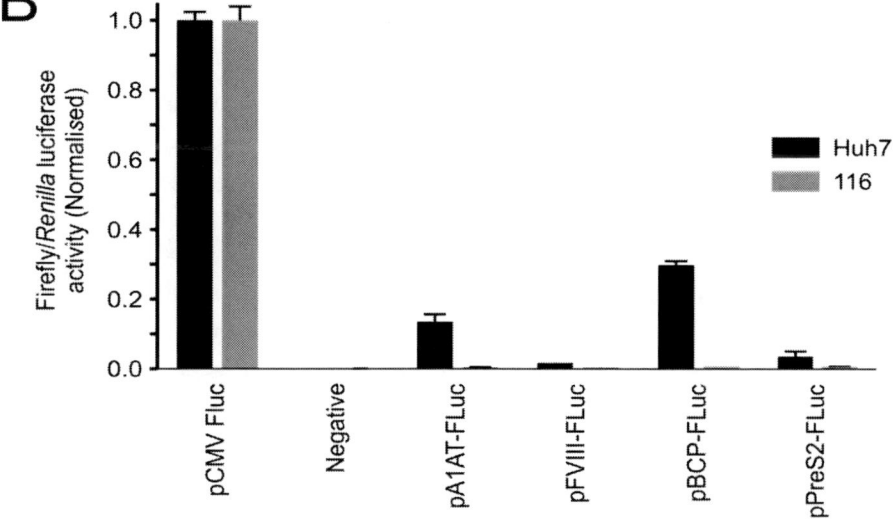

Figure 2. Reporter gene assay to determine liver specificity of Pol II promoters. A. Schematic illustration of the structure of the Pol II expression cassettes encoding the Firefly luciferase reporter. B. Functionality and tissue-specificity of the putative liver-specific promoters was assessed by measuring expression of the Firefly luciferase activity in liver-derived (Huh7) and kidney-derived (116) cells. Bars indicate means of relative Firefly to Renilla luciferase activity in Huh7 (black bars) and 116 (gray bars) cells. Error bars indicate standard error of the mean (SEM).

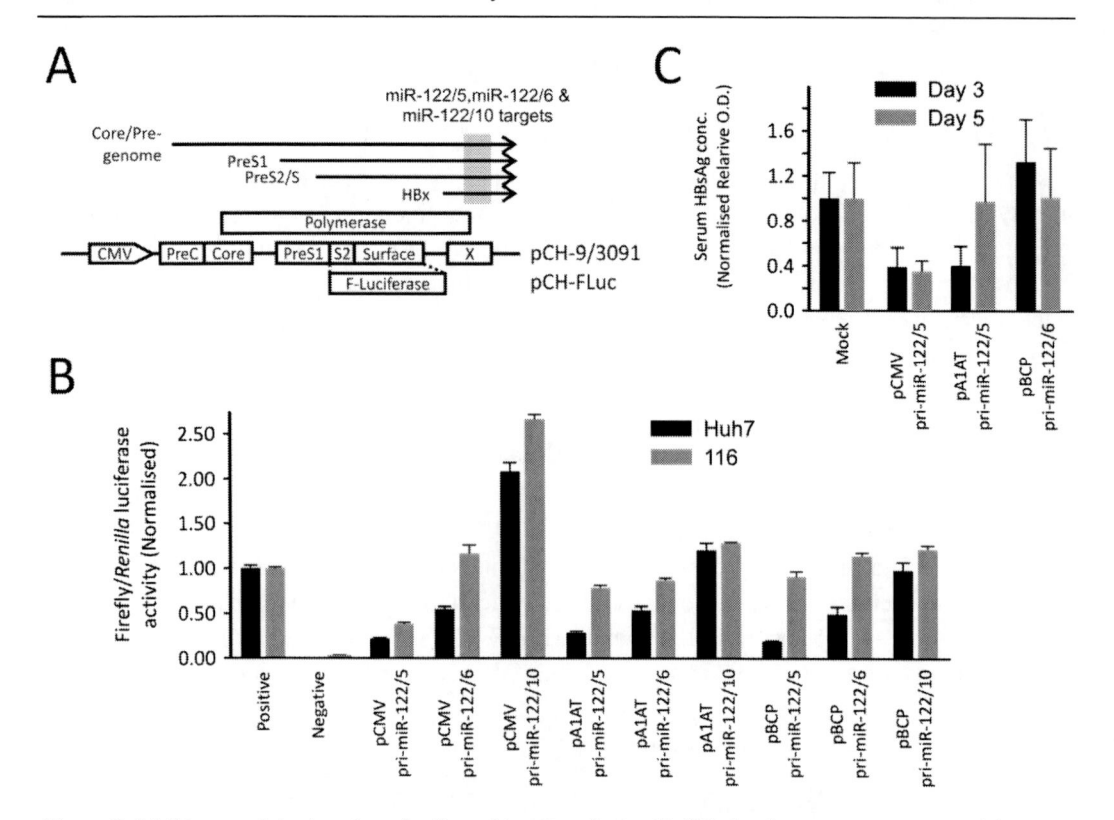

Figure 3. Inhibitory activity in cultured cells and in vivo of pri-miR-122 shuttle sequences expressed from various Pol II transcriptional regulatory elements. A. Organization of the HBV genome with ORFs and sites within the pCH-9/3091 vector that are targets complementary to anti HBV pri-miR-122 shuttles are shown. Four parallel arrows indicate the HBV transcripts, which have common 3′ ends, and include the pri-miR-122/5, pri-miR-122/6 and pri-miR-122/10 targets. The pCH-9/3091-derived pCH-FLuc target vector has the *Firefly luciferase* ORF substituted for the preS2/S HBV sequence. B. Anti HBV efficacy of liver-specific pri-miR-122 shuttles was assessed by co-transfection of Huh7 and 116 cells with pCH-FLuc and the various Pol II expression cassettes. Bars indicate means of relative Firefly to *Renilla* luciferase activities from Huh7 (black bars) and 116 (grey bars) cells. Error bars indicate SEM. C. Silencing of HBV replication *in vivo*. Serum concentration of HBsAg was measured at days 3 and 5 after hydrodynamic injection of mice with pCH-9/3091 and plasmids expressing anti HBV RNAi sequences. Mock injections included the control backbone plasmid containing the CMV promoter without anti HBV RNAi effecters. Each group comprised at least 4 mice and serum was collected 3 and 5 days after injection. HBsAg concentrations were determined by ELISA and mean concentrations of the viral antigen are indicated with SEM.

Pri-miR shuttle sequences driven from the A1AT and HBV core promoters are capable of inhibiting HBV replication in cultured liver-derived cells. Tissue-specific expression achieved by the A1AT and HBV core promoter sequences indicated that these sequences would be useful for expression of anti HBV pri-miR-122 shuttles. To asses this, silencing of the HBV target was initially determined using a reporter gene plasmid (pCH-FLuc) that allowed sensitive and quantitative assessment of HBV targeting in situ. In pCH-FLuc, the preS2/S sequence of pCH-9/3091 [14] was replaced with the Firefly luciferase ORF and the targeted HBx ORF remained intact (Fig. 3A). Results show that pri-miR-122/5 shuttle sequences expressed from the CMV promoter caused knockdown of HBV gene expression in liver- and kidney-derived cell lines (Fig. 3B). However, equivalent knockdown by the A1AT and BCP shuttle vector was only achieved in the liver-derived cell line. Knockdown in the kidney-derived cell line by the tissue-specific pri-miR shuttle vectors was minimal and significantly lower than the silencing achieved by the CMV pri-miR shuttles. Interestingly CMV pri-miR-122/6 did not cause significant silencing in the kidney-derived cell line, which is likely to be a result of lower efficacy of the pri-miR-122/6 sequence. The pri-miR-122/10 shuttle, which is known to be ineffective against HBV [8, 9], was incapable of reporter knockdown irrespective of the promoter from which it was expressed. CMV pri-miR-122/10 seemed to induce HBV expression and is in accordance with previous observations using a similar guide incorporated into U6 short hairpin RNA cassette (U6 shRNA 10) [8]. Collectively these data indicate that anti HBV pri-miR shuttle sequences retain their silencing capabilities when expressed from liver-specific promoters. The silencing activity by A1AT- and BCP-pri-miR-122/5 expression cassettes, which is limited to liver-derived cell lines, prompted investigation of their efficacy in vivo.

A1AT-driven pri-miR shuttle sequences knock down a marker of HBV replication in vivo. Efficacy in vivo of the A1AT- and BCP-driven pri-miR shuttle expression cassettes was assessed using the hydrodynamic injection model of HBV replication. After co injecting animals with an HBV replication competent plasmid together with pri-miR expression cassettes, serum HBsAg concentrations were measured at days 3 and 5 post-injection (Fig. 3C). Assay of this viral antigen is a useful marker of viral replication and may be used to assess HBV knockdown [8]. No silencing of HBV replication was observed in mice that received BCP-driven pri-miR-122 shuttle expression cassettes and may reflect poor activity of this transcriptional control element in murine hepatocytes [20]. Silencing was achieved with pA1AT pri-miR-122/5 and was initially equivalent to that of its CMV promoter-driven counterpart. However, knockdown efficacy with plasmids containing the liver-specific transcription control element to express miR shuttles was not as durable as that achieved with the CMV expression cassettes. This observation may be explained by the relative strengths of the A1AT and CMV promoters. The A1AT promoter is considerably weaker than the CMV promoter (Fig. 2). After hydrodynamic injection, plasmids transfect hepatocytes transiently. With time, a decrease in the intracellular concentration of pA1AT pri-miR-122/5 will result in a lower antiviral RNAi effecter concentration with concomitant diminished antiviral efficacy. Nevertheless this result demonstrates that weaker liver-specific promoters have potential use for therapeutic application and emphasizes the importance of utilizing exogenous RNAi activators that are effective at low concentrations. However, excluding the potentially harmful effects of derangement of the endogenous miR pathway is important.

Liver-specific pri-miR-122 shuttle expression cassettes do not disrupt silencing of a reporter target by endogenous miR-16. To assess whether the liver-specific pri-miR-122 shuttle expression cassettes cause disruption of the miR biogenesis pathway, the previously described miR-16 saturation assay was carried out [10]. This assay is based on the use of miR sponges to derepress miR function [16]. Co-transfection of the control miR-16 sponge plasmid with a plasmid containing a miR-16 downstream of the Renilla *luciferase* ORF (Fig. 4A), caused derepression of reporter activity when compared to mock-transfected cells (Fig. 4B). This result accords with previous observations and verified that the sponge plasmid interfered with endogenous miR-16 silencing of the Renilla *luciferase* target. In contrast, none of the liver-specific pri-miR shuttle sequence expression cassettes derepressed *Renilla* luciferase activity, indicating that they do not disrupt endogenous miR-16 function. This result is not unexpected. The CMV immediate early promoter enhancer is a more powerful transcriptional control element than the liver-specific promoters (Fig. 2), and miR shuttles expressed from the CMV promoter have previously been shown to have little effect on the endogenous miR biogenesis pathway [9, 10]. Lower amounts of pri-miR shuttle sequences should be transcribed from the liver-specific promoters, and are therefore unlikely to outcompete naturally occurring miR for processing by the RNAi processing machinery.

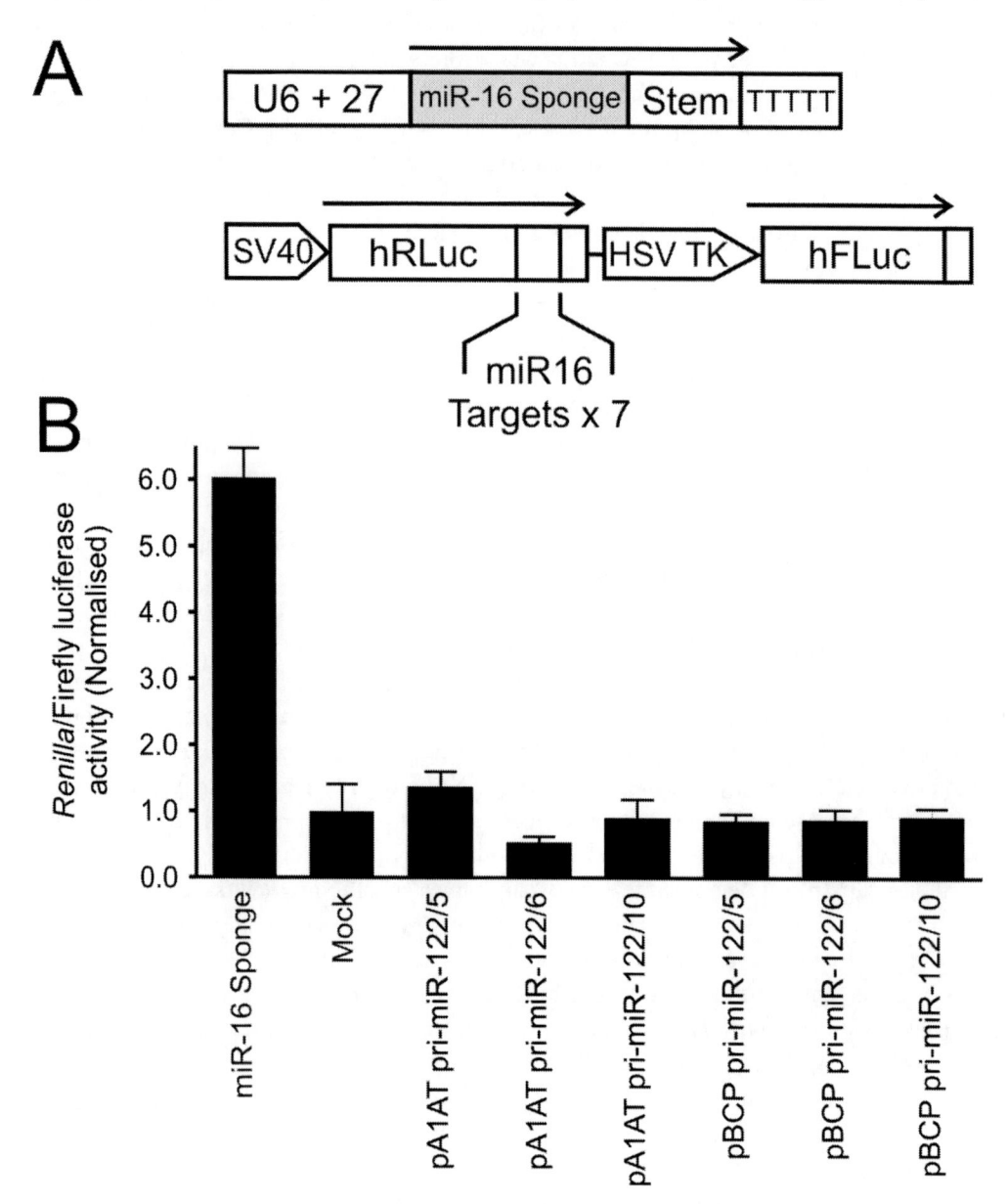

Figure 4. *Assessing saturation of the miR biogenesis pathway by miR-122 shuttles.* A. Schematic representation of the sponge and psiCHECK-miR-16T×7 target plasmids. The dual luciferase reporter target expresses the Renilla *luciferase* transcript with miR-16 target sites within the 3' UTR, and also the untargeted *Firefly luciferase* ORF as normalizing control. B. Huh7 cells were co-transfected with the indicated plasmids and psiCHECK-miR-16T×7 in triplicate and the cells analyzed for luciferase activity 48 hours later. Means of the relative *Renilla* to Firefly luciferase activity are indicated with SEM.

Discussion

Until recently Pol II promoter-regulated RNAi expression cassettes have been characterized by variable efficacy and utility [21-23]. Some of their limitations may be ascribed to difficulties with processing of traditional shRNA sequences that are transcribed from Pol II promoters. The subsequent development of pre- and pri-miR shuttle sequences as RNAi effecters has allowed improved transcription control by exploiting the natural miR biogenesis pathway. pri-miRs are typically transcribed from Pol II regulatory elements and expression of pri-miR shuttle sequences may therefore be subjected to control by these transcription regulatory elements. In support of this, we have previously shown that embedding anti HBV guides within RNAi effecters that mimic naturally occurring pri-miRs enable these sequences to be transcribed from Pol II promoters without compromising silencing activity [9, 10]. In the study reported here, we have extended this observation by demonstrating that miR-122 shuttle sequences may be controlled by liver-specific Pol II promoter elements. However, although efficient and sustained knockdown was achieved in cultured mammalian cells, silencing in vivo was not maintained for longer than 3 days. A likely reason for this observation is that the liver-specific Pol II regulatory elements described here are weaker than the CMV element. Diminished expression of pri-miR shuttles, which accompanies the decreasing number of intrahepatocyte plasmid copies after transient hydrodynamic transfection, is associated with an attenuation of silencing efficacy that is more marked than that observed with the stronger CMV promoter. Improving promoter strength of liver-specific Pol II expression cassettes is therefore a priority of developing this approach. To achieve this, it may be necessary to combine transcription cis elements to optimize promoter strength and liver specificity. By combining a murine alpha-fetoprotein enhancer with a minimal albumin promoter sequence this approach was recently reported to improve expression of an antiviral RNAi activator [24]. The natural propensity for certain miRs to be expressed abundantly in the liver, such as miR-122, indicate that transcription control of these silencing sequences may also be utilized to achieve liver-specific production of anti HBV RNAi activators.

A further refinement of tissue-specific expression of anti HBV miRs would be their production from cassettes that are induced following viral infection. HBV encodes a promiscuous trans activator, HBx, that is capable of indirect induction of transcription through interaction with a variety of trans factors (reviewed in [25]). Incorporating Pol II promoter elements into anti HBV miR expression cassettes that are sensitive and responsive to HBx would be an interesting approach that could further improve regulation of expression of antiviral sequences. An alternative approach is to use regulatory elements that may be controlled by administration or withdrawal of drugs. The recently describe steroid-inducible liver-specific promoter is such an example that could be used to control pri-miR shuttle expression [26].

Tight transcriptional control is important to limit toxic effects of potentially therapeutic exogenous RNAi activators. Unwanted effects, such as silencing of essential cellular sequences and disruption of the natural miR pathway, are more likely to arise as a result of off target and uncontrolled expression of RNAi effecter sequences. To achieve optimal specific silencing, it is also critically important to select RNAi effecter sequences that are potent, that is effective at low concentrations. A therapeutic sequence that is effective against its target at low concentration is less likely to cause unwanted side effects. Expression of potent pri-miR shuttle sequences from regulatable promoters is therefore required to achieve a balance between the required effective concentration of silencing sequences and avoidance of off target effects. The present study advances this approach and further supports the potential therapeutic utility of tissue-specific Pol II-expressed pri-miR shuttle expression cassettes.

Acknowledgments

Work in the authors' laboratory is supported by funding from the Sixth Research Framework Programme of the European Union, Project RIGHT (LSHB-CT-2004-005276), CANSA, the South African National Research Foundation (NRF GUN 68339 and 65495), ESASTAP and the South African Poliomyelitis Research Foundation.

References

[1] WHO. *Hepatitis B virus Fact sheet No. 204 (Revised 2008).* 2000; Available from: *http://www.who.int/mediacentre/factsheets/fs204/en/index.html.*

[2] Lee, W.M., *Hepatitis B virus infection. N Engl J Med*, 1997. 337(24): p. 1733-45.

[3] Arbuthnot, P. and M. Kew, *Hepatitis B virus and hepatocellular carcinoma. Int J Exp Pathol,* 2001. 82(2): p. 77-100.

[4] Chien, Y.C., et al., *Nationwide hepatitis B vaccination program in Taiwan: effectiveness in the 20 years after it was launched. Epidemiol Rev*, 2006. 28: p. 126-35.

[5] Karayiannis, P., *Hepatitis B virus: old, new and future approaches to antiviral treatment. J Antimicrob Chemother*, 2003. 51(4): p. 761-85.

[6] Moolla, N., M. Kew, and P. Arbuthnot, *Regulatory elements of hepatitis B virus transcription. Journal of Viral Hepatitis,* 2002. 9(5): p. 323-331.

[7] Seeger, C., D. Ganem, and H.E. Varmus, *Biochemical and genetic evidence for the hepatitis B virus replication strategy. Science*, 1986. 232(4749): p. 477-84.

[8] Carmona, S., et al., *Effective inhibition of HBV replication in vivo by anti-HBx short hairpin RNAs. Mol Ther*, 2006. 13(2): p. 411-21.

[9] Ely, A., et al., *Expressed anti-HBV primary microRNA shuttles inhibit viral replication efficiently in vitro and in vivo. Mol Ther*, 2008. 16(6): p. 1105-12.

[10] Ely, A., T. Naidoo, and P. Arbuthnot, *Efficient silencing of gene expression with modular trimeric Pol II expression cassettes comprising microRNA shuttles. Nucleic Acids Research*, 2009. 37(13): p. e91.

[11] McGlynn, L.K., et al., *Role of the liver-enriched transcription factor hepatocyte nuclear factor 1 in transcriptional regulation of the factor VIII gene. Mol Cell Biol*, 1996. 16(5): p. 1936-45.

[12] Kalsheker, N., S. Morley, and K. Morgan, *Gene regulation of the serine proteinase inhibitors alpha1-antitrypsin and alpha1-antichymotrypsin. Biochem Soc Trans,* 2002. 30(2): p. 93-8.

[13] Chang, J., et al., *miR-122, a mammalian liver-specific microRNA, is processed from hcr mRNA and may downregulate the high affinity cationic amino acid transporter CAT-1. RNA Biol*, 2004. 1(2): p. 106-13.

[14] Nassal, M., *The arginine-rich domain of the hepatitis B virus core protein is required for pregenome encapsidation and productive viral positive-strand DNA synthesis but not for virus assembly. J Virol,* 1992. 66(7): p. 4107-16.

[15] Passman, M., et al., *In situ demonstration of inhibitory effects of hammerhead ribozymes that are targeted to the hepatitis Bx sequence in cultured cells. Biochem Biophys Res Commun,* 2000. 268(3): p. 728-33.

[16] Ebert, M.S., J.R. Neilson, and P.A. Sharp, *MicroRNA sponges: competitive inhibitors of small RNAs in mammalian cells. Nat Methods,* 2007. 4(9): p. 721-6.

[17] Figueiredo, M.S. and G.G. Brownlee, *cis-acting elements and transcription factors involved in the promoter activity of the human factor VIII gene. J Biol Chem,* 1995. 270(20): p. 11828-38.

[18] Ciliberto, G., L. Dente, and R. Cortese, *Cell-specific expression of a transfected human alpha 1-antitrypsin gene. Cell,* 1985. 41(2): p. 531-40.

[19] Schmidt, E.V., et al., *The cytomegalovirus enhancer: a pan-active control element in transgenic mice. Mol Cell Biol,* 1990. 10(8): p. 4406-11.

[20] 2Rih, J., et al., *The Action of Hepatitis B Virus Enhancer 2-Core Gene Promoter in Non-Viral and Retroviral Vectors for Hepatocyte-Specific Expression. J Biochem Mol Biol,* 1997. 30: p. 269-273.

[21] Peng, Y., J.X. Lu, and X.F. Shen, *shRNA driven by Pol II/T7 dual-promoter system effectively induce cell-specific RNA interference in mammalian cells. Biochem Biophys Res Commun,* 2007. 360(2): p. 496-500.

[22] Yuan, J., et al., *shRNA transcribed by RNA Pol II promoter induce RNA interference in mammalian cell. Mol Biol Rep,* 2006. 33(1): p. 43-9.

[23] Giering, J.C., et al., *Expression of shRNA from a tissue-specific pol II promoter is an effective and safe RNAi therapeutic. Mol Ther,* 2008. 16(9): p. 1630-6.

[24] Snyder, L.L., et al., *Vector design for liver-specific expression of multiple interfering RNAs that target hepatitis B virus transcripts. Antiviral Res,* 2008. 80(1): p. 36-44.

[25] Arbuthnot, P., A. Capovilla, and M. Kew, *Putative role of hepatitis B virus X protein in hepatocarcinogenesis: effects on apoptosis, DNA repair, mitogen-activated protein kinase and JAK/STAT pathways. J Gastroenterol Hepatol,* 2000. 15(4): p. 357-68.

[26] 26.   Wang, L., et al., *Prolonged and inducible transgene expression in the liver using gutless adenovirus: a potential therapy for liver cancer. Gastroenterology,* 2004. 126(1): p. 278-289.

In: Gene Silencing: Theory, Techniques and Applications
Editor: Anthony J.Catalano

ISBN: 978-1-61728-276-8
© 2010 Nova Science Publishers, Inc.

Chapter X

Effective Methods for Selecting siRNA Sequences by Using the Average Silencing Probability and a Hidden Markov Model

Shigeru Takasaki[*]

Toyo University, 1-1-1 Izumino Itakura-machi, Ora-gun Gunma 374-0193, Japan

Abstract

Short interfering RNA (siRNA) has been widely used for studying gene functions in mammalian cells but varies markedly in its gene silencing efficacy. Although many design rules/guidelines for effective siRNAs based on various criteria have been reported recently, there are only a few consistencies among them. This makes it difficult to select effective siRNA sequences in mammalian genes. This chapter first reviews the recently reported siRNA design guidelines and then proposes a new method for selecting effective siRNA sequences from many possible candidates by using the average silencing probability on the basis of a large number of known effective siRNAs. It is different from the previous score-based siRNA design techniques and can predict the probability that a candidate siRNA sequence will be effective. The results of evaluating it by applying it to recently reported effective and ineffective siRNA sequences for various genes indicate that it would be useful for many other genes. The evaluation results indicate that the proposed method would be useful for many other genes. It should therefore be useful for selecting siRNA sequences effective for mammalian genes. The chapter also describes another method using a hidden Markov Model (HMM) to select the optimal functional siRNAs and discusses the frequencies of the combinations for two successive nucleotides as important characteristics of effective siRNA sequences.

[*] Correspondence should be addressed to e-mail: s_takasaki @ toyonet.toyo. ac.jp Phone: +81-276-82-9024, Fax: +81-276-82-9033

Introduction

RNA interference (RNAi) silences gene expression by introducing double-stranded RNA homologous to the target mRNA. It has been widely used for studying gene functions, but many practical obstacles need to be overcome before it becomes an established tool for use in mammalian systems [1–6]. One of the important problems is designing effective siRNA sequences for target genes. The effectiveness of the short interfering RNA (siRNA) responsible for RNA interference varies widely depending on the target sequence positions (sites) selected from the target gene [7,8]. We therefore need useful criteria for gene silencing efficacy when we design siRNA sequences [9,10].

Schwarz *et al.* and Khvorova *et al.* showed that 5' end of the antisense strand might be incorporated into the RNA-induced silencing complex. Strand incorporation may depend on weaker base–pairing, and an A-T terminus may thus lead to more strand incorporation than a G-C terminus [11,12]. Other factors reported to be related to gene silencing efficacy are GC content, point-specific nucleotides, specific motif sequences, and secondary structures of mRNA. Several siRNA design rules/guidelines using efficacy-related factors have been reported [13–17].

Although the effectiveness of siRNA sequences seems to be determined largely by their nucleotide sequences, there are few consistencies among the reported rules/guidelines [18–23]. This implies that they might result in the generation of many candidate sequences, making it difficult to select the effective ones. In addition, the previously reported rules/guidelines cannot estimate the probability that a candidate siRNA will actually silence the target gene. What are therefore needed are not only methods for selecting high-potential siRNA candidates but also methods for estimating the probability that the selected candidates will indeed silence their target genes. Furthermore, there is in RNAi a risk of off-target regulation: a possibility that the siRNA will silence other genes whose sequences are similar to that of the target gene. When we use gene silencing for studying gene functions, we have to first somehow select high-potential siRNA candidate sequences and then eliminate possible off-target ones [24].

This chapter first reviews the recently reported siRNA design guidelines and clarifies their problems. It then describes a prediction method for selecting effective siRNA target sequence from many possible candidate sequences by using the average silencing probability of a large number of siRNA sequences known to be effective. It is quite different from the previous score-based siRNA design techniques and can predict the probability that a candidate siRNA sequence will be effective. The results obtained when applying the method to recently reported effective and ineffective siRNA sequences for various genes showed that it is accurate and thus implies that it would be useful for selecting siRNA sequences silencing many other genes. This chapter also describes another method using a hidden Markov model (HMM) to select the optimal functional siRNAs [25] and discusses the frequencies of the combinations of two successive nucleotides as important characteristics of effective siRNA sequences.

siRNA Sequence Selection Problems

To use RNAi as a biological tool for mammalian cell experiments, we first need to identify target sequences causing gene degradation. They have so far been identified by using a trial-and-error method [3,8], but siRNAs extracted from different regions of the same gene have varied remarkably in their effectiveness. The difficulty of using the trial-and-error method to select target sequences causing gene silencing increases when the coding regions are long, as they are in mammalian cells. This is because the number of candidates increases with the length of the coding region.

The Reported Guidelines for Designing siRNA Sequences

The earliest guidelines for siRNA sequence design were proposed by Elbashir *et al.* [4,8,40]. They suggested that the target mRNA is silenced effectively by siRNA duplexes 21 nucleotides long: 19-nt base-paired sequences with 2-nt overhangs at the 3' ends. Many siRNA design guidelines/rules have been reported since then, and this chapter treats the following five (here designated guidelines G1–G5).

Reynolds *et al.* [18] analyzed 180 siRNAs systematically, targeting every other position of two 197-base regions of firefly luciferase and human cyclophilin B mRNA (90 siRNAs per gene), and reported eight criteria for improving siRNA selection.

Guideline G1:
1. G/C content 30–52%
2. at least 3 As or Ts at positions 15–19
3. absence of internal repeats
4. an A at position 19
5. an A at position 3
6. a T at position 10
7. a base other than G or C at position 19
8. a base other than G at position 13

Ui-Tei *et al.* [19] examined 72 siRNAs targeting six genes and reported four rules for effective siRNA designs.

Guideline G2:
1. an A or T at position 19
2. a G or C at position 1
3. at least five T or A residues from positions 13 to 19
4. no GC stretch more than 9 nt long.

Amarzguioui & Prydz [20] analyzed 46 siRNAs targeting four genes and reported six rules for effective siRNA designs.

Guideline G3:
1. a G or C at position 1
2. an A at position 6
3. a base other than T at position 10
4. a T at position 13
5. a C at position 16
6. an A or T at position 19.

Jagla *et al.* [22] tested 601 siRNAs targeting one exogenous and three endogenous genes and reported four rules.

Guideline G4:
1. an A or T at position 19
2. an A or T at position 10
3. a G or C at position 1
4. more than three A/Ts between positions 13 and 19.

Hsieh *et al.* [21] examined 138 siRNAs targeting 22 genes and reported five position-specific characteristics:

Guideline G5:
1. a T at position 19
2. a C or G at position 11
3. a G at position 16
4. an A at position 13
5. a base other than C at position 6

These guidelines are summarized in Table 1.

Table 1. Effective and ineffective nucleotides specified in the individual guidelines

	Position	1	3	6	10	11	13	16	19
G1	effective		A		T		A/C/T		A/T
G2	effective	G/C							A/T
	ineffective	A/T							G/C
G3	effective	G/C		A			T	C	A/T
	ineffective	T			T				G
G4	effective	G/C			A/T				A/T
G5	effective					C/G	A	G	T
	ineffective			C		A/T			G

Position: nucleotide position from 1 to 19 (5' to 3', cDNA form).
Effective: preferred, ineffective: unpreferred.

Other methods for scoring, screening, and designing functional siRNAs have also been reported recently. Chalk *et al.* [13] reported the following seven rules ("Stockholm rules") based on thermodynamic properties: (1) total hairpin energy < 1, (2) antisense 5' end binding energy < 9, (3) sense 5' end binding energy in range 5 – 9 exclusive, (4) GC between 36% and 53%, (5) middle (7 – 12) binding energy < 13, (6) energy difference < 0, (7) energy difference between −1 and 0. The score of an siRNA candidate is incremented by one for each rule fulfilled and is thus between 0 and 7.

Huesken *et al.* [23] reported a method for screening functional siRNAs by using an artificial neural network. This network was first trained by 2182 randomly selected siRNAs targeted to 34 genes and was used in the design of a genome-wide siRNA collection with two potent siRNAs per gene.

Teramoto *et al.* [14] and Ladunga [34] have reported functional siRNA selection methods using support vector machines (SVMs). Teramoto *et al.* used a generalized string kernel (GSK) combined with a SVM. siRNA sequences were represented as vectors in a multidimensional feature space according to the numbers of subsequences in each siRNA and were classified as effective or ineffective [14]. Ladunga used a SVM with polynomial kernels and constrained optimization models from 572 sequence, thermodynamic, accessibility, and self-hairpin features over 2200 published siRNAs [23,34]. As the key to SVM success is to collect many useful features of effective siRNA sequences, the usefulness of methods using SVMs may depend on the selected siRNAs.

Holen recently reported siRNA rules based on apparent overrepresentation or underrepresentation of certain nucleotides in certain positions of Novartis data set [35]. The criteria for an siRNA candidate depend on the positive and negative scores computed for each position by using scoring table generated by the percentage overrepresentation or underrepresentation of individual nucleotides for each position in the large Novartis data set [23]. Although the method was evaluated by using other reported siRNA sets, which of the candidate siRNAs actually silence genes is not clear. In addition, as the original scores in the scoring table are based on the percentage overrepresentation or underrepresentation of certain nucleotides in certain positions and thus may vary drastically depending on what sets of siRNAs are used. This makes it difficult to evaluate the scores computed for siRNA candidates.

Although secondary structures of siRNA sequences are also thought to be important in predicting siRNA efficacy, there are conflicting results concerning the effects of secondary structures on siRNA functionality. Some studies have suggested that the secondary structure of the siRNA plays a role in determining the efficacy of gene silencing [37–39], but others did not find any correlation between the functionality of the siRNA and the secondary structures of the target mRNA [7,18,20]. This issue therefore requires further study.

The above techniques have also been used to obtain other design rules [42–45], and the features of various siRNA design rules are summarized in Table 2.

Table 2. Features of individual siRNA design rules

	No. of genes	No. of siRNAs	Description	Technique
Reynolds et al. [18]	2	197	Sequence features	
Ui-Tei et al. [19]	6	72	Sequence features	
Amarzguioui et al. [20]	4	46	Sequence features	
Hsieh et al. [21]	22	138	Sequence features	
Hesken et al. [23]	34	2128	Sequence motifs	Neural network
Jagla et al. [22]	4	601	Sequence features	Decision tree
Holen [35]	34	400	Sequence features	Percentage
Saetrom [36]	40	581	Sequence motifs	Genetic programming
Teramoto et al. [14]	2	94	Sequence motifs	Support vector machine
Ladunga [34]	34	2252	Position features	Support vector machine
Calk et al. [13]	92	398	binding energy	Regression tree
Takasaki et al. [30–32]	490	833	Sequence features	Statistics, SOM, RBF

Problems with the Previous Guidelines

Among the problems with the reported guidelines is the problem of inconsistencies with regard to the nucleotide frequencies of each position. Although some guidelines have the same preferred and unpreferred nucleotides at positions 1 and 19, there are few consistencies at other positions (Table 1). These results indicate that though some rules from the guidelines are suitable for getting effective sequences for some genes, they might be unsuitable for others. Since the previous guidelines are based on the analyses of specific genes, it could be inferred that they are not always effective for many other genes. Therefore if these guidelines were used to select sequence candidates for other mammalian genes, many sequences might be selected as candidates. This is because there are mostly long coding regions in mammalian genes but there are only a few consistencies among the previous guidelines. As a result, many candidate sequences might be selected. This is another problem because experimentally evaluating whether the selected sequences provide effective gene degradation is a costly and time-consuming task. To overcome the problems of the previous guidelines, Takasaki et al. recently reported new scoring methods using the statistical and clustering techniques listed in Table 2 [28–32].

Still another problem is that the previously reported methods cannot estimate the probability that a candidate siRNA will actually silence the target gene. Even if a high-scored siRNA were obtained using the reported methods, it would be difficult to estimate the probability that it would actually accomplish the expected gene degradation.

Methods and Materials

Definition of the siRNA Sequence Selection Problem

The problem of selecting target siRNA sequences is to predict whether or not a candidate siRNA sequence for the target mRNA (typically 19 nucleotides $X=X_1X_2......X_{19}$ where X_i is the i-th nucleotide) will result in effective gene silencing: for example, more than 80% silencing at the protein level. The problem of selecting siRNA sequences can therefore be transformed into the problem of finding the degree of gene silencing functionality of a given siRNA candidate X. If individual gene reduction degrees of siRNA candidates were obtained, it might be easy to decide what candidates are appropriate as siRNAs.

It is hypothesized that the evaluation of candidates can be based on the analyses of nucleotide occurrence features at individual positions in the reported effective siRNAs. This is because the effectiveness of siRNA sequences greatly depends on individual nucleotides of the sequences [18–22].

Prediction Analysis Based on the Average Silencing Probability

Many effective gene-silencing siRNA sequences have been reported recently and can be use to predict how new siRNA candidates will function. If the probability of individual nucleotide occurrences at positions from 1 to 19 in the effective siRNA population is obtained, it can be used to calculate the probability that candidate siRNAs will be effective. In addition, if the average probability of a large number of effective siRNA sequences is computed, it could be considered a measure of the potential effectiveness of siRNAs and used to evaluate whether or not a candidate siRNA is likely to silence its target gene. If the probability of the candidate siRNA candidate were greater than the average probability of a large number of effective siRNAs, it would indicate a high likelihood of gene-silencing. To calculate this measure, 833 effective siRNA sequences reported in the literature (PubMed) were collected and nucleotide occurrences at positions from 5' to 3' in the cDNA were summarized. The probability of individual nucleotide occurrence frequencies f_p^N at individual positions can be computed as follows:

$$f_p^N = \frac{\sum_{i=1}^{I}\{A,G,C,T\}}{I}, \tag{1}$$

where N is the kind of nucleotide (A, G, C, or T), p is the position in the cDNA (1, 2, …,19 from 5' to 3'), and I is the number of the effective siRNA sequences (e.g., 833).

Then the probability OF_i of each effective siRNA sequence is calculated in the following way.

$$OF_i = \prod_{p=1}^{19} f_{ip}^N ,$$

(2)

where i is the sequence identification number of the effective siRNAs (i.e., i=1, 2, …, I).

The average sequence probability A_E for the effective siRNAs is therefore computed as follows:

$$A_E = \frac{\sum_{i=1}^{I} OF_i}{I}$$

(3)

A_E could be considered a criterion for candidate siRNA. That is, if the probability of the candidate sequence were greater than A_E, it would indicate a high likelihood of gene-silencing. On the other hand, if the probability of the candidate sequence were remarkably lower than A_E, it would indicate a low likelihood of effectiveness.

siRNA Sequence Selection Based on a Hidden Markov Model

As an siRNA sequence X basically consists of 19 nucleotides, it can be described as X=$X_1 X_2 ……X_{19}$, where X_i indicates the nucleotide A, C, G, or T. Furthermore, this sequence can be expressed as state diagrams of nucleotides A, C, G, and T from the positions 1 to 19 shown in Figure 1. As shown in Figure 1, if the state at position 1 is, for example, the nucleotide C, it can be transmitted to all the states A, C, G, or T at position 2. Likewise, these nucleotide state transitions proceed from the positions 1 to 19. In relations between the state diagrams (top) and the frequency ratios (bottom) as shown in Figure 1, although what states are allocated to the individual positions of effective siRNA sequences are unknown in the intermediate processes, the ratios of the individual nucleotide occurrences are obtained as shown in the bottom of Figure 1. Therefore, the transmission of the individual nucleotides A, C, G, and T from the positions 1 to 19 can be considered a hidden Markov process. If the state diagrams of effective siRNAs were expressed as a hidden Markov model (HMM), the optimal states (nucleotides) for maximizing the state transition probability could be solved as a decoding problem by using the Viterbi algorithm.

Viterbi Algorithm for Selection of the Optimal siRNA Nucleotide

The Viterbi algorithm for selecting the optimal siRNA sequence is expressed as follows:

1. Initialization for individual states i=A, C, G, T.

$$\delta_1 = \prod_i b_i(O_1)$$

(4)

$$\varphi_1(i) = 0$$

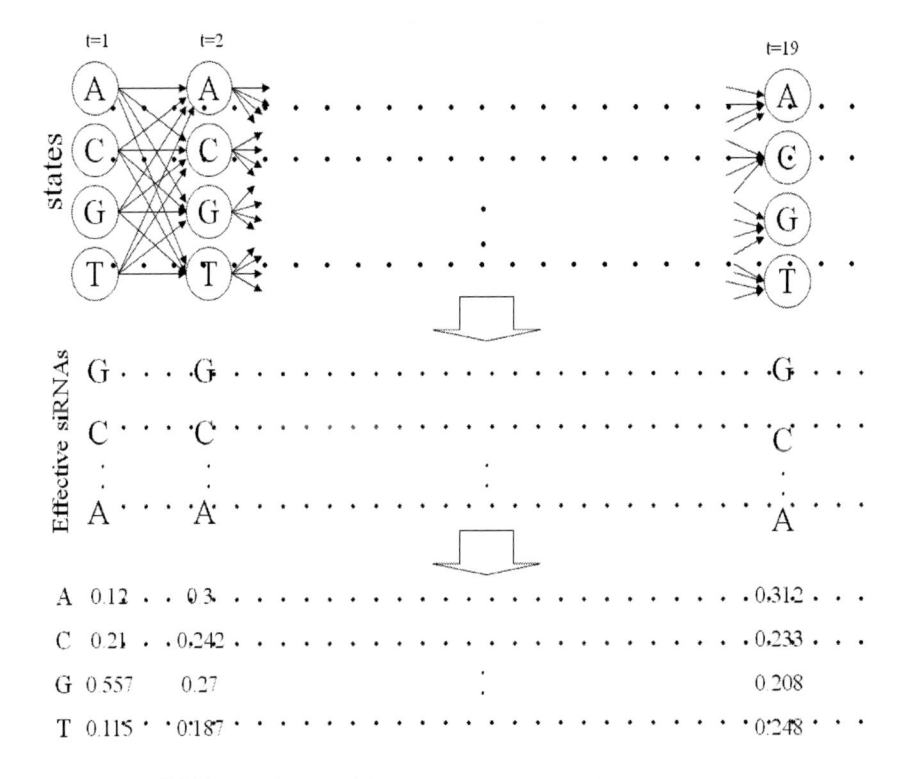

Figure 1. State diagram of hidden Markov model.
A set of 833 effective siRNA sequences from the literature is shown in the middle part of the Figure 1, and the frequency ratios of individual nucleotides at each position in those sequences are shown in the bottom of Figure 1. The ratios of the nucleotides A, C, G, and T at position 2, for example, are respectively 250/833 (=0.3), 202/833 (=0.242), 225/833 (=0.27), and 156/833 (=0.187).

where $\prod_i$ is initial state probability distribution for the state i and

$b_i(o_1)$ is output of the state i at the sequence position 1.

2. Recursive computations for the sequence positions $t = 1, 2,..., 18$ and the individual states $j = $ A, C, G, T.

$$\delta_{t+1}(j) = \max_i \left(\delta_t(i) a_{ij} \right) b_j(O_{t+1}) \tag{5}$$

$$\varphi_{t+1}(j) = \arg \max_i \left(\delta_t(i) a_{ij} \right) \tag{6}$$

where a_{ij} is the state transition probability from the state i to j.

3. Termination of the recursive computations.

$$\hat{P} = \max_i \delta_{19}(i) \tag{7}$$

$$\hat{q}_{19} = \arg \max_{i} \delta_{19}(i) \qquad (8)$$

4. Optimal state generation for sequence positions $t = 18, 17, ..., 1$.

$$\hat{q}_t = \varphi_{t+1}(\hat{q}_{t+1}) \qquad (9)$$

Nucleotide Occurrence Models

Two types of nucleotide occurrences from positions 1 to 19 were assumed. One is that the nucleotides occur independently at individual positions as listed in Table 3(a), and the other is that the occurrence of individual nucleotides at individual positions depends on the nucleotides at other positions. A typical occurrence dependency is, for example, the Markov chain dependency (the simple (first) Markov model). That is, the nucleotide occurrences at the present position depend on the nucleotides at the previous position. The probability of the simple Markov model for the nucleotide at the present position i (i=2, 3, ..., 19) is determined under the condition of the effective nucleotide at the previous position i-1 as listed in Table 3(b). Suppose, for example, that we have the sequence GACTCAACACCGAGTTCAA as a candidate siRNA sequence for some target gene. In this case, the probability of the second nucleotide being A may depend on the first nucleotide being G. From Table 3(b) (the simple Markov Model table) one sees that the probability that A occurs at position 2 under the condition that there is a 0.557 probability that G occurs at position 1 is 0.33. One similarly sees that the probability that the third nucleotide is C is 0.192 under the condition that there is a 0.3 probability that the second nucleotide is A.

Evaluation Criteria Used in the Proposed Method

To make the results estimated using the average silencing probability easily understood, the ratio of the result estimated for a new siRNA candidate to the average sequence probability A_E is considered. This is because the results estimated for the known effective siRNAs could be considered a standard criterion for candidate new siRNAs. This normalized ratio NR is therefore defined as follows:

$$NR = \frac{ER}{A_E}, \qquad (10)$$

where ER is the result estimated by the average silencing probability method and A_E is the average of the probabilities predicted for the known effective siRNAs.

Table 3. Probabilities of individual nucleotide occurrences at each position

(a) Probabilities of independent nucleotide occurrences in 833 effective siRNAs

	1	2	3	4	5	6	7	8	9	10	11	12	13	14	15	16	17	18	19
A	0.12	0.3	0.318	0.218	0.271	0.294	0.247	0.298	0.271	0.229	0.25	0.283	0.259	0.282	0.27	0.232	0.288	0.313	0.312
G	0.557	0.27	0.229	0.291	0.262	0.239	0.304	0.253	0.224	0.257	0.279	0.247	0.255	0.276	0.242	0.287	0.24	0.239	0.208
C	0.208	0.242	0.208	0.294	0.248	0.208	0.263	0.229	0.244	0.275	0.256	0.234	0.247	0.217	0.202	0.251	0.235	0.178	0.233
T	0.115	0.187	0.245	0.197	0.218	0.259	0.186	0.22	0.261	0.239	0.216	0.235	0.239	0.224	0.286	0.23	0.236	0.27	0.248

(b) Probabilities of dependent nucleotide occurrences – the simple Markov model.

First	P_1	Nuc	1'–2	P_2	2'–3	P_3	3'–4	P_4	4'–5	P_5	5'–6	P_6	6'–7	P_7	7'–8	P_8	8'–9	P_9	9'–10
A	0.12	A	0.16	0.3	0.328	0.318	0.2	0.218	0.247	0.271	0.292	0.294	0.229	0.247	0.282	0.298	0.278	0.271	0.204
		G	0.35		0.272		0.336		0.341		0.265		0.363		0.311		0.286		0.323
		C	0.28		0.192		0.264		0.22		0.212		0.257		0.214		0.254		0.257
		T	0.21		0.208		0.2		0.192		0.23		0.151		0.194		0.181		0.217
G	0.557	A	0.33	0.27	0.364	0.229	0.241	0.291	0.314	0.262	0.394	0.239	0.342	0.304	0.316	0.253	0.313	0.224	0.337
		G	0.284		0.169		0.293		0.194		0.206		0.241		0.241		0.218		0.193
		C	0.213		0.227		0.251		0.26		0.22		0.246		0.213		0.223		0.203
		T	0.172		0.24		0.215		0.231		0.179		0.171		0.229		0.246		0.267
C	0.208	A	0.358	0.242	0.371	0.208	0.277	0.294	0.331	0.248	0.329	0.208	0.295	0.263	0.37	0.229	0.33	0.244	0.261
		G	0.116		0.149		0.127		0.171		0.13		0.185		0.137		0.131		0.192
		C	0.295		0.198		0.318		0.245		0.203		0.243		0.224		0.199		0.271
		T	0.231		0.282		0.277		0.253		0.338		0.277		0.269		0.34		0.276
T	0.115	A	0.198	0.187	0.167	0.245	0.172	0.197	0.146	0.218	0.137	0.259	0.144	0.186	0.187	0.22	0.153	0.261	0.134
		G	0.396		0.353		0.368		0.409		0.368		0.389		0.361		0.246		0.304
		C	0.25		0.218		0.353		0.268		0.192		0.301		0.284		0.301		0.359
		T	0.156		0.263		0.108		0.177		0.302		0.167		0.168		0.301		0.203

P_{10}	10'–11	P_{11}	11'–12	P_{12}	12'–13	P_{13}	13'–14	P_{14}	14'–15	P_{15}	15'–16	P_{16}	16'–17	P_{17}	17'–18	P_{18}	18'–19
0.229	0.267	0.25	0.274	0.283	0.195	0.259	0.245	0.282	0.264	0.27	0.218	0.232	0.249	0.288	0.3	0.313	0.3
	0.356		0.284		0.314		0.329		0.306		0.329		0.275		0.288		0.235
	0.257		0.25		0.263		0.236		0.179		0.258		0.218		0.154		0.231
	0.12		0.192		0.229		0.19		0.251		0.196		0.259		0.254		0.235
0.257	0.271	0.279	0.358	0.247	0.364	0.255	0.373	0.276	0.33	0.242	0.267	0.287	0.285	0.24	0.37	0.239	0.387
	0.271		0.263		0.204		0.274		0.226		0.231		0.238		0.2		0.166
	0.285		0.19		0.204		0.193		0.165		0.249		0.247		0.185		0.196
	0.173		0.19		0.228		0.16		0.278		0.151		0.23		0.245		0.236
0.275	0.306	0.256	0.305	0.234	0.282	0.247	0.325	0.217	0.331	0.202	0.222	0.251	0.421	0.235	0.352	0.178	0.412
	0.131		0.155		0.169		0.146		0.16		0.107		0.148		0.133		0.108
	0.262		0.239		0.241		0.214		0.204		0.187		0.167		0.204		0.23
	0.301		0.3		0.308		0.316		0.304		0.231		0.263		0.306		0.23
0.239	0.146	0.216	0.172	0.235	0.204	0.239	0.181	0.224	0.144	0.286	0.151	0.23	0.188	0.236	0.228	0.27	0.183
	0.382		0.294		0.321		0.357		0.262		0.396		0.307		0.325		0.272
	0.216		0.267		0.281		0.226		0.273		0.236		0.313		0.173		0.263
	0.256		0.267		0.194		0.236		0.321		0.276		0.193		0.274		0.277

Table 3. (Continued)

(c) Probabilities of independent nucleotide occurrences in 847 ineffective siRNAs

	1	2	3	4	5	6	7	8	9	10	11	12	13	14	15	16	17	18	19
A	0.312	0.247	0.211	0.247	0.254	0.231	0.273	0.26	0.235	0.295	0.251	0.243	0.226	0.235	0.203	0.261	0.262	0.182	0.084
G	0.185	0.229	0.256	0.262	0.237	0.259	0.236	0.254	0.286	0.279	0.231	0.257	0.323	0.266	0.293	0.26	0.255	0.301	0.319
C	0.215	0.253	0.296	0.285	0.266	0.321	0.283	0.253	0.298	0.242	0.269	0.256	0.247	0.244	0.289	0.269	0.262	0.319	0.426
T	0.288	0.272	0.236	0.207	0.243	0.189	0.208	0.234	0.182	0.184	0.248	0.243	0.204	0.255	0.215	0.21	0.221	0.198	0.171

This NR therefore indicates the gene-silencing potential of the siRNA candidates relative to that of the known effective siRNAs. If $NR \geq 1$, the level of gene silencing expected to be obtained with the siRNA candidate is the same as or higher than level of silencing obtained with the known effective siRNAs. That is, NR indicates that the candidate sequence is likely to silence its target gene. If, on the other hand, $NR < 1$, the gene silencing expected to be obtained with the candidate sequence is lower than the level of silencing obtained with the known effective siRNAs.

Evaluation and Model Generation Data

The recently reported effective and ineffective siRNAs were used as the evaluation data. They are respectively 25 effective and 25 ineffective sequences for human *cyclophilin B* [18]; 38 effective and 24 ineffective sequences for *firefly luciferase (PRL-TK), vimentin, Oct 4, EGFP, ECFP,* and *DsRed* [19]; 21 effective and 25 ineffective sequences for *hTF, mTF, PSK,* and *CSK* [20]; 7 effective and 7 ineffective sequences for the *cyclin B1* [28]; and 12 effective and 12 ineffective sequences for *TC10, UBE2I,* and *CDC34* [23]. These sets of genes are respectively symbolized throughout the present study as MG1, MG2, MG3, MG4, and MG5.

Two kinds of known effective siRNA sequences were used for obtaining frequency ratios of individual nucleotides. One was 833 effective siRNA sequences from 490 different cDNAs in the published references of the PubMed database [5,6,29] and the other was the 636 top-ranked effective siRNA sequences (normalized inhibitory activity >0.832) from 34 genes [23]. 833 effective siRNAs were used to calculate a standard criterion for the effectiveness of siRNAs. The evaluation data were not included in the 833 effective siRNAs. Since it is difficult to select many known ineffective siRNAs from many different genes, the 847 worst-ranked siRNAs (normalized inhibitory activity <0.612) from Huesken *et al.* [23] were used as ineffective siRNA sequences.

Results and Discussion

The proposed method was evaluated by first computing A_E for 833 effective siRNA sequences and then using equation (2) to compute the individual probabilities of the effective and ineffective siRNAs. Since there were ups and downs in the individual ratios of the effective and ineffective siRNAs, the average of them were calculated. The relations between the normalized average ratios of the effective and ineffective siRNAs for the recently reported genes are shown in Figure 2.

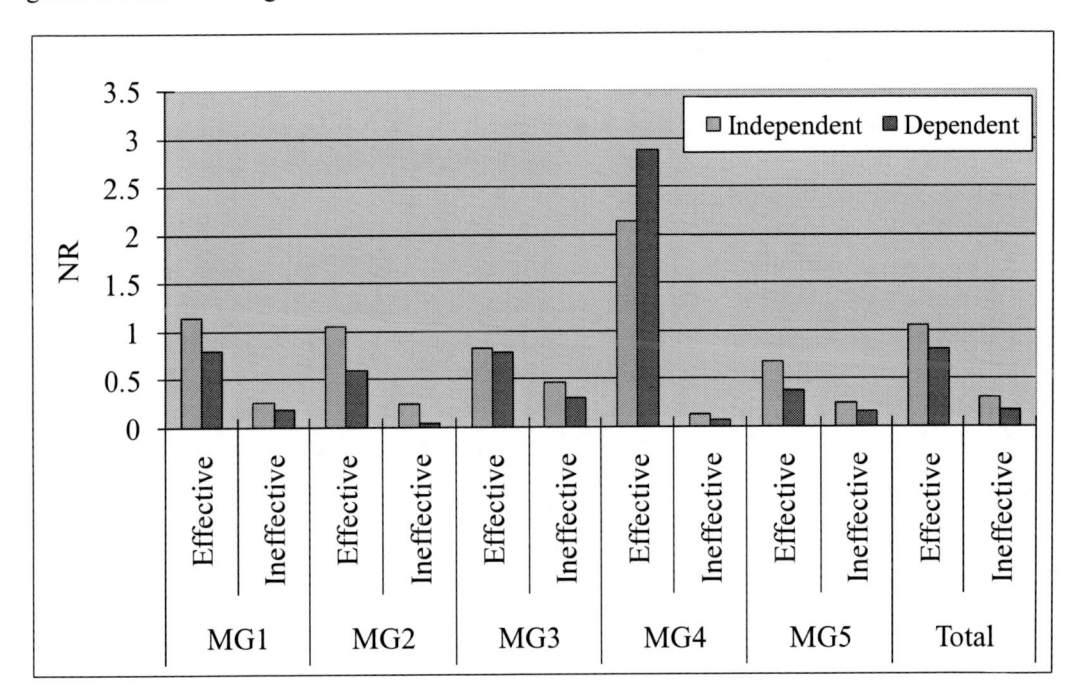

Figure 2. Normalized ratios based on 833 effective siRNAs.

Effective: effective siRNAs. Ineffective: ineffective siRNAs. NR: Normalized ratio calculated by equation (10). Independent: independent occurrences at individual positions, Dependent: the simple Markov model.

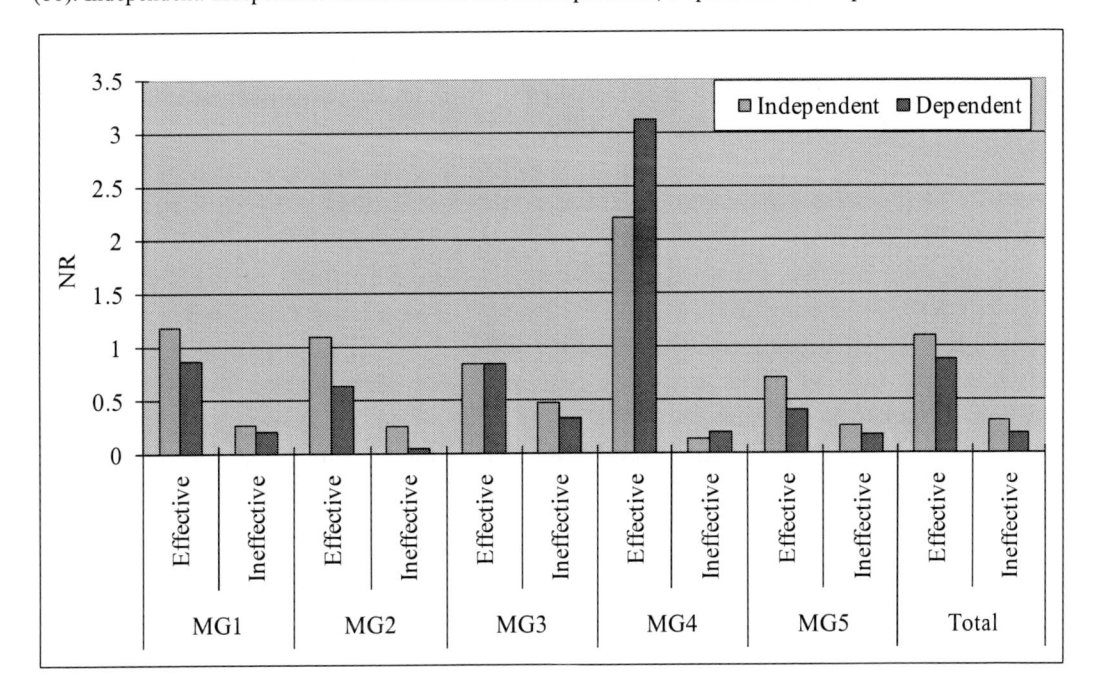

Figure 3. Normalized ratios based on 636 effective siRNAs.

Evaluation Using Nucleotide Frequencies Based on 833 Effective siRNAs

Case 1: Independent Nucleotide Occurrences at Individual Positions

The average normalized ratio *NR* for the MG1 effective siRNAs was 1.14, whereas that for the ineffective ones was 0.26. This indicates that as the *NR* for the sequences of MG1 effective siRNAs are 1.14 times higher than that for the 833 effective siRNAs, it shows the higher level potential in gene-silencing. On the other hand, as the *NR* for the sequences of MG1 ineffective ones shows 0.256 times, i.e., one-fourth compared to the *NR* for the 833 effective siRNAs, it implies one-fourth (low) level potential of gene-silencing. Since the average normalized ratios for MG2 effective and ineffective siRNAs were respectively 1.06 and 0.25, they indicate a similar tendency of MG1. In contrast, the average normalized ratios for MG3 effective and ineffective siRNAs were respectively 0.82 and 0.46. These results indicate that there is no big difference between them (compared to the MG1 and MG2 effective and ineffective siRNAs). That is, the nucleotide frequency characteristics of MG3 effective siRNAs resemble those of MG3 ineffective siRNAs. Although the ratios of the average effective-to-ineffective ratios for MG1 and MG2 are respectively 4.45 (1.14/0.26) and 4.08 (1.06/0.25), the average effective-to-ineffective ratio for MG3 is 1.78 (0.82/0.46). Since the average normalized ratios of MG4 effective and ineffective siRNAs were respectively 2.14 and 0.13, the ratio of the effective to ineffective siRNAs was 16.5 (2.14/0.13). The *NR* of the effective siRNAs for MG4 therefore implies a high likelihood (2.2 times) of gene-silencing compared to that of the 833 effective siRNAs, whereas the *NR* of the ineffective ones show quite low likelihood (0.13 times). On the other hand, since the normalized ratios for MG5 were respectively 0.68 and 0.24, the ratio of the effective and ineffective siRNAs was 2.83. The entire normalized ratio that effective siRNAs for MG1 to MG5 would be effective was 1.06, whereas the entire normalized ratio that the ineffective ones would be effective was 0.297. These evaluation results for the independent nucleotide occurrences indicate that the proposed prediction method based on the effective siRNA sequences is useful for selecting candidate siRNAs for target genes.

Case 2: Dependent Nucleotide Occurrences Based on the Simple Markov Model

As shown in Figure 2, in Case 2 as a whole the average normalized ratios of the effective and ineffective siRNAs for MG1, MG2, MR3, and MG5 were lower than those in Case 1. In contrast, the normalized ratio of MG4 effective siRNAs was higher than that in Case 1 and the normalized ratio of MG4 ineffective ones was lower than that in Case 1. There is, however, a similar tendency in the ratios of the effective-to-ineffective average ratios for MG1, MG2, MG3, and MG5. The average normalized ratio of the MG1 effective siRNAs was 0.79, whereas that of the ineffective ones was 0.19. The average ratio of the effective siRNAs is thus about four times larger than that of the ineffective ones. As the average normalized ratios of the MG2 effective and ineffective siRNAs were respectively 0.59 and 0.04, the average ratio of the effective siRNAs was about 14 times larger than that of the ineffective ones. On the other hand, the average normalized ratios of the MG3 effective and ineffective siRNAs were respectively 0.77 and 0.31. Although the average ratio of the effective siRNAs was only about 2.5 times larger than that of the ineffective ones and this ratio was lower than the corresponding ratios for the MG1 and MG2 siRNAs, there was still a clear difference between the average normalized ratios of the MG3 effective and ineffective siRNAs. Similarly, the normalized ratios of the MG5 effective and ineffective siRNAs were respectively 0.37 and 0.15. Therefore the average ratio of effective siRNAs was approximately 2.5 times larger than that of the ineffective ones. On the other hand, since the average normalized ratios of MG4 (cyclin B1) effective and ineffective siRNAs were respectively 2.88 and 0.08, the difference between them was a remarkably large (36-fold). The *NR* of the effective siRNAs for MG4 therefore indicated the higher likelihood of gene-silencing compared to that of the 833 siRNAs, whereas the *NR* for the ineffective ones showed the quite low likelihood. These evaluation results for the dependent nucleotide occurrences based on the simple Markov model indicate that the proposed prediction method is useful for selecting candidate siRNAs for target genes.

Evaluation Using another Large Number of Known siRNAs

Gene-silencing probabilities were also evaluated using the nucleotide frequencies at individual positions in 636 other effective siRNAs. The independent (Case 1) and dependent (Case 2) nucleotide frequencies at individual positions for other 636 effective probabilities that the effective and ineffective siRNAs would be effective for the reported genes are shown in Figure 3. Although there were ups and downs in *NRs* predicted for MG1 to MG5 using either the 833 or 636 effective siRNAs, the total *NRs* predicted are similar for both cases. That is, the *NRs* based on the 833 effective siRNAs are respectively 1.06 and 0.81 for the independent and dependent cases, and those based on the 636 effective siRNAs are respectively 1.1 and 0.87 for the independent and dependent cases. This implies that the proposed method using the average silencing probabilities could be useful for many other genes.

Evaluation for the HMM

The Viterbi algorithm was carried out for the state diagram of the HMM shown in Figure 1. As a result, the siRNA sequence GAAGAAGAGAGAGAGCAGA was obtained as the optimal nucleotide sequence (i.e., the sequence maximizing the sequence state probability for positions 1 to 19). This result also indicates that the nucleotides G and A might dominate the optimal sequence in reported sets of effective siRNAs.

Table 4. Relations between the maximized and minimized nucleotide sequences and the upper- and lower-level significant nucleotides

	1	2	3	4	5	6	7	8	9	10	11	12	13	14	15	16	17	18	19
Upper-level	G	A	A/T	C/G	C	A/T	G/C	A	C/G	C	C	A/C		A	T/A	T/C	A	T/A	A
Maximized	G	A	A	G	A	A	G	A	G	A	G	A	G	A	G	C	A	G	A
Coincidence	=	=	=	=		=	=	=	=			=		=		=	=		=
Lower-level	A/T	T	G	T/A	T	G/C	T/A	C	G	A/G	T			C	G/C	A	G	C/G	G
Minimized	T	T	T	T	T	A	T	T	A	A	T	C	G	C	G	T	T	C	G
Coincidence	=	=		=	=		=			=	=			=	=			=	=

It is also possible to select individual positional nucleotides for minimizing the sequence state probability. This was done by using the modified Viterbi algorithm, i.e., by changing from maximum to minimum in the equations (4) to (9), and yielded the sequence TTTTTATTAATCGCGTTCG. From the point of gene-silencing by siRNA sequences, the optimal maximized sequence may correspond to the most preferable siRNA sequence in a large number of effective siRNAs. On the other hand, the minimized sequence may correspond to the least preferable one in a large number of effective siRNAs.

These maximized and minimized nucleotide sequences were then compared with the upper- and lower-level significant nucleotides obtained using the previously proposed statistical significance testing for 833 effective siRNA sequences [30]. One sees in Table 4 that the maximized nucleotide obtained using the Viterbi algorithm corresponds to the upper-level nucleotides obtained using the significance testing, and the minimized nucleotide sequence corresponds to the lower-level one obtained using the significance testing. Interestingly, there are many coincidences between the maximized and minimized nucleotides and the upper- and lower-level significant nucleotides. Between the maximized nucleotides and the upper-level ones there are thirteen coincidences (at positions 1, 2, 3, 4, 6, 7, 8, 9, 12, 14, 16, 17, and 19), and between the minimized nucleotides and the lower-level ones there are eleven coincidences (at positions 1, 2, 4, 5, 7, 10, 11, 14, 15, 18, and 19). There are six coincidence positions in both relations: at positions 1, 2, 4, 7, 14, and 19. The positions 1, 2, and 19 correspond to around the 5' and 3' terminal points. This implies that these positions play important roles in gene-silencing.

Evaluation for MG1 to MG5 Based on a Large Number of Ineffective siRNAs

It is also possible to clarify the probability of how siRNA candidates are effective on the basis of a large number of ineffective siRNAs. 847 known siRNAs were selected as ineffective ones (see Methods and Materials). The probabilities of individual nucleotide occurrence frequencies at individual positions are listed in Table 3(c). The relations among NRs of effective and ineffective siRNAs for MG1 to MG5 computed by using the equations (2), (3), and (10) are shown in Figure 4. In the case of using the 847 known ineffective siRNAs, NRs of effective siRNAs for MG1 to MG5 are less than 1, whereas those of ineffective ones are more than 1 as shown in Figure 4. The NR of the total effective siRNAs is 0.67, whereas that of the ineffective ones is 2.37.

Comparing Figure 4 with Figure 2, it is clear that the corresponding NRs of effective and ineffective siRNAs for MG1 to MG5 are respectively reverse relations. This depends on what set of siRNAs, i.e., 833 or 847 siRNAs, is used. There are differences in the nucleotide occurrence frequencies between both sets of siRNAs as shown in Figure 5. Especially, there are big differences at positions 1 and 19. These results are also useful for designing effective siRNA sequences.

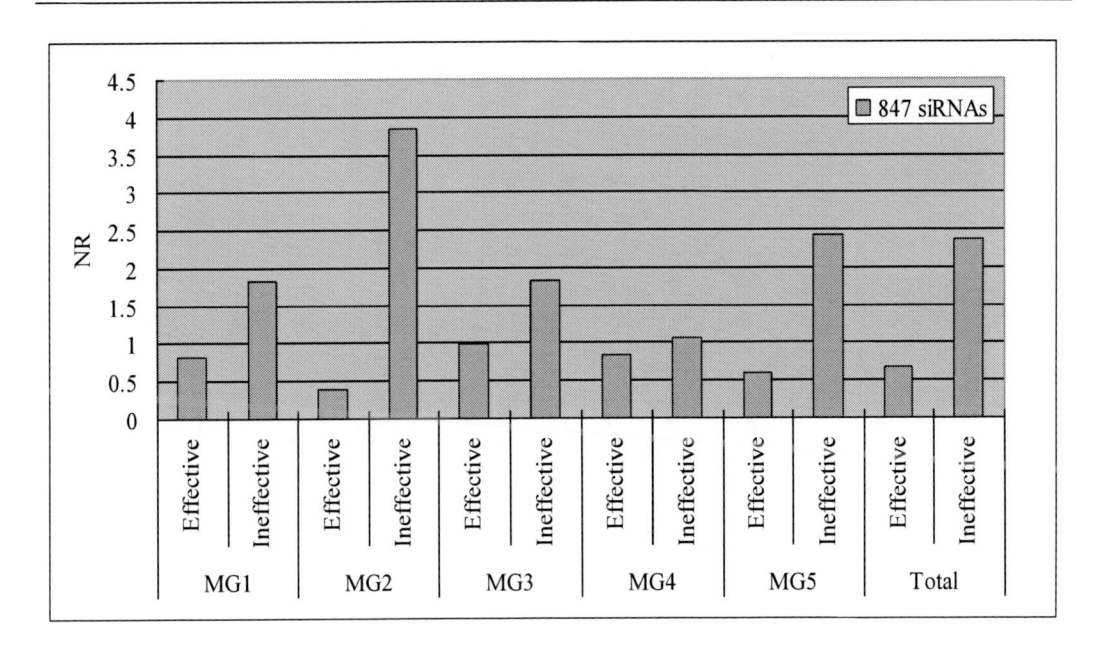

Figure 4. Normarized ratios based on 847 ineffective siRNAs.

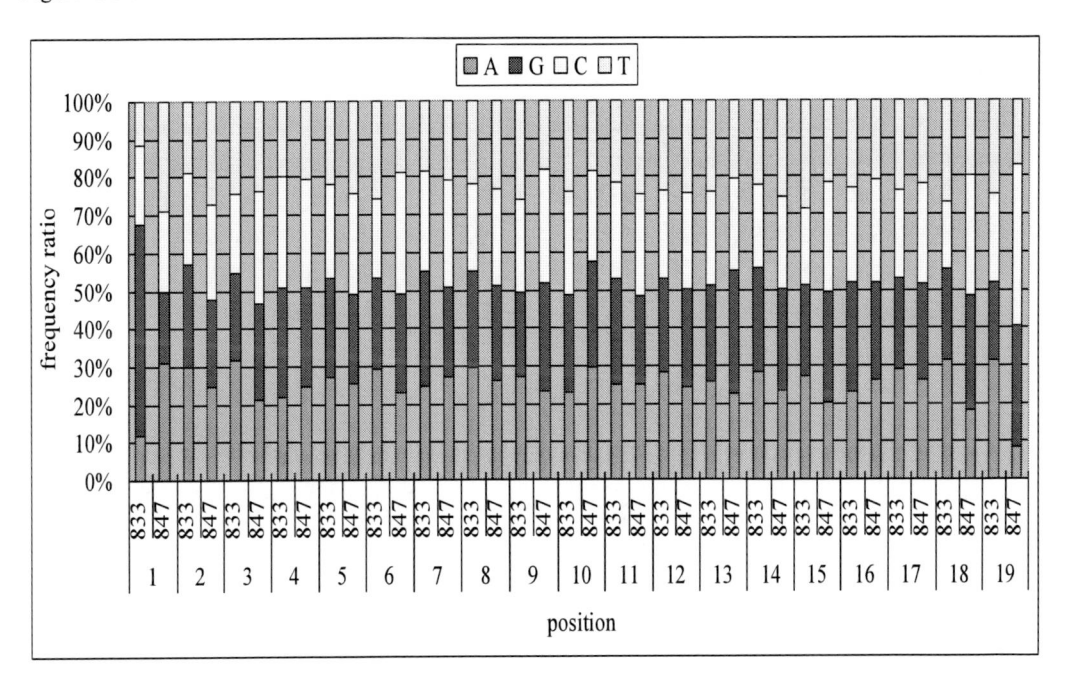

Figure 5. Relations of nucleotide occurrence frequencies between 833 and 847 siRNAs.

Characteristics for the Combinations of Two Successive Nucleotides

From the relations between two successive nucleotides determined by using the first Markov model for 833 known effective siRNAs it is possible to analyze the frequencies of combinations of two successive nucleotides in the sense strand. The relations among the frequency ratios of two successive nucleotides for 833 known effective siRNAs are shown in Figure 6, where it is clear that there are ups and downs in the frequency rations for combinations of two nucleotides. Two-nucleotide combinations with high frequency ratios are TG (34%), GA (33%), CA (33%), and AG (31%), whereas combinations with low ones are CG (15%) and TA (17%).

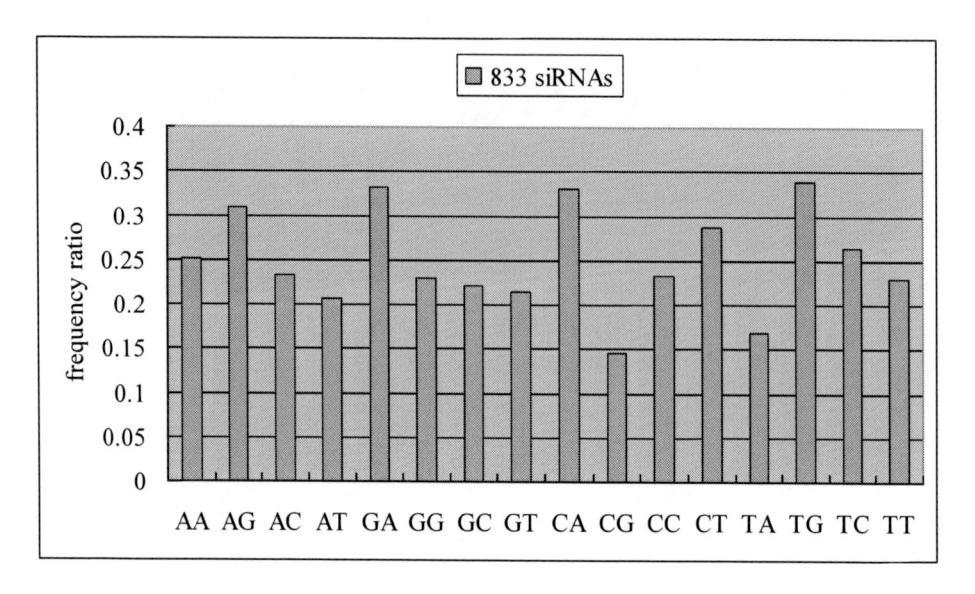

Figure 6. Frequency ratios for combinations of two successive nucleotides in 833 effective siRNAs.

It is also possible to calculate the frequency ratios of two-nucleotide combinations between two successive positions from 5' to 3' of the sense strand. They are shown in Figure7. When designing effective siRNAs for the target genes, it is also necessary to consider these characteristics of the frequencies of two-nucleotide combinations.

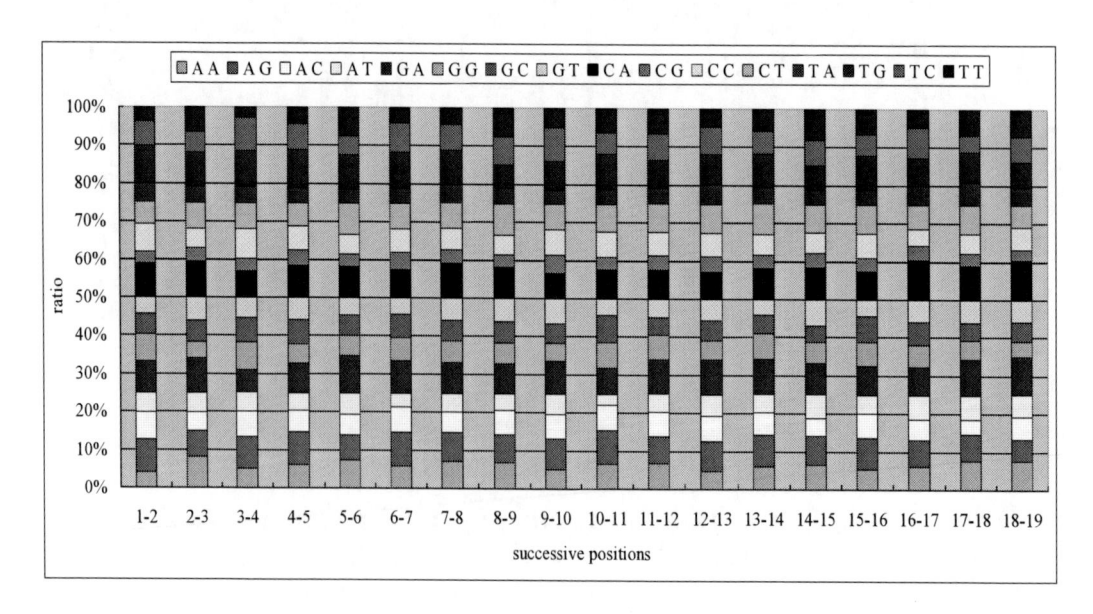

Figure 7. Frequency ratios of two-nucleotide combinations in two successive positions from 5' to 3'.

Comparison with other Reported Methods

The proposed method uses a probability estimation technique for selecting effective siRNA candidates, whereas most of the previous methods use scoring techniques. Although it is not easy to compare the previously reported scoring methods with the proposed average probability technique, the relations between the ratios of the scores for effective and ineffective siRNAs and the probabilities predicted for effective and ineffective siRNAs by using the average probabilities can be compared by analyzing the ROC (Receiver Operating Characteristic) curve based on the true positive fraction (TPF) and the false positive fraction (FPF) [41]. Because the reliability of the ROC curves increases with the numbers of effective and ineffective siRNAs that are used, 833 and 103 (MG1 – MG5) effective siRNAs and 847 and 93 (MG1 – MG5) ineffective siRNAs were adopted (see Methods and Materials). The ROC curves generated using the proposed method and the previously reported scoring methods are shown in Figure 8, where one sees that the curve for the proposed method is similar to those for the methods of Ui-Tei *et al.* [19] and Amarzguioui and Prydz [20]. This indicates that the proposed method distinguishes between effective and ineffective siRNAs as well as the previous top-ranked scoring techniques do.

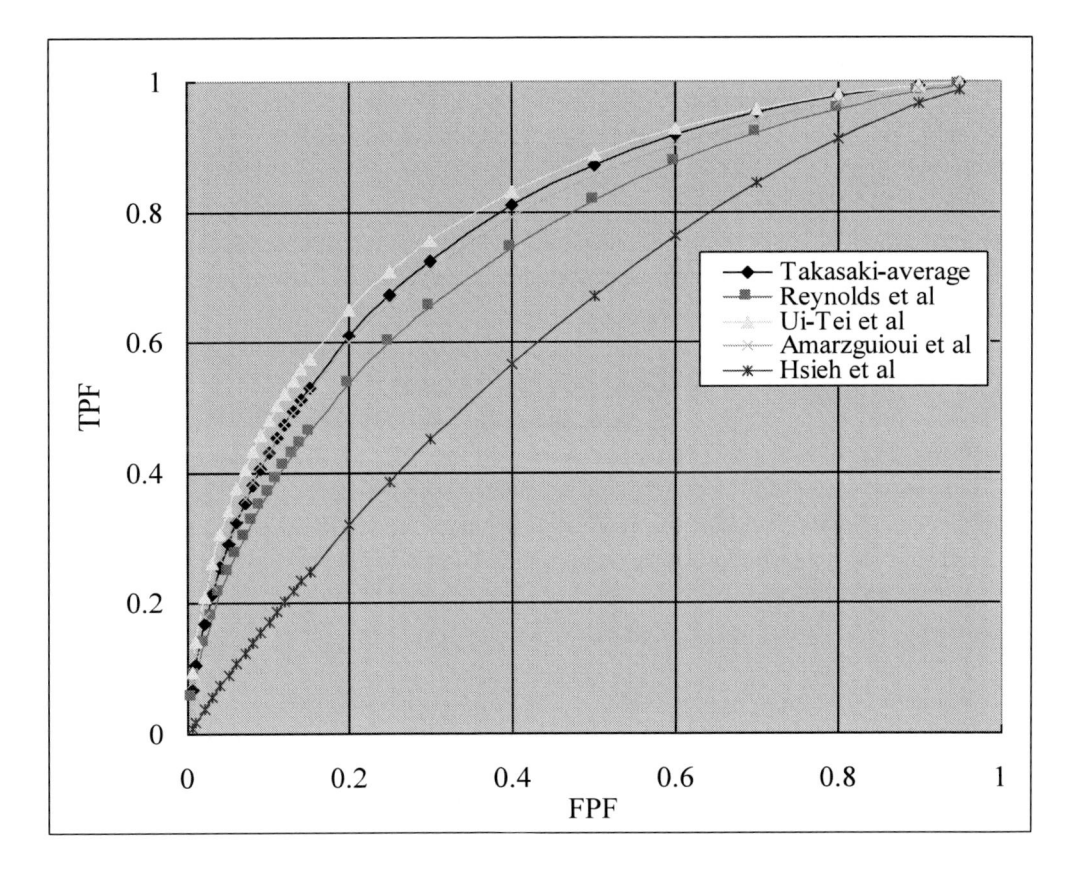

Figure 8. ROC curves comparing the proposed method with previously reported scoring techniques. ROC curves of the individual scoring techniques and the proposed method were generated for 936 effective and 940 ineffective siRNAs [41]. Scores of the effective and ineffective siRNA sequences were computed on the basis of the positional scores of the individual guidelines shown in Saetrom and Snove [33].

Conclusion

This chapter proposed an analytical prediction method using the average silencing probability to select effective siRNA target sequences from many possible candidate sequences. Although the previous scoring methods cannot estimate the probability that a candidate siRNA sequence will actually accomplish the expected gene degradation, the proposed method can. It is therefore quite different from the previous scoring methods. The proposed method was evaluated by applying it to recently reported siRNA sequences effective and ineffective for various genes. The evaluation results indicate that the proposed method would be useful for many other genes. It should therefore be useful for selecting siRNA sequences for mammalian genes. The chapter also described another method using a hidden Markov Model (HMM) to select the optimal functional siRNAs and discussed the frequencies of the combinations for two successive nucleotides as important characteristics of effective siRNA sequences.

References

[1]   Fire, A, Xu. S., Montgomery, M.K., Kostas, S.A., Driver, S.E., and Mello, C.C. (1998) Potent and specific genetic interference by double-stranded RNA in *Caenorhabditis elegans*. *Nature*, 391, 806–811.

[2]   Sharp, P.A. (2001) RNA interference—2001. *Genes Dev.*, 15, 485–490.

[3]   Elbashir, S.M., Harborth, J., Lendeckel, W., Yalcin, A., Weber, K. and Tuschl, T. (2001) Duplexes of 21-nucleotide RNAs mediate RNA interference in mammalian cell culture. *Nature*, 411, 494–498.

[4]   Elbashir, S.M., Lendeckel, W., and Tuschl, T. (2001) RNA interference is mediated by 21- and 22-nucleotide RNAs. *Genes Dev.*, 15, 188–200.

[5]   Dykxhoorn, D.M., Navia, C.D., and Sharp, P.A. (2003) Killing the messenger: short RNAs that silence gene expression. *Nature Review*, 4, 457–467.

[6]   Hannon, G. J. (2002) RNA interference. *Nature*, 418, 244–251.

[7]   Holen, T., Amarzguioui, M., Wiiger, M.T., Babaie, E., and Prydz, H. (2002) Positional effects of short interfering RNAs targeting the human coagulation trigger Tissue Factor. *Nucleic Acids Res.*, 30, 1757–1766.

[8]   Elbashir, S.M., Martinez, J., Patkaniowska, A., Lendeckel, W., and Tuschl, T. (2001) Functional anatomy of siRNAs for mediating efficient RNAi in *Drosophila melanogaster* embryo lysate, *EMBO J.*, 20, 6877–6888.

[9]   Kumar, R., Conklin, D.S., and Mittal, V. (2003) High-throughput selection of effective RNAi probes for gene silencing, *Genome Res.*, 13, 2333–2340.

[10] Mittal, V. (2004) Improving the efficiency of RNA interference in mammals, *Nature Rev. Genetics*, 5, 355–365.

[11] Schwarz, D.S., Hutvagner, G., Du, T., Xu, Z., Aronin N., and Zamore, P.D. (2003) Asymmetry in the assembly of the RNAi enzyme complex. *Cell*, 115, 199–208.

[12] Khvorova, A., Reynolds, A., and Jayasena, S.D. (2003) Functional siRNAs and miRNAs exhibit strand bias. *Cell*, 115, 209–216.

[13] Chalk, A.M., Wahlestedt, C., and Sonnhammer, E.L.L. (2004) Improved and automated prediction of effective siRNA. *Biochem. Biophys. Res. Commun.*, 319, 264–274.

[14] Teramoto, R., Aoki, M., Kimura, T., and Kanaoka, M. (2005) Prediction of siRNA functionality using generalized string kernel and support vector machine. *FEBS Letters*, 579, 2878–2882.

[15] Naito, Y., Yamada, T., Ui-Tei, K., Morishita, S., and Saigo, K. (2004) siDirect: highly effective, target-specific siRNA design software for mammalian RNA interference. *Nucleic Acids Res.*, 32, W124–W129.

[16] Santoyo, J., Vaguerizas, J.M., and Dapozo, J. (2004) Highly specific and accurate selection of siRNAs for high-throughput functional assays. Bioinfomatics, 21, 1376–1382.

[17] Truss, M., Swat, M., Kielbasa, S.M., Schafer, R., Herzed, H. and Hagemeier, C. (2005) HuSiDa – the human siRNA database: an open-access database for published functional siRNA sequences and technical details of efficient transfer into recipient cells. *Nucleic Acids Res.*, 33, D108–D111.

[18] Reynolds, A., Leake, D., Boese, Q., Scaringe, S., Marshall, W.S. and Khvorova, A. (2004) Rational siRNA design for RNA interference. *Nat. Biotech.*, 22, 326–330.

[19] Ui-Tei, K., Naito, Y., Takahashi, F., Haraguchi, T., Ohki-Hamazaki, H., Juni, A., Ueda, R. and Saigou, K. (2004) Guidelines for the selection of highly effective siRNA sequences for mammalian and chick RNA interference. *Nucleic Acids Res.*, 32, 936–948.

[20] Amarzguioui, M., and Prydz, H. (2004) An algorithm for selection of functional siRNA sequences. *Biochem. Biophys. Res. Commun.*, 316, 1050–1058.

[21] Hsieh, A.C., Bo, R., Monola, J., Vazquez, F., Bare, O. Khvorova, A., Scaringe, S., and Sellers, W.R. (2004) A library of siRNA duplexes targeting the phosphoinositide 3-kinase pathway: determinants of gene silencing for use in cell-based screens. *Nucleic Acids Res.*, 32, 893–901.

[22] Jagla, B., Aulner, N., Kelly, P.D. Song, D., Volchuk, A., Zatorski, A., Shum, D., Mayer, T., De Angelis, D.A., Ouerfelli, O., Rutishauser, U., and Rothman, J.E. (2005) Sequence characteristics of functional siRNAs. *RNA*, 11, 864–872.

[23] Huesken, D., Lange, J., Mikanin, C., Weiler, J., Asselbergs, F., Warner, J., Meloon, B., Engel, S., Rosenberg, A., Cohen, D., Labow, M., Reinhardt, M., Natt, F., and Hall, J. (2005) Design of a genome-wide siRNA library using an artificial neural network. *Nat. Biotech,* 23, 995–1001.

[24] Snove, O. Jr., Nedland, M., Fjeldstad, S.H., Humberset, H., Birkeland, O.R. Grunfeld, T., and Saetrom, P.O. (2004) Designing effective siRNAs with off-target control. *Biochem. Biophys. Res. Commun.*, 325, 769–773.

[25] Durbin, R., Eddy, S.R., Krogh, A., and Mitchison, G. (1998) Biological sequence analysis – probabilistic models of proteins and nucleic acids, Cambridge University Press.

[26] Jensen, F.V. (2001) Bayesian Networks and Decision Graphs, Springer.

[27] Takasaki, S., Kawamura, Y., and Konagaya, A. (2006) Selecting effective siRNA sequences by using radial basis function network and decision tree learning, *BMC Bioinformatics*, 7(Suppl 5), S22.

[28] Takasaki, S., Kotani, S., and Konagaya, A. (2004) An effective method for selecting siRNA target sequences in mammalian cells. *Cell Cycle*, 3, 790–795.

[29] Takasaki, S., Kotani, S., and Konagaya, A. (2005) Selecting effective siRNA target sequences for mammalian genes, *RNA Biology*, 2, 21–27.

[30] Takasaki, S., Kawamura, Y., and Konagaya, A. (2006) Selecting effective siRNA sequences based on the self-organizing map and statistical techniques, *Comput. Biol. & Chem.*, 30, 169–178.

[31] Takasaki, S., and Konagaya, A. (2006) Comparative analyses for selecting effective siRNA sequences, *Chem-Bio Informatics Journal*, 6, 69–84.

[32] Takasaki, S., and Kawamura, Y. (2007) Using radial basis function networks and significance testing to select effective siRNA sequences, *Comput. Statistics & Data Analysis*, 51, 6476–6487.

[33] Saetrom, P., and Snove, O. Jr. (2004) A comparison of siRNA efficacy predictors. *Biochem. Biophys. Res. Commun.*, 321, 247–253.

[34] Ladunga, I. (2007) More complete gene silencing by fewer siRNAs: transparent optimized design and biophysical signature, *Nucleic Acids Res.*, 35, 433–440.

[35] Holen, T. (2006) Efficient prediction of siRNAs with siRNArules 1.0: An open-source JAVA approach to siRNA algorithms, *RNA*, 12, 1620–1625.

[36] Saetrom, P. (2004) Predicting the efficacy of short oligonucleotides in antisense and RNAi experiments with boosted genetic programming. *Bioinformtics*, 20, 3055–3063.

[37] Heale, B.S.E., Sifer, H.S., Bowers, C., and Rossi, J.J. (2005) siRNA target site secondary structure predictions using local stable substructures, *Nucleic Acids Res.*, 33, e-30.

[38] Luo, K.Q., and Chang, D.C. (2004) The gene-silencing efficacy of siRNA is strongly dependent on the local structure of mRNA at the target region, *Biochem. Biophys. Res. Commun.*, 318, 303–310.

[39] Bohula, E.A., Salisbury, A.J., Sohail, M., Playford, M.P., Riedemann, J., Southern, E.M., and Macaulay, V.M. (2003) The efficacy of small interfering RNAs targeted to the type I insulin-like growth factor receptor (IGFIR) is influenced by secondary structure in the IGFIR transcript, *J. Biol. Chem.*, 278, 15991–15997.

[40] Elbashir, S.M., Harborth, J. Weber, K., and Tuschl, T. (2002) Analysis of gene function in somatic mammalian cells using small interfering RNAs, *Methods*, 26, 199–213.

[41] Metz, C.E., Herman, B.A., and Roe, C.A. (1998) Statistical comparison of two ROC-curve estimates obtained from partially-paired datasets, *Med. Decis. Making* 18, 110–121.

[42] Shabalina, S.A., Spiridonov, A.N., and Ogurtsov, A.Y. (2006) Computational models with thermodynamic and composition features improve siRNA design, *BMC Bioinformatics*, 7, 65.

[43] Vert, J., Foveau, N., Lajaunie, C., and Vandenbrouck, Y. (2006) An accurate and interpretable model for siRNA efficacy prediction, *BMC Bioinformatics*, 7, 520.

[44] Lu, Z.J., and Mathews, D.H. (2008) Efficient siRNA selection using hybridization thermodynamics, *Nucleic Acids Res.*, 36, 640–647.

[45] Wang, X., Wang, X., Varma, R.K., Beauchamp, L., Magdaleno, S., and Sendera, T.J. (2009) Selection of hyperfunctional siRNAs with improved potency and specificity, *Nucleic Acids Res.*, 37,e152.

In: Gene Silencing: Theory, Techniques and Applications ISBN: 978-1-61728-276-8
Editor: Anthony J. Catalano © 2010 Nova Science Publishers, Inc.

Chapter XI

Small RNA-mediated Gene Silencing for Plant Biotechnology

Ulku Baykal and Zhanyuan Zhang[*]

Plant Transformation Core Facility, Division of Plant Sciences, University of Missouri, Columbia, MO 65211, U.S.A.

Abstract

Small RNA-mediated gene silencing as a natural defense mechanism against viruses, transposons, and other invading nucleic acids or a means of regulating plant endogenous genes is a powerful tool and is being employed to down-regulate the expression of the targeted genes. Such a small RNA-mediated gene silencing has many different applications in a variety of organisms including humans and animals to control disease as a therapeutic agent, as well as plants to alter plant phenotypes. This silencing platform works through RNA-directed degradation or translational repression of target mRNA and has been devised towards a high-throughput approach for the gene suppression. In particular, sequence-specific control of gene expression by these non-coding RNAs has gained a significant amount of importance in plant biotechnology to influence specific plant phenotypes over the past years. It has been demonstrated that crops that were transformed with RNAi constructs, introduced stable modifications to the biochemical pathways. This can open new avenues in the improvement of crop productivity and quality. Here, we review the role of small RNA-directed gene silencing in plant biotechnology. The review will focus on the application of a gene silencing approach mediated by three subclasses of small RNAs for improved oil quality, reduced allergen, virus resistance, and other agronomical traits. The advantages and drawbacks of each gene silencing approach are also discussed with regard to crop improvement.

[*] Correspondence: 1-31 Agriculture Building, University of Missouri, Columbia, MO 65211, USA. Tel: 573-882-6922; Email: zhangzh@missouri.edu

Introduction

RNA interference (RNAi) is a specific gene-silencing process involved in transcript degradation or inhibition of protein synthesis for the regulation of gene expression, control of development and cell defense against invading nucleic acids such as viruses, transposons, or transgenes (Plasterk, 2002; Brodersen and Voinnet 2006). RNAi was accidentally discovered in petunia upon transfer of a chalcone synthase gene to obtain deep purple colored flowers. Instead, the experiments have resulted in white and variegated colored flowers (Napoli et al. 1990). The underlying molecular mechanism of this phenomenon was later unveiled by Fire and Mello's 2006 Nobel Prize-awarded work in *Caenorhabtidis elegans* (Fire et al. 1998). RNAi is now well-analyzed biochemically in both plants and animals and found to be conserved throughout eukaryotes.

The silencing RNA molecules can be synthetic, virus or transcribed nuclear genes. Two kinds of small RNA molecules are the key players in RNAi: small interfering RNAs (siRNAs) from virus, transgene, transposon, or a plant genome loci as well as microRNAs (miRNAs) derived from noncoding sequences of plant genome regions. The siRNAs (~21-24 nucleotides) are produced from double-stranded RNA (dsRNA) molecules by the enzyme DICER (Bernstein et al. 2001). siRNA strands are known as guide and passenger strands. It is the guide strand that incorporates into RNA-induced silencing complex (RISC) and binds to complementary mRNA (target) sequence to induce cleavage by the catalytic unit Argonaute in the RISC (Song et al. 2004; Liu et al. 2004).

RNAi emerges as a powerful tool in genetic engineering of crop plants without introducing new proteins to alter the gene expression to obtain better crop traits with improved quality such as reduced toxins, eliminated allergenic compounds, enhanced nutritive value, abiotic or biotic stress tolerance, etc. It is possible to obtain stable and heritable RNAi phenotype in engineered crops. In the next sections we will discuss the RNAi techniques and their applications as reported in journal publications as sole resources as well as future directions.

Hairpin RNA-mediated Gene Silencing
for Crop Improvement

RNA-silencing (PTGS) mechanism was responsible for the resistance against viruses and depends on the formation of dsRNA, whose antisense strand is complementary to the transcript of a targeted gene (Herr 2005; Brodersen and Voinnet 2006). This discovery led to the introduction of constructs to transgenic plants to produce siRNA species to induce targeted gene silencing efficiently and gain virus resistance. A dsRNA-induced sequence-specific RNA degradation mechanism is used to inhibit specific genes in plants by expressing a DNA molecule made from opposite strands of the gene which are separated by a segment, e.g., an intron sequence. RNA synthesized from this construct can fold to form a double-stranded structure (hairpin).

The use of hairpin RNA (hpRNA) has become very popular for specific gene silencing. Constructs encoding selfcomplementary hpRNA have been shown to be capable of generating post-transcriptional silencing to undetectable levels of the targeted mRNA transcript (Nunes et al. 2006; Wesley et al. 2001). The successful application of this method has been demonstrated in several crop plants.

Decaffeinated Coffee Plants

Caffeine is a secondary metabolite against insect and pest attack in coffee, which has many adverse effects such as caffeine intoxication, anxiety and sleep disorders. The conventional way to obtain decaffeinated coffee is the extraction of caffeine with solvents. The chemical elimination of caffeine is expensive and carries a risk of losing flavor. Three N-methyltransferase enzymes are involved in caffeine biosynthesis in coffee plants, CaXMT1, CaMXMT1 (theobromine synthase) and CaDXMT1 (caffeine synthase), has been cloned and expressed (Uefuji et al. 2003). Ogita et al. (2003) has constructed transgenic coffee plants by silencing the gene expression of theobromine synthase (*CaMXMT1*) using RNAi approach. They have designed hairpin constructs with fragments from the 3' untranslated region of the theobromine synthase mRNA and used these for *Agrobacterium*-mediated transformation of somatic embryos from *Coffea canephora*. The RNAi expression cassette was under the control of CaMV 35S promoter. The caffeine level has been reduced by up to 70% in the leaves and remained to be tested in the beans.

Cottonseed with Improved Fatty Acid Composition and Reduced Gossypol

Cotton is an important crop for its fiber and oilseed. RNAi technology has also been successfully employed in genetic modification of the fatty acid composition of cottonseed oil. A hpRNA-mediated RNAi method was used in cotton to down-regulate two key fatty acid desaturase genes encoding stearoyl-acyl-carrier protein $\Delta 9$-desaturase (ghSAD-1) and oleoyl-phosphatidylcholine $\omega 6$-desaturase (ghFAD2-1) (Liu et al. 2002). Knockdown of these two genes in cotton led to the increase of nutritionally improved high-oleic (HO) and high-stearic (HS) cottonseed oils that are essential fatty acids for human heart health. The cDNAs of *ghSAD-1* or *ghFAD2-1* genes as hairpin under the control of seed specific soybean lectin promoter has been used in the generation of cotton transgenics. The silencing of *ghSAD-1* gene has increased stearic acid levels from 2–3% to 40% and the oleic acid content of the cottonseeds has been elevated to 77% from 15% in seeds of non-transgenics by the silencing of *ghFAD2-1*. A significant decrease has been also observed in plamitic acid content of both *ghSAD-1* and *ghFAD2-1* silenced transgenic lines.

Cottonseed is also rich in dietary protein. However, the ability to use this nutrient-rich resource for food is hampered by the presence of toxic terpenoid product, gossypol. The application of RNAi proved to be successful in the cotton plant (Sunilkumar et al. 2006). RNAi has been used to produce cottonseeds containing lower levels of δ-cadinene synthase, a key enzyme in gossypol production. RNAi-mediated silencing remains confined to the tissues that express the hairpin RNA-encoding transgene in cotton. Their results suggest that the silencing signal from the developing δ-cadinene synthase-suppressed cotton embryo is unlikely to spread and reduce the levels of terpenoids in nontarget tissues, such as the leaves, roots, etc. The tissue specific low-seed-gossypol trait has been achieved with 99% reduction in cotton.

Since gossypol is also produced in vegetative cotton tissues where it protects cotton plants from insects and other pathogens, its targeted suppression in cottonseed is very important for plant defense (Sunilkumar et al. 2006). Indeed, transgenic cotton plants expressing an RNAi construct of the δ-cadinene synthase gene of gossypol synthesis fused to a seed-specific promoter caused seed-specific reduction of this metabolite, while its content in non-seed tissues remained almost unaffected. These cotton plants are thus expected to have similar insect and pathogen resistance to that of wild type cotton, but to produce seeds with higher nutritional value. Gossypol values in the seeds from some of the lines are well below the limit deemed safe for human consumption by United Nations Food and Agriculture Organization and World Health Organization.

Silencing of Omega-3 Fatty Acid Desaturase Gene in Soybean

Soybean is a very important food supply, rich in protein and oil, for both humans and livestock. Modification of the fatty-acid composition of soybean seeds to lower α-linolenic acid (18:3) levels can improve oil stability and flavor, and eliminate the need for hydrogenation. α-linolenic acid is the product of omega-3 fatty-acid desaturase enzyme. This enzyme catalyzes the conversion of linoleic acids (18:2) to α-linolenic acids (18:3) in the polyunsaturated fatty acid biosynthesis pathway during the seed development. *Trans*-fatty acid is generated by unstable fatty acid (e.g., α-linolenic acid) hydrogenization, which is unhealthy to human and animals who consume this fatty acid. It has been proposed that a significant reduction of α-linolenic acids might be possible upon simultaneous down-regulation of three microsomal forms of the omega-3 fatty acid desaturase (*GmFAD3A*, *GmFAD3B*, and *GmFAD3C*) to improve soybean oil quality. A highly conserved region (318-nt long) in all three gene family members has been used as inverted repeats for the construction of intron-spliced hairpin RNAi vector by using glycinin promoter for the seed-specific silencing of the targeted *FAD3* genes (Flores et al. 2008). The silencing of *GmFAD3*s has been remarkably effective with 100% transgenic events displaying *fad3*-mutant phenotypes, i.e., as low as 1-3% of α-linolenic acids in transgenic as compared with 7-10% of α-linolenic acids in non-transgenic dry soybean seeds. Furthermore, such an effective silencing has been demonstrated to be inheritable.

Achievement of Gossypol Intolerance in Cotton Bollworm for Plant Protection

The cotton bollworm (*Helicoverpa armigera*) is a moth whose larval feeding on cotton causes crop damage without getting affected from toxic gossypol. A cytochrome P450 gene (CYP6AE14) expressed in the midgut of cotton bollworm larvae has been identified as the reason behind its gossypol tolerance. Transgenic cotton plants expressing dsRNA of CYP6AE14 was generated to reduce the bollworm damage. Retarded larval growth with reduced expression of CYP6AE14 has been observed upon ingestion of transgenic plant material, which caused the suppression of CYP6AE14 gene. Another application of midgut-expressed gene silencing has also been demonstrated by expressing a glutathione-S-transferase (GST1) dsRNA from *H. armigera* in *Arabidopsis thaliana*. These results are promising for the pest control in field.

Silencing the Major Apple Allergen Mal d 1

Food allergies are a major health concern in industrialized countries. Apples with significantly decreased levels of major apple allergen, Mal d 1, would allow most patients allergic to apples to eat apples without allergic reactions (Puhringer et al. 2000). Mal d 1 belongs to a group of pathogenesis-related protein PR10. To inhibit the expression of Mal d 1 in apple plants by RNA interference, *in vitro*–grown apple plantlets have been transformed with a construct coding for an intron-spliced hairpin RNA containing a Mal d 1–specific inverted repeat sequence separated by a Mal d 1–specific intron sequence. Mal d 1 expression has been successfully reduced by RNAi (Gilissen et al. 2005). The gene-silencing construct was designed on the basis of sequence information of Mal d 1 from Gala apples. Although no detailed information is presently available about the expression of all the individual Mal d 1 genes, the degree of homology with the introduced gene-silencing construct was evidently sufficient for overall silencing of the endogenous Mal d 1 mRNAs.

Peanut Allergen Elimination

Peanuts are used as an ingredient in several food preparations. Peanut contains 7 different allergen proteins, including the most problematic one Ara h 2. Dodo et al. (2008) has demonstrated a reduced Ara h 2 expression in transgenic peanut via RNAi. Transgenic peanut lines have been generated using hpRNA constructs prepared from the 256 bp long coding region of Ara h 2. The allergenicity tests have confirmed the reduced peanut allergy.

Tomato Fruits with Reduced Allergenicity

The efficient silencing of a tomato allergen Lyc e 3 which encodes a nonspecific lipid transfer protein (ns-LTP) 3 has been reported in transgenic tomato plants. Transgenic tomato plants constitutively expressing LTPG1- or LTPG2-specific dsRNA constructs have been generated with efficient silencing of Lyc e 3 (Le et al. 2006). These findings indicate the potential of hairpin RNAi approach in the engineering of hypoallergenic plants to reduce food allergy symptoms.

Rice Dwarf virus Resistance

Rice dwarf virus (RDV) causes severe disease in rice crops. Plants infected with RDV are stunted and fail to produce viable seeds. RDV is transmitted to rice plants by vector insects leafhoppers (*Nephotettix* spp.), after multiplication of the virus in the vector insects. Infection of rice plants by RDV leads to the appearance of white chlorotic spots on leaves and the stunted plant. An internal 500 bp region of the non-structural protein Pns12 of *Rice dwarf virus*, which is one of the early proteins initiate the formation of the viroplasm has been used for the design of hairpin under the control of maize ubiquitin promoter to achieve high viral resistance against RDV. Pns12-specific RNAi constructs have been introduced into rice plants. The resultant transgenic plants have accumulated short interfering RNAs specific to the target virus sequence. The self-fertilized progeny of rice plants with Pns12-specific RNAi constructs has been strongly resistant to viral infection and failed to develop symptoms. Shimizu et al. (2009) suggests that interference with the expression of a protein that is critical for viral replication, such as the viroplasm matrix protein Pns12, might be a practical and effective way to control viral infection in crop plants.

Bean Golden Mosaic virus Resistance

Bean golden mosaic virus (BGMV) belongs to genus *Begomovirus* (family Geminiviridae). Geminivirus diseases, particularly incited by bean golden mosaic, are the most important limitation to the production of common bean and vegetable crops in the tropical lowlands and mid-altitude valleys of Latin America (Morales and Anderson 2001). *Bean golden mosaic virus* (BGMV) is transmitted by the whitefly *Bemisia tabaci* in a persistent, circulative manner, causing the golden mosaic of common bean (*Phaseolus vulgaris* L.). The characteristic symptoms are yellow-green mosaic of leaves, stunted growth, or distorted pods. An intron-spliced hpRNA interference construct has been used for specific silencing of the *AC1* gene product, which acts as a rolling-circle replication initiation factor and is capable of regulating its own expression (Eagle et al. 1994) to inhibit viral gene expression in highly resistant transgenic common bean plants (Bonfim et al. 2007). One transgenic line has showed high resistance and around 93% of its progenies have remained symptomless after exposure to whitefly.

Enhanced Amylose in Wheat

The major nutritional source of plant-derived carbohydrates is starch, which is composed of amylopectin and amylose polysaccharides, synthesized by two competitive pathways. Aiming to increase the relative content of amylose in wheat grains, an RNAi construct has been designed to silence the genes encoding the two starch-branching isozymes (SBEIIa and SBEIIb) of amylopectin synthesis, and was expressed under a seed-specific promoter [a high-molecular-weight glutenin (HMWG) promoter sequence from wheat] (Regina et al. 2006). This resulted in increased grain amylose content to over 70% of the total starch content.

When fed to rats in a diet, the high amylose grains had positive effects on gastrointestinal health, indicating the potential of RNAi technology to improve human health.

Reduced Onion Lachrymatory Factor (LF)

Allium species synthesize a unique set of secondary sulfur metabolites derived from cysteine (Cys). Most notable are the *S*-alk(en)yl-L-Cys sulfoxides, including *S*-2-propenyl-L-cysteine sulfoxide (alliin; 2-PRENCSO) and *trans-S*-1-propenyl-L-cysteine sulfoxide (isoalliin; 1-PRENCSO; Rose et al., 2005). When the tissues of any *Allium* species are wounded, these amino acid derivatives cleaved by the enzyme alliinase (EC 4.4.1.4) to produce sulfenic acids, and volatile sulfur compounds led to the characteristic flavor and bioactivity of the species, causing a typical symptom of tearing. In onion (*Allium cepa*), a crop difficult to transform, Eady et al. (2008) has suppressed the lachrymatory factor synthase (*lfs*) gene using an RNAi construct by using a 512 bp hairpin of the *lfs* gene sequence under the control of CaMV 35S promoter. This reduced lachrymatory synthase activity has produced significantly decreased levels of tear-inducing lachrymatory factor to a negligible amount in wounded onion.

Maize and Soybean Seeds with Reduced Phytic Acid

Maize and soybean contain large amount of phytic acid, which has adverse effect on nutrition and the environment. Phytic acid is essential for germination and seedling formation. Shi et al. (2007) has identified a multidrug resistance-associated protein (MRP4) gene in maize low phytic acid (*lpa1*) mutants and used it for the development of an approach to produce agronomic traits with reduced phytic acid. Since the conventional breeding methods are not successful in the reduction of phytic acid content of crops due to generation of unfavorable traits, RNAi has been employed for reduction of phyic acid by silencing *MRP4* ATP-binding cassette (ABC) transporter specifically in embryo. This transporter gene has been identified in maize low phytic acid (*lpa1*) mutants. The *Ole* and *Glb* promoters were used in the RNAi constructs together with a 5' end fragment of the *MRP4* gene for embryo-specific gene suppression. The highest phytic acid reduction has been reached by *Ole* promoter as 68-87%. The soybean homolog of maize *MRP4* has been identified and used for gene silencing construct driven by the soybean Kunitz trypsin inhibitor 3 (KTI3) promoter. The reduction in soybean phytic acid content has been achieved as 37-90%. The phytic acid content of both maize and soybean has been reduced significantly without causing any abnormalities in the seed traits with a future promise of applicability to other crop plants.

MicroRNA-based Second Generation RNAi Vectors: amiRNAs

Approximately 21-nt miRNA sequences, which are produced enzymatically from longer precursor sequences, are posttranscriptional gene regulators in plants and animals. Plant miRNAs were first discovered in *Arabidopsis thaliana* (Llave et al. 2002; Park et al. 2002; Reinhart et al. 2002). They perform their functions by binding to reverse complementary sequences to direct cleavage or translational inhibition of the target RNA.

Until recently long hpRNAs have been used in RNAi vector construction for transgene-mediated RNAi. These long hpRNAs from conventional RNAi vectors may cause "off-target" effect, i.e., silencing of unintended genes (Jackson et al. 2003). By contrast, miRNAs usually do not cause this off-target effect. It has been demonstrated that structure of miRNA is more important than its sequence in the production of small RNAs (Parizotto et al. 2004; Vaucheret et al. 2004). This finding paves the way for the modification of miRNA sequences to target any gene desired to be silenced. microRNAs structurally different from hpRNAs may offer a better substitute in the development of second-generation RNAi vectors for high-efficiency targeted gene silencing (Llave et al. 2002; Palatnik et al. 2003; Tang et al. 2004). The amiRNA technology exploits endogenous miRNA precursor. Single-stranded miRNAs are initially generated as siRNA-like duplexes whose one strand enters RISC (Khvorova et al. 2003; Zamore 2006) while the other strand is destroyed (Zamore 2006). miRNA-like small RNAs produced by miRNA-based siRNA vectors should display a higher preference for RISC assembly and direct efficient cleavage of their target mRNAs (Tang and Galili 2004).

Transgenic experiments have shown that it was possible to replace the miRNA:miRNA* duplex by an artificial hairpin structure, while maintaining the pattern of matches and mismatches in the foldback without altering miRNA processing. Thus, for functional gene analysis, amiRNAs can be designed to target any gene of interest. Changes in the artificial star sequence are introduced so that the structure of the stem would remain the same as the endogenous structure. The altered sequence is then folded with mfold and the original and altered structures are compared manually (by visual observations). If necessary, further alternations to the artificial star sequence are introduced to maintain the original structure. The DNA sequences corresponding to the artificial star sequences that are used to silence the desired target genes are shown.

miRNAs in plants usually induce cleavage of the target mRNA opposite positions 10 and 11 of the miRNA (Llave et al., 2002). *Arabidopsis* miRNA precursors have been modified to silence endogenous and exogenous target genes in dicatyledonous plants *Arabidopsis*, tomato and tobacco (Alvarez et al. 2006; Niu et al.2006; Schwab et al. 2006; Qu et al. 2007). Because of their specificity and versatility, amiRNA vectors are recognized as second-generation RNAi vectors (Tang et al. 2007). The designing rules are based on the study of extensive miRNA duplex structure and have been proved to be effective in gene silencing *in vitro* and *in vivo*. Software tools (http://wmd3.weigelworld.org/cgi-bin/webapp.cgi) and protocols for the design of *Arabidopsis* and rice amiRNA constructs can be easily adapted to other crops. Artificial miRNAs (amiRNAs), designed to target one or several genes of interest, provide a new and highly specific approach for effective post-transcriptional gene silencing (PTGS) in plants.

Artificial MicroRNA-Mediated Virus Resistance in Plants

RNA silencing in plants is a natural defense system against foreign genetic elements including viruses. The virus-specific siRNAs were shown to be a hallmark of the acquired virus resistance. Engineering endogenous miRNAs for antiviral strategies has shown to possess great potential in crop plant improvement. The first case study of using miRNA against virus was the gained resistance to *Cucumber mosaic virus* (CMV). A plant expression vector (p35SmiR2bprec) harboring the CaMV 35S promoter driving miRNA precursor sequence was designed to generate an artificial miRNA targeting the viral suppressor 2b of CMV (Qu et al. 2007). Transgenic tobacco plants expressing the *2b*-specific miRNA (miR2b) has inhibited multiplication of the CMV and exhibited resistance to CMV effectively. The presence of strong correlation between virus resistance and the expression level of the miRNA has suggested that the miRNA-mediated viral silencing is an effective approach against CMV infection. A recent report has showed high resistance in *Arabidopsis* to two other plant RNA viruses by expressing virus-specific miRNAs (Niu et al. 2006). Therefore, miRNA-mediated approach would be useful to engineer crop plants resistant to a wide range of viruses. The resistance level conferred by the transgenic miRNA is well correlated to the miRNA expression level. Comparison of the anti-CMV effect of the artificial miRNA to that of a short hairpin RNA-derived small RNA targeting the same site has revealed that the miRNA approach was superior to the approach using short hairpin RNA both in transient assays and in transgenic plants. By targeting the strain-specific sequences or sequences conserved in different strains, virus resistance with strain specificity or against a broad spectrum of virus strains could be achieved (Qu et al. 2007).

The second case study is the miRNA-mediated resistance against turnip yellow mosaic virus (TYMV) and turnip mosaic virus (TuMV). These two viruses can infect Brassica and non-Brassica hosts inflicting significant economic damages worldwide (Prod'homme et al. 2003; Tomimura et al. 2003). A report has demonstrated the modification of an *Arabidopsis thaliana* miR159 precursor to express artificial miRNAs (amiRNAs) targeting viral mRNA sequences encoding two gene-silencing suppressors, P69 of TYMV and HC-Pro of TuMV (Niu et al. 2006). Transgenic *A. thaliana* plants expressing amiR-P69[159] and amiR-HC-Pro[159] are specifically resistant to TYMV and TuMV, respectively. Expression of amiR-TuCP[159] targeting TuMV coat protein sequences also confers specific TuMV resistance. However, transgenic plants that express both amiR-P69[159] and amiR-HC-Pro[159] from a dimeric pre-amiR-P69[159]/amiR-HC-Pro[159] transgene are resistant to both viruses. The virus resistance trait is displayed at the cell level and is hereditable. This report has pointed out the applicability of obtaining broad-spectrum multiple viral resistances by designing RNAi vectors comprise several amiRNAs in crop plants.

amiRNA in Rice

Rice is a very important crop and model plant for the monocots. Warthmann et al. (2008) employed an amiRNA-based strategy for both japonica and indica types of cultivated rice, *Orayza sativa*, as enabling tools for the breeding and functional genomics. Their method is based on a vector derived from an endogenous rice miRNA precursor, osa-miR528 with a possibility to expand to other monocot crops. amiRNA constructs targeting three different genes *Phytoene desaturase* (*Pds*, Os03g08570; mutations cause albino phenotype), *Spotted leaf 11* (*Spl11*, Os12g38210; inactivation result in spontaneous lesion formation in the absence of pathogens), and *Elongated uppermost internode1* (*Eui1/CPY714D1*, Os05g40384; loss of function: elongation of the uppermost internode at heading stage). Upon constitutive expression of these amiRNAs in the varieties Nipponbare (japonica) and IR64 (indica), the targeted genes are down-regulated by amiRNA-guided cleavage of the transcripts, resulting in the expected mutant phenotypes. The effects are highly specific to the target gene, the transgenes are stably inherited and they remain effective in the progeny. They have obtained different mutant phenotypes by highly specific amiRNA-directed gene silencing with stable inheritance.

These results not only show that amiRNAs can efficiently trigger gene silencing in a monocot crop, but also that amiRNAs can effectively modulate agronomically important traits in varieties used in modern breeding programs. This approach is useful for candidate gene validation, comparative functional genomics between different varieties, and for improvement of agronomic performance and nutritional value.

Artificial *Trans*-Acting siRNAs as Effective Gene Silencing Tool

Trans-acting small interfering RNAs (tasiRNAs) are a class of endogenous small RNAs that are generated from four families of noncoding *TAS* gene-derived transcripts for posttranscriptional gene regulation. *Trans*-acting siRNAs (tasiRNAs) involves both 21-nt siRNAs and miRNAs. tasiRNAs differ from conventional siRNAs in that they target genes in *trans* (Peragine et al. 2004; Vazquez et al. 2004; Allen et al. 2005). tasiRNA generation depends on several proteins, including suppressor of gene silencing 3 (SGS3), RNA-dependent RNA polymerase 6 (RDR6) and DICER-like 4 (DCL4) (Peragine et al. 2004; Vazquez et al. 2004; Allen et al. 2005; Gasciolli et al. 2005; Xie et al. 2005; Yoshikawa et al. 2005).

TasiRNA formation occurs in phase with the miRNA-guided cleavage site. The tasiRNA productions are triggered by miR173 downstream from the cleavage site in *TAS1* and *TAS2*, by miR390 upstream from the miR390-guided cleavage site in *TAS3* (Allen et al. 2005), or by miR828 downstream from the cleavage site in *TAS4* (Rajagopalan et al. 2006). Of these *TAS* loci, *TAS1* is unique to *Arabidopsis*. The production of tasiRNAs from *TAS3* in *A. thaliana* is dependent on the presence of both a downstream, cleavable miR390 target site and an upstream, non-cleavable miR390 target site (Axtell et al. 2006; Montgomery et al. 2008a). Demonstration of tasiRNAs formation from chimeric transcripts containing indispensable miR173 or miR390 target sites has specified these trigger sequences as the minimal requirement for artificial tasiRNA formation (Montgomery et al. 2008a and b; Felippes and Weigel 2009).

As a tool for gene silencing, artificial tasiRNA offer several advantages over other RNAi methods. Although amiRNA constructs are also very predictable and lack of off-target effects (Allen et al, 2005; Ossowski et al, 2008; Mallory et al, 2004; Schwab et al, 2006 and Xie et al, 2005), the artificial tasiRNA approach is more practical for stacking multiple functional small RNA sequences into a single construct. Natural *TAS* RNA produces limited amount of tasiRNA, many more artificial tasiRNA could be generated from a single construct. The ability to generate multiple artificial tasiRNA enables silencing of multiple transcripts or transcript families from a single construct. This feature is particularly suitable for complex metabolic engineering in crop plants.

Artificial tasiRNAs in Plant Gene Silencing

The engineering of *TAS* locus to silence the targeted gene has been demonstrated successfully as a useful tool for genetic manipulation of plant (Talmor-Neiman et al. 2006; Montgomery et al. 2008; Felippes and Weigel, 2009). Recently, de la Luz Gutiérrez-Nava et al. (2008) has engineered *TAS1c* locus to silence the FAD2 gene expression in *Arabidopsis* plant by replacing a single or multiple copies of endogenous siRNA with an artificial siRNA targeting *FAD2* (*siFAD2*), or by substituting the five endogenous siRNAs with a 210-bp fragment of the FAD2 gene. These three attempts have resulted in consistent and very efficient silencing of *FAD2*. Furthermore, it has been shown that the silencing effect was inherited. The authors also anticipate the engineered tasiRNA genes as a valuable resource for functional genomics for gene function and epistasis analysis due to the full loss-of function phenotypes.

Several parameters were found to be important to influence silencing efficiency by tasiRNA. Manipulating construct promoters could also produce incomplete silencing or spatially and temporally restricted silencing that would be useful for studying essential genes or moderating gene activity. Substituting miR173 in *TAS1a* (Felippe and Weigel, 2009) and *TAS1c* (Carrington; de la Luz Gutiérrez-Nava et al. 2008) or miR390 in *TAS3* (Carrington; Felippe and Weigel, 2009) with other miRNA target sites has significantly reduced the silencing ability of the artificial siRNAs. Thus the involvement of an undiscovered mediator, which is possibly recruited by these specific miRNA for the tasiRNA biogenesis, has been suggested (Montgomery et al. 2008a and b; Felippes and Weigel 2009). One attractive possibility for silencing genes in a tissue-specific or temporally-specific manner would be to engineer *TAS* constructs triggered by miRNAs that are expressed in specific tissues or at specific times in development. The expression patterns of trigger miRNAs could be used to confer spatial or temporal specificity on *TAS*-induced silencing.

Conclusion

The results reviewed herein demonstrate that targeted gene silencing can be used to modulate biosynthetic pathways in a specific tissue to obtain a desired phenotype that is not possible by traditional breeding. The new discoveries in RNAi strategies open up a new frontier in the use of genetic manipulation to enhance global food supply with greater safety and fewer consumer concerns. Thus, an approach based on the removal of naturally occurring toxic compounds from the edible portion of the plant not only improves food safety but also provides an additional and potentially extraordinary means to meet the nutritional requirements of the growing world population. These more advanced gene-silencing platforms will also become better tools for crop functional genomics studies and testing research hypothesis for the plant biology studies in general.

References

Allen, E., Xie, Z., Gustafson, A. M., and Carrington, J. C. (2005). microRNA-directed phasing during trans-acting siRNA biogenesis in plants. *Cell, 121*, 207-221.

Alvarez, J. P., Pekker, I., Goldshmidt, A., Blum, E., Amsellem, Z., and Eshed, Y. (2006). Endogenous and synthetic microRNAs stimulate simultaneous, efficient, and localized regulation of multiple targets in diverse species. *Plant Cell, 18*, 1134-1151.

Bernstein, E., Caudy, A. A., Hammond, S. M., and Hannon, G. J. (2001). Role for a bidentate ribonuclease in the initiation step of RNA interference. *Nature, 409*, 363-6.

Bonfim, K., Faria, J. C., Nogueira, E. O., Mendes, E. A., and Aragão, F. J. (2007). RNAi-mediated resistance to *Bean golden mosaic virus* in genetically engineered common bean (*Phaseolus vulgaris*). *Molecular Plant-Microbe Interactions, 20*, 717-726.

Brodersen, P., and Voinnet, O. (2006). The diversity of RNA silencing pathways in plants. *Trends in Genetics, 22*, 268-280.

Crowe, T. C., Seligman, S. A., and Copeland, L. (2000). Inhibition of enzymic digestion of amylose by free fatty acids in vitro contributes to resistant starch formation. *Journal of Nutrition, 130*, 2006-2008.

de la Luz Gutiérrez-Nava, M., Aukerman, M. J., Sakai, H., Tingey, S. V., and Williams, R. W. (2008). Artificial trans-acting siRNAs confer consistent and effective gene silencing. *Plant Physiology, 147*, 543-551.

Dodo, H. W., Konan, K. N., Chen, F. C., Egnin, M., and Viquez, O. M. (2008). Alleviating peanut allergy using genetic engineering: the silencing of the immunodominant allergen Ara h 2 leads to its significant reduction and a decrease in peanut allergenicity. *Plant Biotechnology Journal, 6*, 135-145.

Eady, C. C., Kamoi, T., Kato, M., Porter, N. G., Davis, S., Shaw, M., Kamoi, A., and Imai, S. (2008). Silencing Onion Lachrymatory Factor Synthase Causesa Significant Change in the Sulfur Secondary Metabolite Profile. *Plant Physiology, 147*, 2096-2106.

Eagle, P. A., Orozco, B. M., and Hanley-Bowdoin, L. (1994). A DNA sequence required for geminivirus replication also mediates transcriptional regulation. *Plant Cell 6*, 1157-1170.

Felippes, F. F., and Weigel, D (2009). Triggering the formation of tasiRNAs in Arabidopsis thaliana: the role of microRNA miR173. *EMBO Reports 10*, 264-270.

Fire, A., Xu, S., Montgomery, M. K., Kostas, S. A., Driver, S. E., and Mello, C. C. (1998). Potent and specific genetic interference by double-stranded RNA in *Caenorhabditis elegans*. *Nature, 391*, 806-811.

Flores, T., Karpova, O., Su, X., Zeng, P., Bilyeu, K., Sleper, D. A., Nguyen, H. T., and Zhang, Z. J. (2008). Silencing of GmFAD3 gene by siRNA leads to low alpha-linolenic acids (18:3) of *fad3*-mutant phenotype in soybean [*Glycine max* (Merr.)]. *Transgenic Research, 17*, 839-850.

Gilissen, L. J., Bolhaar, S. T., Matos, C. I., Rouwendal, G. J., Boone, M. J., Krens, F. A., Zuidmeer, L., Van Leeuwen, A., Akkerdaas, J., Hoffmann-Sommergruber, K., Knulst, A. C., Bosch, D., Van de Weg, W. E., and Van Ree, R. (2005). Silencing the major apple allergen Mal d 1 by using the RNA interference approach. *Journal of Allergy and Clinical Immunology, 115*, 364-369.

Herr, A. J., Jensen, M. B., Dalmay, T., and Baulcombe, D. C. (2005). RNA polymerase IV directs silencing of endogenous DNA. *Science, 308*, 118-120.

Jackson, A. L., Bartz, S. R., Schelter, J., Kobayashi, S. V., Burchard, J., Mao, M., Li, B., Cavet, G., and Linsley, P. S. (2003). Expression profiling reveals off-target gene regulation by RNAi. *Nature Biotechnology, 21*, 635-937.

Jones-Rhoades, M. W., Bartel, D. P., and Bartel, B. (2006). MicroRNAs and their regulatory roles in plants. *Annual Review of Plant Biology, 57*, 19-53.

Khvorova, A., Reynolds, A., and Jayasena, S. D. (2003). Functional siRNAs and miRNAs exhibit strand bias. *Cell, 115*, 209-216.

Langland, J. O., Jin, S., Jacobs, B. L., and Roth, D. A. (1995). Identification of a plant-encoded analog of PKR, the mammalian double-stranded RNA-dependent protein kinase. *Plant Physiology, 108*, 1259-1267.

Le, L., Lorenz, Y., Scheurer, S., Fötisch, K., Enrique, E., Bartra, J., Biemelt, S., Vieths, S., and Sonnewald, U. (2006). Design of tomato fruits with reduced allergenicity by dsRNAi-mediated inhibition of ns- LTP (Lyc e 3) expression. *Plant Biotechnology, 4*, 231-242.

Liu, J., Carmell, M. A., Rivas, V. F., Marsden, C. G., Thomson, J. M., Song, J. J., Hammond, S. M., Joshua-Tor, L. J., and Hannon, G. J. (2004). Argonaute2 is the catalytic engine of mammalian RNAi. *Science, 305,* 1437-1441.

Liu, Q., Singh, S. P., and Green, A. G. (2002). High-stearic and high-oleic cottonseed oils produced by hairpin RNA-mediated post-transcriptional gene silencing. *Plant Physiology, 129*, 1732-1743.

Llave, C., Kasschau, K. D., Rector, M. A., and Carrington, J. C. (2002). Endogenous and silencing-associated small RNAs in plants. *Plant Cell, 14*, 1605-1619.

Mallory, A. C., Reinhart, B. J., Jones-Rhoades, M. W., Tang, G., Zamore, P. D., Barton, M. K., and Bartel, D. P. (2004). MicroRNA control of PHABULOSA in leaf development: importance of pairing to the microRNA 5′ region. *EMBO Journal 23*, 3356-3364.

Mao, Y. B., Cai, W. J., Wang, J. W., Hong, G. J., Tao, X. Y., Wang, L. J., Huang, Y. P., and Chen, X. Y. (2007). Silencing a cotton bollworm P450 monooxygenase gene by plant-mediated RNAi impairs larval tolerance of gossypol. *Nature Biotechnology, 25*, 1307-1313.

Montgomery, T. A., Howell, M. D., Cuperus, J. T., Li, D., Hansen, J. E., Alexander, A. L., Chapman, E. J., Fahlgren, N., Allen, E., Carrington, J. C. (2008a). Specificity of ARGONAUTE7–miR390 interaction and dual functionality in TAS3 trans-acting siRNA formation. *Cell, 133*, 128-141.

Montgomery, T. A., Yoo, S. J., Fahlgren, N., Gilbert, S. D., Howell, M. D., Sullivan, C. M., Alexander, A., Nguyen, G., Allen, E., Ahn, J. H., and Carrington, J. C. (2008b). AGO1-miR173 complex initiates phased siRNA formation in plants. *Proceedings of the National Academy of Sciences of the United States, 105*, 20055-20062.

Morales, F. J., and Anderson, P. K. (2001). The emergence and dissemination of whitefly-transmitted geminiviruses in Latin America. *Archives of Virology, 146*, 415-441.

Napoli, C., Lemieux, C., and Jorgensen, R. (1990). Introduction of a Chimeric Chalcone Synthase Gene into Petunia Results in Reversible Co-Suppression of Homologous Genes in trans. *Plant Cell, 2*, 279-289.

Niu, Q. W., Lin, S. S., Reyes, J. L., Chen, K. C., Wu, H. W., Yeh, S. D., and Chua, N. H. (2006). Expression of artificial microRNAs in transgenic Arabidopsis thaliana confers virus resistance. *Nature Biotechnology, 24*, 1420-1428.

Nunes, A. C., Vianna, G. R., Cuneo, F., Amaya-Farfán, J., de Capdeville, G., Rech, E. L., and Aragão, F. J. (2006). RNAi-mediated silencing of the myo-inositol-1-phosphate synthase gene (GmMIPS1) in transgenic soybean inhibited seed development and reduced phytate content. Planta *224*, 125-132.

Ogita S, Uefuji H, Yamaguchi Y, Koizumi N, Sano H. (2003). Producing decaffeinated coffee plants. *Nature, 423*, 823.

Ossowski, S., Schwab, R., and Weigel, D. (2008). Gene silencing in plants using artificial microRNAs and other small RNAs. *Plant Journal, 53*, 674-690.

Palatnik, J. F., Allen, E., Wu, X., Schommer, C., Schwab, R., Carrington, J. C., and Weigel, D. (2003). Control of leaf morphogenesis by microRNAs. *Nature, 425*, 257-263.

Parizotto, E.A., Dunoyer, P., Rahm, N., Himber, C., and Voinnet, O. (2004). In vivo investigation of the transcription, processing, endonucleolytic activity, and functional relevance of the spatial distribution of a plant miRNA. *Genes and Development, 18*, 2237-2242.

Park, W., Li, J., Song, R., Messing, J., and Chen, X. (2002). CARPEL FACTORY, a Dicer homolog, and HEN1, a novel protein, act in microRNA metabolism in Arabidopsis thaliana. *Current Biology, 12*, 1484-1495.

Plasterk, R.H. (2002). RNA silencing: the genome's immune system. *Science, 296*, 1263-1265.

Prod'homme, D., Jakubiec, A., Tournier, V., Drugeon, G., and Jupin, I. (2003). Targeting of the Turnip yellow mosaic virus 66K replication protein to the chloroplast envelope is mediated by the 140K protein. *Journal of Virology, 77*, 9124-9135.

Puhringer, H., Moll, D., Hoffmann-Sommergruber, K., Watillon, B., Katinger, H., and Machado, M. L. D. (2000). The promoter of an apple Ypr10 gene, encoding the major allergen Mal d 1, is stress- and pathogen-inducible. *Plant Science, 152*, 35-50.

Rajagopalan, R., Vaucheret, H., Trejo, J., Bartel, D. P. (2006). A diverse and evolutionarily fluid set of microRNAs in *Arabidopsis thaliana. Genes and Development, 20*, 3407-3425.

Qu, J., Ye, J., and Fang, R. (2007). Artificial miRNA-mediated virus resistance in plants. *Journal of Virology 81*, 6690-6699.

Regina, A., Bird, A., Topping, D., Bowden, S., Freeman, J., Barsby, T., Kosar-Hashemi, B., Li, Z., Rahman, S., and Morell, M. (2006). High-amylose wheat generated by RNA interference improves indices of large-bowel health in rats. *Proceedings of the National Academy of Sciences of the United States, 103*, 3546-3551.

Reinhart, B. J., Weinstein, E. G., Rhoades, M. W., Bartel, B., and Bartel, D. P. (2002). microRNAs in plants. *Genes and Development, 16*, 1616-1626.

Schwab, R., Ossowski, S., Riester, M., Warthmann, N., and Weigel, D. (2006). Highly specific gene silencing by artificial microRNAs in *Arabidopsis. Plant Cell, 18*, 1121-1133.

Shi, J., Wang, H., Schellin, K., Li, B., Faller, M., Stoop, J. M., Meeley, R. B., Ertl, D. S., Ranch, J. P., and Glassman, K. (2007). Embryo-specific silencing of a transporter reduces phytic acid content of maize and soybean seeds. *Nature Biotechnology, 25*, 930-937.

Shimizu, T., Yoshii, M., Wei, T., Hirochika, H., and Omura, T. (2009). Silencing by RNAi of the gene for Pns12, a viroplasm matrix protein of Rice dwarf virus, results in strong resistance of transgenic rice plants to the virus. *Plant Biotechnology Journal, 7*, 24-32.

Song, J. J., Smith, S. K., Hannon, G. J., and Joshua-Tor, L. (2004). Crystal structure of Argonaute and its implications for RISC slicer activity. *Science, 305*, 1434-1437

Sunilkumar, G., Campbell, L., Puckhaber, L., Stipanovic, R., and Rathore, K. (2006). Engineering cottonseed for use in humannutrition by tissue-specific reduction of toxic gossypol. *Proceedings of the National Academy of Sciences of the United States, 103*, 18054-18059.

Talmor-Neiman, M., Stav, R., Klipcan, L., Buxdorf, K., Baulcombe, D. C., and Arazi, T. (2006). Identification of trans-acting siRNAs in moss and an RNA-dependent RNA polymerase required for their biogenesis. *Plant Journal, 48*, 511-521.

Tang, G., Galili, G., and Zhuang, X. (2007). RNAi and microRNA: breakthrough technologies for the improvement of plant nutritional value and metabolic engineering. *Metabolomics, 3*, 357-369.

Tang, G., Reinhart, B. J., Bartel, D. P., and Zamore, P. D. (2003). A biochemical framework for RNA silencing in plants. *Genes and Development, 17*, 49-63.

Tomimura, K., Gibbs, A. J., Jenner, C. E., Walsh, J. A., and Ohshima, K. (2003). The phylogeny of Turnip mosaic virus; comparisons of 38 genomic sequences reveal a Eurasian origin and a recent 'emergence' in East Asia. *Molecular Ecology, 12*, 2099-2111.

Uefuji, H., Ogita, S., Yamaguchi, Y., Koizumi, and N., Sano H. (2003). Molecular cloning and functional characterization of three distinct N-methyltransferases involved in the caffeine biosynthetic pathway in coffee plants. *Plant Physiology, 132*, 372-80.

Vaucheret, H., Vazquez, F., Crété, P., and Bartel, D. P. (2004). The action of ARGONAUTE1 in the miRNA pathway and its regulation by the miRNA pathway are crucial for plant development. *Genes and Development, 18*, 1187-1197.

Warthmann, N., Chen, H, Ossowski, S., Weigel, D., and Hervé, P. (2008). Highly specific gene silencing by artificial miRNAs in rice. *PLoS ONE, 3*, e1829.

Wesley, S. V., Helliwell, C. A., Smith, N. A., Wang, M. B., Rouse, D. T., Liu, Q., Gooding, P. S., Singh, S. P., Abbott, D., Stoutjesdijk, P. A., Robinson, S. P., Gleave, A. P., Green, A. G., Waterhouse, P. M. (2001). Construct design for efficient, effective and high-throughput gene silencing in plants. *Plant Journal, 27*, 581-90.

Xie, Z., Allen, E., Fahlgren, N., Calamar, A., Givan, S. A., and Carrington, J. C. (2005) Expression of Arabidopsis miRNA genes. *Plant Physiology, 138*, 2145-2154.

Xu, P., Zhang, Y., Kang, L., Roossinck, M. J., and Mysore, K. S. (2006). Computational estimation and experimental verification of off-target silencing during posttranscriptional gene silencing in plants. *Plant Physiology, 142*, 429-440.

Zhang, C., and Ghabrial, S.A. (2006). Development of *Bean pod mottle virus*-based vectors for stable protein expression and sequence-specific virus-induced gene silencing in soybean. *Virology, 344*, 401-411.

Zhu, X., and Galili, G. (2003). Increased lysine synthesis coupled with a knockout of its catabolism synergistically boosts lysine content and also transregulates the metabolism of other amino acids in Arabidopsis seeds. *Plant Cell, 15*, 845-853.

In: Gene Silencing: Theory, Techniques and Applications
Editor: Anthony J. Catalano

ISBN: 978-1-61728-276-8
© 2010 Nova Science Publishers, Inc.

Chapter XII

Post Transcriptional Gene Silencing Methods for Functional Characterization of Abiotic Stress Responsive Genes in Plants

Muthappa Senthil-Kumar[1,2] *and Makarla Udayakumar*[1]

[1] Department of Crop Physiology, University of Agricultural Sciences,
GKVK, Bangalore 560 065 India

[2] Present address: Plant Biology Division, The Samuel Roberts Noble Foundation,
2510 Sam Noble Pky, Ardmore, OK 73402 USA

Abstract

Several abiotic stress specific functional and regulatory genes have been cloned, and a number of EST databases representing stress specific genes are available for many plant species. These sequences have to be translated into functional information, necessitating the need for potential functional genomic approaches. Post transcriptional gene silencing (PTGS) is one of approaches to characterize functional relevance of stress responsive genes. Virus-induced gene silencing (VIGS) and developing stable gene knock down plants using hairpin RNA interference (hpRNAi) constructs (referred here as RNAi) are two important PTGS methods. Over a period, these methods are becoming integral part of plant stress functional genomics. Among these two methods, use of VIGS for characterizing abiotic stress responsive genes is still an emerging approach while RNAi has been widely used.

This review is focused on VIGS vector resources, brief methodology of VIGS and application of gene silencing to identify/characterize genes involved in drought-, salinity- oxidative-, high light-, and nutrient-stress management. VIGS can be used as fast forward

---

[1] Corresponding author: Muthappa Senthil-Kumar, Plant Biology Division, Noble Foundation, 2510 Sam Noble Pky, Ardmore, OK 73402 USA, Email: skmuthappa@noble.org , or senthilphy@yahoo.com.

genetic screening method to identify genes involved in stress tolerance and also an effective reverse genetic tool to validate the relevance of genes identified from high-throughput screening. Further, VIGS can be effectively integrated with abiotic stress imposition and response of gene silenced plants can be quantified using suitable techniques. We describe here an comprehensive approach to silence large number of cDNA clones and characterize the silenced plants under abiotic stresses. We also discussed application of other PTGS based methods like RNAi and artificial micro RNA (amiRNA) in abiotic stress functional genomics.

We propose that PTGS is an useful technology for translational genomics to assign function to large number of abiotic stress responsive genes. Even with their current limitations, gene silencing techniques are set to revolutionize plant abiotic stress functional genomics. Limitations and future directions for these techniques are also briefly discussed.

Introduction

Plants have acquired various stress tolerance mechanisms involving physiological and biochemical changes that result in adaptation to abiotic stresses. Recent advances in genome-wide analyses have revealed complex regulatory networks that control global gene expression, protein modification, and metabolite composition [1, 2]. Genomic technologies now provide high throughput integrated approaches for transcript and protein analysis and useful for novel gene discovery [1]. Microarray studies show that at least several hundred genes are involved in response to drought or salt stress in plants [1, 3]. A careful analysis of the transcription profiles should reveal not only individual stress-activated genes but also their pathways. Over a period genomic information has been made available through several databases. The microarray expression database (MAEDA) developed by the RIKEN group (http://rarge.gsc.riken.go.jp), the database resource for analysis of signal transduction in cells (DRASTIC) developed by the Scottish crop research institutive, the database compiled by the stress genomics consortium (www.stress-genomics.org) and the website www.plantstress.com are providing useful information for abiotic stress genomics. However, understanding functional relevance of each and every gene identified from these approaches is a mammoth task ahead. Hence a powerful technology that can speedup translating sequence information into a functional information is necessary. Gene down-regulation is one of the approaches to achieve this goal.

PTGS-Based Methods are Integral Part of Plant Abiotic Stress Functional Genomics

The functional relevance of abiotic stress-responsive genes can be elucidated by gene down-regulation. Some of widely used gene down-regulation approaches include chemical, irradiation, transposon, T-DNA insertional mutagenesis, and posttranscriptional gene silencing (PTGS)-based methods [4-6]. Chemical and radiation mutagenesis approaches are very powerful and convenient to generate mutants. However, it is more difficult and time consuming to clone the genes from such mutants, as this involves map-based cloning or transcript profiling. T-DNA insertion mutagenesis causes loss of gene function by ectopic activation of neighboring genes or disruption of coding sequence or un-translated regions (UTR). This technique is cumbersome, time consuming and suffers from the difficulty of defining the number of insertions. Even transposon tagging has many of the same limitations.

Down-regulation of endogenous genes by PTGS is one of the best approach to characterize abiotic stress responsive gene function in plants. PTGS is a sequence-specific gene-silencing mechanism initiated by the introduction of double-stranded RNA (dsRNA), homologous in sequence to the silenced gene, which triggers degradation of mRNA. Three PTGS methods namely RNA interference (RNAi), virus-induced gene silencing (VIGS) and artificial micro RNA (amiRNA) are commonly used. Among these methods RNAi has been widely used as a tool for exploring stress responsive gene function in crop species [7] especially to study the abiotic stress tolerance [8]. Apart from RNAi, VIGS has been emerging as a suitable method to study genes from many angiosperms where genome sequencing projects are underway or completed [9]. Here we review how VIGS has been used to assess abiotic stress responsive gene function and also the available methodological elements, such as vectors, inoculation procedures, and analysis of silenced phenotypes under stress.

Mechanism of VIGS

VIGS is a plant defense mechanism that can protect plants against invading viruses. Virus vector technology exploits this RNA defense. Upon inoculation of VIGS vectors into plants, the inserted gene fragment is amplified along with viral RNA by the viral replication system, spreads systemically in the infected plants. The dsRNA formed during virus replication are targeted by DICER-like enzymes and cleaved into small interfering RNA (siRNA). In turn, the siRNA molecule serves as a template for degradation of complementary endogenous mRNAs, through the RNA-induced silencing complex (RISC) [10]. Thus, if the insert is from a host gene, the antisense strand of siRNAs would target RISC to the corresponding host mRNA and cleaves it, leading to loss of function in the encoded protein [11].

VIGS Vectors

Current understanding about the phenomenon of PTGS in plant defense against viruses has enabled researchers to modify virus genomes as VIGS vectors suitable for different plant species. Silencing of endogenous plant genes require cloning of gene fragments into the virus without compromising viral replication and movement. Initially VIGS vector was developed by modifying RNA viruses such as *Tobacco mosaic virus* (TMV) [12]. These studies used plant gene triggers that act as visible markers (*PDS*, phytoene desaturase and *ChlH*, H subunit of Mg protoporphyrin chelatase) for gene silencing. RNA viruses, such as TMV, *Potato virus-X* (PVX), and *Tobacco rattle virus* (TRV) are widely used for gene silencing. However, not all RNA virus derived vectors are useful as VIGS vectors because many viruses, such as *Tobacco etch virus* (TEV) have potent anti-silencing proteins that directly interfere with host silencing machinery. At a later stage DNA viruses were also engineered as VIGS vectors despite their size constraints for movement between plant cells. For example, certain Gemini viruses like *Cabbage leaf curl geminivirus* (CbLCV) has great potential as a VIGS vector. It has been shown to infect and trigger silencing of transgenes and endogenous genes in model plant *Arabidopsis* [12]. The list of viral vectors presently used for VIGS are presented in recent reviews [9, 13].

RNA viruses replicate in the cytoplasm and DNA viruses replicate in plant nuclei using host DNA replication machinery. Both types of viruses induce diffusible, homology-dependent systemic silencing of endogenous genes [12]. However, the extent of silencing spread and the severity of viral symptoms can vary significantly in different host plants and host/virus combinations.

Methodology of VIGS

For analysis of plant genes using VIGS, a plant gene fragment (reverse genetics approach) or a cDNA library (fast-forward genetics approach) is cloned into VIGS vector and inoculated into plants [14]. Inoculation can be done by either smearing *in vitro* transcripts of infectious RNA or Agrobacterium-mediated delivery (spray or syringe infiltration or Agrodrench) of T-DNA containing the virus or bombardment by biolistic or DNA abrasion or sap inoculation [13]. Agroinoculation is one of the widely used methods for delivery of VIGS vectors into plant cells [15]. The protocol used in our lab for Agroinoculation of *Tobacco Rattle Virus* (TRV)-VIGS vector in *Nicotiana benthamiana* is as follows. *Agrobacterium tumefaciens* strains (GV2260) separately containing pTRV1 and pTRV2 or its derivatives are grown at 28°C in Luria-Bertani (LB) medium containing appropriate antibiotics. The cells are harvested from cultures grown overnight and re-suspended in the Agrobacterium induction buffer (10 mM MES, pH 5.5, and 200 μM acetosyringone) to a final absorbance [optical density (OD) at 600 nm] of 0.2-0.4, and incubated for 2-4 h at room temperature in a shaker. For leaf infiltration, each *A. tumefaciens* strain containing pTRV1 and pTRV2 or its derivatives are mixed in a 1 : 1 ratio in MES buffer (pH 5.5) and infiltrated into lower leaves using a 1-ml needleless syringe and the infiltrated plants are maintained under a temperature range of 21–23°C for effective viral infection and spread. Infiltration of 21-d-old *N. benthamiana* plants induces effective silencing and maintenance of these plants at 20 ± 3°C provides maximum efficiency of gene silencing. A video demonstration of this protocol has been recently published [16].

Gene silenced plants are subjected to a suitable abiotic stress to evaluate their performance. For each experiment two subsets of plants are maintained. One set of plants from each treatment (wild-type, vector control and gene silenced) is maintained under non-stress conditions and other set of plants is subjected to stress. Stress effects can be quantified using suitable physiological, biochemical and molecular techniques and results are compared with performance of non-stressed but silenced plants and vector controls [17-19].

Apart from characterization of abiotic stress genes, VIGS has been widely used for characterizing gene function in different cellular processes associated with basic cell function, metabolic pathways, evolutionary and developmental biology and plant–microbe interactions in diverse plant species (solanaceous plants, a few legumes, brassicaceae and a few monocots) [13]. Several studies have also demonstrated the utility of VIGS method for characterizing organ-specific genes in flowers, fruits, leaves and roots [13]. Detailed protocols and information about methodology of VIGS are also covered in some of earlier manuscripts [6, 11, 14].

Application of VIGS to Characterize Genes Involved in Various Abiotic Stresses

Ever since the development of VIGS protocols in *N. benthamiana*, tomato and Arabidopsis, several studies used VIGS to down-regulate genes that are related to abiotic stress tolerance. Over a period development of VIGS vector from various viruses expanded the application of VIGS to several crop plants that are often susceptible to abiotic stresses. Using the presently available resources the target stress responsive gene can be effectively down-regulated and gene function can be studied. We review below some of applications of VIGS in characterizing genes under various abiotic stresses.

High Light and Oxidative Stress

Plants are sensitive to photo-oxidative stress that occurs due to excess photon energy supplied to chloroplast. The oxidative stress damage is enhanced when plants are also exposed to other abiotic stresses like extreme temperature or water deficit along with high light. Plants adopt several strategies to combat high-light induced oxidative stress. Some of the tolerance mechanisms include enhanced production of free radical scavenging enzymes, accumulation of anthocyanins and red-ox regulation [20]. Using VIGS, the oxidative stress tolerance mechanism in plants has been further explored. We review here four such studies. First study- one of stress responsive gene, aconitase was silenced in *N. benthamiana* using PVX-VIGS and leaf discs from gene silenced plants were stressed using a herbicide compound, paraquat. The results showed that ion leakage in silenced plant was significantly lower than that from PVX vector control plant [21]. Consistent with these results even the Arabidopsis knockout mutants of this gene also conferred enhanced tolerance to reactive oxygen species (ROS) [21]. These results indicate that down-regulation of aconitase enhances oxidative stress tolerance. Second study- a study by Chang-Sook and colleagues provided evidences for the involvement of two prohibitin genes in oxidative stress tolerance. Silencing of genes that encode subunits of prohibitin, namely *NbPHB1* and *NbPHB2* in *N. benthamiana* using TRV-VIGS resulted in a 10- to 20-fold higher production of ROS and induced premature leaf senescence [22]. Also, the silenced plants were hyper susceptible to various oxidative stress-inducing reagents, including H_2O_2, paraquat, antimycin A and salicylic acid. These results suggest that prohibitins play a crucial role in protection against oxidative stress in plant cells [22]. Similarly, *N. benthamiana* plants silenced with *PDS* gene also reported to have low photo system-II (PS-II) quantum yield and did not show photosynthesis [23]. Third study- mitochondria-associated hexokinases that are previously implicated in cell death control are recently demonstrated for their involvement in oxidative stress response. A hexokinase gene, *NbHxk1*, silenced *N. benthamiana* plants showed increased susceptibility to paraquat-induced oxidative stress [24]. Further, results from this study also demonstrated that higher levels of hexokinase can confer improved resistance to photo-oxidative stress [24].

Fourth study- physiological role of a ran-binding-protein encoding gene, *NbRanBP1*, in controlling ROS production was studied using VIGS in *N. benthamiana*. *NbRanBP1* gene silenced plants showed reduced mitochondrial membrane potential and excessive production of ROS indicating the role of this gene in oxidative stress tolerance [25].

UV Stress

UV-B radiation is a key environmental signal that initiates diverse responses in plants that affect metabolism, development, and viability. Many effects of UV-B involve differential regulation of gene expression [26]. Components of UV-B induced signaling in plants have been analyzed in detail and the functional relevance of genes involved are being characterized. UVc photons are highly energetic and trigger biochemical responses [18]. For instance, exposure of tobacco leaves to UVc light induces the accumulation of salicylic acid (SA). VIGS has been used to characterize the genes involved in UV-stress responses. Catinot and colleagues have analyzed the importance of isochorismate for SA biosynthesis in *N. benthamiana* by silencing the isochorismate synthase (*ICS*) gene using TRV-VIGS [27]. Their study showed that *ICS* gene silencing by VIGS leads to a strong suppression of SA accumulation in response to UV-induced abiotic stress, demonstrating the relevance of isochorismate pathway and SA biosynthesis.

Shading Stress

In addition to providing a key energy resource for photosynthesis, light signals provide plants grown under natural conditions with important spatial and temporal information about their surrounding environment. Little is known about how plants sense and adapt to environmental stresses like shading that compromise photosynthesis and respiration and deplete energy supplies. The molecular mechanisms underlying physiological responses of plants to low light or shading and dark-light cycle are complex [28]. Studies showed changes in dark-light cycle rhythm and low light adaptation is mainly mediated by photoreceptors and other regulatory genes [28]. We review here the function of two protein kinases in dark-light cycle. Functional role of Arabidopsis protein kinases KIN10 and KIN11 in plant survival under reduced light or dark has been studied by using TRV-VIGS [29]. Simultaneous silencing of both *KIN10* and *KIN11* genes abrogated the transcriptional switch in darkness and stress signaling, and impaired starch mobilization at night. This study clearly indicated the role of these two genes in energy balance, growth and survival of Arabidopsis in reduced light conditions [29]. Further, activity of nitrate reductase (NR) is coordinated with the changes of environmental conditions such as light-dark transitions by its post-translational modifications involving protein phosphorylation. Subsequent binding of 14-3-3 proteins to the NR phosphorylated under darkness is needed for its inactivation. 14-3-3s are ubiquitous binding proteins conserved among eukaryotes [30]. NR activity repression under dark condition is lost in leaves of *N. benthamiana* plants silenced with *Nb14-3-3a* and *Nb14-3-3b* genes [31]. Physiological down-regulation of NR activity under darkness or shade should be important to avoid the accumulation of toxic nitrite in the cells. NR activity is also important for maintaining photosynthesis. In higher plants, NR is inactivated by the phosphorylation of a conserved Ser residue and binding of 14-3-3 proteins in the presence of divalent cations or polyamines. It is likely that changes in plant metabolism during shading or reduced light significantly interfere with dark-light cycle. Hence, studies related to dark-light cycle can be indirectly useful to understand this stress. Apart from VIGS, RNAi has been used to manipulate plant leaf area, erectness of leaf and other agronomical traits like plant height [32]. These traits are important in shade avoidance among field crops, as they can regulate excessive biomass and leaf area index, preventing self shading and shading of its neighbors.

Water-Deficit Stress

Plant response to water-deficit stress are complex. Stress-specific genes encodes products that directly protect the plant cell against stresses and regulate gene expression and signal transduction during water-deficit. VIGS is one of several approaches used to understand relevance of such stress regulated genes. Silencing of late embryogenesis abundant 4 (*lea4*), a known stress responsive gene involved in desiccation-stress tolerance, resulted in susceptibility of tomato plants to water-deficit stress [19]. Similarly, silencing of a transcription factor, WRKY in *N. benthamiana* plants resulted in reduced photosynthesis, less efficiency of PS-II reaction centre, and exhibited photoinhibition under water-deficit stress. Results from this study indicated the role of WRKY in basic physiological processes under stress [17]. Silencing of drought induced *Lea5*, *HSP20* and *HSP70* genes that belongs to broad group of chaperone proteins in *N. benthamiana* altered response of silenced plants to water-deficit stress. *HSP70* gene silenced plants showed higher pheophytin levels under drought stress. Besides this, down-regulation of Jumonji (JMJC), bHLH, and zinc finger regulatory genes made *N. benthamiana* plants tolerant to drought [3, 33]. These genes were further validated for their function using other gene silencing methods, proving that VIGS is an efficient method for characterizing the relevance of genes involved in water-deficit stress [8].

Flooding or Hypoxia Stress

Flooding or water-logging causes yield reduction in crop plants. A major component of flooding stress is the lack of oxygen available to submerged tissues (hypoxia stress). Genes altered during this stress are associated with cell wall growth and modification, tetrapyrrole synthesis, hormone response, starch or sugar metabolism and nitrogen metabolism [34]. While differential regulation of protein and metabolite levels have been studied in response to low oxygen stress, little research has been done on characterizing genes involved in these processes. *KIN10* and *KIN11* genes were silenced in Arabidopsis using TRV-VIGS and the response of protoplasts obtained from silenced plants to hypoxia has been studied [29]. This study provided clues about the role of these two protein kinases during hypoxia. However, VIGS has not yet been fully exploited to study the genes involved in hypoxia stress.

Salinity Stress

Dunaliella salina, a unicellular green alga is able to withstand extreme salt concentrations. Studying the function of genes in *D. salina* could provide valuable information about the molecular mechanisms of halotolerance. Since VIGS protocols are not available in this algae, relevance of its genes can be studied by silencing their homologs in *N. benthamiana* by VIGS. Efficient silencing of a DEAD box helicase gene from this algae in *N. benthamiana* has demonstrated the possibility to use a sequence from *D. salina* to induce silencing in *N. benthamiana* [35]. Although, this study does not provide direct experimental evidence for studying gene under salinity, this has demonstrated such a possibility in future. Till date VIGS has not been used to directly study the relevance of genes in salinity tolerance. However, other gene down-regulation approaches like RNAi are widely used to study salinity tolerance [36].

Nutrient Deficiency and Toxicity Stress

Expression of large numbers of genes have been shown to be up- or down-regulated in response to mineral nutrient deficiency. For example, 115 genes were up-regulated by nitrate re-supply in nitrogen deficient tomato plants, while 195 genes exhibited significant changes in expression patterns in response to alterations of the nutritional status of P, K, or Fe. Hence, it is highly desirable to find out the key genes involved in plant responses to nutrition stresses [37]. VIGS is potentially an attractive reverse-genetics tool for studies of mineral nutrition-related genes in roots [18]. Ferric chelate reductase gene (*FRO1*) silencing in tomato using *Tomato yellow leaf curl China virus*-based VIGS vector resulted in reduced ferric chelate reductase activity in root cells. This study demonstrated that coinciding nutritional stress treatment and maximum silencing period is important. For example, *14-3-3* genes are known to be up-regulated within 2h of removal of Fe from the soil or plant growth medium [37]. Therefore, in this study Fe deficiency treatment was given to *FRO1* gene silenced tomato plants at a time when the positive control plants silenced with marker gene '*Su*' showed visible yellowing phenotype in leaves [18]. These results clearly demonstrated that VIGS can be employed to investigate gene function associated with plant nutrient deficiency in roots. Further, Kang and colleagues examined the *in vivo* effects of sulfite reductase (SiR) deficiency on chloroplast development in *N. benthamiana*. VIGS of *SiR* resulted in leaf yellowing and growth retardation phenotypes [38]. This study and others showed the dual functions of *SiR*, acting as a sulfur assimilation enzyme and as a chloroplast nucleoid binding protein. Further, VIGS has been effectively used to silence a nodule inception gene (*Nin*) in pea and their relevance in Nitrogen fixation has been demonstrated [39]. Also, role of 14-3-3a and Nb14-3-3b proteins in inactivation of NR activity under darkness in *N. benthamiana* to prevent nitrite toxicity has been shown [31]. Furthermore, functions of KIN10/KIN11 in plant protection during nutrient or sugar deprivation conditions has been shown by VIGS in Arabidopsis [29].

Wound Stress

Plant induces several genes in response to wound injury. Wounding leads to the release of systemin and also induces mapkinase (MAPK)-activating signals, such as ROS, lipases and oligosaccharide elicitors. Here we review two studies showing the relevance of a lipase gene and MAPK in wound signaling. By using VIGS, involvement of a *Capsicum annuum* lipase (*GDSL-lipase 1*) gene in signaling pathway during wound responses has been shown [40]. Further, TRV-VIGS-mediated co-silencing of two MAPK genes, *LeMPK1* and *LeMPK2*, in tomato resulted in decreased wound-induced accumulation of jasmonic acid (JA) while JA levels increased strongly in control plants [41]. This study clearly showed that LeMPKs function upstream of JA and are required for expression of a subset of wound-response genes.

Assessing Relevance of Genes in High-, Low-Temperature and Freezing Stress Using RNAi

Plants possess several intriguingly complex mechanisms to tolerate temperature extremes. We review the application of gene silencing to characterize genes involved in all the three temperature extremes. First, heat-stress-associated 32-kD protein, Hsa32, has been shown to be required for maintenance of acquired thermotolerance, an important feature for thermotolerance of plants [42]. Second, the C-repeat-binding factor (CBF) role in cold acclimation has also been demonstrated by gene silencing. In this study, CBF1 and CBF3 have been shown to impart additive effect to induce the CBF regulon to induce cold acclimation [43]. Third, downregulation of *Populus tomentosa* Delta-12 fatty acid desaturase gene (*PtFAD2*) in the hybrid poplar revealed the importance of this gene in freezing stress tolerance [44]. This study showed that the level of polyunsaturated fatty acid (PUFA) is essential for cold acclimation and changes in PUFA levels can lead to the alteration of freezing tolerance in poplar trees. These studies clearly indicated that gene silencing can be a tool for assessing the relevance of plant genes associated with temperature extremes.

Advantages of VIGS to Study Abiotic Stress Responsive Gene Function

(a) Organ specific gene silencing is possible by VIGS. For example, silencing in tomato flowers and fruits can be achieved by inoculating VIGS vectors shortly before flower bud initiation [13]. Such method does not affect the target gene in already developed leaves, stem and roots, thus facilitating target organ specificity. This is highly useful to study the relevance of target gene in anthesis, pollen viability and seed development during drought or temperature extremes. (b) Similarly, VIGS can be done at a particular growth phase of the plant, thus facilitating the analysis of relevance of stress responsive gene in vegetative and/or reproductive growth phase and also during stress or recovery period. (c) Unlike RNAi, VIGS can be applied to wide range of crop plants. This can be done by two means. First, by establishing suitable VIGS protocols for the target species. Second, by using recalcitrant crop genes as heterologous silencing trigger and study gene function in VIGS amenable model plant and indirectly infer the gene function [19]. (d) En-mass cloning of stress responsive cDNA into gateway cloning ready VIGS vectors hastens forward genetic screens. For example, stress specific cDNA library can be constructed for a specific abiotic stress in specific VIGS vector and individual gene knock down plants can be obtained. These plants can be subjected to stress and forward genetic screening can be done to identify relevant gene. (e) VIGS can be used to silence multiple gene family members using a single trigger sequence in VIGS vector. This allows researchers to study redundant gene function during stress. This is important as many of abiotic stresses tolerant mechanisms are multigene regulated and functional genomics based on type of mechanism (controlled by group of genes) is more relevant rather than by studying individual gene. (f) Simultaneous silencing of more than one gene (even distantly related) in a plant is possible by VIGS [29]. This approach can also be used to evaluate partner gene combinations that can play role in more than one abiotic stress tolerance or multistress tolerance. (g) VIGS can be used in conjunction with RNAi or mutant or overexpression studies. For example, a particular gene cluster combination speculated to be involved in abiotic stress tolerance can be evaluated by transiently silencing the second target gene using VIGS in a RNAi- or mutant- or gene overexpressing- plant. Such an approach is suitable alternate to development of double knockout mutants using conventional crossing which is time consuming. (h) Apart from these, VIGS can be applied to characterize genes that may cause embryo lethality or seedling mortality when mutagenesis or stable RNAi are used. Also, VIGS protocols for a particular species (or even genera) can be applied independent of the plant genetic background, i.e. across any cultivar, variety or sub species.

Challenges in Using VIGS and Ways to Overcome the Same

(a) *Interference of VIGS with natural PTGS machinery in plants*. Recent studies indicate that post-transcriptional regulations of gene expression play an important role in how plants respond to abiotic stresses [1, 45]. Virus, siRNA and other components of PTGS pathway may influence the gene regulation under abiotic stress. However, such cross talks can be accounted to certain extent by having appropriate positive- and vector- control plants. (b) *Virus infection itself can alter abiotic stress tolerance* [46]. Use of proper vector control can neutralize possible changes that occur due to influence of VIGS vector inoculation. (c) *RNA-dependent RNA polymerases (RDRs) play role in RNA silencing and natural gene regulation*. RDR1, a potentially important enzyme in PTGS, has been shown to be regulated during biotic and abiotic stresses [47]. However, the role of RDR during VIGS is very minimal as the silencing occurs mainly due to antiviral defense of the plant. (d) *Off-target silencing*. Unintended gene silencing of close homologous gene is possible in some cases. This can be minimized by taking adequate care while selecting 'trigger' sequence for VIGS vector. (e) *Change in environmental conditions during stress treatment may affect silencing of gene*. Influence of change in environmental conditions during abiotic stress (e.g. temperature-, water-deficit-, and high light-stress) impositions on VIGS is a possibility [23]. However, by adopting a change of green house environmental factors in a phased manner, i.e. by initially maintaining optimum environmental conditions favorable for VIGS and later changing the conditions favoring stress imposition, will help to achieve optimum silencing [4]. We have also detailed the difficulties of using VIGS and suitable remedies in our earlier manuscript [4].

New Approaches for Gene Silencing

The amiRNAs designed to target one or several genes of interest provide a highly specific approach for effective PTGS in plants and *Chlamydomonas reinhardtii* [48, 49]. amiRNAs can effectively modulate agronomically important traits in varieties used in modern breeding programs and this has been demonstrated in rice [49]. Software tools and a protocol for the design of rice amiRNA constructs are available. Suitability of this approach has been demonstrated for candidate gene validation, comparative functional genomics between different varieties, and for improvement of agronomic performance [49]. Model algae like Chlamydomonas can be easily used for high-throughput abiotic stress functional genomics by silencing genes using amiRNA constructs.

Apart from this overexpression of full-length sense strand cDNA using strong promoters can induce co-suppression in plants (called sense RNAi; [50]). This will be highly useful to knockout gene function in plants. Unlike RNAi, that needs cloning of trigger sequence in sense and antisense orientation, the constructs for sense overexpression can be easily made and cDNA library for specific abiotic stress responsive genes can be constructed in the sense RNAi vector. Screening of such library by forward genetics will be useful. We believe that these newly emerging gene silencing methods will have great potential for plant stress functional genomics in future.

Conclusion

VIGS is considered specifically useful to directly assess abiotic stress tolerance in crop plants, which is not always possible with mutants, as genetic resources are minimal in crop plants. Over a period, VIGS protocols have been standardized for functional characterization of genes involved in drought-, oxidative-, mineral nutrient deficiency-, and other abiotic-stresses. Our group is in the process of developing VIGS protocols for salinity-, freezing- and high temperature-stresses. Apart from VIGS, RNAi and other recently developed PTGS approaches are also powerful tools for studying relevance of abiotic stress responsive genes. Based on the information available till date, we propose a model to show the application of PTGS methods in gene function analysis under stress (Figure 1). Taken together, PTGS methods became an indispensible tools for functional genomics of abiotic stress tolerance.

Acknowledgments

RNAi and VIGS projects in MU lab is supported by projects from Department of Science and Technology (DST) and Department of Biotechnology (DBT), New Delhi. SM was supported by a research grant from International Foundation for Science (IFS), Sweden (C/4066-1/Ac/17341R). We thank Dr. Hema Ramanna for critical reading of the manuscript.

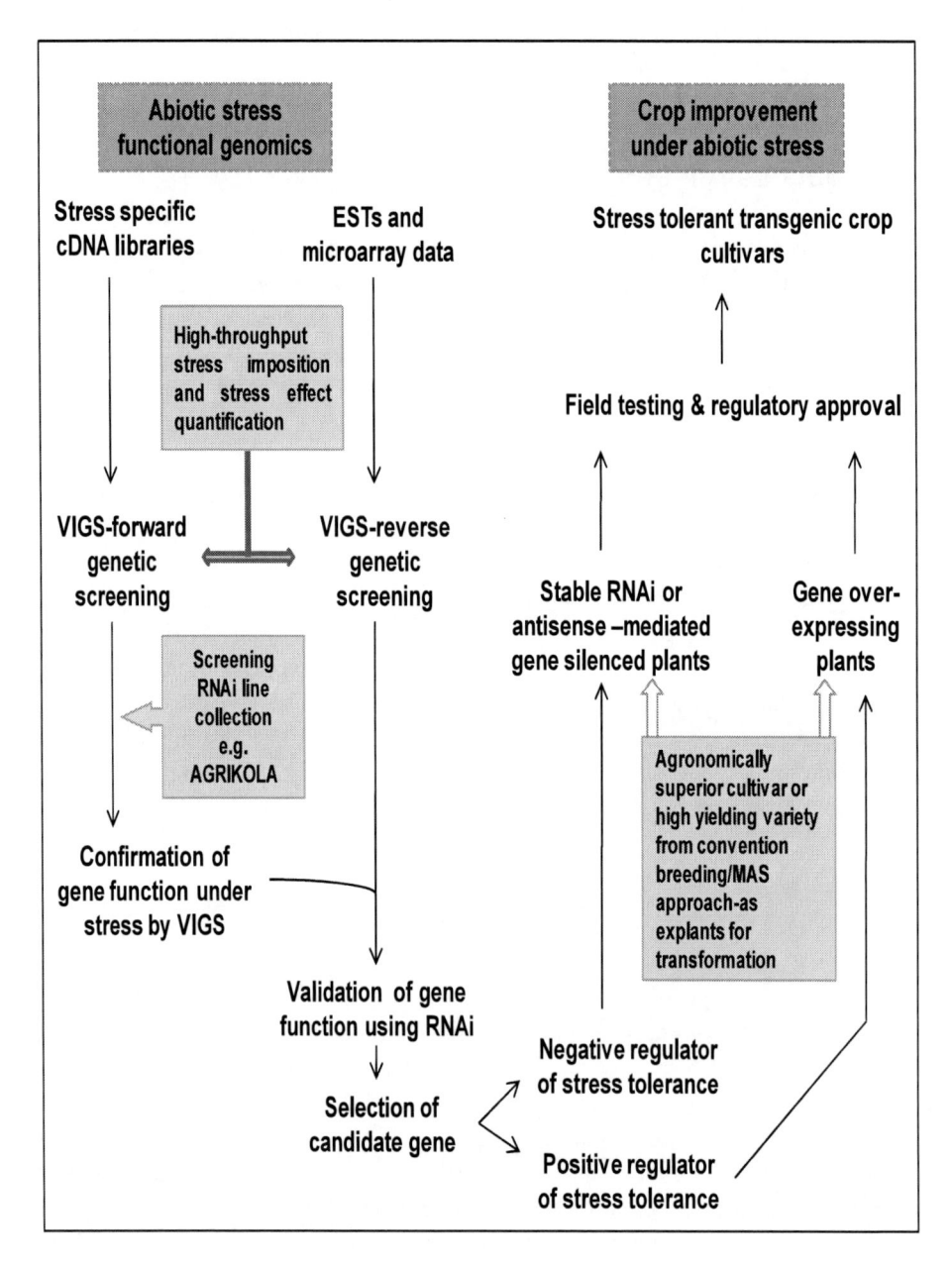

Figure 1. Application of PTGS methods for functional characterization of abiotic stress responsive genes and development of transgenic stress tolerant crop plants. VIGS can be used as both forward and reverse genetic tool. Abiotic stress related genes that are known to impart stress tolerance can be identified by VIGS-based screening. Apart from VIGS, RNAi line collections (http://maizegdb.org/; http://chromdb.org/; https://mtrnai.msi.umn.edu/) can also be used for forward genetic screening. To perform this screening crop specific stress imposition and stress effect quantification protocols are needed. The candidate genes identified from functional genomics screen can be used for developing transgenic crop plants tolerant to stress. If needed suitable explants for genetic transformation can be used from a high yielding commercial variety or germplasm line, thus incorporating high yielding trait over and above the stress tolerance in the resultant transgenic crop plant. MAS, marker assisted selection; AGRIKOLA, Arabidopsis genomic RNAi knock-out line analysis .

References

[1]   Urano, K., Kurihara, Y., Seki, M. and Shinozaki, K. (2010). 'Omics' analyses of regulatory networks in plant abiotic stress responses. *Current Opinion in Plant Biology* doi:10.1016/j.pbi.2009.12.006

[2]   Yamaguchi-Shinozaki, K. and Shinozaki, K. (2006). Transcriptional regulatory networks in cellular responses and tolerance to dehydration and cold stresses. *Annual Review of Plant Biology* 57, 781-803.

[3]   Govind, G., Harshavardhan, V., Patricia, J., Dhanalakshmi, R., Senthil Kumar, M., Sreenivasulu, N. and Udayakumar, M. (2009). Identification and functional validation of a unique set of drought induced genes preferentially expressed in response to gradual water stress in peanut. *Molecular Genetics and Genomics* 281, 607-607.

[4]   Senthil-Kumar, M., Rame Gowda, H. V., Hema, R., Mysore, K. S. and Udayakumar, M. (2008). Virus-induced gene silencing and its application in characterizing genes involved in water-deficit-stress tolerance. *Journal of Plant Physiology* 165, 1404-1421.

[5]   Shubha, V. and Akhilesh, K. T. (2007). Emerging trends in the functional genomics of the abiotic stress response in crop plants. *Plant Biotechnology Journal* 5, 361-380.

[6]   Tessa, M. B.-S., Jeffrey, C. A., Gregory, B. M. and Dinesh-Kumar, S. P. (2004). Applications and advantages of virus-induced gene silencing for gene function studies in plants. *The Plant Journal* 39, 734-746.

[7]   Travella, S. and Keller, B. (2008) Down-regulation of gene expression by RNA-induced gene silencing, in *Transgenic Wheat, Barley and Oats: Production and Characterization Protocols* (Huw D. Jones, and Shewry, P. R., Eds.), pp 185-199, Springer, Secaucus.

[8]   Senthil-Kumar, M., Hema, R., Suryachandra, T. R., Ramegowda, H. V., Gopalakrishna, R., Rama, N., Udayakumar, M. and Mysore, K. S. (2010). Functional characterization of three water deficit stress-induced genes in tobacco and Arabidopsis: An approach based on gene down regulation. *Plant Physiology and Biochemistry* 48, 35-44.

[9]   Becker, A. and Lange, M. (2010). VIGS - genomics goes functional. *Trends in Plant Science* 15, 1-4.

[10] Waterhouse, P. M., Wang, M.-B. and Lough, T. (2001). Gene silencing as an adaptive defence against viruses. *Nature* 411, 834-842.

[11] Lu, R., Martin-Hernandez, A. M., Peart, J. R., Malcuit, I. and Baulcombe, D. C. (2003). Virus-induced gene silencing in plants. *Methods* 30, 296-303.

[12] Robertson, D. (2004). VIGS vectors for gene silencing: many targets, many tools. *Annual Review of Plant Biology* 55, 495-519.

[13] Senthil-Kumar, M., Anand, A., Uppalapati, S. R. and Mysore, K. S. (2008). Virus-induced gene silencing and its applications. *CAB Reviews: Perspectives in Agriculture, Veterinary Science, Nutrition and Natural Resources* 3, doi:10.1079/PAVSNNR20083011.

[14] Dinesh-Kumar, S. P., Anandalakshmi, R., Marathe, R., Schiff, M. and Liu, Y. (2003) Virus-induced gene silencing, in *Plant Functional Genomics*, pp 287-293.

[15] Vaghchhipawala, Z. E. and Mysore, K. S. (2008) Agroinoculation: A simple procedure for systemic infection of plants with viruses, in *Plant Virology Protocols*, pp 555-562.

[16] Velasquez, A., Chakravarthy, S. and Martin, G. (2009). Virus-induced gene silencing (VIGS) in *Nicotiana benthamiana* and tomato. *Journal of visualized experiments* 28, doi: 10.3791/1292.

[17] Archana, K., Rama, N., Mamrutha, H. and Nataraja, K. N. (2009). Down-regulation of an abiotic stress related *Nicotiana benthamiana* WRKY transcription factor induces physiological abnormalities. *Indian Journal of Biotechnology* 8, 53-60.

[18] He, X., Jin, C., Li, G., You, G., Zhou, X. and Zheng, S. (2008). Use of the modified viral satellite DNA vector to silence mineral nutrition-related genes in plants: silencing of the tomato ferric chelate reductase gene, FRO1, as an example. *Science in China Series C: Life Sciences* 51, 402-409.

[19] Senthil-Kumar, M. and Udayakumar, M. (2006). High-throughput virus-induced gene-silencing approach to assess the functional relevance of a moisture stress-induced cDNA homologous to lea4. *Journal of Experimental Botany* 57, 2291-2302.

[20] Ishikawa, T. and Shigeoka, S. (2008). Recent advances in ascorbate biosynthesis and the physiological significance of ascorbate peroxidase in photosynthesizing organisms. *Bioscience, Biotechnology, and Biochemistry* 72, 1143-1154.

[21] Moeder, W., del Pozo, O., Navarre, D., Martin, G. and Klessig, D. (2007). Aconitase plays a role in regulating resistance to oxidative stress and cell death in Arabidopsis and *Nicotiana benthamiana*. *Plant Molecular Biology* 63, 273-287.

[22] Chang Sook, A., Jeong Hee, L., Hwang, A. R., Woo Taek, K. and Hyun-Sook, P. (2006). Prohibitin is involved in mitochondrial biogenesis in plants. *The Plant Journal* 46, 658-667.

[23] Nethra, P., Nataraja, K. N., Rama, N. and Udayakumar, M. (2006). Standardization of environmental conditions for induction and retention of post-transcriptional gene silencing using tobacco rattle virus vector. *Current Science* 90, 431-435.

[24] Sarowar, S., Lee, J., Ahn, E. and Pai, H. (2008). A role of hexokinases in plant resistance to oxidative stress and pathogen infection. *Journal of Plant Biology* 51, 341-346.

[25] Cho, H. K., Park, J. A. and Pai, H. S. (2008). Physiological function of NbRanBP1 in *Nicotiana benthamiana*. *Molecules and Cells* 26, 270-277.

[26] Jenkins, G. I. (2009). Signal transduction in responses to UV-B radiation. *Annual Review of Plant Biology* 60, 407-431.

[27] Catinot, J., Buchala, A., Abou-Mansour, E. and Métraux, J.-P. (2008). Salicylic acid production in response to biotic and abiotic stress depends on isochorismate in *Nicotiana benthamiana*. *FEBS Letters* 582, 473-478.

[28] Keara, A. F. (2008). Shade avoidance. *New Phytologist* 179, 930-944.

[29] Baena-Gonzalez, E., Rolland, F., Thevelein, J. M. and Sheen, J. (2007). A central integrator of transcription networks in plant stress and energy signalling. *Nature* 448, 938-942.

[30] Yan, J., He, C., Wang, J., Mao, Z., Holaday, S. A., Allen, R. D. and Zhang, H. (2004). Overexpression of the Arabidopsis 14-3-3 protein GF14l in cotton leads to a "Stay-Green" phenotype and improves stress tolerance under moderate drought conditions. *Plant Cell Physiology* 45, 1007-1014.

[31] Hirano, T., Ito, A., Berberich, T., Terauchi, R. and Saitoh, H. (2007). Virus-induced gene silencing of 14-3-3 genes abrogates dark repression of nitrate reductase activity in *Nicotiana benthamiana. Molecular Genetics and Genomics* 278, 125-133.

[32] Kusaba, M. (2004). RNA interference in crop plants. *Current Opinion in Biotechnology* 15, 139-143.

[33] Senthil-Kumar, M., Govind, G., Kang, L., Mysore, K. and Udayakumar, M. (2007). Functional characterization of *Nicotiana benthamiana* homologs of peanut water deficit-induced genes by virus-induced gene silencing. *Planta* 225, 523-539.

[34] Christianson, J. A., Llewellyn, D. J., Dennis, E. S. and Wilson, I. W. (2010). Global gene expression responses to waterlogging in roots and leaves of cotton (*Gossypium hirsutum* L.). *Plant Cell Physiology* 51, 21-37.

[35] Howes, T. and Kumagai, M. H. (2005). Virus-induced gene silencing in *Nicotiana benthamiana* using a DEAD-box helicase sequence derived from *Dunaliella salina. The Journal of Young Investigators* 12.

[36] Oh, D.-H., Leidi, E., Zhang, Q., Hwang, S.-M., Li, Y., Quintero, F. J., Jiang, X., D'Urzo, M. P., Lee, S. Y., Zhao, Y., Bahk, J. D., Bressan, R. A., Yun, D.-J., Pardo, J. M. and Bohnert, H. J. (2009). Loss of halophytism by interference with SOS1 expression. *Plant Physiology* 151, 210-222.

[37] Wang, Y.-H., Garvin, D. F. and Kochian, L. V. (2002). Rapid induction of regulatory and transporter genes in response to phosphorus, potassium, and iron deficiencies in tomato roots. Evidence for cross talk and root/rhizosphere-mediated signals. *Plant Physiology* 130, 1361-1370.

[38] Kang, Y.-W., Lee, J.-Y., Jeon, Y., Cheong, G.-W., Kim, M. and Pai, H.-S. (2010). *In vivo* effects of NbSiR silencing on chloroplast development in *Nicotiana benthamiana. Plant Molecular Biology*, 10.1007/s11103-11009-19593-11108.

[39] Constantin, G. D., Gronlund, M., Johansen, I. E., Stougaard, J. and Lund, O. S. (2008). Virus-induced gene silencing (VIGS) as a reverse genetic tool to study development of symbiotic root nodules. *Molecular Plant-Microbe Interactions* 21, 720-727.

[40] Kim, K.-J., Lim, J. H., Kim, M. J., Kim, T., Chung, H. M. and Paek, K.-H. (2008). GDSL-lipase1 (CaGL1) contributes to wound stress resistance by modulation of CaPR-4 expression in hot pepper. *Biochemical and Biophysical Research Communications* 374, 693-698.

[41] Kandoth, P. K., Ranf, S., Pancholi, S. S., Jayanty, S., Walla, M. D., Miller, W., Howe, G. A., Lincoln, D. E. and Stratmann, J. W. (2007). Tomato MAPKs LeMPK1, LeMPK2, and LeMPK3 function in the systemin-mediated defense response against herbivorous insects. *Proceedings of the National Academy of Sciences* 104, 12205-12210.

[42] Charng, Y.-y., Liu, H.-c., Liu, N.-y., Hsu, F.-c. and Ko, S.-s. (2006). Arabidopsis Hsa32, a novel heat shock protein, is essential for acquired thermotolerance during long recovery after acclimation. *Plant Physiology* 140, 1297-1305.

[43] Novillo, F., Medina, J. n. and Salinas, J. (2007). Arabidopsis CBF1 and CBF3 have a different function than CBF2 in cold acclimation and define different gene classes in the CBF regulon. *Proceedings of the National Academy of Sciences* 104, 21002-21007.

[44] Zhou, Z., Wang, M.-J., Zhao, S.-T., Hu, J.-J. and Lu, M.-Z. (2010). Changes in freezing tolerance in hybrid poplar caused by up- and down-regulation of PtFAD2 gene expression. *Transgenic Research*, doi:10.1007/s11248-11009-19349-x.

[45] Floris, M., Mahgoub, H., Lanet, E., Robaglia, C. and Menand, B. (2009). Post-transcriptional regulation of gene expression in plants during abiotic stress. *International Journal of Molecular Sciences* 10, 3168-3185.

[46] Ping, X., Fang, C., Jonathan, P. M., Tracy, F., Lloyd, W. S. and Marilyn, J. R. (2008). Virus infection improves drought tolerance. *New Phytologist* 180, 911-921.

[47] Liu, Y., Gao, Q., Wu, B., Ai, T. and Guo, X. (2009). NgRDR1, an RNA-dependent RNA polymerase isolated from *Nicotiana glutinosa*, was involved in biotic and abiotic stresses. *Plant Physiology and Biochemistry* 47, 359-368.

[48] Attila, M., Andrew, B., Eva, T., Frank, S., Shantanu, K., Stephan, O., Detlef, W. and Baulcombe, D. (2009). Highly specific gene silencing by artificial microRNAs in the unicellular alga *Chlamydomonas reinhardtii*. *The Plant Journal* 58, 165-174.

[49] Warthmann, N., Chen, H., Ossowski, S., Weigel, D. and Herve, P. (2008). Highly specific gene silencing by artificial miRNAs in rice. *PLoS ONE* 3, e1829.

[50] Jorgensen, R. A., Ma, C. and Doetsch, N. (2005) Functional genomics by sense-RNAi: A forward genetic approach for cell-type-targeted mutagenesis, in *Plant Genomics in China VI*, Kunming, China.

In: Gene Silencing: Theory, Techniques and Applications
Editor: Anthony J. Catalano

ISBN: 978-1-61728-276-8
© 2010 Nova Science Publishers, Inc.

Chapter XIII

Genome-wide Identification and Analysis of miRNAs Complementary to Upstream Sequences of mRNA Transcription Start Sites

Kenta Narikawa[1], Kenji Nishi[2], Yuki Naito[2], Minami Mazda[2] and Kumiko Ui-Tei[1,2,*]

[1]Department of Computational Biology, Graduate School of Frontier Sciences, University of Tokyo, 5-1-5 Kashiwanoha, Kashiwa-shi, Chiba-ken 277-8561, Japan
[2]Department of Biophysics and Biochemistry, Graduate School of Science, University of Tokyo, 2-11-16 Yayoi, Bunkyo-ku, Tokyo 113-0032, Japan

Abstract

Small regulatory RNAs including short interfering RNAs (siRNAs) and microRNAs (miRNAs) are crucial regulators of gene expression at the posttranscriptional level. Recently, additional roles for small RNAs in gene activation and suppression at the transcriptional level were reported; these RNAs were shown to have sequences that closely or completely match to their respective promoter regions. However, no global analysis for identifying target sequences for miRNAs in the promoter region have been carried out in the human genome.

We performed a genome-wide search for upstream sequences of mRNA transcription start sites where miRNAs are capable of hybridizing with high complementarity. We identified 219 sites in the 10-kb upstream regions of transcription start sites with complete complementarity to 94 human mature miRNAs. Furthermore, the mismatched positions and nucleotides in near-completely matched sites were highly biased, and most

---

* Email address: ktei@bi.s.u-tokyo.ac.jp

of them appear to be possible target sites of miRNAs. The expression of downstream genes of miRNA target sites were examined following transfection of each miRNA into three different human cell lines. The results indicate that miRNAs dynamically modulates gene expression depending on the downstream genes and the cell type.

Introduction

Small regulatory RNAs including small interfering RNAs (siRNAs) and microRNAs (miRNAs) regulate gene expression at the posttranscriptional level in various organisms [1,2]. RNA silencing encompasses a group of mechanistically related pathways. However, the biogenesis and mechanism of action differ between siRNA and miRNA. In the siRNA-mediated RNA interference (RNAi) pathway, long double-stranded RNAs (dsRNAs) are processed by the cytoplasmic RNase III enzyme Dicer into 21–23-nucleotide (nt) siRNAs. The siRNA guide strand that is incorporated into the RNA-induced silencing complex (RISC) closely or completely matches their target sequences, and a core component of RISC, Argonaute (Ago) protein, mediates the sequence-specific cleavage of target mRNAs [3,4]. In the miRNA pathway, miRNAs are transcribed by RNA polymerases II and III [5]. This generates primary miRNA (pri-miRNA) transcripts, which are further processed to stem-loop-structured precursor miRNAs (pre-miRNAs) by the nuclear RNase III Drosha as part of a microprocessor complex with DGCR8 [6-8]. The pre-miRNA is exported into the cytoplasm by the Exportin-5/Ran-GTP complex [9]. Dicer cleaves the cytoplasmic pre-miRNA hairpin into the mature miRNA, which enters the miRNA-containing ribonucleoprotein (miRNP) and is capable of binding to partially complementary sequences within the 3′ UTR of the target mRNA.

In the past few years, additional roles for small non-coding RNAs to direct both gene activation and suppression at the transcriptional level in nucleus of human cells have been reported. Exogenous siRNAs with sequence complementarity to promoter regions have been shown to induce transcriptional gene silencing (TGS) [10-15]. Meanwhile, siRNA sequences with promoter complementarity also direct transcriptional activation in a process referred to as RNA activation (RNAa) [16-18]. While little is known about the mechanisms and biological effects of this aberrant siRNA behaviour, studies regarding the mechanisms of TGS and RNAa have provided some insight. TGS induced by promoter-directed siRNAs has been shown to require Ago proteins [12,13] and promoter-associated RNAs that span the targeted loci [15,19]. RNAa was shown to be associated with the 5′ end of the siRNA guide strand and Ago2 [17] and antisense transcripts [18,19]. However, siRNA targeted to the HIV-1 LTR promoter was also demonstrated to induce gene activation via an indiscriminate off-target effect [20].

Both TGS and RNAa have been shown to be inducible by endogenous miRNAs that completely or nearly completely match promoters in human cells [21,22]. miR-373 has highly complementary target sequences to the promoters of E-cadherin and CSDC2 (cold shock domain-containing protein C2), and it induces the expression of both genes [22]. Furthermore, miR-320 is encoded within the promoter region of the cell cycle gene POLR3D in the antisense orientation, and transcriptionally silences POLR3D when associated with Ago1, Polycomb-group (PcG) component EZH2, and trimethyl histone H3 lysine 27 (H3K27me3) [21]. These results suggest that endogenous miRNAs sharing complementary sequences to promoter regions regulate gene expression at the level of transcription. In the human genome, ~700 miRNA genes are distributed in single or clustered fashion and comprise independent transcription units in intergenic regions or are embedded within introns of protein-coding genes [23]. While genome-wide computational and experimental searches for predicting miRNA regulatory targets in the 3′ UTRs but not in the upstream regions have become major challenges [24], but their intended regions have been mainly 3′ UTRs. To date, no global analyses have been carried out to identify miRNAs targeting upstream regions of mRNA coding regions. We performed genome-wide analyses searching for upstream sequences of transcription start sites (TSSs) where miRNAs share complementarity. The changes in the expression of downstream genes due to the miRNAs were examined.

Materials and Methods

Search for Upstream Regions Containing Sequences Complementary to miRNAs

The transcriptional start sites were assumed to be the 5′-terminus of the mRNAs registered in the RefSeq database in this study. The 10-kb sequences upstream of the TSSs were retrieved by matching of the 5′-terminus of mRNA sequences registered in the RefSeq release 27 database (http://www.ncbi.nlm.nih.gov/RefSeq/) [25] with human genome sequences from the UCSC Genome Browser (http://genome.ucsc.edu) [26]. Then, the 10-kb upstream sequences of human mRNA and the miRNA sequences retrieved from the miRBase release 10.0 (http://microrna.sanger.ac.uk/) [27] were aligned, and regions with perfect complementarity in both directions were selected. In addition, 1-kb upstream regions that contained incomplete complementary sequences (i.e., 1- or 2-bp mismatches) with miRNAs were also selected.

Cell Culture

Human carcinoma cell line of the uterine cervix, HeLa, and prostatic carcinoma cell line, PC-3 (Health Science Research Resources Bank, Japan), were cultured in Dulbecco's Modified Eagle's Medium (DMEM; Invitrogen) at 37°C. Media for both cell lines were supplemented with 10 % heat-inactivated fetal bovine serum (FBS; Mitsubishi Kagaku) and antibiotics (10 units/ml of penicillin (Meiji) and 50 µg/ml of streptomycin (Meiji)). Human MCF-7 breast cancer cell line (Health Science Research Resources Bank, Japan) was cultured in Modified Eagle's Medium (MEM; Invitrogen) supplemented with 10 % heat-inactivated FBS, 100 mM sodium pyruvate (Invitrogen) and 4 mg/ml insulin (Invitrogen).

Plasmid Construction and RNA Silencing Assay

All plasmids constructed are derivatives of psiCHECK-1 (Promega, Madison, Wisconsin). Chemically synthesized double-stranded oligonucleotides containing 81-bp were inserted into the corresponding restriction enzyme sites of psiCHECK-1 to generate psiCHECK-miR-target. Each oligonucleotide included three tandem repeats of an identical 25-bp containing completely matched target sequences of miR-378, miR-548c-5p, miR-548d-5p, miR-572, miR-574-5p and miR-606 miRNAs and cohesive XhoI/EcoRI ends (Table 1). The inserted miRNA targets were expressed as part of the 3'_UTR of the *Renilla luciferase* mRNA in the transfected cells. Cells in each well of a 24-well culture plate were transfected simultaneously with one of the psiCHECK-miR-target (100 ng), a pGL3-Control (0.5 µg; Promega) and synthetic miRNA (0.005, 0.05, and 0.5 µg; Sigma). Cells were harvested 24 h after transfection and relative *luc* activity (*Renilla luciferase* activity/firefly *luciferase* activity) was determined using the Dual-Luciferase Reporter Assay System (Promega). Unrelated siRNA against the firefly *luciferase* gene (5'-CGUACGCGGAAUACUUCGAUU-3' and 5'-UCGAAGUAUUCCGCGUACGUG-3') was used as a miRNA control (dsControl).

**Table 1. Oligonucleotide sequences inserted
into psiCHECK-miR-target constructs**

oligo (-SS)	sequence (5' ⇒ 3')
378-SS	tcgaGGCCTTCTGACTCCAAGTCCAGTGCGGCCTTCTGACTCCAAGTCCAGTGCGGCCTTCTGACTCCAAGTCCAGTGC
548c-5p-SS	tcgaTGGCAAAAACCGCAATTACTTTTGCTGGCAAAAACCGCAATTACTTTTGCTGGCAAAAACCGCAATTACTTTTGC
548d-5p-SS	tcgaTGGCAAAAACCACAATTACTTTTGCTGGCAAAAACCACAATTACTTTTGCTGGCAAAAACCACAATTACTTTTGC
572-SS	tcgaGGCTGGGCCACCGCCGAGCGGACGAGGCTGGGCCACCGCCGAGCGGACGAGGCTGGGCCACCGCCGAGCGGACGA
574-5p-SS	tcgaCACACACTCACACACACACACTCACCACACACTCACACACACACACTCACCACACACTCACACACACACACTCAC
606-SS	tcgaGTATCTTTGATTTTCAGTAGTTTACGTATCTTTGATTTTCAGTAGTTTACGTATCTTTGATTTTCAGTAGTTTAC

oligo (-AS)	sequence (5' ⇒ 3')
378-AS	aattGCACTGGACTTGGAGTCAGAAGGCCGCACTGGACTTGGAGTCAGAAGGCCGCACTGGACTTGGAGTCAGAAGGCC
548c-5p-AS	aattGCAAAAGTAATTGCGGTTTTTGCCAGCAAAAGTAATTGCGGTTTTTGCCAGCAAAAGTAATTGCGGTTTTTGCCA
548d-5p-AS	aattGCAAAAGTAATTGTGGTTTTTGCCAGCAAAAGTAATTGTGGTTTTTGCCAGCAAAAGTAATTGTGGTTTTTGCCA
572-AS	aattTCGTCCGCTCGGCGGTGGCCCAGCCTCGTCCGCTCGGCGGTGGCCCAGCCTCGTCCGCTCGGCGGTGGCCCAGCC
574-5p-AS	aattGTGAGTGTGTGTGTGTGAGTGTGTGGTGAGTGTGTGTGTGTGAGTGTGTGGTGAGTGTGTGTGTGTGAGTGTGTG
606-AS	aattGTAAACTACTGAAAATCAAAGATACGTAAACTACTGAAAATCAAAGATACGTAAACTACTGAAAATCAAAGATAC

AS and SS, respectively, indicate antisense and sense strand sequences of three copies of each mRNA target sequences. Both strands were annealed and inserted into the psiCHECK vector.

Table 2. Stem‑loop RT primers specific for each miRNA

name	sequence (5' ⇒ 3')
st21	gtcgtatccagtgcagggtccgaggtattcgcactggatacgactcaaca
st378	gtcgtatccagtgcagggtccgaggtattcgcactggatacgacccttct
st548c-5p	gtcgtatccagtgcagggtccgaggtattcgcactggatacgacggcaaa
st548d-5p	gtcgtatccagtgcagggtccgaggtattcgcactggatacgacggcaaa
st572	gtcgtatccagtgcagggtccgaggtattcgcactggatacgactgggcc
st574-5p	gtcgtatccagtgcagggtccgaggtattcgcactggatacgacacacac
st606	gtcgtatccagtgcagggtccgaggtattcgcactggatacgacatcttt

st21, 378, 548c-5p, 548d-5p, 572, 574-5p and 606 are RT primers for miR-21, miR-378, miR-548c-5p, miR-548d-5p, miR-572, miR-574-5p, and miR-606, respectively.

Table 3. PCR primers used for each miRNA

name	sequence (5' ⇒ 3')
stRT-21fp	gcccgctagcttatcagactgatg
stRT-378fp	gccgccactggacttggagtc
stRT-548cfp	gccgcaaaagtaattgcggt
stRT-548dfp	gccgcgaaaagtaattgtggt
stRT-572fp	gccgcggtccgctcggcggtg
stRT-574fp	gcccgctgagtgtgtgtgtgtgagt
stRT-606fp	gcccgcaaactactgaaaatga
stRT-rp	gtgcagggtccgaggt

stRT-21fp, stRT-378fp, stRT-548c-5pfp, stRT-548d-5pfp, stRT-572fp, stRT-574-5pfp and stRT-606fp are forward primers for PCR of miR-21, miR-378, miR-548c-5p, miR-548d-5p, miR-572, miR-574-5p and miR-606, respectively. stRT-rp is a universal reverse primer.

Table 4. PCR primers for downstream genes

gene name	fp sequence (5' ⇒ 3')	rp sequence (5' ⇒ 3')	PCR product (bp)
OSBPL10	catgctggtggtgtactctgct	cgggagcttggagcactctt	144
RNPS1	tagggctccttcacctaccaaa	cgatccttctcactcgactcct	102
OSBPL9	agcgtccatcttccctacca	tgctactcggtggtgaatgg	105
POLR3A	gacctggagttgccgtgttt	tgatgtggcagcaggttttg	88
ANXA3	gcatctcatggtggccccta	cttcatttgcctgcttgtcc	136
CDC23	ctgcttattgcgggccttac	gcatccatatcctgggcatc	149
RNF6	agattatcggcttatgagagacca	ttccttgacgccatctaacc	94
C20orf43	ctcttgggaaggcagcatct	ctggaggtcatcgtgcttgt	125
actin	cacactgtgcccatctacga	gccatctcttgctcgaagtc	203

PCR primers are indicated by the gene names. fp and rp represent forward and reverse primers, respectively. Expected size of PCR product is shown in the right colume.

Detection of Endogenous miRNAs by RT-PCR

Endogenous miRNAs were detected by RT-PCR according to the procedure reported by Chen et al [28]. Briefly, 1 ml of the HeLa, PC-3, or MCF-7 cell suspensions (containing 1 x

10^5 cells/ml) was inoculated into each well of a 24-well plate 24 h prior to transfection. Cells were transfected without or with chemically synthesized miRNAs (500 ng/well) using Lipofectamine 2000 (Invitrogen, Carlsbad, California). The cells transfected with synthesized miRNAs were used as miRNA detection control. Total RNA was purified 3 days after transfection using the mirVana miRNA Isolation Kit (Ambion, Austin, Texas) according to the manufacturer's protocol. The reverse transcriptase reactions contained total RNA (0.25 µg), 50 nmoles of stem–loop RT primer specific for each miRNA (Table 2), $1\times$ RT buffer (RETROscript Kit; Ambion), 0.25 mM dNTPs, 3.33 U/µl of MultiScribe reverse transcriptase (Applied Biosystems, Foster City, California), and 0.25 U/µl of RNase OUT (Invitrogen). The mixture was incubated at 16 °C for 30 min, 42 °C for 30 min, 85 °C for 5 min and then held at 4 °C. The PCR reaction was performed using AmpliTaq Gold PCR Master Mix (Applied Biosystems) for each miRNA using universal reverse primers (Table 3). The mixture was incubated at 95 °C for 10 min, followed by 40 cycles at 95 °C for 15 s, and 60 °C for 1 min. The reactions were fractionated on a 3 % agarose gel with TBE.

Detection of Downstream Gene Expression by Real Time RT-PCR

Using Lipofectamine 2000 (Invitrogen), HeLa, PC-3 and MCF-7 cells were cotransfected with or without 500 ng/well chemically synthesized miRNA. Total RNA was treated with RQ1 DNase (Promega) and purified using RNeasy Mini Kit (QIAGEN) 3 days after transfection. Reverse transcriptase reactions were carried out using total RNA (0.25 µg) with Anchored-oligo(dT)$_{18}$ primer by Transcriptor High Fidelity cDNA Synthesis Kit (Roche) according to the manufacturer's protocol. Real time PCR was performed using FastStart Universal SYBR Green Master (Roche) with forward and reverse primers specific for OSBPL10, RNPS1, OSBPL9, POLR3A, ANXA3, CDC23, RNF6, C20orf43 and actin genes (Table 4). The mixture was incubated at 95 °C for 10 min, followed by 40 cycles of 95 °C for 15 sec and 60 °C for 1 min and monitored by ABI PRISM 7000 (Applied Biosystems). The expression level of each mRNA was first normalized to the amount of actin and then to the siRNA against the firefly *luciferase* (dsControl) transfection control.

Results

Search for miRNAs Having Perfect Complementarity within Upstream Sequences of mRNA Coding Regions

We identified 219 sites in the 10-kb upstream regions of TSSs with complete complementarity to 94 human mature miRNAs (Tables 5 and 6). In this study, the 5'-end of mRNA was assumed to be TSS (position +1). More than one site was included in some of the upstream sequences, and overlapping transcriptional variants were counted at several sites; the 219 sites corresponded to the upstream sequences of 155 genes (Table 5). Among these sites, 153 sites were complementary to pre-miRNAs which generate 88 mature miRNAs registered in the miRBase, indicating that these miRNAs are transcribed from the upstream regions as origins. As shown for miR-320 [21], these miRNAs may have multiple regulatory functions if the antisense transcript is transcribed at the same position. Except for these cases, the 66 sites of 50 genes (labelled with blue in Table 6) appeared perfect complementarity to the miRNAs (miR-378, miR-548c-5p, miR-548d-5p, miR-572, miR-574-5p and miR-606) transcribed from the different location other than the 10-kb upstream regions.

Table 5. Results of the search for miRNAs with complementarity to upstream sequences of transcription start sites

	miRNAs	Complementary sites***	Genes
Total number	722	24,854	18,578
Perfect match containing miRNA transcription sites	94* (20**)	219* (36**)	155* (26**)
Perfect match	6* (4**)	66* (13**)	50* (10**)
1-bp mismatch	10**	69**	53**
2-bp mismatch	44**	4453**	763**

The search was carried out using 722 mature miRNA sequences against 24,854 sequences 10-kb upstream of 18,578 genes transcription start sites. * indicates the outcome of the 10-kb upstream sequences; ** denotes the 1-kb upstream sequences. *** Note that some of the upstream sequences have multiple complementary sites. Details of completely complementary and 1-bp mismatched sites are shown in Tables 6 and 7.

Table 6. List of miRNAs and genes with perfectly matched sequences in the 10-kb upstream regions

let-7i	gene	ID	position		
UGAGGUAGUAGUUUGUGCUGUU	C12orf61	NM_175895	-257	+	origin

let-7i*	gene	ID	position		
CUGCGCAAGCUACUGCCUUGCU	C12orf61	NM_175895	-313	+	origin

miR-1	gene	ID	position		
UGGAAUGUAAAGAAGUAUGUAU	C20orf200	NM_152757	-2827	+	origin

miR-7	gene	ID	position		
UGGAAGACUAGUGAUUUUGUUGU	ISG20L1	NM_022767	-9502	-	origin

miR-7-2*	gene	ID	position		
CAACAAAUCCCAGUCUACCUAA	ISG20L1	NM_022767	-9462	-	origin

miR-10a	gene	ID	position		
UACCCUGUAGAUCCGAAUUUGUG	HOXB3	NM_002146	-5478	-	origin
	HOXB4	NM_024015	-1545	-	origin

Table 6. (Continued)

miR-10a*	gene	ID	position		
CAAAUUCGUAUCUAGGGGAAUA	HOXB3	NM_002146	-5437	-	origin
	HOXB4	NM_024015	-1504	-	origin

miR-10b	gene	ID	position		
UACCCUGUAGAACCGAAUUUGUG	HOXD4	NM_014621	-1056	-	origin

miR-10b*	gene	ID	position		
ACAGAUUCGAUUCUAGGGGAAU	HOXD4	NM_014621	-1017	-	origin

miR-15b	gene	ID	position		
UAGCAGCACAUCAUGGUUUACA	IFT80	NM_020800	-5075	+	origin

miR-15b*	gene	ID	position		
CGAAUCAUUAUUUGCUGCUCUA	IFT80	NM_020800	-5113	+	origin

miR-16	gene	ID	position		
UAGCAGCACGUAAAUAUUGGCG	IFT80	NM_020800	-5222	+	origin

miR-16-2*	gene	ID	position		
CCAAUAUUACUGUGCUGCUUUA	IFT80	NM_020800	-5265	+	origin

miR-22	gene	ID	position		
AAGCUGCCAGUUGAAGAACUGU	WDR81	NM_152348	-2601	+	origin

miR-22*	gene	ID	position		
AGUUCUUCAGUGGCAAGCUUUA	WDR81	NM_152348	-2563	+	origin

miR-25	gene	ID	position		
CAUUGCACUUGUCUCGGUCUGA	AP4M1	NM_004722	-7915	+	origin

miR-25*	gene	ID	position		
AGGCGGAGACUUGGGCAAUUG	AP4M1	NM_004722	-7877	+	origin

miR-26a	gene	ID	position		
UUCAAGUAAUCCAGGAUAGGCU	AVIL	NM_006576	-8610	-	origin

miR-26a-2*	gene	ID	position		
CCUAUUCUUGAUUACUUGUUUC	AVIL	NM_006576	-8572	-	origin

Table 6. (Continued)

miR-34b	gene	ID	position		
CAAUCACUAACUCCACUGCCAU	BTG4	NM_017589	-648	+	origin
	FLJ46266	NM_001100388	-1798	-	origin
	FLJ46266	NM_207430	-1798	-	origin

miR-34b*	gene	ID	position		
UAGGCAGUGUCAUUAGCUGAUUG	BTG4	NM_017589	-611	+	origin
	FLJ46266	NM_001100388	-1835	-	origin
	FLJ46266	NM_207430	-1835	-	origin

miR-34c-3p	gene	ID	position		
AAUCACUAACCACACGGCCAGG	BTG4	NM_017589	-1145	+	origin
	FLJ46266	NM_001100388	-1301	-	origin
	FLJ46266	NM_207430	-1301	-	origin

miR-34c-5p	gene	ID	position		
AGGCAGUGUAGUUAGCUGAUUGC	BTG4	NM_017589	-1112	+	origin
	FLJ46266	NM_001100388	-1334	-	origin
	FLJ46266	NM_207430	-1334	-	origin

miR-92b	gene	ID	position		
UAUUGCACUCGUCCCGGCCUCC	MUC1	NM_001018016	-2328	+	origin
	MUC1	NM_001018017	-2328	+	origin
	MUC1	NM_001044390	-2328	+	origin
	MUC1	NM_001044391	-2328	+	origin
	MUC1	NM_001044392	-2328	+	origin
	MUC1	NM_001044393	-2328	+	origin
	MUC1	NM_002456	-2328	+	origin

miR-92b*	gene	ID	position		
AGGGACGGGACGCGGUGCAGUG	MUC1	NM_001018016	-2287	+	origin
	MUC1	NM_001018017	-2287	+	origin
	MUC1	NM_001044390	-2287	+	origin
	MUC1	NM_001044391	-2287	+	origin
	MUC1	NM_001044392	-2287	+	origin
	MUC1	NM_001044393	-2287	+	origin
	MUC1	NM_002456	-2287	+	origin

miR-93	gene	ID	position		
CAAAGUGCUGUUCGUGCAGGUAG	AP4M1	NM_004722	-7670	+	origin

miR-93*	gene	ID	position		
ACUGCUGAGCUAGCACUUCCCG	AP4M1	NM_004722	-7709	+	origin

Table 6. (Continued)

miR-106b	gene	ID	position		
UAAAGUGCUGACAGUGCAGAU	AP4M1	NM_004722	-7444	+	origin

miR-106b*	gene	ID	position		
CCGCACUGUGGGUACUUGCUGC	AP4M1	NM_004722	-7484	+	origin

miR-132	gene	ID	position		
UAACAGUCUACAGCCAUGGUCG	HIC1	NM_001098202	-6360	+	origin
	HIC1	NM_006497	-5149	+	origin

miR-132*	gene	ID	position		
ACCGUGGCUUUCGAUUGUUACU	HIC1	NM_001098202	-6324	+	origin
	HIC1	NM_006497	-5113	+	origin

miR-138	gene	ID	position		
AGCUGGUGUUGUGAAUCAGGCCG	SLC12A3	NM_000339	-6703	-	origin

miR-138-2*	gene	ID	position		
GCUAUUUCACGACACCAGGGUU	SLC12A3	NM_000339	-6656	-	origin

miR-141	gene	ID	position		
UAACACUGUCUGGUAAAGAUGG	EMG1	NM_006331	-6676	-	origin

miR-141*	gene	ID	position		
CAUCUUCCAGUACAGUGUUGGA	EMG1	NM_006331	-6718	-	origin

miR-142-3p	gene	ID	position		
UGUAGUGUUUCCUACUUUAUGGA	BZRAP1	NM_004758	-3150	-	origin

miR-142-5p	gene	ID	position		
CAUAAAGUAGAAAGCACUACU	BZRAP1	NM_004758	-3186	-	origin

miR-146b-3p	gene	ID	position		
UGCCCUGUGGACUCAGUUCUGG	CUEDC2	NM_024040	-3890	+	origin

miR-146b-5p	gene	ID	position		
UGAGAACUGAAUUCCAUAGGCU	CUEDC2	NM_024040	-3854	+	origin

miR-181c	gene	ID	position		
AACAUUCAACCUGUCGGUGAGU	NANOS3	NM_001098622	-2411	-	origin

miR-181c*	gene	ID	position		
AACCAUCGACCGUUGAGUGGAC	NANOS3	NM_001098622	-2373	-	origin

Table 6. (Continued)

miR-181d	gene	ID	position		
AACAUUCAUUGUUGUCGGUGGGU	NANOS3	NM_001098622	-2226	-	origin

miR-190b	gene	ID	position		
UGAUAUGUUUGAUAUUGGGUU	TPM3	NM_152263	-1600	-	origin

miR-191	gene	ID	position		
CAACGGAAUCCCAAAAGCAGCUG	C3orf60	NM_199069	-947	+	origin
	C3orf60	NM_199070	-947	+	origin
	C3orf60	NM_199073	-491	+	origin
	C3orf60	NM_199417	-947	+	origin
	DALRD3	NM_001009996	-2110	-	origin

miR-191*	gene	ID	position		
GCUGCGCUUGGAUUUCGUCCCC	C3orf60	NM_199069	-989	+	origin
	C3orf60	NM_199070	-989	+	origin
	C3orf60	NM_199073	-533	+	origin
	C3orf60	NM_199417	-989	+	origin
	DALRD3	NM_001009996	-2068	-	origin

miR-195	gene	ID	position		
UAGCAGCACAGAAAUAUUGGC	BCL6B	NM_181844	-5363	+	origin

miR-195*	gene	ID	position		
CCAAUAUUGGCUGUGCUGCUCC	BCL6B	NM_181844	-5401	+	origin

miR-196a	gene	ID	position		
UAGGUAGUUUCAUGUUGUUGGG	HOXB9	NM_024017	-6080	-	origin
	HOXC9	NM_006897	-8331	-	origin

miR-196a*	gene	ID	position		
CGGCAACAAGAAACUGCCUGAG	HOXC9	NM_006897	-8294	-	origin

miR-196b	gene	ID	position		
UAGGUAGUUUCCUGUUGUUGGG	HOXA9	NM_152739	-4019	-	origin

miR-200c	gene	ID	position		
UAAUACUGCCGGGUAAUGAUGGA	EMG1	NM_006331	-7089	-	origin

miR-200c*	gene	ID	position		
CGUCUUACCCAGCAGUGUUUGG	EMG1	NM_006331	-7128	-	origin

Table 6. (Continued)

miR-208b	gene	ID	position		
AUAAGACGAACAAAAGGUUUGU	MYH6	NM_002471	-9745	-	origin

miR-210	gene	ID	position		
CUGUGCGUGUGACAGCGGCUGA	C11orf35	NM_173573	-7354	-	origin
	KIAA1542	NM_020901	-8353	+	origin

miR-212	gene	ID	position		
UAACAGUCUCCAGUCACGGCC	HIC1	NM_001098202	-6000	+	origin
	HIC1	NM_006497	-4789	+	origin

miR-219-1-3p	gene	ID	position		
AGAGUUGAGUCUGGACGUCCCG	RING1	NM_002931	-613	-	origin
	RXRB	NM_021976	-7237	+	origin
	RXRB	NM_021976	-7241	+	origin
	RXRB	NM_021976	-7245	+	origin

miR-219-5p	gene	ID	position		
UGAUUGUCCAAACGCAAUUCU	RING1	NM_002931	-654	-	origin
	RXRB	NM_021976	-7196	+	origin
	RXRB	NM_021976	-7200	+	origin
	RXRB	NM_021976	-7204	+	origin

miR-301a	gene	ID	position		
CAGUGCAAUAGUAUUGUCAAAGC	PRR11	NM_018304	-4561	+	origin

miR-320	gene	ID	position		
AAAAGCUGGGUUGAGAGGGCGA	POLR3D	NM_001722	-110	+	origin

miR-324-3p	gene	ID	position		
ACUGCCCCAGGUGCUGCUGG	DLG4	NM_001365	-3589	-	origin

miR-324-5p	gene	ID	position		
CGCAUCCCCUAGGGCAUUGGUGU	DLG4	NM_001365	-3626	-	origin

miR-345	gene	ID	position		
GCUGACUCCUAGUCCAGGGCUC	SLC25A29	NM_001039355	-1353	+	origin

miR-375	gene	ID	position		
UUUGUUCGUUCGGCUCGCGUGA	CRYBA2	NM_005209	-8270	-	origin
	CRYBA2	NM_057093	-8264	-	origin
	CRYBA2	NM_057094	-8264	-	origin

Table 6. (Continued)

miR-378	gene	ID	position		
ACUGGACUUGGAGUCAGAAGG	OSBPL10	NM_017784	-4582	-	

miR-425	gene	ID	position		
AAUGACACGAUCACUCCCGUUGA	C3orf60	NM_199069	-1420	+	origin
	C3orf60	NM_199070	-1420	+	origin
	C3orf60	NM_199073	-964	+	origin
	C3orf60	NM_199074	-254	+	origin
	C3orf60	NM_199417	-1420	+	origin
	DALRD3	NM_001009996	-1637	-	origin

miR-425*	gene	ID	position		
AUCGGGAAUGUCGUGUCCGCCC	C3orf60	NM_199069	-1461	+	origin
	C3orf60	NM_199070	-1461	+	origin
	C3orf60	NM_199073	-1005	+	origin
	C3orf60	NM_199074	-295	+	origin
	C3orf60	NM_199417	-1461	+	origin
	DALRD3	NM_001009996	-1596	-	origin

miR-483-3p	gene	ID	position		
UCACUCCUCUCCUCCCGUCUU	IGF2AS	NM_016412	-6345	+	origin

miR-483-5p	gene	ID	position		
AAGACGGGAGGAAAGAAGGGAG	IGF2AS	NM_016412	-6305	+	origin

miR-484	gene	ID	position		
UCAGGCUCAGUCCCCUCCCGAU	KIAA0430	NM_014647	-149	+	origin
	NDE1	NM_017668	-6946	-	origin

miR-497	gene	ID	position		
CAGCAGCACACUGUGGUUUGU	BCL6B	NM_181844	-5051	+	origin

miR-497*	gene	ID	position		
CAAACCACACUGUGGUGUUAGA	BCL6B	NM_181844	-5091	+	origin

miR-548c-5p	gene	ID	position		
AAAAGUAAUUGCGGUUUUUGCC	ABTB1	NM_032548	-8217	-	
	ABTB1	NM_172027	-8217	-	
	ABTB1	NM_172028	-8217	-	
	ANXA3	NM_005139	-248	+	
	C1orf94	NM_032884	-1735	+	
	C20orf43	NM_016407	-8342	-	
	CDC23	NM_004661	-6363	+	
	CDC42EP4	NM_012121	-3863	-	
	EBI2	NM_004951	-8486	+	

Table 6. (Continued)

miR-548c-5p	gene	ID	position		
	FGF23	NM_020638	-9861	+	
	FOXC2	NM_005251	-8776	+	
	GNB5	NM_006578	-3189	-	
	GPR141	NM_181791	-8087	+	
	LBP	NM_004139	-6809	-	
	MANBAL	NM_001003897	-4841	-	
	MANBAL	NM_022077	-4841	-	
	MTHFSD	NM_022764	-3325	-	
	NLGN4X	NM_020742	-9292	-	
	NLGN4X	NM_181332	-8474	-	
	NR0B1	NM_000475	-5731	+	
	NSUN5	NM_018044	-6978	+	
	NSUN5	NM_148956	-6978	+	
	OR2T8	NM_001005522	-7046	+	
	PDE6C	NM_006204	-3974	-	
	PLAC1	NM_021796	-1244	+	
	PSTPIP2	NM_024430	-798	+	
	RBP4	NM_006744	-7378	+	
	RNF6	NM_005977	-7624	-	
	RNF6	NM_183043	-8154	-	
	RNF6	NM_183044	-8292	-	
	UBQLN2	NM_013444	-1751	+	
	ZNF331	NM_001079907	-9126	-	
	ZNF418	NM_133460	-732	+	

miR-548d-5p	gene	ID	position		
AAAAGUAAUUGUGGUUUUUGCC	ADAM23	NM_003812	-7057	-	
	ALDH1L2	NM_001034173	-914	+	
	C1orf94	NM_032884	-1790	-	
	C8B	NM_000066	-333	-	
	CLEC12A	NM_138337	-525	+	
	CLEC12A	NM_201623	-525	+	
	FSTL5	NM_020116	-9873	+	
	GC	NM_000583	-6013	+	
	GJB3	NM_001005752	-3678	+	
	GJB3	NM_024009	-2557	+	
	MAPKAPK5	NM_003668	-2843	+	
	MAPKAPK5	NM_139078	-2843	+	
	NANOS1	NM_199461	-9261	-	
	NLRP4	NM_134444	-8220	+	
	POLR3A	NM_007055	-8654	-	
	SOX30	NM_007017	-621	+	
	SOX30	NM_178424	-621	+	
	SPAG11B	NM_058206	-5668	-	
	SPAG11B	NM_058207	-5668	-	
	TAS2R39	NM_176881	-2594	+	
	WNT5A	NM_003392	-7212	+	
	WWP2	NM_199424	-2131	-	

miR-565	gene	ID	position		
GGCUGGCUCGCGAUGUCUGUUU	SACM1L	NM_014016	-443	+	origin

miR-572	gene	ID	position		
GUCCGCUCGGCGGUGGCCCA	RNPS1	NM_006711	-24	-	

miR-574-5p	gene	ID	position		
UGAGUGUGUGUGUGUGAGUGUGU	JPH2	NM_020433	-1016	+	
	JPH2	NM_175913	-1016	+	
	LOC645037 (GAGE2C)	NM_001098411	-6003	+	
	OPN4	NM_001030015	-283	+	
	OPN4	NM_033282	-283	+	
	PRKCE	NM_005400	-815	+	

miR-580	gene	ID	position		
UUGAGAAUGAUGAAUCAUUAGG	SKP2	NM_005983	-4159	+	origin
	SKP2	NM_032637	-4159	+	origin

miR-584	gene	ID	position		
UUAUGGUUUGCCUGGGACUGAG	NM_001101342	NM_001101342	-923	+	origin

miR-596	gene	ID	position		
AAGCCUGCCCGGCUCCUCGGG	ARHGEF10	NM_014629	-6737	-	origin

miR-606	gene	ID	position		
AAACUACUGAAAAUCAAAGAU	OSBPL9	NM_148904	-5101	+	
	OSBPL9	NM_148905	-5101	+	
	OSBPL9	NM_148907	-5131	+	

miR-607	gene	ID	position		
GUUCAAAUCCAGAUCUAUAAC	LCOR	NM_032440	-4276	+	origin

miR-611	gene	ID	position		
GCGAGGACCCCUCGGGGUCUGAC	FEN1	NM_004111	-156	+	origin

miR-616	gene	ID	position		
AGUCAUUGGAGGGUUUGAGCAG	MBD6	NM_052897	-3672	+	origin

miR-616*	gene	ID	position		
ACUCAAAACCCUUCAGUGACUU	MBD6	NM_052897	-3632	+	origin

Table 6. (Continued)

miR-632	gene	ID	position		
GUGUCUGCUUCCUGUGGGA	C17orf75	NM_022344	-7999	+	origin

miR-636	gene	ID	position		
UGUGCUUGCUCGUCCCGCCCGCA	JMJD6	NM_001081461	-9689	-	origin
	JMJD6	NM_015167	-9689	-	origin
	MFSD11	NM_024311	-1213	+	origin

miR-649	gene	ID	position		
AAACCUGUGUUGUUCAAGAGUC	SLC7A4	NM_004173	-1654	-	origin

miR-658	gene	ID	position		
GGCGGAGGGAAGUAGGUCCGUUG GU	EIF3EIP	NM_016091	-5052	+	origin

miR-659	gene	ID	position		
CUUGGUUCAGGGAGGGUCCCCA	ANKRD54	NM_138797	-3418	-	origin
	EIF3EIP	NM_016091	-1649	+	origin

miR-661	gene	ID	position		
UGCCUGGGUCUCUGGCCUGCGCGU	PLEC1	NM_201381	-492	-	origin
	PLEC1	NM_201382	-1332	-	origin
	PLEC1	NM_201383	-2705	-	origin
	PLEC1	NM_201384	-5639	-	origin

miR-801	gene	ID	position		
GAUUGCUCUGCGUGCGGAAUCGAC	TAF12	NM_005644	-5568	+	origin

miR-923	gene	ID	position		
GUCAGCGGAGGAAAAGAAACU	NLE1	NM_001014445	-8906	-	origin
	NLE1	NM_018096	-8906	-	origin

miR-935	gene	ID	position		
CCAGUUACCGCUUCCGCUACCGC	CACNG6	NM_031897	-9926	-	origin
	CACNG6	NM_145814	-9926	-	origin
	CACNG6	NM_145815	-9926	-	origin

miR-940	gene	ID	position		
AAGGCAGGGCCCCCGCUCCCC	RNPS1	NM_006711	-4010	+	origin
	RNPS1	NM_080594	-3693	+	origin

Mature miRNAs and genes with perfectly matched sequences in regions 10-kb upstream of the TSSs. The nucleotide sequence below each miRNA designation shows the mature miRNA sequence. The gene name indicates the gene located downstream of the miRNA complementary site. ID indicates the accession ID of each transcript. Position indicates the nucleotide position relative to the 5'-end of the miRNA counted from the transcription start site (+1). The +/= shows the orientation of the miRNA target sequences; this symbol indicates the relative direction with respect to downstream mRNA transcription (+/= = same/opposite). In the most right colume, "origin" indicates the locus from which own miRNA is originally transcribed. Blue indicates the sites appearing perfectly complementarity to the miRNAs transcribed from the different locations.

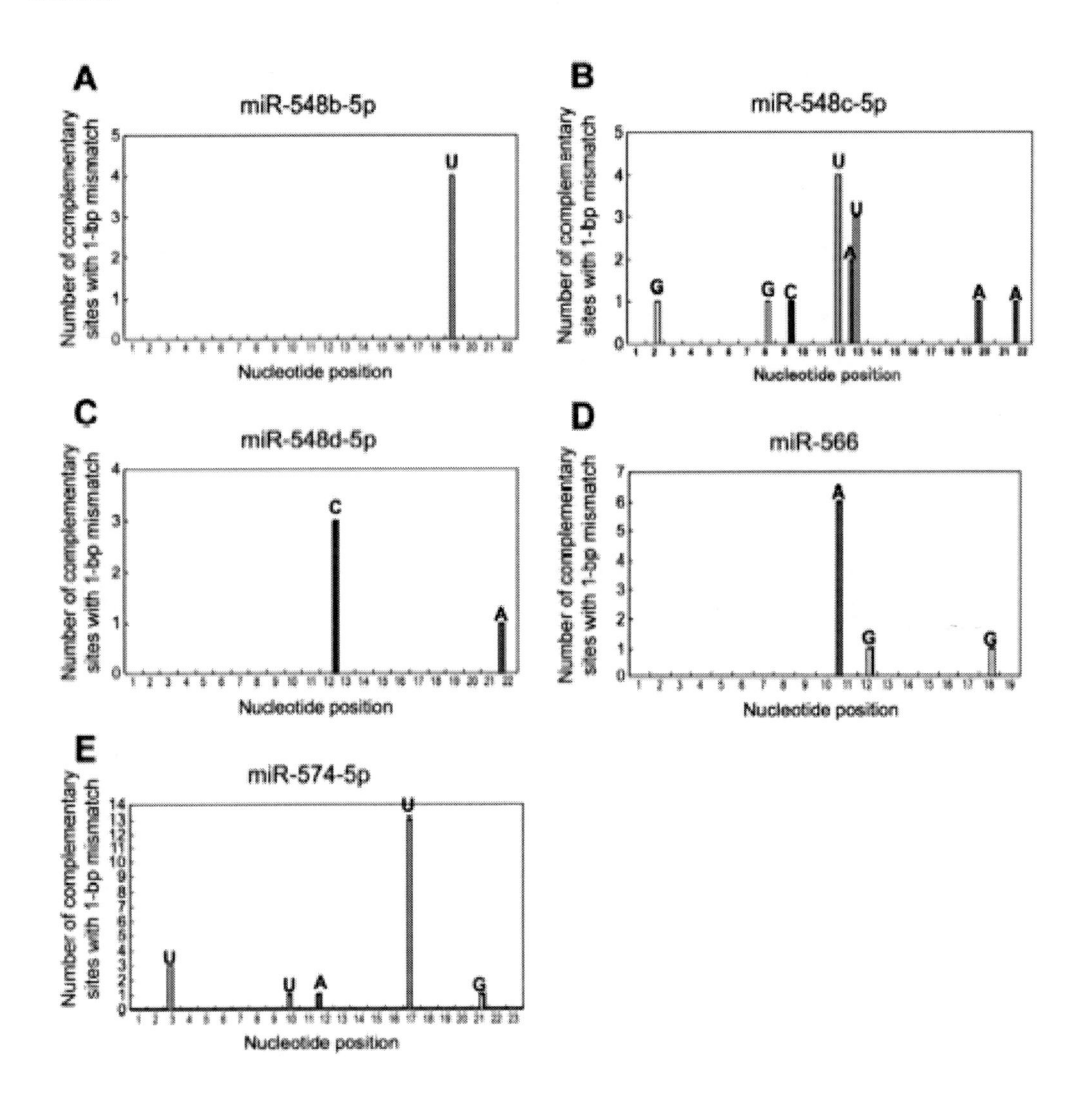

Figure 1. Positions containing 1-bp mismatches in 1-kb upstream regions of mRNA coding regions.

The mismatched positions in miR-548b-5p (A), miR-548c-5p (B), miR-548d-5p (C), miR-566 (D) and miR-574-5p (E) with target genes within the 1-kb upstream regions of the mRNA coding regions are shown from the 5'-end (+1) of the guide strand. The red bar indicates miRNA that is perfectly complementary to the target sequence when the nucleotide is changed to A. Green bar, U; Yellow, G; Blue, C.

Search for and Analysis of miRNAs with Complementary Sequences Containing 1- and 2-bp Mismatches within 1-kb of the Upstream Sequences

In a study on RNAi, target mRNAs carrying a few base mismatches with siRNAs have been demonstrated to be silenced to various degrees according to the position and the identity of the mismatched nucleotide [29, 30]. Furthermore, miRNAs recognize target mRNA by the "seed sequence" positioned between nucleotides 2–8 from the miRNA 5′-end [31-35], but not by the entire sequence. In RNAi, the identities of off-targets are also determined by complementarity to the seed regions of the introduced siRNA guide strand [36-38]. In a study on RNAa, a sequence located at position –645 relative to the TSS in the E-cadherin promoter was reported to function, although it has a few base mismatches with miR-373 [22]. Then, we searched for 1- and 2-bp mismatched positions and nucleotides of human miRNAs with the 1-kb of the upstream sequences.

We found 69 sites that had 1-bp mismatched sequences to 10 miRNAs in the 1-kb upstream regions of 53 genes (Tables 5 and 7). The complementary sites of five miRNAs (miR-548b-5p, miR-548c-5p, miR-548d-5p, miR-566 and miR-574-5p) recognized multiple 1-bp mismatched sites; the mismatched positions and nucleotides of these miRNAs are summarized in Figure 1. The major mismatched sites of miR-548b-5p and miR-574-5p were identical at nucleotide positions 19 and 17, respectively, from the 5′-end of each miRNA (Figure 1A and E). When G is changed to U at position 19 in miR-548b-5p or A is changed to U at position 17 in miR-574-5p, the majority of these miRNAs can pair with perfect complementarity, indicating mismatched nucleotides are remarkably characteristic. The major mismatched nucleotides in miR-548c-5p, miR-548d-5p and miR-566 were found at position 11 or 12 corresponding to the central region of the miRNAs (Figure 1B–D). In total, 87 % of 1-bp mismatched nucleotides were found in the non-seed region, and the remaining 13 % of them were mismatched in the seed region. Thus, the mismatched positions and nucleotides were highly biased. Total of 4453 target sites containing 2-bp mismatches with 44 miRNAs were found (Table 5). Among them, miR-548c-3p, miR-548d-5p, miR-566 and miR-574-5p had more than ten mismatched target sites (Figure 2). The major mismatched positions in the target sequences were unevenly positioned at 11 and 18 in miR-566 (212 of 276 sites (77 %), Figure 2A). In the cases of miR-548d-5p and miR-548c-3p, the mismatched nucleotides were concentrated at position 12 and 7, respectively (Figure 2B and C). The miR-574-5p (5′-UGAGUGUGUGUGUGUGAGUGUGU-3′) is highly GU repeated sequence without the nucleotides at positions 3 and 17 where Us are changed to As. The 3753 of 4060 mismatched target sites (92 %) of miR-574-5p showed simultaneous nucleotide changes from As to Us at positions 3 and 17 (Figure 2D), suggesting that these sites might be highly distributed GU repeated sequences. In total, 2-bp mismatched positions and nucleotides are also remarkably biased similar to the 1-bp mismatched sites. In total, only 12.0 % of 2-bp mismatched nucleotide positions were found in the seed region, and the remaining 88.0 % of them were mismatched in the non-seed region, except for the miR-574-5p mismatched sites.

Changes in Gene Expression by Targeting Upstream Sequences with miRNA

Six miRNAs, which transcribed from the different locations other than the upstream regions, were found to be completely complementary at 66 sites in the 10-kb upstream regions of 50 genes (Tables 5 and 6). To investigate whether miRNAs complementary to upstream sequences of mRNAs are general regulators of downstream genes, chemically synthesized miRNAs were transfected into the cells and changes in the expression of downstream genes were measured by real-time PCR. Similar to endogenous miRNAs, the synthetic mature miRNAs were annealed with opposite strand RNAs to form duplex RNAs (Figure 3A).

To investigate the effects of synthetic miRNAs against the completely matched sequences in the 3′ UTRs, psiCHECK-miR-target constructs containing three copies of completely matched target sequences in the 3′ UTR of the *Renilla luciferase* gene were made (Figure 3B). The synthetic miRNA and the corresponding psiCHECK-miR-target construct were transfected simultaneously into HeLa cells, and *luciferase* activity was measured. All of six miRNAs (miR-378, miR-548c-5p, miR-548d-5p, miR-572, miR-574-5p, and miR-606) reduced *Renilla luciferase* activity (Figure 3C), indicating that these synthetic miRNAs can repress gene expression when completely matched sequences are contained in the 3′ UTRs.

Next, the endogenous expression of six miRNAs labelled with blue in Table 6 were examined by RT-PCR in human HeLa, PC-3, and MCF-7 cells. For detection of miRNAs by RT-PCR, we used miRNA-specific stem–loop primers for reverse transcription according to the method reported by Chen et al. [28]. The synthetic miRNAs transfected into HeLa cells were detected using stem–loop RT primers and PCR primers specific for each miRNA. Endogenous miR-378 and miR-574-5p were detected in all three cell lines, but miR-548c-5p, miR-548d-5p, miR-572 and miR-606 were not detected in any of the cell lines tested (Figure 4).

We examined the downstream gene expression changes caused by the introduction of miRNAs; the expression of 8 genes that had sequences corresponding to the transfected miRNAs in their upstream regions were determined by real-time RT-PCR (Figure 5). The expression of OSBPL10 was reduced significantly in two of three cell lines by transfection with miR-378 (Figure 5A). The expression levels of four genes, RNF6, RNPS1, C20orf43 and ANXA3 were lowered in one of three or two cell lines by miR-548c-5p, miR-572, miR-548c-5p and miR-548c-5p, respectively (Figure 5B-E). The expression levels of OSBPL9 and CDC23 genes were significantly increased by transfection of miR-606 and miR-548c-5p, respectively, in one of three cell lines (Figure 5F,G). Thus, the mRNA expression was differentially regulated in each cell type. Furthermore, the effects of miRNA transfection varied depending on the downstream genes. In PC-3 cells, RNF6, C20orf43 were reduced by miR-548c-5p transfection, but ANXA3 and the other genes (GNB5, MANBAL, and RBP4) containing perfect complementary sites with miR-548c-5p in the upstream regions were not reduced (Figure 5, data not shown). POLR3A was decreased with miR-548d-5p, but the expression levels of the other genes (MAPKAPK5 and GJB3) were not changed in PC-3 cells (Figure 5, data not shown). In this study, when mRNA levels were significantly increased, the changes were not greater than ~30 % in any instances. Similarly, miRNA transfection resulted in ~40 % downregulation of mRNAs in all cases of reduction. Although the rate of change of gene expression was not so abundant, the effect of miRNA was obviously dose-dependent (Figure 6). Note that miR-378 and miR-574-5p are endogenously expressed in all three cell lines, which could lead to an underestimation of their effects on gene expressions.

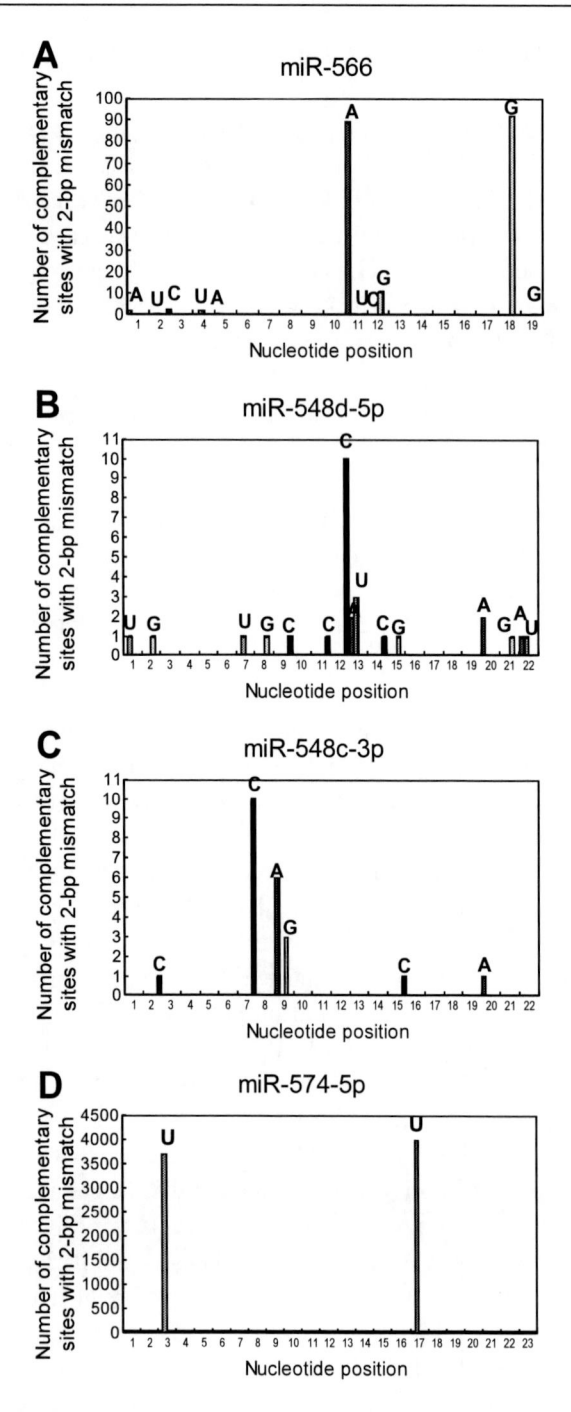

Figure 2. Positions containing 2-bp mismatches in 1-kb upstream regions of mRNA coding regions. The mismatched positions in miR-566 (A), miR-548d-5p (B), miR-548c-3p (C) and miR-574-5p (D) with target sites within the 1-kb upstream sequences of the mRNA coding regions are shown from the 5'-end (+1) of the guide strand. The red bar indicates miRNA that is perfectly complementary to the target sequence when the nucleotide is changed to A. Green bar, U; Yellow, G; Blue, C.

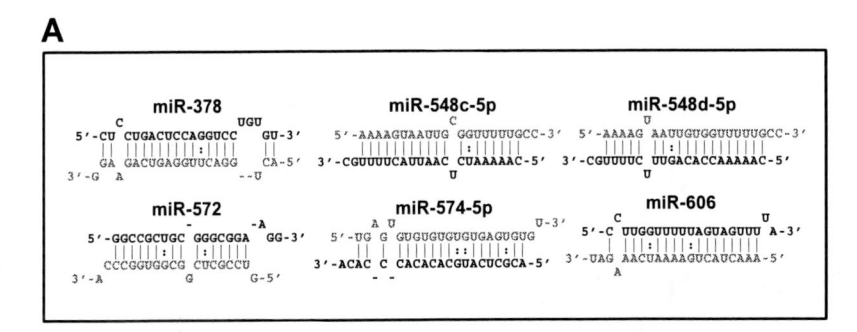

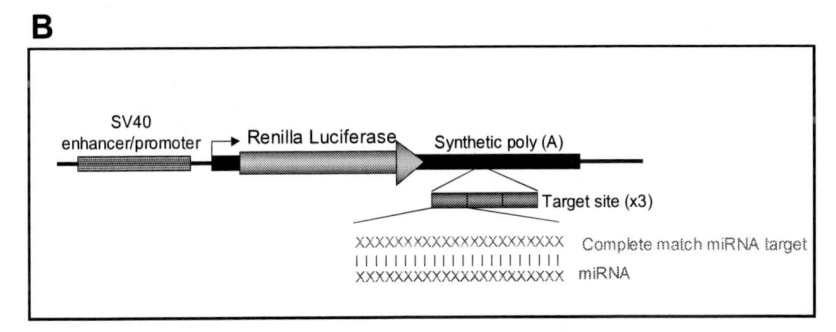

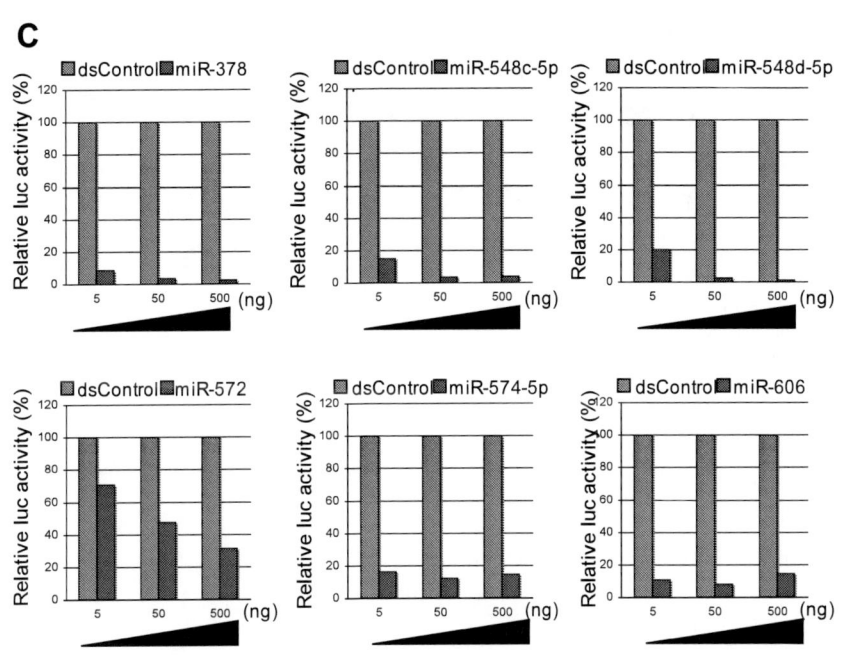

Figure 3. Effects of miRNAs against completely matched targets in 3' UTRs.
(A) Chemically synthesized miRNAs; the red characters indicate mature miRNAs. (B) The schematic structure of the target sequence expression construct, psiCHECK-miR-target, which contains three copies of matched target sequences in the 3' UTR of the *Renilla luciferase* gene. (C) Silencing activity of synthetic miRNAs; the activity was monitored using psiCHECK-miR-target constructs. The silencing activity is shown as the percentage of relative *luc* activity normalized to the dsControl transfection (100 %).

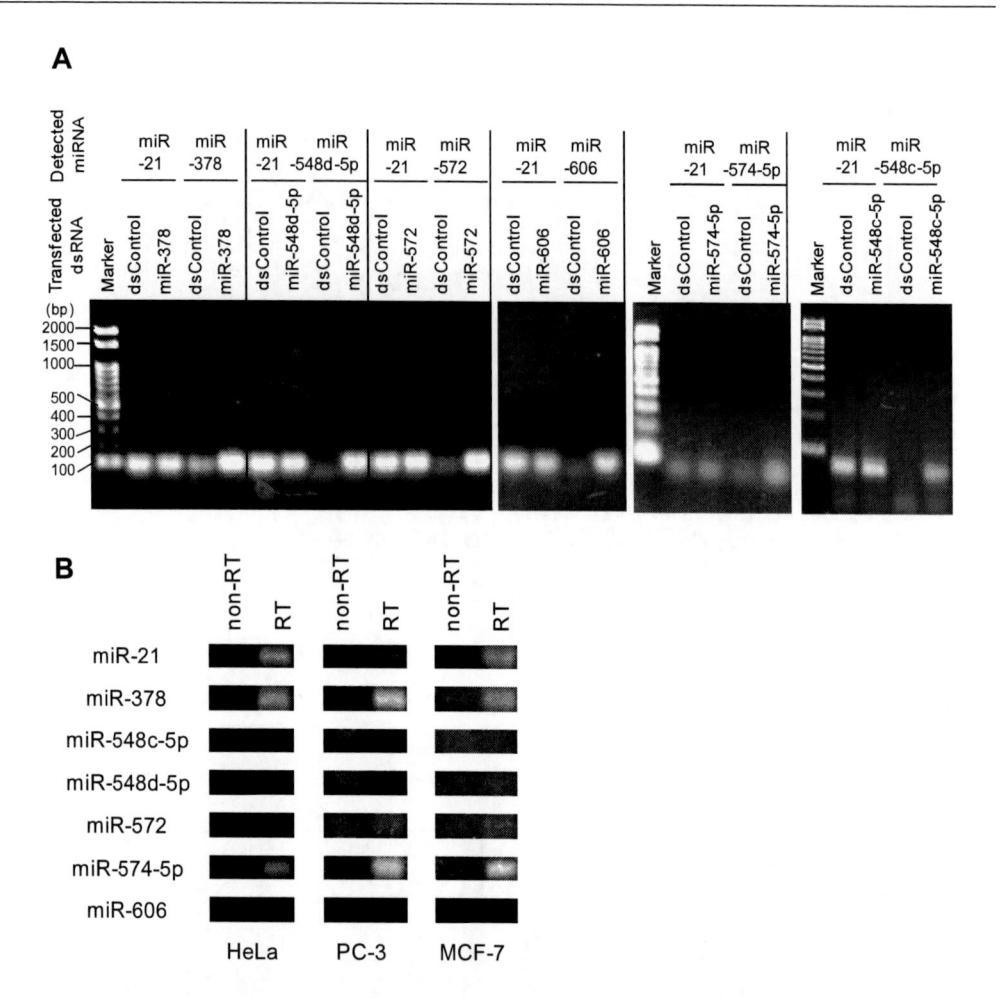

Figure 4. Detection of endogenous miRNAs by RT-PCR.
(A) Detection of miRNAs by stem-loop RT primers specific for each miRNA. The synthetic miRNAs were transfected into HeLa cells, and each miRNA was detected with specific stem-loop RT primers. miR-21 was used as transfection and detection control. Expression of each endogenous miRNA was detected when dsControl alone was transfected and intended miRNA was detected by specific primers for RT-PCR. Expected sizes of each miRNA is 64, 61, 60, 61, 61, 65 and 62 bp for miR-21, miR-378, miR-548c-5p, miR-548d-5p, miR-572, miR-574-5p and miR-606, respectively. (B) Endogenous miRNAs detected using stem-loop RT-PCR in HeLa, PC-3, and MCF-7 cells.

Table 7. List of miRNAs and genes with 1-bp mismatched complementary sequences within the 1-kb upstream regions

miR-190	gene	ID	position	sub	sequence (5' ⇒ 3')	
UGAUAUGUUUG AUAUAUUAGGU	RNF212	NM_194439	-974	7-U	UGAUAUUUUUGA UAUAUUAGGU	+

miR-297	gene	ID	position	sub	sequence (5' ⇒ 3')	
AUGUAUGUGUG CAUGUGCAUG	C12orf54	NM_152319	-151	15-A	AUGUAUGUGUGC AUAUGCAUG	+

miR-548b-5p	gene	ID	position	sub	sequence (5' ⇒ 3')	
AAAAGUAAUUG UGGUUUUUGGCC	ALDH1L2	NM_001034173	-914	19-U	AAAAGUAAUUGU GGUUUUUGCC	+
	C8B	NM_000066	-333	19-U	AAAAGUAAUUGU GGUUUUUGCC	-
	CLEC12A	NM_138337	-525	19-U	AAAAGUAAUUGU GGUUUUUGCC	+
	CLEC12A	NM_201623	-525	19-U	AAAAGUAAUUGU GGUUUUUGCC	+
	SOX30	NM_007017	-621	19-U	AAAAGUAAUUGU GGUUUUUGCC	+
	SOX30	NM_178424	-621	19-U	AAAAGUAAUUGU GGUUUUUGCC	+

miR-548c-5p	gene	ID	position	sub	sequence (5' ⇒ 3')	
AAAAGUAAUUG CGGUUUUUGCC	ACTR10	NM_018477	-735	13-U	AAAAGUAAUUGC UGUUUUUGCC	+
	ALDH1L2	NM_001034173	-914	12-U	AAAAGUAAUUGU GGUUUUUGCC	+
	AOX1	NM_001159	-468	20-A	AAAAGUAAUUGC GGUUUUUACC	+
	C5orf32	NM_032412	-715	13-U	AAAAGUAAUUGC UGUUUUUGCC	-
	C8B	NM_000066	-333	12-U	AAAAGUAAUUGU GGUUUUUGCC	-
	CCDC90A	NM_001031713	-913	8-G	AAAAGUAGUUGC GGUUUUUGCC	-
	CLEC12A	NM_138337	-525	12-U	AAAAGUAAUUGU GGUUUUUGCC	+
	CLEC12A	NM_201623	-525	12-U	AAAAGUAAUUGU GGUUUUUGCC	+
	LPO	NM_006151	-167	22-A	AAAAGUAAUUGC GGUUUUUGCA	-
	MS4A4A	NM_148975	-446	13-A	AAAAGUAAUUGC AGUUUUUGCC	+
	OR10W1	NM_207374	-104	13-A	AAAAGUAAUUGC AGUUUUUGCC	-
	PGM2L1	NM_173582	-789	13-U	AAAAGUAAUUGC UGUUUUUGCC	+
	PSMA1	NM_002786	-454	9-C	AAAAGUAACUGC GGUUUUUGCC	+
	PTER	NM_001001484	-887	2-G	AGAAGUAAUUGC GGUUUUUGCC	-
	PTER	NM_030664	-887	2-G	AGAAGUAAUUGC GGUUUUUGCC	-

Table 7. (Continued)

miR-548c-5p	gene	ID	position	sub	sequence (5' ⇒ 3')	
	SOX30	NM_007017	-621	12-U	AAAAGUAAUUGU GGUUUUUGCC	+
	SOX30	NM_178424	-621	12-U	AAAAGUAAUUGU GGUUUUUGCC	+

miR-548d-5p	gene	ID	position	sub	sequence (5' ⇒ 3')	
AAAAGUAAUUG UGGUUUUUGCC	ANXA3	NM_005139	-248	12-C	AAAAGUAAUUGC GGUUUUUGCC	+
	C5orf32	NM_032412	-658	22-A	AAAAGUAAUUGU GGUUUUUGCA	+
	PSTPIP2	NM_024430	-798	12-C	AAAAGUAAUUGC GGUUUUUGCC	+
	ZNF418	NM_133460	-732	12-C	AAAAGUAAUUGC GGUUUUUGCC	+

miR-566	gene	ID	position	sub	sequence (5' ⇒ 3')	
GGGCGCCUGUG AUCCCAAC	C14orf2	NM_004894	-979	11-A	GGGCGCCUGUAAU CCCAAC	+
	C20orf24	NM_018840	-980	18-G	GGGCGCCUGUGAU CCCAGC	-
	C20orf24	NM_199483	-980	18-G	GGGCGCCUGUGAU CCCAGC	-
	CT45-1	NM_001017417	-267	12-G	GGGCGCCUGUGGU CCCAAC	+
	GSTCD	NM_001031720	-444	11-A	GGGCGCCUGUAAU CCCAAC	+
	PRRG2	NM_000951	-788	11-A	GGGCGCCUGUAAU CCCAAC	+
	RASL11A	NM_206827	-104	11-A	GGGCGCCUGUAAU CCCAAC	+
	RLN3	NM_080864	-591	11-A	GGGCGCCUGUAAU CCCAAC	-
	ZNF594	NM_032530	-466	11-A	GGGCGCCUGUAAU CCCAAC	-

miR-574-5p	gene	ID	position	sub	sequence (5' ⇒ 3')	
UGAGUGUGUGU GUGUGAGUGUG U	BAHCC1	NM_001080519	-40	17-U	UGAGUGUGUGUG UGUGUGUGUGU	-
	C4orf31	NM_024574	-206	17-U	UGAGUGUGUGUG UGUGUGUGUGU	-
	CALN1	NM_031468	-899	21-G	UGAGUGUGUGUG UGUGAGUGGGU	-
	CD300A	NM_007261	-496	12-A	UGAGUGUGUGUA UGUGAGUGUGU	-
	CD300A	NM_007261	-556	10-U	UGAGUGUGUUUG UGUGAGUGUGU	-
	EMP2	NM_001424	-645	17-U	UGAGUGUGUGUG UGUGUGUGUGU	-

Table 7. (Continued)

miR-574-5p	gene	ID	position	sub	sequence (5' ⇒ 3')	
	ERC1	NM_015064	-302	3-U	UGUGUGUGUGUG UGUGAGUGUGU	-
	ERC1	NM_178037	-302	3-U	UGUGUGUGUGUG UGUGAGUGUGU	-
	ERC1	NM_178038	-302	3-U	UGUGUGUGUGUG UGUGAGUGUGU	-
	ERC1	NM_178039	-302	3-U	UGUGUGUGUGUG UGUGAGUGUGU	-
	ERC1	NM_178040	-302	3-U	UGUGUGUGUGUG UGUGAGUGUGU	-
	GDF5	NM_000557	-159	17-U	UGAGUGUGUGUG UGUGUGUGUGU	-
	GLT25D2	NM_015101	-23	17-U	UGAGUGUGUGUG UGUGUGUGUGU	+
	MAP3K13	NM_004721	-197	17-U	UGAGUGUGUGUG UGUGUGUGUGU	+
	MUC16	NM_024690	-194	17-U	UGAGUGUGUGUG UGUGUGUGUGU	+
	OPN4	NM_00103001 5	-154	17-U	UGAGUGUGUGUG UGUGUGUGUGU	+
	OPN4	NM_033282	-154	17-U	UGAGUGUGUGUG UGUGUGUGUGU	+
	OR1N2	NM_00100445 7	-423	17-U	UGAGUGUGUGUG UGUGUGUGUGU	-
	PI3	NM_002638	-801	3-U	UGUGUGUGUGUG UGUGAGUGUGU	+
	SORCS1	NM_00101303 1	-756	17-U	UGAGUGUGUGUG UGUGUGUGUGU	+
	SORCS1	NM_052918	-756	17-U	UGAGUGUGUGUG UGUGUGUGUGU	+
	SYT14	NM_153262	-659	17-U	UGAGUGUGUGUG UGUGUGUGUGU	-
	TRPA1	NM_007332	-273	17-U	UGAGUGUGUGUG UGUGUGUGUGU	+
	VCX	NM_013452	-143	17-U	UGAGUGUGUGUG UGUGUGUGUGU	+
	VCX3A	NM_016379	-142	3-U	UGUGUGUGUGUG UGUGAGUGUGU	+

miR-646	gene	ID	position	sub	sequence (5' ⇒ 3')	
AAGCAGCUGCCU CUGAGGC	PRELID2	NM_138492	-419	2-G	AGGCAGCUGCCU CUGAGGC	+
	PRELID2	NM_182960	-451	2-G	AGGCAGCUGCCU CUGAGGC	+
	PRELID2	NM_205846	-451	2-G	AGGCAGCUGCCU CUGAGGC	+

Table 7. (Continued)

miR-649	gene	ID	position	sub	sequence (5' ⇒ 3')	
AAACCUGUGUU GUUCAAGAGUC	ZNF211	NM_006385	-289	19-U	AAACCUGUGUUG UUCAAGUGUC	-
	ZNF211	NM_198855	-289	19-U	AAACCUGUGUUG UUCAAGUGUC	-

miR-940	gene	ID	position	sub	sequence (5' ⇒ 3')	
AAGGCAGGGCCC CCGCUCCCC	GABRA2	NM_000807	-249	15-C	AAGGCAGGGCCC CCCCUCCCC	+

Mature miRNAs and genes with complementary sequences containing 1-bp mismatches in the 1-kb upstream sequences of the TSSs. The nucleotide sequence below each miRNA designation shows the mature miRNA sequence. The gene name indicates the gene located downstream of the miRNA complementary site. ID indicates the accession ID of each transcript. Position indicates the nucleotide position relative to the 5'-end of the miRNA counted from the transcription start site (+1). "Sub" represents the mismatched position and nucleotide in miRNA. The red characters in the miRNA sequences show sequences mismatched. The +/= shows the orientation of the miRNA target sequences; this symbol indicates the relative direction with respect to downstream mRNA transcription (+/= = same/opposite).

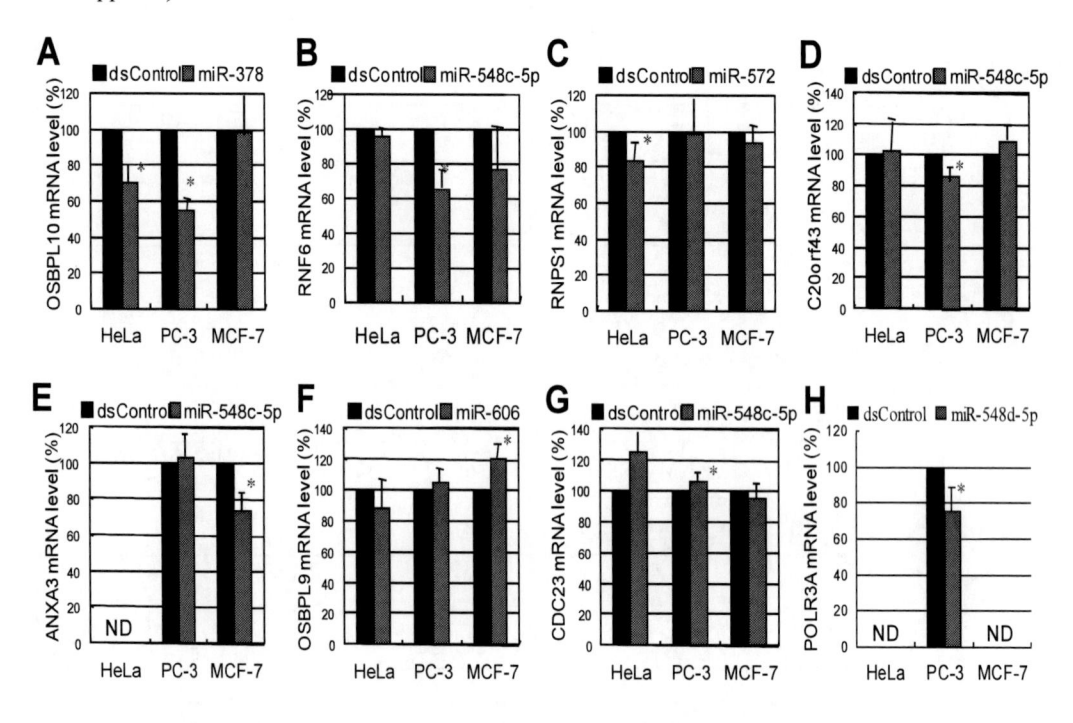

Figure 5. Changes in mRNA expression levels after transfection of miRNAs.
Downstream mRNAs were detected by real-time RT-PCR in HeLa, PC-3, and MCF-7 cells. The combinations of downstream mRNA/miRNA were as follow: (A) OSBPL10/miR-378, (B) RNF6/miR-548c-5p, (C) RNPS1/miR-572, (D) C20orf43/miR-548c-5p, (E) ANXA3/miR-548c-5p, (F) OSBPL9/miR-606, (G) CDC23/miR-548c-5p, and (H) POLR3A/miR-548d-5p. The expression level of each mRNA was normalized to the dsControl, which is presented as 100 %. The values are shown as the means ± SD from at least three independent experiments. Statistical significance was calculated using paired two-tailed Student's t-test; *P < 0.05. ND, not determined.

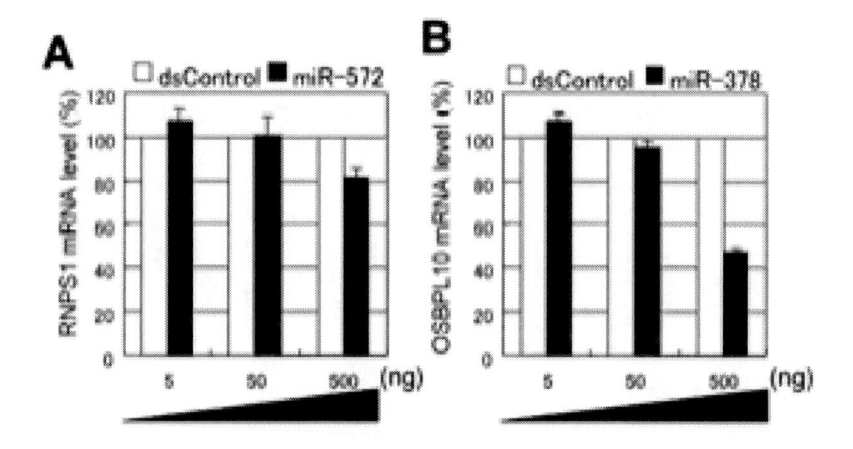

Figure 6. Dose-dependent effect of miRNA on gene expression. (A) Downstream RNPS1 mRNA was detected by real-time RT-PCR in HeLa cells transfected with miR-572 (5, 50 and 500 ng/well). (B) Downstream OSBPL10 mRNA was detected by real-time RT-PCR in PC-3 cells transfected with miR-378 (5, 50 and 500 ng/well). The expression level of each mRNA was normalized to the dsControl, which is presented as 100 %. The values are shown as the means ± SD.

Discussion

This study is the first report of global identification of miRNAs that contain completely or near-completely (i.e., 1- or 2-bp mismatched) complementary sequences to sites in the 10-kb or 1-kb upstream regions of human mRNAs. In total, 66 sites within 10-kb upstream of the coding regions were found to be completely complementary with miRNAs transcribed from the different locations (Tables 5 and 6). Although each of miR-378, miR-572, and miR-606 has only one complementary site, miR-548c-5p, miR-548d-5p and miR-574-5p appeared to be complementary with the upstream sequences of 26, 17, and 4 genes, respectively. The miR-548c-5p and miR-548d-5p miRNAs are members of a family of human miRNA genes derived from short miniature inverted-repeat transposable elements (MITEs) duplicated in many genomic regions, and are capable of forming highly stable hairpin loop recognized by the RNAi enzymatic machinery and processed to form 22 bp mature miRNA sequences [39]. Thus, the many complementary sites of miR-548 family in the upstream regions might represent evolutionary specific regulation of gene expression related to species diversification.

The mismatched positions and types of nucleotides spanning the miRNAs appeared to be highly biased based on searches for sequences containing 1- and 2-bp mismatches with mature miRNAs in the 1-kb upstream region (Figure 1 and 2). The seed sequences of miRNAs are the primary region for 3′ UTR target recognition [31-35]. In RNAi, the identities of off-targets are determined by complementarity to the seed regions of the introduced siRNA [36-38]. In this study, among human miRNAs, 10 and 44 miRNAs were revealed to have 1- and 2-bp mismatched sites, respectively, in the 1-kb upstream region of TSSs (Table 5). The search result of 1-bp mismatched target sites of miRNAs showed that 87 % of 1-bp mismatched sites and 88 % of 2-bp mismatched sites were located in the non-seed regions, except for the target sites of miR-574-5p. The results suggest that these miRNAs are able to act on their upstream target sites with perfect complementarity in the seed region, if the mechanism of transcriptional regulation on the upstream region is similar to that of the known miRNA function. In RNAi, mismatches at the central region of the siRNA guide strand have been shown to be highly sensitive to knockdown efficiency, but the mismatches in the other site are the most tolerable [29, 30]. In both 1-bp and 2-bp mismatched site analyses, miR-548d-5p and miR-566 showed major mismatched site at positions 12 and 11 (Figs. 1 and 2), respectively, indicating that these miRNAs can act on the upstream regions through the known miRNA machinery but not RNAi machinery. In an analogous fashion, miR-548c-5p also has major mismatched site at position 12 (Figure 1B). In contract, major mismatched site of miR-548c-3p was position 7, and the miR-574-5p has mismatches at positions 3 and 17 (Figure 2). The positions 7 and 3 are involved in the seed region, suggesting that these miRNAs can not act through a known miRNA machinery.

We next evaluated the effects of the miRNAs on gene expression. Eight combinations of miRNAs that matched the upstream sites of mRNAs positioned -24 to -8654-bp relative to the TSS were used. Reductions in mRNA levels were observed in 6 of 8 combinations (Figure 5). In most cases, the reduction was observed in one of three cell lines. In only one combination, reduction was observed in two cell lines simultaneously (Figure 5A). The result indicates that the reducing effect of each miRNA is different depending on cell type. None of the miRNA sequences completely matched the mRNA coding regions. However, two complementary sites of the seed sequence of miR-548d-3p, an opposite strand of miR-548d-5p, were found in the 3' UTR and CDS of POLR3A gene (Table 8). This suggests that the decrease in POLR3A mRNA by miR-548d-5p is a seed-dependent silencing activity. Furthermore, two seed sequence complementary sites of miR-378 were found in the CDS of OSBPL10 mRNA (Table 8). Recently, the CDS was reported to comprise miRNA and siRNA targets [38,40,41]. Thus, the decrease in OSBPL10 mRNA after transfection of miR-378 could also be a seed-dependent silencing effect that targets the OSBPL10 CDS in HeLa and PC-3 cells, although such reduction was not observed in MCF-7 cells (Figure 5A). The increases in mRNA levels were observed in two of the miRNA/mRNA combinations in only one of three cell lines (Figure 5F, G), indicating that the machinery for increasing mRNA expression is also highly cell type-specific. These results show that miRNA can both silence and activate gene expression; these effects were largely dependent on cell type and downstream gene. In this study, the indirect interaction of miRNAs through the regulation of some transcription factor(s) that control the expression of the downstream genes is hardly distinguishable from the direct interaction of miRNAs with the upstream regions. Further detailed analyses might reveal the mechanism of transcriptional regulation (e.g., RNAa, TGS) targeted the upstream region of mRNAs.

Table 8. Seed complementary sites in the mRNA coding regions

A

gene / miRNA	Region in mRNA		
	5'UTR	CDS	3'UTR
OSBPL10 / miR-378	0	$2^{1), 2)}$	0
RNF6 / miR-548c-5p	0	0	0
CDC23 / miR-548c-5p	0	0	0

A

gene / miRNA	Region in mRNA		
	5'UTR	CDS	3'UTR
OSBPL9 / miR-606	0	0	0
RNPS1 / miR-572	0	0	0
C20orf43 / miR-548c-5p	0	0	0
ANXA3 / miR-548c-5p	0	0	0
P OLR3A / miR-548d-5p	0	0	0

1)

mRNA   5'-ACCAUGGAGAGGGCAGUCCAGGGC-

miRNA 3'

```
III  :   IIIIIII
```

2)

mRNA   5'-AGCAGGGACCCAUGGAGUCCAGGA-

miRNA 3'

```
 I      I : :IIIIIII
```

B

gene / miRNA	Region in mRNA		
	5'UTR	CDS	3'UTR
OSBPL10 / miR-378*	0	0	0
RNF6 / miR-548c-3p	0	0	0
CDC23 / miR-548c-3p	0	0	0
OSBPL9 / miR-606 opposite strand	0	0	0
RNPS1 / miR-572 opposite strand	0	0	0
C20orf43 / miR-548c-3p	0	0	0
ANXA3 / miR-548c-3p	0	0	0
POLR3A / miR-548d-3p	0	1[1]	1[2]

1)

mRNA   5'-AUACGCAGAUGAAAAGGUUUUUGA-3'

miRNA   : III :   IIIIIIII

                3'-CGUUUUCUUUGACACCAAAAAC-5'

2)

mRNA   5'-CAAUGGGGAUGAGAUGGUUUUUUG-3'

miRNA   :::II :   IIIIIIII

The number of seed complementary sites of the mature miRNA guide strand (A) and opposite strand (B) in the 5′ UTR, CDS and 3′ UTR of each mRNA which expression level was changed by the corresponding miRNA. Two seed complementary sites were observed in CDS of OSBPL10 (A). The opposite strand of miR-548c-5p, miR-548c-3p, showed seed complementarity in CDS and 3′ UTR of POLR3A (B). The pairing pattern of miRNA and mRNA in the seed complementary regions are shown below each table. Seven nucleotides at nucleotide positions 2-8 from 5′ end of each miRNA were perfectly matched.

Conclusion

In the present study, we identified miRNAs and their complementary sites in upstream sequences of mRNA coding regions. Based on our results, perfectly matched complementary sites and 1- and 2-bp mismatched sites appear to be potential targets for miRNA. Using 8 combinations of miRNAs and downstream mRNAs, we found that miRNAs are able to both silence and promote the expression of downstream genes and that these changes are largely gene- and cell type-dependent.

Acknowledgments

This work was partially supported by grants from the Ministry of Education, Culture, Sports, Science and Technology of Japan and Core Research Project for Private University: matching fund subsidy to K.U.-T.

References

[1]   Filipowicz, W., Jaskiewicz, L., Kolb, F. A. & Pillai, R. S. (2005) Post-transcriptional gene silencing by siRNAs and miRNAs. *Curr. Opin. Struct. Biol.,* 15, 331-341.

[2]   Zaratiegui, M., Irvine, D.V. & Martienssen, R. A. (2007) Noncoding RNAs and gene silencing. *Cell,* 128, 763-776.

[3]   Song, J. J., Smith, S. K., Hannon, G. J. & Joshua-Tor, J. (2004) Crystal structure of Argonaute and its implications for RISC slicer activity. *Science,* 305, 1434-1437.

[4]   Liu, J., Carmell, M. A., Rivas, F. V., Marsden, C. G., Thomson, J. M., Song, J.J., Hammond, S. M., Joshua-Tor, J. & Hannon, G. F. (2004) Argonaute2 is the catalytic engine of mammalian RNAi. *Science,* 205, 1437-1441.

[5]   Borchert, G. M., Lanier, W. & Davidson, B. L. (2006) RNA polymerase III transcribes human microRNAs. *Nat. Struct. Mol. Biol.,* 13, 1097-1101.

[6]   Lee, R. C., Ahn, C., Han, J., Choi, H., Kim, J., Yim, J., Lee, J., Provost, P., Radmark, O., Kim, S. & Kim, V. N. (2003) The nuclear RNase III Drosha initiates microRNA processing. *Nature,* 425, 415-419.

[7]   Han, J., Lee, Y., Yeom, K. H., Kim, Y. K., Jin, H. & Kim, V. N. (2004) The Drosha-DGCR8 complex in primary microRNA processing. *Genes Dev.,* 18, 3016-3027.

[8]   Denli, A. M., Tops, B. B., Plasterk, R. H., Ketting, R. F. & Hannon, G. J. (2004) Processing of primary microRNAs by the Microprocessor complex. *Nature,* 432, 231-235.

[9]   Yi, R., Qin, Y., Macara, I. G. & Cullen, B. R. (2003) Exportin-5 mediates the nuclear export of pre-microRNAs and short hairpin RNAs. *Genes Dev.,* 17, 3011-3016.

[10] Morris, K. V., Chan, A. W. L., Jacobsen, S. E. & Looney, D. F. (2004) Small interfering RNA-induced transcriptional gene silencing in human cells. *Science,* 305, 1289-1292.

[11] Ting, A. H., Schuebel, K. E., Herman, J. G. & Baylin, S. B. (2005) Short double-stranded RNA induces transcriptional gene silencing in human cancer cells in the absence of DNA methylation. *Nat. Genet.,* 37, 906-910.

[12] Janowski, B. A., Huffman, K. E., Schwartz, J. C., Ram, R., Nordsell, R., Shames, D. S., Minna, J. D. & Corey, D. R. (2006) Involvement of Ago1 and Ago2 in mammalian transcriptional silencing. *Nat. Struct. Mol. Biol.,* 13, 787-792.

[13] Kim, D. H., Willeneuve, L. M., Morris, K. V. & Rossi, J. J. (2006) Argonaute-1 directs siRNA-mediated transcriptional gene silencing in human cells. *Nat. Struct. Mol. Biol.,* 13, 793-797.

[14] Weinberg, M. S., Villneuve, L. M., Ehsani, A., Amarzguioui, M., Aagaard, L., Chen, Z. X., Riggs, A. D., Rossi, J. J. & Morris, K. V. (2006) The antisense strand of small interfering RNAs directs histone methylation and transcriptional gene silencing in human cells. *RNA,* 12, 256-262.

[15] Han, J., Kim, D. & Morris, K. V. (2007) Promoter-associated RNA is required for RNA-directed transcriptional gene silencing in human cells. *Proc. Natl. Acad. Sci., USA,* 104, 12422-12427.

[16] Li, L. C., Okino, S. T., Zhao, H., Pookot, D., Place, R. F., Urakami, S., Enokida, H. & Dahiya, R. (2006) Small dsRNAs induce transcriptional activation in human cells. *Proc. Natl. Acad. Sci., USA,* 103, 17337-17342.

[17] Janowski, B. A., Younger, S. T., Hardy, D. B., Ram, R., Huffman, K. D. & Corey, D. R. (2007) Activating gene expression in mammalian cells with promoter-targeted duplex RNAs. *Nat. Chem. Biol.,* 3, 166-173.

[18] Schwartz, J. C., Younger, S. T., Nguyen, N. B., Hardy, D. B., Monia, B. P., Corey, D. R. & Janowski, B. A. (2008) Antisense transcripts are targets for activating small RNAs. *Nat. Struct. Mol. Biol.,* 15, 842-848.

[19] Morris, K. V., Santoso, S., Turner, A. M., Pastori, C. & Hawkins, P. G. (2008) Bidirectional transcription directs both transcriptional gene activation and suppression in human cells. *PLoS Genet.,* 4, 1-9.

[20] Weinberg, M. S., Barichievy, S., Schaffer, L., Han, J. & Morris, K. V. (2007) An RNA targeted to the HIV-1 LTR promoter modulates indiscriminate off-target gene activation. *Nucleic Acids Res.,* 35, 7303-7312.

[21] Kim, D. H., Sætrom, P., Snøve, O. Jr. & Rossi, J. J. (2008) MicroRNA-directed transcriptional gene silencing in mammalian cells. *Proc. Natl. Acad. Sci., USA,* 105, 16230-16235.

[22] Place, R. F., Li, L. C., Rookot, D., Noonan, E. J. & Dahiya, R. (2008) MicroRNA-373 induces expression of genes with complementary promoter sequences. *Proc. Natl. Acad. Sci., USA,* 105, 1608-1613.

[23] Rodriguez, A., Griffiths-Jones, S., Ashurst, J. L. & Bradley, A. (2004) Identification of mammalian microRNA host genes and transcription units. *Genome Res.,* 14, 1902-1910.

[24] Bartel, D. P. (2009) MicroRNAs: target recognition and regulatory functions. *Cell,* 136, 215-233.

[25] Sayers, E. W., Barrett, T., Benson, D. A., Bryant, S. H., Canese, K., Chetvernin, V., Church, D. M., DiCuccio, M., Edgar, R., Federhen, S., Feolo, M., Geer, L. Y., Helmberg, W., Kapustin, Y., Landsman, D., Lipman, D. J., Madden, T. L., Maglott, D. R., Miller, V., Mizrachi, I., Ostell, J., Pruitt, K. D., Schuler, G. D., Sequeira, E., Sherry,

S. T., Shumway, M., Sirotkin, K., Souvorov, A., Starchenko, G., Tatusova, T. A., Wagner, L. & Yaschenko, E., Ye, J. (2009) Database resources of the National Center for Biotechnology Information. *Nucleic Acids Res., 37*, D5-15.

[26] Kuhn, R. M., Karolchik, D., Zweig, A. S., Trumbower, H., Thomas, D. J., Thakkapallayil, A., Sugnet, C. W., Stanke, M., Smith, K. E., Siepel, A., Rosenbloom, K. R., Rhead, B., Raney, B. J., Pohl, A., Pedersen, J. S., Hsu, F., Hinrichs, A. S., Harte, R. A., Diekhans, M., Clawson, H., Bejerano, G., Barber, G. P., Baertsch, R., Haussler, D. & Kent, W. J. (2007) The UCSC genome browser database: update 2007. *Nucleic Acids Res., 35*, D668-73.

[27] Griffiths-Jones, S., Saini, H. K., van Dongen, S. & Enright, A. J. (2008) miRBase tools for microRNA genomics. *Nucleic Acids Res., 36*, D154-158.

[28] Chen, C., Ridzon, D. A., Broomer, A. J., Zhou, Z., Lee, D. H., Nguyen, J. T., Barbisin, M., Xu, N. L., Mahuvakar, V. R., Andersen, M. R., Lao, K. Q., Livak, K. L. & Guegler, K. J. (2005) Real-time quantification of microRNAs by stem-loop RT-PCR. *Nucleic Acids Res., 33*, e179.

[29] Du, Q., Thonberg, H., Zhang, H.-Y., Wahlestedt, C. & Liang, Z. (2005) A systematic analysis of the silencing effects of an active siRNA at all single-nucleotide mismatched target sites. *Nucleic Acids Res., 33*, 1671-1677.

[30] Dahlgren, C., Zhang, H.-Y., Du, Q., Grahn, M., Norstedt, G., Wahlestedt, C. & Liang, Z. (2008) Analysis of siRNA specificity on targets with double-nucleotide mismatches. *Nucleic Acids Res., 36*, e53.

[31] Lewis, B. P., Shih, I. H., Jones-Rhoades, M. W., Bartel, D. P. & Burge, C. B. (2003) Prediction of mammalian microRNA targets. *Cell, 115*, 787-798.

[32] Bartel, D. P. (2004) MicroRNAs: genomics, biogenesis, mechanism, and function. Cell, 116, 281-297.

[33] Lewis, B. P., Burge, C. B. & Bartel, D. P. (2005) Conserved seed pairing, often flanked by adenosines, indicates that thousand of human genes are microRNA targets. *Cell, 120*, 15-20.

[34] Lim, L. P., Lau, N. C., Garrett-Engele, P., Grimson, A., Schelter, J. M., Castle, J., Bartel, D. P., Linsley, P. S. & Johnson, J. M. (2005) Microarray analysis shows that some microRNAs downregulate large numbers of target mRNAs. *Nature, 433*, 769-773.

[35] Grimson, A., Farh, K. K., Johnston, W. K., Garrett-Engele, P., Lim, L. P. & Bartel, D. P. (2007) MicroRNA targeting specificity in mammals: determinants beyond seed pairing. *Mol. Cell, 27*, 91-105.

[36] Jackson, A. L., Burchard, J., Schelter, J., Chau, B. N., Cleary, M., Lim, L. & Linskey, P. S. (2006) Widespread siRNA "off-target" transcript silencing mediated by seed region sequence complementarity. *RNA, 12*, 1179-1187.

[37] Birmingham, A., Anderson, E. M., Reynolds, A., Ilsley-Tyree, D., Leake, D., Fedorov, Y., Baskerville, S., Maksimova, E., Robinson, K., Karpilow, J., Marshall, W. S. & Khvorova, A. (2006) 3'UTR seed matches, but not overall identity, are associated with RNAi off-targets. *Nat. Methods, 3*, 199-204.

[38] Ui-Tei, K., Naito, Y., Nishi, K., Juni, A. & Saigo, K. (2008) Thermodynamic stability and Watson-Crick base pairing in the seed duplex are major determinants of the efficiency of the siRNA-based off-target effect. *Nucleic Acids Res., 36*, 7100-7109.

[39] Piriyapongsa, J. & Jordan, K. (2007) Family of human microRNA genes from miniature inverted-repeat transposable elements. *PLoS ONE,* 2, e203.

[40] Duursma, A. M., Kedde, M., Schrier, M., Sage, C. & Agami, R. (2008) miR-148 targets human DNMT3b protein coding region. *RNA,* 14, 872-877.

[41] Tay, Y., Zhang, J., Thomson, A. M., Lim, B. & Rigoutsos, I. (2008) MicroRNAs to Nanog, Oct4 and Sox2 coding regions modulate embryonic stem cell differentiation. *Nature,* 455, 1124-1128.

In: Gene Silencing: Theory, Techniques and Applications   ISBN: 978-1-61728-276-8
Editor: Anthony J. Catalano   © 2010 Nova Science Publishers, Inc.

Chapter XIV

Gene Silencing:
Theory, Techniques and Applications

Prashant Mohanpuria, Vinay Kumar, Monika Mahajan,
Hasan Mohammad and Sudesh Kumar Yadav[*]
Biotechnology Division, Institute of Himalayan Bioresource Technology, CSIR,
Palampur-176061 (HP), India

Abstract

Gene silencing is an exciting field of functional genomics. It has been used as a research tool to discover or validate the functions of genes. It involves short sequence of nucleic acid that can bind to RNA of the gene and interferes the process of its expression. It is diverse in occurrence as well as in applications. This phenomenon occurs from nematodes to fungi and can cause gene silencing in plants, animals and human beings. The core aspects of the mechanisms and functions of gene silencing include co-suppression, RNA-mediated virus resistance and RNA-directed DNA methylation (RdDM). The applications of gene silencing cover a wide spectrum in plants, from designer flower colors to plant-produced medical therapeutics. These functions are achieved by two types of approach such as protection of the plant against attack and fine-tuning of metabolic pathways. RNA-mediated gene control mechanism has already provided new platforms for developing molecular tools for gene function studies and crop improvements. We are now exploring this technology for commercially focused applications in plants. Here, we review the theory of gene silencing discovery and the mechanism of this technique in plants. Further, we discuss the potential use of this technique in plant science particularly in crop improvements.

[*] Corresponding author: skyt@rediffmail.com; sudeshkumar@ihbt.res.in

1. Gene Silencing Theory

Gene silencing is related to the reduced expression phenomenon that results in no expression or very low expression of a gene or a RNA sequence. Gene silencing in plants occurs at both transcriptional (TGS) and post-transcriptional levels (PTGS) [1]. In TGS, mRNA synthesis of a particular gene is greatly reduced or absent altogether. While in PTGS, mature mRNA or mRNA precursors are synthesized but degraded rapidly or improperly processed. Transcriptional gene silencing (TGS) involves gene disruption, RNA-directed DNA methylation (RdDM) and chromatin modifications. In former, the physical structure of a gene is disrupted by the insertion of a piece of DNA of known sequence such as T-DNA or transposon elements, and treatment with mutagens or irradiation. RdDM has been highly elaborated in the plant kingdom. It is the first example of an RNA-mediated epigenetic modification which reinforces the prevailing concept of 'homology-dependent gene silencing' where DNA–DNA, RNA–DNA and RNA–RNA interactions are considered as potential triggers for gene silencing [2]. It requires conventional DNA methyltransferases, histone modifying enzymes and some other specific proteins. Either small interfering RNA (siRNA) processed from double stranded RNA or dsRNA itself directs methylation of cytosine residues of promoter region as well as open reading frame [3-5]. However, methylation of the promoter region alone is sufficient for producing stable and heritable silencing of a transgene.

Post-transcriptional gene silencing (PTGS) was first observed in plants during 1990. It was initially referred to as 'co-suppression' because the introduction of chimeric chalcone synthase gene (CHS) resulted in blocking of anthocyanin biosynthesis by co-inhibiting endogenous as well as introduced CHS gene expression in petunia. Thus, many of transgenic petunia plants produced flowers with variegated or developed white sectors [6]. But the real cause of this suppression phenomenon was discovered later and was found to be due to dsRNA. Further, coat protein mediated protection (CPMP) was discovered in 1995, where the virus resistance was conferred by the translatable as well untranslatable coat protein transgenes in plants [7]. Later on, a number of gene silencing phenomena that inactivate genes at the post transcriptional level were identified. Antisense RNA suppression is an early form of PTGS employed by plant scientists. It involves the base pairing of antisense RNA strand with its complementary sense RNA strand (mRNA) of the corresponding target gene intended to silence [8, 9]. This complementary base pairing prevents the translation of the target mRNA because of the inability of ribosomes to bind with it [9, 10] and also makes mRNA susceptible to endonucleases [11]. This antisense approach stimulated a great deal of interest in functional genomics at various laboratories around the world. But later, it was found that this approach was not very specific and stable to target a gene expression.

In 1998, Fire and Mello reported the dsRNA as the real inducer of gene suppression in nematode worms and termed the phenomenon as RNA interference (RNAi) [12]. It is generally believed that RNAi originally evolved as a defense mechanism against invasive nucleic acids, including viruses and transposons. In this phenomenon, RNA itself is responsible for the sequence specific degradation of a target mRNA. It requires the introduction of homologous dsRNA to specifically target a gene product, resulting in complete loss or reduction in appreciable amount of the expression of the target gene. The complex process of RNAi involves ATP dependent two highly conserved enzyme systems. The first one is Dicer in animals and Dicer-like (DCL) elements in plants. The second one is

RISC (RNA-Induced Silencing Complex). Dicer family members are large, multidomain proteins that contain a putative RNA helicase domain (at N-terminal), a PAZ (Piwi/Argonaute/Zwille) protein-protein interaction domain, two tandem ribonuclease III (RNase III) and one or two dsRNA-binding domains (at C-terminal) [13, 14]. RISC consists of both protein and RNA [15]. The Argonaute proteins are the major protein component responsible for small RNA binding as well as ribonuclease activity to cut target RNA [5]. It also contains a PAZ domain. There are still other components of RISC yet to be identified along with their functions. RISC utilizes small functional RNA molecules like siRNA or micro RNA (miRNA) to target the complementary mRNAs. It is thought that there is protein-protein interaction between Dicer and RISC through the PAZ domain which enables these small RNA fragments (siRNA or miRNA) to enter into the RNAi pathway. Partial evidence has suggested that RISC degrades the sense strand and only utilizes the antisense strand of these small RNA molecules to search for its complementary base pairing with the target mRNA.

RNAi mechanism is conserved in a wide range of eukaryotes including plants, fungi, nematodes, protozoa, insects, and vertebrates which reflects its ancient evolutionary origin to protect the genome of a species. However, this mechanism has not been discovered in prokaryotes. This mechanism in plants has numerous distinctive features, including signal amplification, short- and long-range signaling, and TGS responses [16]. PTGS based RNAi has been employed in various plant systems due to its specificity and stability to target a gene function.

2. Gene Silencing Techniques

In gene silencing, small functional RNA molecules of 20-30 nucleotides play major role. These small RNAs are either produced endogenously or exogenously by different ways. These molecules are very small but their impacts are very big. Thus such RNAs may be considered as powerful entity. Based upon the origin of small RNAs, they are classified into two small interfering RNA (siRNA) and micro RNA (miRNA). In the succeeding sections, we have discussed the origin of these small RNAs and their role in gene silencing.

2.1 Small Interfering RNA (siRNA)

Gene silencing through RNAi is triggered by long exogenously supplied dsRNA that produces siRNAs into cells [17]. The dsRNAs are produced from replication of plant RNA viruses, transgenic inverted repeats, and amplified through RNA-dependent RNA polymerase (RdRp) in plants. The siRNAs can be produced *in vitro* from dsRNA or intron-containing hairpin RNA (hpRNA) constructs. The hpRNA is a type of dsRNA derived from a long piece of single stranded RNA containing inverted repeats and a loop connecting between them [18]. It is expected that siRNAs derived from hpRNA may produce 100% gene silencing [19]. In case of plants, siRNAs are generated by RNase III- like enzyme called DICER-LIKE (DCL) [20]. Plant produces two distinct classes of siRNAs with 2-nucleotide 3' overhangs *i.e.* short (21-23 nucleotides) and long (24-26 nucleotides) siRNAs [21]. In plants and worms, amplification of the silencing signal and cell-to-cell RNAi spreading has been observed [22]. The siRNA can also be used as primer for the generation of new dsRNA by RdRp. The new dsRNA subsequently serve as target for the DCL and are processed into secondary siRNAs [23, 24]. The movement of dsRNA or siRNA between plant cells can take place through plasmodesmata and/or via the vascular system [12, 25]. The mature single stranded siRNA is loaded into RISC containing Argonaute proteins which direct the mRNA cleavage or translational inhibition of the target gene [26]. We are also working on gene silencing in plant systems. The hpRNA strategy has been employed successfully for the first time in a woody, perennial and difficult plant system tea for functional analysis of glutathione synthetase gene [27]. It has also been employed to silence caffeine synthase gene in tea for the purpose of reducing the caffeine contents [28].

2.2. Micro RNA (miRNA)

RNAi is also triggered by small non-coding RNAs known as miRNAs [17]. The miRNAs are chemically and functionally similar to siRNAs but are produced endogenously from miRNA genes present in the genome. The miRNAs are a large family of small regulatory RNA molecules with 21-24nt in length. These are produced from larger precursors that are transcribed from non-protein-coding genes in most of the plants and animals [29]. Mostly miRNAs are evolutionary conserved. For the first time, lin-4 miRNA was described in *Caenorhabditis elegans* during 1993 by Lee and colleagues in the Victor Ambros lab [30]. Later, the term 'microRNA' was coined in 2001 [31]. Majority of the plant miRNA genes exist as independent transcriptional units, which are transcribed by RNA polymerase II into primary transcripts (pri-miRNAs) of 1-2kb or longer. The pri-miRNAs are processed into precursor miRNAs (pre-miRNAs) with a specific hairpin shaped stem-loop structure. The processing of these precursor molecules is accomplished by nuclear-localized RNase III enzyme DICER-LIKE 1 (DCL1) in case of plants [32]. Further, the processed molecule is exported from nucleus to cytoplasm by a miRNA transporter, HASTY (HST). After methylation of mature miRNA, it joins with Argonaute to form RISC that can subsequently target specific mRNAs [33]. The plant miRNAs generally target their mRNAs through perfect or near-perfect complementarity, directing the post-transcriptional regulation of gene expression by mediating either mRNA cleavage or translational repression [34, 35].

Recently, these small molecules have been reported to play multiple roles in plant development, signal transduction, protein degradation and apoptosis [36]. The miRNAs act as regulators of transcriptional factors of various important plant genes [37] and are reported to involve in plant responses to biotic and abiotic stresses [38]. So far, a large number of miRNAs have been identified from various plant species but still many other plant species are yet to identify with the conserved or novel miRNA sequences. Identifying miRNAs and their target genes is therefore seem very important to understand their functions and the underlying control mechanism in the plant metabolic pathways for plant improvements. The mechanism of RNAi involving formation of siRNA and miRNA in plants has been shown in figure 1.

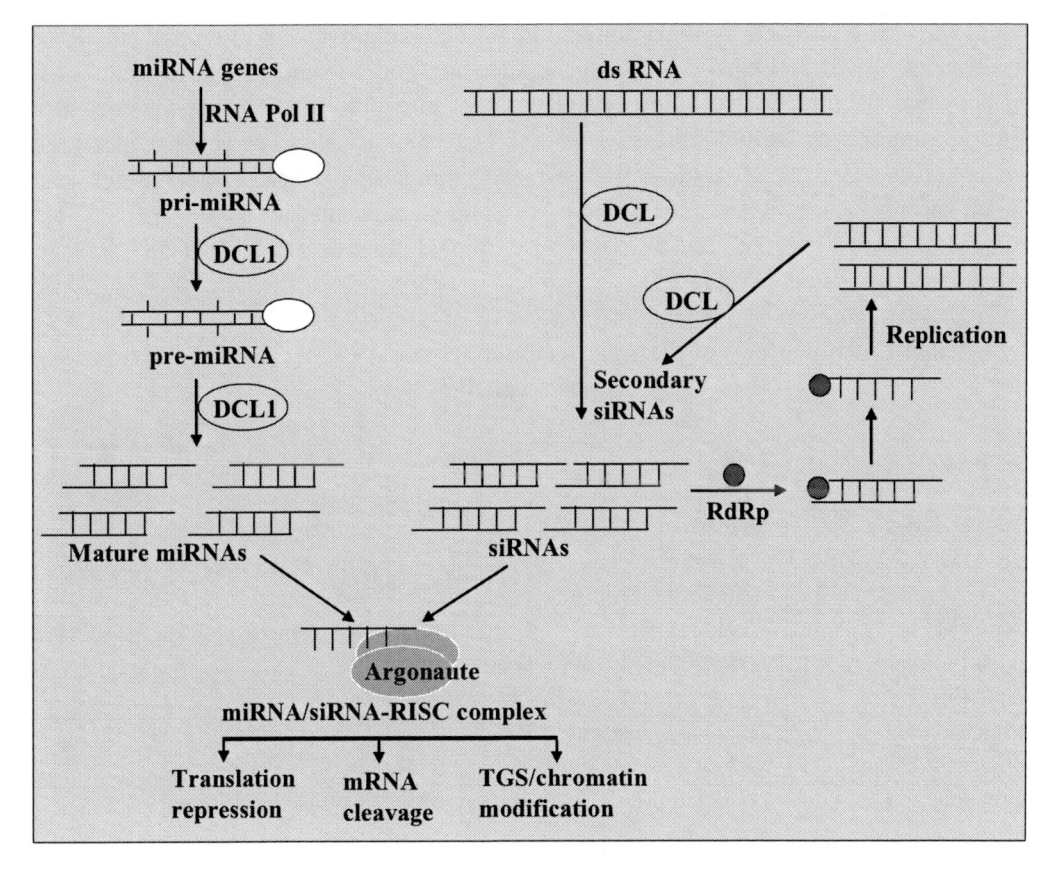

Figure 1. Representative model of gene silencing mediated by RNA interference (RNAi) in plants. RNAi mechanism is initiated by 21-26 nt siRNAs which are processed from long dsRNAs through RNase III (DCL; Dicer-like). While 21-24 nt miRNAs are transcribed from miRNA gene present in the genome. siRNAs can also be used as primers for the generation of new dsRNA by RNA-dependent RNA polymerase (RdRp). The new dsRNA can subsequently serve as target for the DCL and are processed into secondary siRNAs for amplification of silencing signals. In plants nuclear-localized DCL1 appears to be responsible for processing of both pri (primary) - and pre (precursor) -miRNAs to generate mature miRNAs. The mature single stranded miRNA or siRNA are loaded into RNA induced silencing complex (RISC) to direct mRNA cleavage or translational repression or chromatin modifications.

3. Applications of Gene Silencing in Crop Improvement

Gene silencing mediated by RNA interference (RNAi) is one of the most exciting discoveries in the field of genomics as an efficient tool for PTGS. Stable transgenic plants have been produced in which the transgene constructs generate dsRNA or hpRNA. These constructs are delivered to plants using *Agrobacterium* or biolistic mediated approaches. RNAi mechanism has been reported to occur naturally in most of the eukaryotes including plants. Plants are an important natural resource of numerous products, including food, fiber, wood, oils, dyes, secondary metabolites, and pharmaceuticals. However, these valuable plant products are produced in small amounts in limited tissues. But RNAi provides opportunity for metabolic engineering of plants to remove several un-useful or deleterious substances and to produce the useful products in larger amount. For genetic improvement of crop plants, RNAi has advantages over antisense-mediated gene silencing and co-suppression, in terms of its efficiency and stability. It also offers advantages over mutation-based reverse genetics in its ability to suppress transgene expression in multigene families in a regulated manner. The silencing efficiency of hpRNA and antisense based approaches has been compared in a range of plant species. The hpRNA strategy increases the gene silencing by 90-100% [18]. Some examples documenting the use of post-transcriptional gene silencing (PTGS) for crop improvements are listed in Table 1.

Table 1. Applications of post-transcriptional gene silencing (PTGS) in crop improvement

Target gene	Trait	Host plant	Potential benefits	Ref. no.
ghSAD-1 and ghFAD2-1	To increase stearic acid and oleic acid contents of seed oil	Cotton	Cooking applications without the need for hydrogenation	52
CaMXMT1 (Theobromine synthase)	To reduce caffeine content	Coffee beans plants	Decaffeinated coffee	40
Glutelin genes	To reduce glutelin content	Rice	Consumer health	41
Zein gene	To enrich the protein content	Maize	Increased nutritional value	51
CYP79D1 and CYP79D2 genes	To reduce the linamarin content	Cassava	Removal of deleterious substance	39
DET1 gene	To increase carotenoid and flavonoid contents	Tomato	Increased nutritional value	54
Mal-d1 gene family	To reduce allergenicity	Apple	Consumer health	44
Lol-p1 and Lol-p2 genes	To reduce allergenicity	Ryegrass	Hypo-allergic ryegrass	43
Gene encoding SBE	To increase amylose content	Maize	Improved maize quality	50
ACR2	To increase arsenic uptake	*Arabidopsis*	Phytoremediation of soils	57
SBEIIa and SBEIIb genes	To increase amylose content	Wheat	Improved wheat quality	55
SBE1 and SBE2 genes	To increase amylose content	Potato	Improved potato quality	56
FEH genes	To suppress insulin degradation	Chicory	Improved chicory quality	49

Table 1. (Continued)

Target gene	Trait	Host plant	Potential benefits	Ref. no.
Delta-cadinene synthase gene	To reduce toxic gossypol	Cotton	Remove deleterious substance	47
LTPG1 and LTPG2 genes	To reduce allergenicity	Tomato	Consumer health	45
LFY gene	To down regulate LFY gene	Chinese cabbage	Increased productivity	58
LeACO1 gene	To reduce ethylene biosynthesis	Tomato	Longer shelf life (slower ripening)	53
ORF IV of RTBV	To increase the resistance against RTBV	Rice	Virus resistance	59
Lachrymatory synthase gene	Breakdown of trans-s-1-propenyl-L- cysteine sulfoxide	Onion	Remove deleterious substance	46
Ara-h2 gene	To reduce allergenicity	Peanut	Consumer health	42
LKR/SDH gene	To suppress lysine catabolism	Corn	Improved lysine content	48

RNAi as an efficient gene silencing tool holds up the potential in genetic improvement of crop plants, particularly targeting the reduction in the levels of natural toxin compounds. Cassava (*Manihot esculenta* Crantz.) is a major source of calories. But it contains potentially toxic levels of the cyanogenic glucoside, linamarin. Therefore, short-cut processing techniques yield toxic food products. Using a leaf-specific promoter to drive the antisense expression of the cytochrome P450 genes (CYP79D1/CYP79D2), catalyzing the first-important step in linamarin synthesis resulted in their reduced expressions along with 60%-94% and 99% reduction in leaf and root linamarin contents respectively [39]. Similarly, transgenic coffee plants expressing the hpRNA constructs containing *CaMXMT1* sequences, a cDNA encoding one of the genes involved in the caffeine biosynthetic pathway were produced. They showed a marked reduction of 30-80% and 50-70% in their theobromine and caffeine contents, respectively. These findings confirmed theobromine as an important intermediate in major caffeine biosynthetic pathway. In addition, feasibility of decaffeinated coffee for caffeine sensitive people was accomplished [40]. As a relief to kidney patients which are unable to digest glutelin, a low glutelin rice variety (LGC-1) was developed using RNAi approach [41].

RNAi has also shown promise in the development of hypoallergenic food crops. Peanut allergy is due to the presence of immunodominant Ara-h2 proteins. It is one of the most life-threatening food allergies and one of the serious challenges before the peanut and food industries. RNAi was employed for making transgenic peanuts with reduced Ara-h2 content and for this an inverted repeat cassette was prepared from a coding region of Ara-h2. Thus, the feasibility of reducing peanut allergy was demonstrated [42]. Similarly, proteins Lol-p1 and Lol-p2 are the main allergens in ryegrass pollen. These proteins belong to two major classes of grass pollen allergens to which over 90% of pollen-allergic patients are sensitive. Using antisense constructs, transgenic plants were generated with reduced levels of these pollen allergens in the most important worldwide cultivated ryegrass species, *Lolium perenne* and *Lolium multiflorum*. Thus, development of hypoallergenic ryegrass cultivars could be possible [43]. As the apple allergy is dominated by IgE antibodies against Mal-d1 protein antigen in areas where birch pollen is endemic. To combat against this, *in vitro* grown apple plantlets were transformed with a construct coding for an intron-spliced hairpin RNA containing a Mal-d1–specific inverted repeat sequence separated by a Mal-d1–specific intron sequence. The apple transformants were reported for significant reduction in *in vivo* allergenicity [44]. Later on, hypoallergenic tomatoes (*Lycopersicon esculentum*) were also produced. Tomato contains Lyc-e3 allergen that encodes a nonspecific lipid transfer protein (ns-LTP). Transgenic tomato plants constitutively expressing LTPG1- or LTPG2-specific dsRNAi constructs were created which resulted in suppression of Lyc-e3 accumulations [45]. Recently, tear-inducing lachrymatory factor synthase gene was also suppressed through RNAi in transgenic onion (*Allium cepa*) and resulted in the reduction of lachrymatory synthase activity up to 1,544-fold [46].

Cotton is a nutritionally and economically important fiber crop in developed countries and also in many developing countries where malnutrition and starvation are prevalent. The cotton seeds are rich in dietary protein but unsuitable for human consumption because of its toxic terpenoid product, gossypol. RNAi was used to produce cotton stocks with seeds containing lower levels of delta-cadinene synthase, a key enzyme in gossypol production. This does not affect the enzyme production in other parts of the plant, where gossypol is important in preventing damage from plant pests [47]. This targeted genetic modification applied to an underutilized agricultural byproduct, provides a mechanism to open up a new source of nutrition for hundreds of millions of people.

RNAi has been used in the upregulation of important steps in a biosynthetic pathway by blocking their catabolism. In plants, lysine catabolism is thought to be controlled by a bifunctional enzyme, lysine ketoglutarate reductase/saccharopine dehydrogenase (LKR/SDH). The suppression of LKR/SDH gene in maize kernels resulted in an increase in free lysine content in its developing endosperm as well as embryos. This also led to increase in lysine content of mature kernels [48]. Similarly, transgenic chicory plants were generated by suppression of fructan 1-exohydrolase (1-FEH) gene, responsible for insulin degradation in its roots. This prevents the degradation of insulin and enhances its accumulation in roots [49].

The nutritional quality of several edible crops has also been improved using RNAi. Maize quality was improved by silencing the gene encoding starch branch enzyme (SBE). For this, gene segment of SBE of maize was employed to design inverted repeat construct (pCJSBE2b) for transformation studies. The total starch content of transgenic maize did not change compared to the control plants. However, up to 50% increase in amylose content was observed in transgenics [50]. In maize, the main protein component of seed storage proteins is alpha-zeins. These are the major determinants of nutritional imbalance when maize is used as the sole food source. For improving the nutritional quality of alpha-zein rich maize, *opaque-2* (*o2*) mutation was employed in breeding varieties. However, this *o2* worked in a recessive fashion by affecting the expression of 22-kD alpha-zeins, as well as additional endosperm gene functions. RNAi constructs designed from a 22-kD zein gene were used to transform maize. Thus, dominant opaque phenotype with suppressed storage protein genes without interrupting O2 synthesis was produced and this illustrated an approach for creating more nutritious crop plants [51]. Further, the fatty acid composition of cotton oil was also genetically modified. Two key fatty acid desaturase genes encoding stearoyl-acyl-carrier protein D9-desaturase and oleoyl-phosphatidylcholine x6-desaturase were down regulated using hpRNA construct transformation in cotton. Knockdown of these two genes led to the increase of nutritionally improved high-oleic (HO) and high-stearic (HS) cottonseed oils. These are essential fatty acids for human heart health [52]. In order to prolong the ripening of tomato fruit and to increase its shelf life, the expression of tomato LeACO1 gene (1-aminocyclopropane-1-carboxylic acid oxidase 1) in the transgenic tomato was suppressed using co-suppression as a gene silencing tool. It was resulted in the decrease of endogenous ethylene biosynthesis and increased the storage ability of tomato fruits [53]. A fruit-specific promoter with RNAi has been used to suppress an endogenous photomorphogenesis regulatory gene DE-ETHIOLATED1 (DET1) expression. It was resulted in improved carotenoid and flavonoid levels in tomato fruits with minimal effects on plant growth and other fruit quality parameters [54]. Further, grain amylose content was increased to over 70% of the total starch content in wheat by silencing the genes encoding two starch-branching isozymes of amylopectin synthesis. These high amylose grains were fed to rats in a diet as a whole meal which had resulted in positive effects on indices of gastrointestinal health. [55]. Similarly, by inhibiting Sbe1 and Sbe2 genes (coding for starch branching enzymes) using a potato granule-bound starch synthase promoter, more than 50% high-amylose potato lines were produced [56]. These elucidate the potential of RNAi technology to improve the quality of food crops.

Silencing of ACR2 gene encoding an arsenic reductase in *Arabidopsis* resulted in plants that could accumulate 10- to 16-fold more arsenic in shoots (350-500ppm) and retain less arsenic in roots compared with wild-type plants. By reducing the expression of ACR2 homologs in tree, shrub, and grass species could play a vital role in the phytoremediation of environmental arsenic contamination [57].

A somewhat different approach was adopted to improve productivity of Chinese cabbage (*Brassica rapa* L. ssp. pekinensis). LEAFY (LFY) gene plays an important role in determining plant flowering mainly by controlling the timing of phase transition. The LFY gene expression was down-regulated using RNAi to delay Chinese cabbage bolting and flowering. Due to late transition to the reproductive phase, the resulted transgenic plant showed higher productivity in terms of more branching and more leaves. By inhibiting LFY gene expression, bolting could be delayed in a cold-sensitive long-day (LD) condition and thus, late flowering of Chinese cabbage can be utilized as a good genetic resource for the breeding late-bolting Chinese cabbage [58].

Use of RNAi has also been implicated in various crops to protect them against virus infection. Rice tungro is a viral disease seriously affecting rice production in South and Southeast Asia. It is caused by the simultaneous infection of Rice tungro bacilliform virus (RTBV) and Rice tungro spherical virus (RTSV). For the control of RTBV infection, transgenic rice plants were produced, expressing DNA encoding ORF IV of RTBV, both in sense as well as in anti-sense orientation, resulting in the formation of double-stranded (ds) RNA. The resulted transgenic lines showed resistance against RTBV [59].

4. Conclusion and Future Perspectives

Gene silencing is an efficient tool for functional genomics. Although tissue or organ-specific hpRNA based RNAi constructs have been employed for several gene silencing applications in various important crops. Still the uses of finely tuned RNAi-based gene silencing vectors that minimize the 'off-target' effects are to be awaited. Recently, the newly discovered miRNAs have been shown as important regulatory factors which play multiple roles in biological processes, including development, cell proliferation, apoptosis, and stress responses in plants. The future strategy of gene silencing would be based on new-generation RNAi constructs (miRNA constructs) with high silencing efficiency, accuracy, and fewer side effects in plants. Still there is unexplored possibility of silencing undesirable or pathogen genes in agriculturally valuable crops. Thus, the full potential of RNAi for crop improvement has yet to be realized.

Acknowledgments

Authors are thankful to Dr. P. S. Ahuja, Director, IHBT for his valuable suggestions and guidance to write this article. We would like to acknowledge the financial grants in our laboratory from Council of Scientific and Industrial Research (CSIR) and Department of Science and Technology (DST), Govt. of India.

References

[1]   Fagard, M., and Vaucheret, H. (2000). (Trans)gene silencing in plants: How many mechanisms? *Annu Rev Plant Physiol Plant Mol Biol* 51, 167-194.

[2]   Matzke, M., Kanno, T., Huettel, B., Jaligot, E., Mette, F., Kreil, P., Daxinger, L., Rovina, P., Aufsatz, W., and Matzke, M. (2005). RNA-directed DNA methylation, in: P. Meyer (Ed.), Plant Epigenetics, Annual Plant Reviews, vol. 19, Blackwell Publishing, Oxford, pp. 69-105.

[3]   Matzke, M., Matzke, A. J., and Kooter, J. M. (2001). RNA: guiding gene silencing. *Science* 293, 1080-1083.

[4]   Agrawal, N., Dasaradhi, P. V., Mohmmed, A., Malhotra, P., Bhatnagar, R. K., and Mukherjee, S. K. (2003). RNA interference: biology, mechanism, and applications. *Microbiol Mol Biol Rev* 67, 657-685.

[5]   Huettel, B., Kanno, T., Daxinger, L., Bucher, E., van der Winden, J., Matzke, A. J., and Matzke, M. (2007). RNA-directed DNA methylation mediated by DRD1 and Pol IVb: a versatile pathway for transcriptional gene silencing in plants. *Biochim Biophys Acta* 1769, 358-374.

[6]   Napoli, C., Lemieux, C., and Jorgensen, R. (1990). Introduction of a chimeric chalcone synthase gene into Petunia results in reversible co-suppression of homologous genes in trans. *Plant Cell* 2, 279-289.

[7]   Watson, J. M., Fusaro, A. F., Wang, M., and Waterhouse, P. M. (2005). RNA silencing platforms in plants. *FEBS Lett* 579, 5982-5987.

[8]   Knee, R., and Murphy, P. R. (1997). Regulation of gene expression by natural antisense RNA transcripts. *Neurochem Int* 31, 379-392.

[9]   Brantl, S. (2002). Antisense-RNA regulation and RNA interference. *Biochim Biophys Acta* 1575, 15-25.

[10]   Arenz, C., and Schepers, U. (2003). RNA interference: from an ancient mechanism to a state of the art therapeutic application? *Naturwissenschaften* 90, 345-359.

[11]   Kimball, S. R. (2002). Regulation of global and specific mRNA translation by amino acids. *J Nutr* 132, 883-886.

[12]   Hannon, G. J. (2002). RNA interference. *Nature* 418, 244-251.

[13]   Yu, H., and Kumar, P. P. (2003). Post-transcriptional gene silencing in plants by RNA. *Plant Cell Rep* 22, 167-174.

[14]   Tang, G., Reinhart, B. J., Bartel, D. P., and Zamore, P. D. (2003). A biochemical framework for RNA silencing in plants. *Genes Dev* 17, 49-63.

[15]   Hammond, S. M., Bernstein, E., Beach, D., and Hannon, G. J. (2000). An RNA-directed nuclease mediates post-transcriptional gene silencing in Drosophila cells. *Nature* 404, 293-296.

[16]   Qi, Y., and Hannon, G. J. (2005). Uncovering RNAi mechanisms in plants: biochemistry enters the foray. *FEBS Lett* 579, 5899-5903.

[17]   Rana, T. M. (2007). Illuminating the silence: understanding the structure and function of small RNAs. *Nat Rev Mol Cell Biol* 8, 23-36.

[18]   Wesley, S. V., Helliwell, C. A., Smith, N. A., Wang, M. B., Rouse, D. T., Liu, Q., Gooding, P. S., Singh, S. P., Abbott, D., Stoutjesdijk, P. A., Robinson, S. P., Gleave, A. P., Green, A. G., and Waterhouse, P. M. (2001). Construct design for efficient, effective and high-throughput gene silencing in plants. *Plant J* 27, 581-590.

[19]   Smith, N. A., Singh, S. P., Wang, M. B., Stoutjesdijk, P. A., Green, A. G., and Waterhouse, P. M. (2000). Total silencing by intron-spliced hairpin RNAs. *Nature* 407, 319-320.

[20]   Margis, R., Fusaro, A. F., Smith, N. A., Curtin, S. J., Watson, J. M., Finnegan, E. J., and Waterhouse, P. M. (2006). The evolution and diversification of Dicers in plants. *FEBS Lett* 580, 2442-2450.

[21]   Zamore, P. D. (2004). Plant RNAi: How a viral silencing suppressor inactivates siRNA. *Curr Biol* 14, R198-200.

[22]   Stanislawska, J., and Olszewski, W. L. (2005). RNA interference--significance and applications. *Arch Immunol Ther Exp (Warsz)* 53, 39-46.

[23]   Ahlquist, P. (2002). RNA-dependent RNA polymerases, viruses, and RNA silencing. *Science* 296, 1270-1273.

[24]   Martinez, J., Patkaniowska, A., Urlaub, H., Luhrmann, R., and Tuschl, T. (2002). Single-stranded antisense siRNAs guide target RNA cleavage in RNAi. *Cell* 110, 563-574.

[25] Szweykowska-Kulinska, Z., Jarmolowski, A., and Figlerowicz, M. (2003). RNA interference and its role in the regulation of eukaryotic gene expression. *Acta Biochim Pol* 50, 217-229.

[26] Bartel, D. P. (2004). MicroRNAs genomics, biogenesis, mechanism, and function. *Cell* 116, 281-297.

[27] Mohanpuria, P., Rana, N. K., and Yadav, S. K. (2008). Transient RNAi based gene silencing of glutathione synthetase reduces GSH levels in somatic embryos of *Camellia sinensis* L. *Biol Plant* 52, 381-384.

[28] Mohanpuria, P., Ahuja, P. S., Yadav, S. K. (2007). Expression analysis of caffeine synthase and its transient silencing reduces caffeine levels in somatic embryos of *Camellia sinensis* (L.) O. Kuntze (abstract no P20). In: Abstracts: Indo-French conference RNAi in Genome Control. December 12-14, CCMB, Hyderabad, India.

[29] Zhou, X., Jeker, L. T., Fife, B. T., Zhu, S., Anderson, M. S., McManus, M. T., and Bluestone, J. A. (2008). Selective miRNA disruption in T reg cells leads to uncontrolled autoimmunity. *J Exp Med* 205, 1983-1991.

[30] Lee, R. C., Feinbaum, R. L., and Ambros, V. (1993). The *C. elegans* heterochronic gene lin-4 encodes small RNAs with antisense complementarity to lin-14. *Cell* 75, 843-854.

[31] Ruvkun, G. (2001). Molecular biology. Glimpses of a tiny RNA world. *Science* 294, 797-799.

[32] Kurihara, Y., and Watanabe, Y. (2004). Arabidopsis micro-RNA biogenesis through Dicer-like 1 protein functions. *Proc Natl Acad Sci* USA 101, 12753-12758.

[33] Jones-Rhoades, M. W., Bartel, D. P., and Bartel, B. (2006). MicroRNAs and their regulatory roles in plants. *Annu Rev Plant Biol* 57, 19-53.

[34] Rhoades, M. W., Reinhart, B. J., Lim, L. P., Burge, C. B., Bartel, B., and Bartel, D. P. (2002). Prediction of plant microRNA targets. *Cell* 110, 513-520.

[35] Yao, Y., Guo, G., Ni, Z., Sunkar, R., Du, J., Zhu, J. K., and Sun, Q. (2007). Cloning and characterization of microRNAs from wheat (Triticum aestivum L.). *Genome Biol* 8, R96.

[36] Xie, W., Brown, W. T., and Denman, R. B. (2008). Translational regulation by non-protein-coding RNAs: different targets, common themes. *Biochem Biophys Res Commun* 373, 462-466.

[37] Zhang, B., Pan, X., Cannon, C. H., Cobb, G. P., and Anderson, T. A. (2006). Conservation and divergence of plant microRNA genes. *Plant J* 46, 243-259.

[38] Sanan-Mishra, N., Kumar, V., Sopory, S. K., and Mukherjee, S. K. (2009). Cloning and validation of novel miRNA from basmati rice indicates cross talk between abiotic and biotic stresses. *Mol Genet Genomics* 282, 463-474.

[39] Siritunga, D., and Sayre, R. T. (2003). Generation of cyanogen-free transgenic cassava. *Planta* 217, 367-373.

[40] Ogita, S., Uefuji, H., Yamaguchi, Y., Koizumi, N., and Sano, H. (2003). Producing decaffeinated coffee plants. *Nature* 423, 823.

[41] Kusaba, M., Miyahara, K., Iida, S., Fukuoka, H., Takano, T., Sassa, H., Nishimura, M., and Nishio, T. (2003). Low glutelin content1: a dominant mutation that suppresses the glutelin multigene family via RNA silencing in rice. *Plant Cell* 15, 1455-1467.

[42] Dodo, H. W., Konan, K. N., Chen, F. C., Egnin, M., and Viquez, O. M. (2008). Alleviating peanut allergy using genetic engineering: the silencing of the

immunodominant allergen Ara-h2 leads to its significant reduction and a decrease in peanut allergenicity. *Plant Biotechnol J* 6, 135-145.

[43] Petrovska, N., Wu, X., Donato, R., Wang, Z., Ong, E. J., Jones, E., Forster, J., Emmerling, M., Sidoli, A., Hehir, R. O., and Spangenberg, G. (2004). Transgenic ryegrasses (*Lolium* spp.) with down-regulation of main pollen allergens. *Mol Breed* 14, 489-501.

[44] Gilissen, L. J., Bolhaar, S. T., Matos, C. I., Rouwendal, G. J., Boone, M. J., Krens, F. A., Zuidmeer, L., Van Leeuwen, A., Akkerdaas, J., Hoffmann-Sommergruber, K., Knulst, A. C., Bosch, D., Van de Weg, W. E., and Van Ree, R. (2005). Silencing the major apple allergen Mal-d1 by using the RNA interference approach. *J Allergy Clin Immunol* 115, 364-369.

[45] Le, L. Q., Mahler, V., Lorenz, Y., Scheurer, S., Biemelt, S., Vieths, S., and Sonnewald, U. (2006). Reduced allergenicity of tomato fruits harvested from Lyc-e1-silenced transgenic tomato plants. *J Allergy Clin Immunol* 118, 1176-1183.

[46] Eady, C. C., Kamoi, T., Kato, M., Porter, N. G., Davis, S., Shaw, M., Kamoi, A., and Imai, S. (2008). Silencing onion lachrymatory factor synthase causes a significant change in the sulfur secondary metabolite profile. *Plant Physiol* 147, 2096-2106.

[47] Sunilkumar, G., Campbell, L. M., Puckhaber, L., Stipanovic, R. D., and Rathore, K. S. (2006). Engineering cottonseed for use in human nutrition by tissue-specific reduction of toxic gossypol. *Proc Natl Acad Sci* USA 103, 18054-18059.

[48] Reyes, A. R., Bonin, C. P., Houmard, N. M., Huang, S., and Malvar, T. M. (2009). Genetic manipulation of lysine catabolism in maize kernels. *Plant Mol Biol* 69, 81-89.

[49] Asad, M. (2006). RNA interference (RNAi) as a tool to engineer high nutritional value in chicory (*Chicorium intybus*). *Commun Agric Appl Biol Sci* 71, 75-78.

[50] Chai, X. J., Wang, P. W., Guan, S. Y., and Xu, Y. W. (2005). Reducing the maize amylopectin content through RNA interference manipulation. *Zhi Wu Sheng Li Yu Fen Zi Sheng Wu Xue Xue Bao* 31, 625-630.

[51] Segal, G., Song, R., and Messing, J. (2003). A new opaque variant of maize by a single dominant RNA-interference-inducing transgene. *Genetics* 165, 387-397.

[52] Liu, Q., Singh, S. P., and Green, A. G. (2002). High-stearic and High-oleic cottonseed oils produced by hairpin RNA-mediated post-transcriptional gene silencing. *Plant Physiol* 129, 1732-1743.

[53] Hu, Z. L., Chen, X. Q., Chen, G. P., LÜ, L. J., and Donald, G. (2007). The influence of co-suppressing tomato 1-aminocyclopropane-1-carboxylic acid oxidase I on the expression of fruit ripening-related and pathogenesis-related protein genes. *Agri Sci China* 6, 406-413.

[54] Davuluri, G. R., van Tuinen, A., Fraser, P. D., Manfredonia, A., Newman, R., Burgess, D., Brummell, D. A., King, S. R., Palys, J., Uhlig, J., Bramley, P. M., Pennings, H. M., and Bowler, C. (2005). Fruit-specific RNAi-mediated suppression of DET1 enhances carotenoid and flavonoid content in tomatoes. *Nat Biotechnol* 23, 890-895.

[55] Regina, A., Bird, A., Topping, D., Bowden, S., Freeman, J., Barsby, T., Kosar-Hashemi, B., Li, Z., Rahman, S., and Morell, M. (2006). High-amylose wheat generated by RNA interference improves indices of large-bowel health in rats. *Proc Natl Acad Sci* USA 103, 3546-3551.

[56] Andersson, M., Melander, M., Pojmark, P., Larsson, H., Bulow, L., and Hofvander, P. (2006). Targeted gene suppression by RNA interference: an efficient method for production of high-amylose potato lines. *J Biotechnol* 123, 137-148.

[57] Dhankher, O. P., Rosen, B. P., McKinney, E. C., and Meagher, R. B. (2006). Hyperaccumulation of arsenic in the shoots of Arabidopsis silenced for arsenate reductase (ACR2). *Proc Natl Acad Sci* USA 103, 5413-5418.

[58] Xia, G. Q., Zhu, J. Y., He, Q. W., Zhao, S. Y., and Wang, C. H. (2007). Late-bolting transgenic Chinese cabbage obtained by RNA interference technique. *Zhi Wu Sheng Li Yu Fen Zi Sheng Wu Xue Xue Bao* 33, 411-416.

[59] Tyagi, H., Rajasubramaniam, S., Rajam, M. V., and Dasgupta, I. (2008). RNA-interference in rice against Rice tungro bacilliform virus results in its decreased accumulation in inoculated rice plants. *Trans Res* 17, 897-904.

In: Gene Silencing: Theory, Techniques and Applications
Editor: Anthony J. Catalano

ISBN: 978-1-61728-276-8
© 2010 Nova Science Publishers, Inc.

Chapter XV

RNA Interference Mediated by Cell-Penetrating Peptides

Cheng-Yi Lee, Betty Revon Liu, Ya-Hui Wang, Yu-Wun Hou, Jyh-Ching Chou and Han-Jung Lee[*]

Department of Natural Resources and Environmental Studies, National Dong Hwa University, Hualien 97401, Taiwan

Abstract

RNA interference (RNAi) has been utilized in a variety of applications to target specific gene silencing mediated by small-interfering RNA (siRNA) over the last few years. Cell-penetrating peptides (CPPs) were proven to be able to traverse cell membranes and deliver biological macromolecules into living cells. Here, we provided an efficient and safe method for the delivery of siRNA into mammalian cells mediated by CPPs noncovalently. We first established a GC-EGFP cell line stably expressing enhanced green fluorescent protein (EGFP) from human gastric cells. CPPs were demonstrated to interact with and deliver siRNA into GC-EGFP cells, and the internalized dsRNA tended to localize in the perinuclear region within cells. The sulforhodamine B (SRB) assay further confirmed CPPs were nontoxic to cell viability. Finally, our results showed that siRNA fulfilled its targeted *egfp* gene silencing. In the future, CPPs may provide a useful and nontoxic tool for the delivery of siRNA into mammalian cells.

[*] Corresponding author: E-mail: hjlee@mail.ndhu.edu.tw

Abbreviations

AID	arginine-rich intracellular delivery
CPP	cell-penetrating peptide
DMSO	dimethyl sulfoxide
dsRNA	double-stranded RNA
EGFP	enhanced green fluorescent protein
N/P	nitrogen/phosphate
PBS	phosphate buffered saline
PTD	protein transduction domain
PTGS	post-transcriptional gene silencing
RNAi	RNA interference
siRNA	short interfering RNA
SR9	synthetic nona-arginine
SRB	sulforhodamine B
Tat	*trans*-activator of transcription
TRBP	transactivating response RNA-binding protein

Introduction

RNA interference (RNAi) was coined after the discovery of double-stranded RNA (dsRNA) injected into the nematode *Caenorhabditis elegans* [1], leading to specific silencing of genes highly homologous in sequence to the delivered dsRNA. RNAi, also called post-transcriptional gene silencing (PTGS) in plants [2], is a process in which a short interfering RNA (siRNA) triggers the silencing of gene expression in a sequence-specific manner. The siRNA which composed of 21-25 nucleotides is intracellularly generated from endogenous or exogenous dsRNA cleaved by a ribonuclease III-type enzyme Dicer, its cofactor transactivating response RNA-binding protein (TRBP) and the Argonaute family of proteins [3]. The siRNA duplexes undergo 5' phosphorylation and are associated with the RNA-induced silencing complex (RISC) [4, 5]. Activated RISC and the antisense strand complementary to the targeted mRNA interact with the mRNA target. Single site-specific cleavage of the mRNA target then occurs, and the target mRNA is ultimately degraded.

The *trans*-activator of transcription (Tat), a protein with 86 amino acids of the human immunodeficiency virus type 1, was observed to enter into mammalian cells freely [6, 7]. The basic region of 47–57 amino acids of Tat protein, called protein transduction domain (PTD) or cell-penetrating peptide (CPP), was proven that can rapidly translocate through the plasma membrane and accumulate in the cell nucleus [8]. More observations have been reported that CPPs for cellular internalization are usually rich in basic amino acids, including arginine-rich peptides [9]. CPPs can serve as efficient carriers to deliver numerous cargoes, such as DNAs, RNAs, proteins, liposomes and nanoparticles into cells both *in vitro* and *in vivo* [10]. This peptide-mediated cellular delivery has many desirable features, such as ease of preparation, safety, efficiency and noncytotoxicity.

In this study, we present evidence to show that synthetic nona-arginine (SR9) peptides can efficiently deliver siRNA into mammalian cells noncovalently. The enhanced green fluorescent protein (*egfp*)-targeted siRNA results in *egfp*-specific inhibition of gene expression after delivery in cells.

Materials and Methods

Preparation of Peptide, SiRNA and Stable Transfectants

SR9 peptide containing the nona-arginine sequence was synthesized as previously described [11]. Both the sense (5'-GGCAAGCUGACCCUGAAGUUCTT-3') and antisense (5'-GAACUUCAGGGUCAGCUUGCCTT-3') sequences of the siRNA duplexes targeted for the *egfp* gene were described previously [12]. Human gastric carcinoma (GC) cells (American Type Culture Collection, CRL-1739) were cultured as previously described [12]. The pEGFP-N1 plasmid (Clontech) was used for the establishment of the GC-EGFP cell line stably expressing EGFP.

CPP-Mediated SiRNA Delivery

Two hundred nM of EGFP siRNA duplexes were pre-mixed with or without (as a negative control) SR9 peptides at a nitrogen/phosphate (N/P) ratio of 12 in phosphate buffered saline (PBS) to a final volume of 80 µl and incubated for 15 min at room temperature. Cells were washed with PBS and treated with SR9-plus-siRNA mixtures for 10 min in a 37°C incubator. After 48 hours of culture, medium was removed, and cells were washed twice with 1 ml of PBS. Cellular images were collected by the TCS SL confocal microscope system (Leica) and quantified by the UN-SCAN-IT software (Silk Scientific).

Cytotoxic Assay

To determine cell viability relative to the control group after treatment of SR9 peptides, the sulforhodamine B (SRB) assay was conducted as previously described [13].

Statistical Analysis

Results were represented as means ± standard errors of measurement. Statistical comparisons of the control with the treated group were performed by the Student's *t*-test. The level of significance was set at $P < 0.05$ (*) or 0.01 (**).

Results

In this study, we employed SR9 peptides to deliver siRNA into cells by a noncovalent manner. Because fluorescence is simple to be monitored, we established the GC-EGFP cell line which stably expresses EGFP [12]. GC-EGFP cells were treated with 200 nM of EGFP-targeted siRNA alone or siRNA pre-mixed with SR9 peptides at N/P ratio of 12. After incubation of 48 hours, inhibition of EGFP expression in cells was monitored by a confocal microscope (Figure 1). The green fluorescence of cells treated with siRNA alone (Figure 1A) was normalized to the control level (100%). However, when cells treated with SR9-plus-siRNA complexes (Figure 1B), the fluorescent intensity was attenuated ~43% due to gene silencing (Figure 1C). This result demonstrated that SR9 peptides were able to deliver siRNA into cells, causing silencing of EGFP expression in GC-EGFP cells.

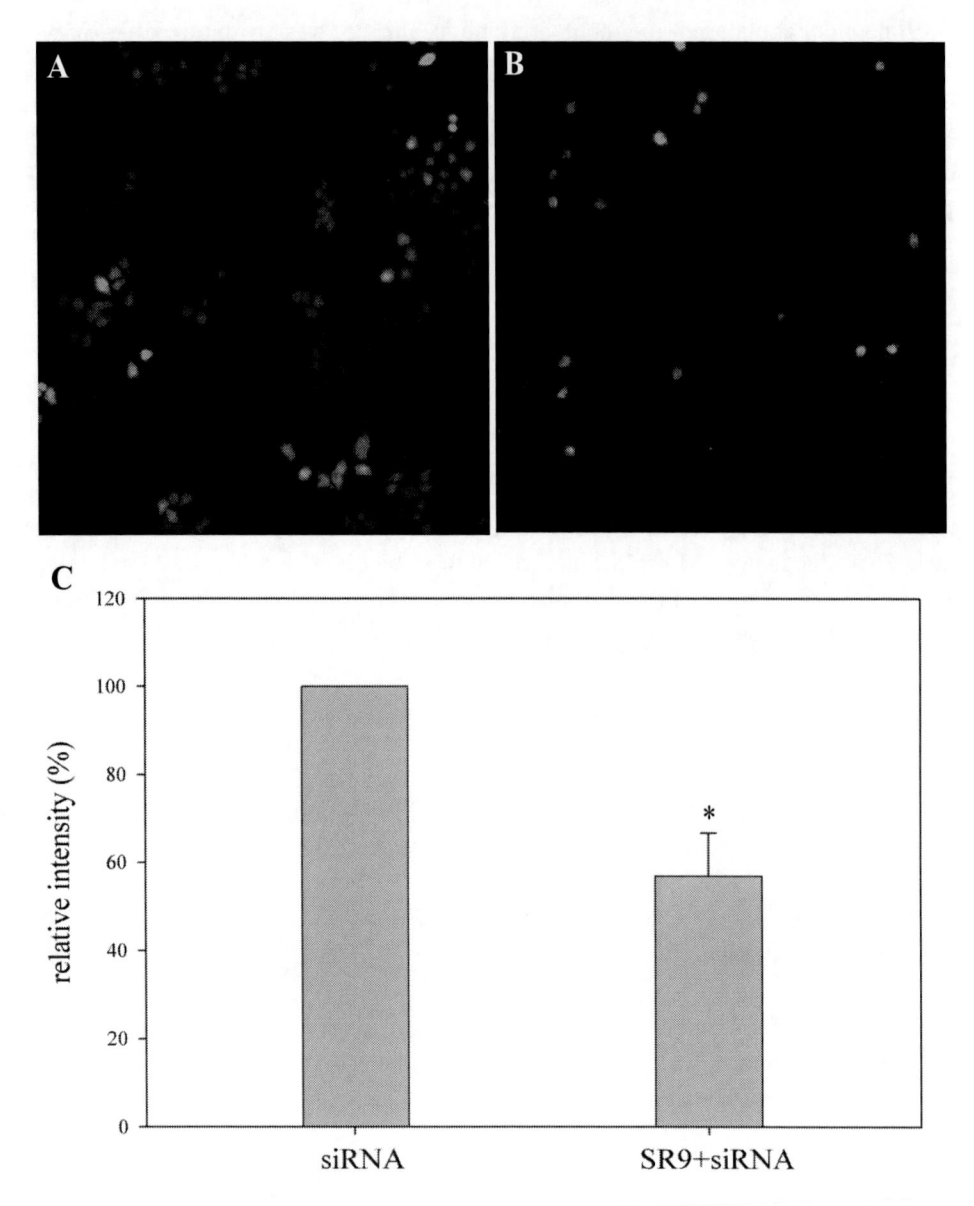

Figure 1. Intracellular delivery of siRNA mediated by SR9 peptides. GC-EGFP cells were treated with siRNA alone (A) or SR9 (N/P = 12) mixed with siRNA (B) and observed by a confocal microscope. (C) Quantification of fluorescence.

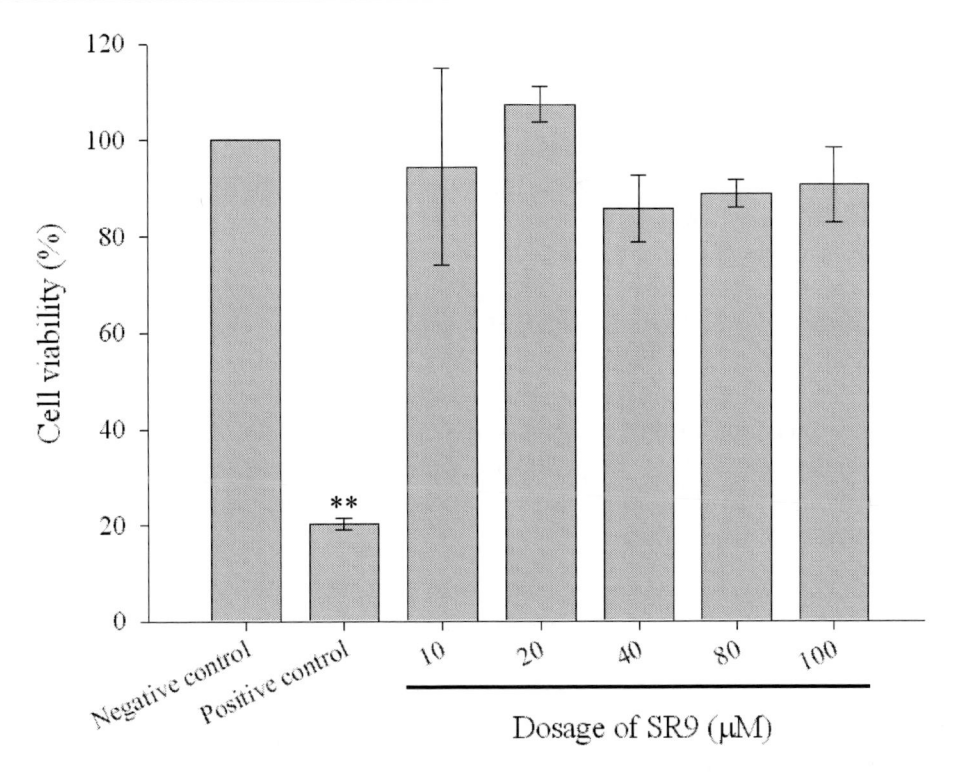

Figure 2. Cell viability of SR9 peptide treatment. Human A549 lung carcinoma cells (American Type Culture Collection, CCL-185) were treated with PBS as a negative control, 100% DMSO as a positive control or 10–100 μM of SR9 peptide.

To investigate the cytotoxicity caused by SR9 peptides, cell viability was determined by the SRB assay (Figure 2). Human A549 cells were treated with PBS as a negative control, dimethyl sulfoxide (DMSO) as a positive control or different concentrations of SR9 peptide. Total cell number was calculated and analyzed after 24 hours. Cells treated with various concentrations, even up to 100 μM, of SR9 peptide did not result in any cytotoxicity. In great contrast, cells treated with 100% DMSO exhibited dramatically reduction in cell viability. This result indicated that no detectable cytotoxicity caused by SR9 peptides in human cells.

Discussion

RNAi-mediated gene-specific silencing is a promising strategy in biological research and gene therapy. One of the primary barriers for siRNA action is to cross the plasma membrane. CPP-mediated siRNA delivery system is one of the nonviral-mediated siRNA delivery methods. This efficient method has attracted much attention from our laboratory [12] and others [14–21]. In the present study, we demonstrated that nontoxic SR9 peptides, members of CPPs, are able to deliver siRNA into mammalian cells, and the internalized siRNA ultimately causes RNAi. Free of toxicity is an important issue before CPPs can be widely applied in biomedical and therapeutic studies *in vivo*. As shown in Figure 2, we applied up to 500 times more dosage of SR9 peptide used in Figure 1 and tested cytotoxicity of SR9 peptide. These results confirmed that SR9 peptides up to 100 μM were nontoxic to cell viability.

Our results showed that SR9 peptides are able to deliver siRNA into GC-EGFP cells. Our previous study proved that siRNA tends to localized in the perinuclear region within cells [12]. This result agrees previous reports that localizations of siRNA [22, 23], Dicer [24], RISC [23] and RNAi activity [21, 25] are restricted to the cytoplasm. Recently, Argonaute2, the central protein component of RISC, was reported to reside in cytoplasmic processing bodies [26]. Specific gene silencing using RNAi has demonstrated huge potentials as a therapeutic strategy for eliminating pathogenic gene expression. Despite the early promise and excitement of gene-specific silencing, several critical problems remain to be solved before widespread clinical applications. These include, but not limited to, off-target effects, toxicity due to saturation of the endogenous RNAi functions, limited duration of silencing and effective targeted delivery [27]. Finally, this CPP-mediated siRNA delivery system presented here may provide a useful, easy and nontoxic tool for the delivery of exogenous siRNA in RNAi studies in the future.

Conclusion

In the present study, a stable transfectant, called GC-EGFP cell line, over-expressing EGFP in human gastric cells was established. Nontoxic CPPs can efficiently deliver *egfp*-targeted siRNA into GC-EGFP cells noncovalently. The internalized siRNA delivered by CPPs results in *egfp*-specific inhibition of gene expression in GC-EGFP cells.

Acknowledgments

This work was supported by the National Science Council (NSC 97-2621-B-259-003-MY3 to H.J.L.), Taiwan.

References

[1] Fire, A., Xu, S., Montgomery, M. K., Kostas, S. A., Driver, S. E. & Mello, C. C. (1998). Potent and specific genetic interference by double-stranded RNA in *Caenorhabditis elegans*. *Nature*, *391*, 806–811.

[2] Baulcombe, D. C. (1996). RNA as a target and an initiator of post-transcriptional gene silencing in transgenic plants. *Plant Mol. Biol.*, *32*, 79–88.

[3] Tiemann, K. & Rossi, J. J. (2009). RNAi-based therapeutics–current status, challenges and prospects. *EMBO Mol. Med. 1*, 142–151.

[4] Hammond, S. M., Bernstein, E., Beach, D. & Hannon, G. J. (2000). An RNA-directed nuclease mediates post-transcriptional gene silencing in *Drosophila* cells. *Nature*, *404*, 293–296.

[5]   Nykanen, A., Haley, B. & Zamore, P. D. (2001). ATP requirements and small interfering RNA structure in the RNA interference pathway. *Cell*, *107*, 309–321.

[6]   Green M. & Leowenstein, P. (1988). Autonomous functional domains of chemically synthesized human immunodeficiency virus tat trans-activator protein. *Cell*, *55*, 1179–1188.

[7]   Frankel, A. D. & Pabo, C. O. (1988). Cellular uptake of the tat protein from human immunodeficiency virus. *Cell*, *55*, 1189–1193.

[8]   Vives, E., Brodin, P. & Lebleu, B. (1997). A truncated HIV-1 Tat protein basic domain rapidly translocates through the plasma membrane and accumulates in the cell nucleus. *J. Biol. Chem.*, *272*, 16010–16017.

[9]   Futaki, S. (2002). Arginine-rich peptides: potential for intracellular delivery of macromolecules and the mystery of the translocation mechanisms. *Int. J. Pharm.*, *245*, 1–7.

[10] Schwartz, J. J. & Zhang, S. (2000). Peptide-mediated cellular delivery. *Curr. Opin. Mol. Ther.*, *2*, 162–167.

[11] Chang, M., Chou, J. C., Chen, C. P., Liu, B. R. & Lee, H. J. (2007). Noncovalent protein transduction in plant cells by macropinocytosis. *New Phytol.*, *174*, 46–56.

[12] Wang, Y. H., Hou, Y. W. & Lee, H. J. (2007). An intracellular delivery method for siRNA by an arginine-rich peptide. *J. Biochem. Biiophys. Methods*, *70*, 579–586.

[13] Hu, J. W., Liu, B. R., Wu, C. Y., Lu, S. W. & Lee, H. J. (2009). Protein transport in human cells mediated by covalently and noncovalently conjugated arginine-rich intracellular delivery peptides. *Peptides*, *30*, 1669–1678.

[14] Unnamalai, N., Kang, B. G. & Lee, W. S. (2004). Cationic oligopeptide-mediated delivery of dsRNA for post-transcriptional gene silencing in plant cells. *FEBS Lett.*, *566*, 307–310.

[15] Read, M. L., Singh, S., Ahmed, Z., Stevenson, M., Briggs, S. S., Oupicky, D., Barrett, L.B., Spice, R., Kendall, M., Berry, M., Preece, J. A., Logan, A. & Seymour, L. W. (2005). A versatile reducible polycation-based system for efficient delivery of a broad range of nucleic acids. *Nucl. Acids Res.*, *33*, e86.

[16] Simeoni, F., Morris, M. C., Heitz, F. & Divita, G. (2005). Peptide-based strategy for siRNA delivery into mammalian cells. Methods Mol. Biol., 309, 251–260.

[17] Veldhoen, S., Laufer, S. D., Trampe, A. & Restle, T. (2006). Cellular delivery of small interfering RNA by a non-covalently attached cell-penetrating peptide: quantitative analysis of uptake and biological effect. *Nucl. Acids Res.*, *34*, 6561–6573.

[18] Meade, B. R. & Dowdy, S. F. (2007). Exogenous siRNA delivery using peptide transduction domains/cell penetrating peptides. *Adv. Drug Deliv. Rev.*, *59*, 134–140.

[19] Endoh, T., Sisido, M. & Ohtsuki, T. (2008). Cellular siRNA delivery mediated by a cell-permeant RNA-binding protein and photoinduced RNA interference. *Bioconjug. Chem. 19*, 1017–1024.

[20] Law, M., Jafari, M. & Chen, P. (2008). Physicochemical characterization of siRNA-peptide complexes. *Biotechnol. Prog. 24*, 957–963.

[21] Endoh, T. & Ohtsuki, T. (2009). Cellular siRNA delivery using cell-penetrating peptides modified for endosomal escape. *Adv. Drug Deliv. Rev.*, *61*, 704–709.

[22] Kawasaki, H. & Taira, K. (2003). Short hairpin type of dsRNAs that are controlled by tRNA(Val) promoter significantly induce RNAi-mediated gene silencing in the cytoplasm of human cells. *Nucl. Acids Res.*, *31*, 700–707.

[23] Kawakami, S. & Hashida, M. (2007). Targeted delivery systems of small interfering RNA by systemic administration. *Drug Metab. Pharmacokinet.*, *22*, 142–151.

[24] Billy, E., Brondani, V., Zhang, H., Muller, U. & Filipowicz, W. (2001). Specific interference with gene expression induced by long, double-stranded RNA in mouse embryonal teratocarcinoma cell lines. *Proc. Natl. Acad. Sci. U.S.A.*, *98*, 14428–14433.

[25] Zeng, Y. & Cullen, B. R. (2002). RNA interference in human cells is restricted to the cytoplasm. *RNA*, *8*, 855–860.

[26] Zhou, H., Yang, L., Li, H., Li, L. & Chen, J. (2009). Residues that affect human Argonaute2 concentration in cytoplasmic processing bodies. *Biochem. Biiophys. Res. Commun.*, *378*, 620–624.

[27] Sibley, C. R., Seow, Y. & Wood, M. J. (2010). Novel RNA-based strategies for therapeutic gene silencing. *Mol. Ther.*, *18*, 466–476.

In: Gene Silencing: Theory, Techniques and Applications ISBN: 978-1-61728-276-8
Editor: Anthony J.Catalano © 2010 Nova Science Publishers, Inc.

Chapter XVI

Asymmetric RNA Duplexes as Next Generation RNAi Inducers

Dong-ki Lee[1,*], Soyoun Kim[2], Andrew C. Keates[3] and Chiang J. Li[2,3,†]

[1]Global Research Laboratory for RNAi Medicine, Department of Chemistry and BK21 School of Chemical Materials Science, Sungkyunkwan University, Suwon, Korea

[2]Department of Biomedical Engineering, Dongguk University, Seoul, Korea

[3]Skip Ackerman Center for Molecular Therapeutics, Beth Israel Deaconess Medical Center, Harvard Medical School, Boston, Massachusetts, USA

Abstract

RNA interference (RNAi) has become an indispensible technology for biomedical research and promises to usher in a brand new class of therapeutics that work by silencing disease genes. Until recently, the paradigm for gene silencing in mammalian cells has relied on a small symmetrical RNA structures containing a 19-base-pair duplex with 2 nucleotide overhangs at each 3' end: the standard siRNA structure. The standard siRNA scaffold is based on structures generated by Dicer digestion of a double stranded RNA, and is considered to be the fundamental template for designing RNAi inducers. In fact, early studies suggested there was only very limited flexibility regarding the length and symmetry of the siRNA structure in order to maintain optimal gene silencing. Recent studies, however, have demonstrated that gene silencing siRNAs with duplex lengths shorter than 19 bp or asymmetric structures can trigger specific gene silencing in mammalian cells. Importantly, asymmetric siRNA structures can ameliorate several sequence-independent, nonspecific effects triggered by the canonical siRNA structure. These findings demonstrate the structural flexibility of RNAi inducers in mammalian cells.

[*] Correspondence: Dong-ki Lee, dklee@skku.edu
[†] Chiang J. Li, cli@bidmc.harvard.edu

Introduction

RNA interference (RNAi), discovered by Fire and Mello in 1998, is an endogenous post-transcriptional gene silencing mechanism conserved in diverse eukaryotic organisms [1]. Since its discovery, this highly potent and specific gene silencing mechanism has been rapidly adopted by the research community as an important functional genomics tool. More importantly, RNAi has unlimited potential for the development of novel therapeutics for a wide variety of diseases. However, the execution of RNAi in mammalian cells proved difficult initially as the long dsRNA typically used to trigger RNAi in nematodes and flies activated potent antiviral responses in mammalian cells that resulted in non-specific mRNA downregulation and inhibition of protein synthesis. RNAi-mediated gene silencing in mammalian cells only became possible after the discovery of small interfering RNA (siRNA)-mediated target gene silencing in mammalian cells [2,3]. Since then, siRNA-mediated gene silencing has become one of the most exciting and promising biotechnologies for both basic research and the development of therapeutic applications. Compared to other existing oligonucleotide-based gene targeting methods such as antisense and ribozyme technology, siRNA shows significantly improved efficacy while maintaining the sequence specificity. This improved efficacy is most likely due to the fact that RNAi is a naturally existing cellular mechanism present in almost all eukaryotic cell types. At present, a number of therapeutic siRNA-based drug candidates are undergoing clinical trials, with many more in advanced preclinical development [4].

The canonical siRNA architecture consists of a symmetric short RNA duplex with a 19-bp central duplex region and 2-nt overhangs at each 3' end: the so-called 19+2 siRNA structure. This scaffold was designed to mimic the structures typically generated by Dicer processing of long dsRNA [5]. Moreover, initial structure-activity relationship studies conducted in Drosophila cell lysates indicated that the standard siRNA architecture was most active whereas deviations from this structure, such as changes to the duplex length or removal of 3'-overhangs, resulted in loss of gene silencing activity [6].

Despite the initial promise of the specific siRNA-mediated gene silencing, researchers quickly realized that the canonical siRNA structure triggered several non-specific responses when introduced in cells and animals [4]. Some of the major non-specific effects triggered by the standard siRNA scaffold include: (1) off-target gene silencing triggered by incorporation of the sense strand into RNA-induced silencing complex (RISC), or incomplete base paring of the siRNA antisense strand with non-target mRNA; (2) activation of non-specific innate immune responses by pattern recognition receptors (e.g. Toll-like receptors and RIG-I-like cytoplasmic helicases); (3) and saturation of the RNAi machinery, and hence the inhibition of endogenous microRNA processing, by excess exogenous siRNA. Unfortunately, these nonspecific effects significantly hamper the development of siRNA as a specific tool for functional genomics studies and as a therapeutic modality. In response to these problems, several studies have investigated the use of chemically modified siRNA as a means to circumvent some of the nonspecific effects of 19+2 siRNA. Chemical modification of siRNA, however, is often associated with unfavorable side effects such as increased toxicity and reduced silencing efficiency. Thus, an important area of current research is to discover and characterize novel RNAi inducers that overcome the non-specific effects generated by the canonical siRNA structure.

As outlined above, the general consensus in the RNAi field was that the canonical siRNA architecture represented the optimal structure for siRNA-mediated gene silencing. Nevertheless, ongoing efforts have focused on the identification of RNAi inducers that are distinct from the canonical siRNA structure. Here, we summarize several recent findings that challenge the central dogma regarding the 19+2 siRNA architecture, and propose novel structures with enhanced functionalities.

Asymmetric siRNA as Novel RNAi Inducing Architectures

As originally proposed by Elbashir et al. [6], the optimal siRNA structure for triggering RNAi was thought to be a symmetric dsRNA with 19-bp duplex region and 2 nt-overhang at each 3' end of the duplex. Moreover, according to this study, removal of the 3' overhang structures, or shortening of the duplex region significantly reduced gene silencing activity.

Recently, however, Chu et al. have reported that siRNA containing a 16-bp central duplex region with two nucleotide overhangs can induce RNAi in human cells [7]. These authors reported that the 16-bp siRNA was more effective at knocking down CDK9 gene expression than the canonical 19 bp siRNA. However, the conclusion that 16-bp siRNA is more potent than 19 bp siRNA requires further validation, since the 5' seed sequences of two siRNA were different, and the comparison was only carried out on one gene and one gene site. Thus, a side-by-side comparison of gene silencing potency of each siRNA against a number of genes needs to be carried out before a generalization. The idea that siRNA duplexes may not require 19-bp for gene silencing was also observed in a study by Li et al. which used a bacteria-based screening method to identify potent siRNA targeting the MVP and EGFP genes [8]. These results raised the possibility that RISC assembly and activation during RNAi may not require the use of canonical siRNA containing a 19-bp duplex region.

In addition to the duplex length, the necessity for symmetric 3' overhang structures in traditional siRNA has also been investigated. In particular, Sano et al. reported the importance of scaffold structures with asymmetric 3' overhangs on gene silencing activity and strand selection bias [9]. These authors showed that siRNA containing a 2 nt overhang at the 3' end of the antisense strand, but not the sense strand, are more effective than the cannonical siRNA, due to preferential loading of the antisense strand into the RISC.

As a significant departure from the canonical siRNA architecture and its variations, the novel asymmetric RNA duplex scaffolds for gene silencing have been proposed and developed by Sun et al. [10] and Chang et al. [11], and termed aiRNA (asymmetric interfering RNA) and asiRNA (asymmetric shorter-duplex siRNA) respectively (Figure 1). The aiRNA architecture consists of a 21-nt antisense strand and a 15-nt sense strand that together generate an asymmetric RNA duplex structure containing a 15-bp RNA duplex region with 5' and 3' antisense overhangs. asiRNA, on the other hand, consists of a 19 to 21-nt antisense and 16 nt sense strands that generate a 16-bp RNA duplex region with 3 to 5-nt long 3' antisense overhang. Both asymmetric structures have been shown to mediate sequence-specific cleavage of target mRNA between bases 10 and 11 relative to the 5' end of the antisense strand, indicating that they act via the canonical RNAi mechanism. The RISC dependency of asymmetric RNA duplexes was further supported by use of the RISC incorporation assay [10] and Ago2-dependency test [11]. Notably, the gene silencing potency of aiRNAs was shown to be superior to the corresponding standard siRNA.

The most exciting aspect of these asymmetric RNAi inducers is their potential to overcome several non-specific effects triggered by traditional siRNA structures. Both aiRNA and asiRNA markedly reduced sense strand-mediated off target silencing, likely due to the combined effect of the shortened sense strands in each molecule as well as their asymmetric overhang structure. In addition, asiRNA was shown to have reduced saturation of the endogenous RNAi machinery when compared to standard siRNA. Interestingly, the reduction in RNAi pathway saturation was dependent on the length of asiRNA duplex. aiRNA and asiRNA are also less likely to trigger non-specific activation of TLR3. A recent report has demonstrated that intraocular injection of conventional 19+2 siRNA, including chemically modified siRNAs, can suppress angiogenesis via non-specific activation of TLR3, and that this effect was dependent upon the length of the siRNA duplex [12]. The 15-16 bp long duplex inherent to aiRNA and asiRNA, however, is outside the range for non-specific TLR3 activation by dsRNAs [12].

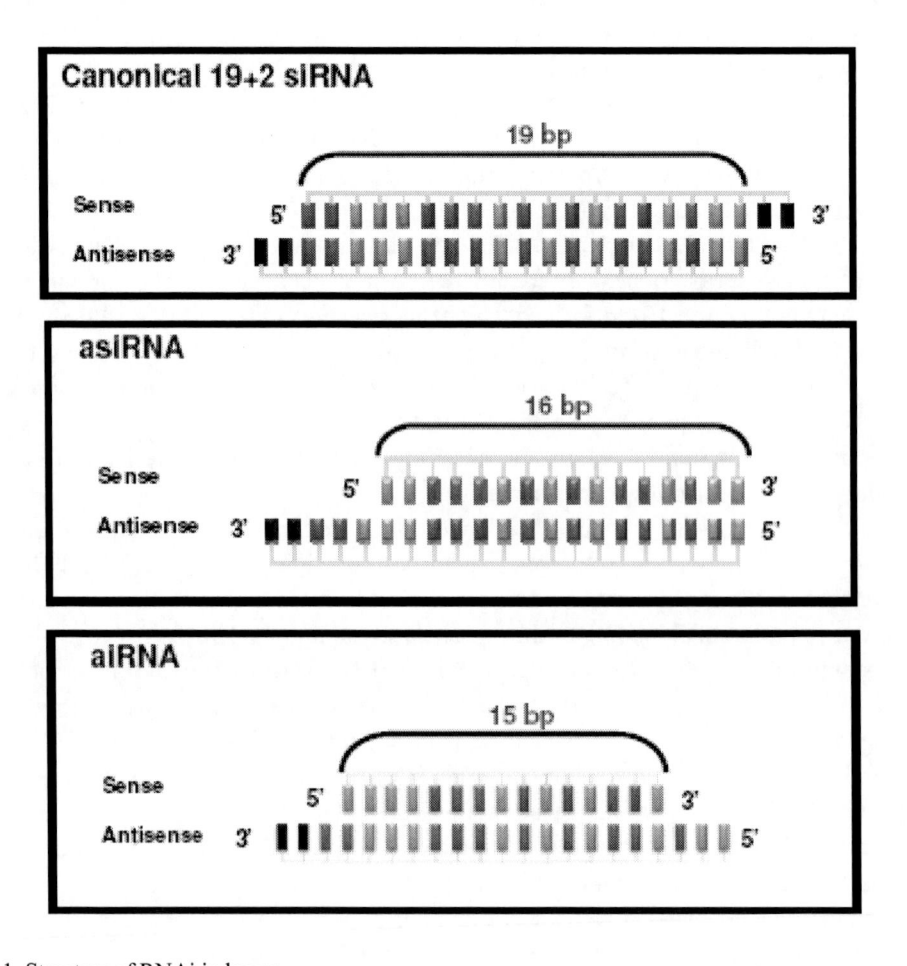

Figure 1. Structure of RNAi inducers.

Conclusion and Perspective

The recent discovery and characterization of novel asymmetric RNAi inducers demonstrates that, in contrast to the original findings of Elbashir et al., the fundamental gene silencing scaffold is indeed quite flexible. More importantly, these second-generation RNAi inducers provide significant improvements compared to the traditional symmetric 19+2 siRNA structure. In particular, asymmetric RNAi technologies have improved gene silencing efficacy and markedly reduced non-specific effects such as sense strand off-target silencing and innate immune response activation. Given the numerous advantages of asymmetric RNAi inducers we believe that these new structures have the potential to significantly improve the study of gene function in vitro. Moreover, the use of asymmetric RNAi inducers may also overcome some of the key problems associated with the development of RNAi-based therapeutics.

Acknowledgments

We would like to thank our colleagues Drs. Arthur B. Pardee, Harry A. Rogoff, Xiangao Sun, colleagues at the Skip Ackerman Center for Molecular Therapeutics, and Biomolecular Therapeutics Laboratory at SKKU for helpful discussions and advice on the study of asymmetric RNAi inducers. We thank NIH and Global Research Laboratory of Korean Ministry of Education, Science and Technology for supports.

References

[1]   Fire, A. et al. (1998) Potent and specific genetic interference by double-stranded RNA in Caenorhabditis elegans. *Nature* 391 (6669), 806-811.

[2]   Elbashir, S.M. et al. (2001) Duplexes of 21-nucleotide RNAs mediate RNA interference in cultured mammalian cells. *Nature* 411 (6836), 494-498.

[3]   Zamore, P.D. et al. (2000) RNAi: double-stranded RNA directs the ATP-dependent cleavage of mRNA at 21 to 23 nucleotide intervals. *Cell* 101 (1), 25-33.

[4]   Tiemann, K. and Rossi, J.J. (2009) RNAi-based therapeutics-current status, challenges and prospects. *EMBO Mol Med* 1 (3), 142-151.

[5]   Elbashir, S.M. et al. (2001) RNA interference is mediated by 21- and 22-nucleotide RNAs. *Genes Dev* 15 (2), 188-200.

[6]   Elbashir, S.M. et al. (2001) Functional anatomy of siRNAs for mediating efficient RNAi in Drosophila melanogaster embryo lysate. *EMBO J* 20 (23), 6877-6888.

[7]   Chu, C.Y. and Rana, T.M. (2008) Potent RNAi by short RNA triggers. *RNA* 14 (9), 1714-1719.

[8]   Li, Z. et al. (2009) Forward and robust selection of the most potent and noncellular toxic siRNAs from RNAi libraries. *Nucleic Acids Res* 37 (1), e8.

[9]   Sano, M. et al. (2008) Expression of long anti-HIV-1 hairpin RNAs for the generation of multiple siRNAs: advantages and limitations. *Mol Ther* 16 (1), 170-177.

[10] Sun, X. et al. (2008) Asymmetric RNA duplexes mediate RNA interference in mammalian cells. *Nat Biotechnol* 26 (12), 1379-1382.

[11] Chang, C.I. et al. (2009) Asymmetric shorter-duplex siRNA structures trigger efficient gene silencing with reduced nonspecific effects. *Mol Ther* 17 (4), 725-732.

[12] Kleinman, M.E. et al. (2008) Sequence- and target-independent angiogenesis suppression by siRNA via TLR3. *Nature* 452 (7187), 591-597.

Index

D

E

F

N

O

P

Q

R

S

U

V

W

X

Y

Z